THE OVARY

SECOND EDITION

Volume II

Physiology

Contributors

E. C. Amoroso
P. M. F. Bishop
Georgiana M. Bonser
J. T. Eayrs
P. Eckstein
A. Glass
J. Herbert
K. C. Highnam
J. W. Jull
P. L. Krohn
J. S. Perry
A. Raynaud
I. W. Rowlands
Heidi H. Swanson
Barbara J. Weir

THE OVARY

SECOND EDITION

Volume II

Physiology

EDITED BY **Professor Lord Zuckerman**

Zoological Society of London
London, England
and
University of East Anglia
Norwich, Norfolk
England

AND **Barbara J. Weir**

Wellcome Institute of Comparative Physiology
Zoological Society of London
London, England
and
Journal of Reproduction and Fertility
Cambridge, England

ACADEMIC PRESS   *New York   San Francisco   London   1977*

A Subsidiary of Harcourt Brace Jovanovich, Publishers

ACADEMIC PRESS, INC.
111 Fifth Avenue, New York, New York 10003

United Kingdom Edition published by
ACADEMIC PRESS, INC. (LONDON) LTD.
24/28 Oval Road, London NW1

Library of Congress Cataloging in Publication Data

Zuckerman, Professor Lord   (date)   ed.
    The ovary.

    Includes bibliographies and index.
    1.   Ovaries.   I.   Title.   [DNLM: 1.   Ovary.
QL876 096]
QP261.Z8   1976   599'.01'66   76-13955
ISBN 0–12–782602–5   (v. 2)

Contents

1 *Control of Ovarian Development in Invertebrates*

K. C. HIGHNAM

2 *The Development of Estrogen-Sensitive Tissues of the Genital Tract and the Mammary Gland*

A. RAYNAUD

List of Contributors

Numbers in parentheses indicate the pages on which the authors' contributions begin.

E. C. AMOROSO (315), A. R. C. Institute of Animal Physiology, Babraham, Cambridge, England

P. M. F. BISHOP* (185), Guy's Hospital Medical School, London, England

GEORGIANA M. BONSER** (129), Cancer Research Annexe, University of Leeds, Leeds, England

J. T. EAYRS (399), Department of Anatomy, Medical School, University of Birmingham, Birmingham, England

P. ECKSTEIN (275), Department of Anatomy, Medical School, University of Birmingham, Birmingham, England

A. GLASS (399), Department of Anatomy, Medical School, University of Birmingham, Birmingham, England

J. HERBERT (457), Department of Anatomy, University of Cambridge, Cambridge, England

K. C. HIGHNAM (1), Department of Zoology, University of Sheffield, Sheffield, England

J. W. JULL† (129), Cancer Research Center, University of British Columbia, Canada

P. L. KROHN (101), La Forêt, St. Mary, Jersey, Channel Islands, Great Britain

J. S. PERRY (315), A. R. C. Institute of Animal Physiology, Babraham, Cambridge, England

A. RAYNAUD (63), Laboratoire Pasteur, Sannois, France

I. W. ROWLANDS‡ (217), Wellcome Institute of Comparative Physiology, Zoological Society of London, Regents Park, London, England

* Present address: Moorhaven, Bovey Tracey, Devon, England.
** Present address: 10 Elmete Court, Leeds, England
† Deceased.
‡ Present address: Department of Anatomy, University of Cambridge, Cambridge, England.

HEIDI H. SWANSON (399), Department of Anatomy, Medical School, University of Birmingham, Birmingham, England

PARBARA J. WEIR* (217), Wellcome Institute of Comparative Physiology, Zoological Society of London, Regents Park, London, England

* Present address: Journal of Reproduction and Fertility, 7 Downing Place, Cambridge, England.

Preface

Insofar as its speed is necessarily that of its slowest member, the completion of a book by several hands is somewhat like the voyage of a convoy of ships. This does not however make the enterprise any the less valuable. The first edition of "The Ovary" appeared over fifteen years ago, and it was time the reviews it incorporated were brought up to date. The slowness with which some chapters arrived was not, however, the only reason for the delay in the appearance of this new edition. The editors could perhaps have tried to be more demanding of their contributors than they were. As it turned out, however, illness held up some chapters, and one, which was all but completed, had to be restarted when its main author was killed in an air crash. In 1961 I found myself constrained to apologize for the tardy appearance of the original edition, and I therefore do so again for this new one, not only to those authors who were first with their contributions and who have had to wait longest to see them in print, but also to our patient publishers, Academic Press, and to the scientific public for whom the work is designed.

Like any book of reviews, it is obvious that the chapters of this new edition will fail to mention some papers that may have appeared in the past three to four years. This, however, detracts little from their value. Immediately before the texts were sent to the publisher every author was invited to update, given she or he so wished, what had been written. No doubt there is a lesson to be drawn from the fact that few felt this necessary. A review does not gain in value if it merely catalogues the names of authors who have written about the subject with a brief reference to the summaries of their published papers. For a review is nothing if it is not critical, and has little merit if its purpose is not to focus on such generalizations as are justified by the pieces of information on which it is based. In the ideal, the updating of a review should be an exercise in which the validity of any general proposition that has already been defined is examined in the light of new experimental data and in which new hypotheses are formulated where they are called for by new findings. This, of course, is the personal view of an editor who knows full well that it is not necessarily accepted universally.

I have heard it said that with the enormous growth of the world's scientific effort over the past two to three decades budding scientists are often advised that when they survey the literature which bears on the problems they are investigating there is little point in going back more than ten years, so rapidly does new observation overlay what is already known. This is something which I am sure many who belong to an older generation of scientists greatly regret. Sometimes it results in what is already established being "rediscovered." Often new findings are treated out of all proportion to the major generalizations to which they relate. What is more—and this applies not only to the fields of science with which "The Ovary" deals—the vast expansion of scientific activity over recent years has inevitably resulted in resources becoming available not only for what is called "big science" but for a wider range of enquiry in "small science" than would ever have been possible in more penurious times. When editing a work such as the present one, it is difficult to avoid the impression that on occasion an experiment or set of experiments has been undertaken merely because a professor or supervisor has had to provide a theme for a postgraduate's thesis. One also finds that new techniques have defined topics for experiment without the kind of critical preliminary evaluation of the limitations of new methods of enquiry in relation to the central questions which it is hoped they will help elucidate. With the vast growth in experimental work—and inevitably, therefore, the occasional dilution of its quality—"controls" also sometimes appear inadequate. In spite of these general observations, I have, however, no hesitation in saying that there has been a considerable increase in our knowledge of the ovary over the period since the appearance of the first edition.

Dr. Anita Mandl and Professor Peter Eckstein collaborated with me in the editing of the first edition but were unable to help in this, Dr. Mandl because she had retired from academic life and Professor Eckstein because of the heavy load of other work which he has since assumed. Fortunately I was able to recruit as coeditor Dr. Barbara Weir, to whom my thanks, as well as that of the contributors, are due, as they also are to Academic Press.

S. Zuckerman

Acknowledgments

The help of the librarians and their staff of the Royal Society of Medicine (Mr. R. Wade), Wellcome History of Science Library (Mr. L. Symons) and of the Zoological Society of London (Mr. R. A. Fish) in checking references is greatly appreciated.

Preface to the First Edition

To the best of my knowledge no book on the normal ovary has appeared in English since the publication, in 1929, of Professor A. S. Parkes' monograph entitled "The Internal Secretion of the Ovary". The greater length of the present work, the object of which is to provide a detailed account of the principal aspects of ovarian development, structure and function as understood today, reflects the vigour with which researches on these subjects have been pursued over the past thirty years.

An almost unlimited number of topics could have been regarded as falling within the scope of the review. Since there was a necessary limit to size, the two volumes it constitutes cannot, accordingly, be claimed to exhaust the subject they were designed to cover. To whatever extent arbitrariness marks the fields dealt with, the treatise also partakes of a characteristic common to all scientific reviews, and one which reflects the fact that the content, pattern and emphasis of different fields of knowledge are always in a state of change.

The original intention was to publish the work in a single volume. When it became necessary to allocate the material to two, some rearrangement of chapters was called for, and the original sequence of topics which had been planned was changed. In the main, those chapters which relate to what might be called the natural history of the ovary are now included in Volume I, while information derived from more experimental and chemical studies is assembled in Volume II. The two volumes overlap to some extent, as do certain topics, but so far as possible this has been dealt with by means of cross references.

I am deeply grateful to the many contributors for their generous assent to my invitation to participate in what has proved a lengthier and more arduous task than I, and perhaps they, anticipated at the outset. The authors of the various chapters are of course individually responsible for the content and bibliographic references as well as the style and accuracy of their contributions.

When manuscripts started to arrive, I had to turn to two of my colleagues, Dr. Anita Mandl and Dr. Peter Eckstein, for assistance in the work of editing, and of arranging for the translations of those chapters which were submitted in French. I am deeply grateful for their help, as I am also to the Academic Press for its tolerance during the long period in which this review has been in train. My thanks are also due to Miss Heather Paterson for her able help in preparing manuscripts and checking proofs, to Mr. L. T. Morton for compiling the Subject Index, and to the Academic Press for constructing the Author Index.

The delays which are inevitably associated with the production of a lengthy treatise have meant that a number of contributions appear less up-to-date in print than they did in typescript. Even though references to papers published in last year's scientific journals may be lacking, I nonetheless believe that at the moment the two volumes provide a more comprehensive picture of the whole subject than can be found in any other single work.

August, 1961　　　　　　　　　　　　　　　　*S. Zuckerman*

Contents of Other Volumes

THE OVARY

SECOND EDITION

Volume II

Physiology

1

Control of Ovarian Development in Invertebrates

K. C. Highnam

I. INTRODUCTION

Although there are indications that gonadal differentiation and egg development are controlled by hormones in coelenterates, nemerteans, flatworms, and ascidians (Brien, 1963; Bierne, 1964, 1966, 1968, 1970; Dodd,

1955; Sengel and Kiery, 1962; Sengel and Georges, 1966; Dawson and Hisaw, 1964; Bouchard-Madrelle, 1967), it is in the annelids, molluscs, crustaceans, insects, and echinoderms that the processes involved have been examined in detail. Only these major invertebrate phyla will therefore be dealt with here.

As in vertebrates, the primordial germ cells in invertebrates segregate early during embryonic development, although in molluscs, particularly, the actual timing varies between species (Raven, 1958). In the major invertebrate phyla, the primordial germ cells migrate to their final positions where they become associated with mesodermal elements to form the definitive gonad. In insects, it is likely that the mesodermal parts of the ovaries are induced by the germ cells (Poulson and Waterhouse, 1960; Davis, 1967), although an independent origin of the mesodermal gonad has been suggested (Geigy, 1931; Aboim, 1945; Hathaway and Selman, 1961).

In hermaphrodite forms, the primordial germ cells can become either male or female, or develop into nurse cells. Even in gonochoristic crustaceans, the cells are equipotent for maleness or femaleness until the time of sexual differentiation. In the insects, the primordial germ cells are usually considered to be genetically male or female, although those of the glowworm, *Lampyris noctiluca,* are similar to those of crustaceans in being equipotent until differentiation is induced by the presence or absence of an endocrine factor (Naisse, 1963, 1965). Echinoderm primordial germ cells are also likely to be equipotent (Delavault, 1966).

Regardless of the way in which the primordial germ cells differentiate, once they become determined as female cells their subsequent history is similar in all animals. The cells become oogonia, which enter a phase of rapid multiplication, sometimes with a fixed number of divisions, so that each primary oogonium produces a predetermined number of secondary oogonia. These transform into oocytes, which initially grow relatively slowly, then rapidly during the period of vitellogenesis. The fully grown oocyte leaves the gonad and is ready for fertilization, either immediately or after having undergone its maturation divisions.

The process of oocyte development must be related to the growth and development of the animal, and also to the environmental situation, so that eggs are fertilized and laid in conditions suitable for the embryonic and post-embryonic development of the following generation. It is consequently not surprising that mechanisms have evolved to synchronize reproductive development with somatic growth and with environmental cycles. Moreover, the basic similarity of oocyte development in all animals imposes limits on the ways in which controls can operate. In general, it might be expected that primordial germ cell differentiation, oogonial proliferation, vitellogenesis, oocyte maturation, and ovulation or spawning are all likely stages in egg production at which control mechanisms could operate. As the following sections show, one or more of these stages are subject to control by hormones in certain invertebrate phyla. The variety in form and structure of the animals necessarily implies differences in the sources and chemical nature of the hormones involved in the control of oogenesis.

II. ANNELIDA

A. Polychaeta

The sexes are separate in almost all polychaete species. The germ cells are derived from cells of the coelomic epithelium, frequently in the septal walls or blood vessel sheaths. The ovaries thus formed may occur in many body segments, although in some species they are confined to the more posterior parts of the body, particularly in those creatures which undergo an epitokous metamorphosis or which bud off sexual individuals posteriorly. Oocytes, initially surrounded by follicle cells, are often liberated into the coelom to grow and lay down yolk. In some species, oocyte maturation occurs after spawning and sperm entry.

1. The Endocrine System

Groups of neurosecretory cells are present in the supraesophageal ganglia (brain) and in the ganglia of the ventral chain (Gabe, 1966). In *Nephtys californiensis* (Clark, 1959) and *Perinereis cultrifera* (Bobin and Durchon, 1952), axon tracts containing neurosecretion extend ventrally through the brain to terminate in the vicinity of the dorsal blood vessel. In *Platynereis dumerilii,* two major groups of neurosecretory cells in the brain are considered to be the source of hormone(s) involved in reproduction (Hauenschild, 1964). In polynoids, a large midsagittal fiber tract at the level of the optic commissure passes ventrally to end on the surface of the brain; the tract derives from anterior and posterior roots, the latter originating as bilaterally symmetrical branches converging on the midsagittal tract to form a "Y" (Baskin, 1971). The arms of the "Y" are formed, at least in part, of axons from neurosecretory cells (Korn, 1959; Baskin, 1971).

A cerebrovascular complex beneath the brain was previously thought to be associated with the release of neurosecretory products in nereids and nephtyids (Bobin and Durchon, 1952; Clark, 1956, 1959). This structure is considerably more complex than was originally thought and is called the infracerebral gland (Golding *et al.,* 1968); there are different kinds of neurosecretory axon terminals and some are closely associated with apparently secretory epithelial cells (Dhainaut-Courtois, 1966a,b, 1968a,b; Golding *et al.*, 1968; Golding, 1970; Baskin, 1970). In polynoids, there is a similar epithelial–neurosecretory complex in the ventral region of the brain, associated with a coelomic sinus and a blood vessel (Baskin, 1971).

The infracerebral gland fulfills one or more endocrine functions (Dhainaut-Courtois, 1968a; Golding *et al.*, 1968), although this view is based more upon ultrastructural and cytological indications of the secretory activities of different cell types than upon definitive experiments (Dhainaut-Courtois, 1968a; Baskin, 1970). However, in *Nereis limnicola*, the dorsal and ventral parts of the brain (the ventral part containing the infracerebral gland) are separately not as effective as a complete brain in controlling maturation, which suggests that the cerebral neurosecretory cells together with the infracerebral gland form an integrated system whose integrity is essential for the secretion at normal levels of its hormone(s) (Baskin and Golding, 1970).

2. Control of Oocyte Development

There is now good evidence that oocyte growth and vitellogenesis in nereids is inhibited during the juvenile life of the animals by a hormone produced by neurosecretory cells in the supraesophageal ganglion (Hauenschild, 1965, 1966; Durchon, 1962, 1967a; Clark, 1962, 1966, 1969). Removal of the brain results in a rapid enlargement of the oocytes, but the development of cytoplasmic inclusions, including yolk granules, and the metachromatic system, which is the precursor of acid mucopolysaccharides in the oocytes, is abnormal, and the oocytes do not attain their natural size (Clark and Ruston, 1963; Hauenschild, 1965, 1966; Durchon, 1967a; Dhainaut and Porchet, 1967; Dhainaut, 1970a,b; Porchet, 1970). Synthesis of RNA is very much greater than normal following brain removal. In addition, the development of perinuclear annular lamellae, which are continuous with the endoplasmic reticulum, and which may be associated with the excessive synthesis of RNA, is grossly abnormal (Durchon, 1967a; Dhainaut, 1966a). The functional evolution of the Golgi apparatus is also a major characteristic of oocyte growth, incorporating amino acids and transferring products to the cytoplasm (Dhainaut, 1967), as well as metabolizing sugars for mucopolysaccharide formation (Dhainaut, 1968).

The accelerated development of oocytes following brain removal could suggest not only the withdrawal of an inhibitory hormone but also that of a hormone which actually promotes vitellogenesis. However, when a brain from a full grown but immature worm is implanted for a period of time into a decerebrate fragment of another individual, or when a brain from a very young posttrochophore individual is implanted into an isolated parapodium, oocyte growth and vitellogenesis are normal (Hauenschild, 1964, 1965, 1966). It seems clear, then, that oocyte growth and vitellogenesis in normal nereids results from the *gradual* withdrawal of a single "inhibitory" hormone. In high concentration, as in juvenile individuals, the hormone is inhibitory; in reducing concentrations, as when the individuals attain "puberty" with oocytes of a certain size, the hormone stimulates and controls the normal development of intracellular inclusions within the oocytes (Porchet, 1970). The final stages of vitellogenesis can take place in the complete absence of the hormone. Using a standard bioassay technique, the brains of *Nereis diversicolor* females with oocytes 100 to 150 μm in diameter are five times less inhibiting than those of juvenile forms, or females with oocytes 5 μm in diameter (Durchon and Porchet, 1970).

Although the abnormal oocytes resulting from sudden withdrawal of the brain hormone can be fertilized, further development is abnormal (Choquet, 1962; Clark and Ruston, 1963). In the viviparous *Nereis limnicola,* which is a self-fertilizing hermaphrodite, the brain exerts inhibitory endocrine control over gamete development as in other nereids (Baskin and Golding, 1970). The normal, mature oocytes resemble the abnormal oocytes obtained by decerebration in the related, but oviparous, species *Nereis diversicolor* (Clark and Ruston, 1963). The precociously mature oocytes in *Nereis limnicola* are capable of fertilization and further development, as determined by the presence of normal larval stages in the coelom following the operation (Baskin and Golding, 1970). It would seem that the development of viviparity in *Nereis limnicola,* transferring to the coelom the function of providing nutrients and other factors necessary for growth of the embryos and relieving the oocyte of intense vitellogenic metabolism, can account for this difference in control between *Nereis limnicola* and other nereids (Baskin and Golding, 1970).

Before vitellogenesis, the brain hormone does not merely inhibit oocyte development (Golding, 1972). *Perinereis cultrifera* lives for 3 years, and oocytes in the second summer animals are simply refractory to the effects of decapitation, whereas oocytes in the second and third winter animals rapidly degenerate after decapitation if they are less than 120 μm in diameter (Dhainaut and Porchet, 1967). Worms with an oocyte diameter of about 120 μm can be considered to be pubertal, and their brains are weakly inhibitory. When such brains are implanted into young animals, they induce the precocious development of inclusions within the oocytes (Porchet, 1970). In *Nereis grubei,* similarly, oocytes less than 50 μm in diameter are resorbed after decapitation (Schroeder, 1971). *Nereis grubei* lives for 2 years and the oocytes of this size class are probably equivalent to the second and third winter oocytes of *Perinereis cultrifera* (Schroeder, 1971). The brain hormone is thus necessary for the normal development of the oocytes, including the production of cytoplasmic inclusions which precedes the stage of rapid vitellogenesis.

In *Nephtys hombergi,* breeding worms have a higher level of free sugars in the coelomic fluid together with alcohol-soluble nitrogenous material than do juveniles, and can maintain these levels despite starvation. Decerebration of nonbreeding individuals brings about similar changes (Clark, 1964). In *Nereis diversicolor,* decerebration profoundly affects the oxygen consumption of the operated individuals (Dhainaut, 1966b). Although changes in the somatic tissues during reproduction are undoubtedly mediated by a mechanism similar to that affecting oocyte growth and vitellogenesis (see below), it is not implausible to suggest that the infracerebral gland may affect other bodily processes (Baskin and Golding, 1970; Golding *et al.,* 1968), although the relationships may not necessarily be direct (Baskin, 1970).

3. Somatic Maturation

In many nereids, oocyte growth and vitellogenesis are accompanied by somatic metamorphosis into an epitokous swimming form, the heteronereis (Clark, 1961). The epitokous metamorphosis involves enlargement and change in form of the parapodia, the secretion of new oar-shaped chaetae, replacement of much of the musculature, enlargement of the eyes, and multiplication of coelomocytes. Even in species which reproduce in the atokous form, somatic changes accompany gametic development, although these are not as dramatic as in epitokous forms (Dales, 1950). In *Nereis diversicolor,* coelomocytes fill the coelomic spaces during the early stages of sexual development, disappearing later as the coelom becomes filled with mature oocytes. The musculature is histolyzed and differentiated, and the body wall is thin and fragile at the time of spawning.

The somatic changes associated with oocyte development are inhibited in juvenile forms by a hormone from the brain (Durchon, 1952, 1967a; Hauenschild, 1965, 1966; Clark, 1966, 1969). Sudden withdrawal of the hormone following brain removal initiates the metamorphosis processes, but these are not completed and the worm dies; reimplantation of the brain for a period into decerebrate fragments, or transplantation of a small brain from very young animals into individual parapodia, enables metamorphosis to proceed to completion (Hauenschild, 1964, 1966). There seems little doubt that, as in the control of oocyte development, high concentrations of brain hormone are inhibitory to metamorphosis, but that a declining titer of hormone promotes somatic metamorphosis. Because of the similarity in effect of brain hormone upon both oocyte development and somatic metamorphosis, it is argued that the same hormone controls both processes.

The viviparous *Nereis limnicola,* like the oviparous *Nereis diversicolor,* reproduces in the atokous form. In both species, somatic changes involving variations in coelomocyte number and histolysis of the body musculature occurs (Smith, 1950; Baskin and Golding, 1970). However, in *Nereis limnicola* the somatic changes are not synchronous with maturation of the gametes, being related instead to the release of the larvae (Baskin and Golding, 1970). This might suggest that separate inhibitory hormones controlling gametic and somatic maturation are withdrawn at different times in *Nereis limnicola,* and by inference, are withdrawn simultaneously in *Nereis diversicolor* and other nereids. A more plausible explanation, however, is that a single inhibitory hormone is present in all nereids but that in *Nereis limnicola* the gametic and somatic tissues have developed different sensitivities to the hormone in relation to viviparity—either the gametes have become less sensitive to the hormone compared with other nereids, or the somatic tissues have adapted to respond to lower levels of the hormone (Baskin and Golding, 1970).

4. Control of Oocyte Development in Other Polychaetes

In the syllids, oocyte development is associated with the proliferation of new segments to form reproductive stolons, often with the formation of a new prostomium by modification of an existing adult segment, which separates from the parent body. The resemblance of stolon formation in syllids to the epitokous transformation in nereids is superficial, the former probably being developed from an originally asexual method of reproduction (Clark, 1961). The essential difference between the two groups is underlined by the fact that in syllids, although gamete development and stolon formation are under inhibitory endocrine control as in the nereids, the source of the factor is the pharyngeal region of the gut, and not the brain (Durchon, 1959; Durchon and Wissocq, 1964; Hauenschild, 1959). There are also indications that gonadal hormones are present in the syllids; the development of the secondary sexual characters in *Autolytus* is abnormal following gonadal X irradiation (Durchon, 1967a).

In *Arenicola marina,* the endocrine control of oocyte development is completely unlike that in the nereids and syllids; a hormone from the brain *promotes* oocyte maturation after the completion of vitellogenesis (Howie, 1963, 1966). Without the hormone, the oocytes remain in the arrested prophase of the first maturation division. The hormone can therefore be compared with "meiosis-inducing" substance of starfish ovaries (Section VI,C).

Maturation of oocytes in *Arenicola marina* begins in the coelom and precedes spawning by only a matter of hours. This explains why coelomic eggs are unfertilizable unless taken from spent or partially spent worms or from worms kept in the laboratory until they are about to spawn (Howie, 1961b). Spawning is automatic after maturation (Howie, 1963); mature eggs injected into a nonspawning worm will pass through the nephromixia, although unripe eggs cannot do so. However, since a few unripe eggs are spawned after the injection of tissue extracts from spawning worms, it is possible that the hormone has some effect upon the nephromixia (Howie, 1961b). The maturation hormone is common to both sexes, and in the male causes the breakdown of the sperm morulae so that they can be passed through the nephromixia (Howie, 1961a). The maturation hormone is absent from both male and female *Arenicola marina* in the period prior to breeding, but, when extractable during the breeding season, it occurs in equivalent amounts in both sexes (Howie, 1966). Unlike other polychaetes, *Arenicola marina* is characterized by a scarcity of stainable neurosecretory cells in the brain, and although the midbrain does contain concentrations of presumed neurosecretory cells, only the posterior lobes of the brain will induce oocyte maturation and spawning (Howie, 1966). Moreover, during the breeding season, all oocytes are in much the same stage of development, and although there is no direct evidence that coelomic gametes inhibit the release of a gonad-stimulating hormone, it is suggested that a feedback exists between the gametes and the brain to prevent the rapid proliferation of oogonia and spermatogonia (Howie and McLeneghan, 1965). Thus the earliest stages of oogenesis in *Arenicola marina* could be influenced by hormones, as in some gastropods and cephalopods (Section III,C,2 and 3) and arthropods (Section IV,C,1).

5. Coordination of Reproductive Development in Polychaetes

a. Coordination with Somatic Growth. In nereids, it is very likely that the hormone from the cerebral neurosecretory system which inhibits oocyte development and the epitokal metamorphosis in those species in which it occurs, is the same as that which promotes growth and regeneration in juvenile forms (Clark, 1966, 1969: Hauenschild and Fisher, 1962; Hauenschild, 1965). The relationship between somatic growth and reproductive development is thus determined very simply by the presence or absence of the cerebral hormone. The incompatibility between growth and reproduction may be due to more than the competing demands of the two stages; physiological adaptation and the gain in plasticity conferred by a long life (Clark, 1962), although clearly a mechanism which precludes demands for energy by competing processes, must have survival value (Farner, 1962).

The contrasting endocrine mechanisms found in nereids and other polychaete families may be related to different patterns of reproductive biology. Nereids and the stolons of syllids are monotelic, death following closely upon breeding, whereas many other polychaetes are polytelic, with repeating breeding cycles (Golding, 1974).

b. Coordination with the Environment. In *Platynereis dumerilii,* the breeding season extends from March to October in the Mediterranean, and the worms swarm on the surface with a monthly periodicity, maximum swarming occurring at the time of the new moon (Hauenschild, 1955, 1956). Oocyte development and metamorphosis take about 2 weeks to complete, and withdrawal of the brain hormone is associated with the increased "photoperiod" which begins at the time of the full moon (Hauenschild, 1960, 1965). Swarming and spawning are also synchronized by the production of a dialyzable, low molecular weight pheromonal compound by the individual worms (Boilly-Marer, 1969).

Arenicola marina, on the other hand, has a restricted breeding season of about 3 weeks duration, initiated by the first major fall in air temperature during low tide in the autumn (Howie, 1959), and this presumably causes the release of the oocyte maturation hormone. There is greater likelihood of synchrony in maturation between individuals when the terminal rather than the initial stages of development are hormonally controlled (Clark, 1965, 1969).

B. Oligochaeta

1. General Considerations

All oligochaetes are hermaphrodite, and most species possess a single pair of ovaries, although these may be duplicated in a few species. The ovaries are attached to a septum and each consists of a basal zone, a zone of proliferation, and a distal region filled with vitellogenic oocytes surrounded by flattened cells. The oocytes are shed from the distal region and pass to the exterior through ciliated coelomoducts.

2. Endocrine System

Neurosecretory cells are present in the brain and ventral ganglia (Schmid, 1947; Hubl, 1953, 1956; Herlant-Meewis, 1956, 1957). No specific neurohemal organs are present, but axon terminals are associated with the extensive system of blood capillaries within the brain (Herlant-Meewis, 1956, 1957; Hubl, 1953). No epithelial endocrine organs have been described.

3. Control of Reproduction

In *Lumbricus terrestris,* the number and content of material of specific cerebral neurosecretory cells increase with approaching maturation. They also vary with the annual reproductive cycle and after gonadectomy (Hubl, 1953). A similar relationship between supposed neurosecretory cell activity and reproductive development is found in *Eisenia foetida* (Herlant-Meewis, 1956). Extirpation of the cerebral ganglia from *Eisenia foetida* causes the worms to lose about one-third of their weight, and prevents egg laying for weeks in the majority of the operated animals (Herlant-Meewis, 1957). Extirpation of the subesophageal ganglion interrupts egg laying for 2 to 15 weeks, removal of the whole circumesophageal collar interrupts egg laying for 7 to 17 weeks, and removal of the ventral nerve chain between segments four to six prevents egg production for 3 to 8 weeks (Herlant-Meewis, 1957). Egg laying also stops in sham-operated worms, but in the majority of individuals recommences after 1 or 2 weeks. Together with observations on the histology of neurosecretory cells within the brain and subesophageal ganglion, these results are interpreted to mean that the brain exerts a positive effect upon egg laying in *Eisenia foetida,* but that the effect is indirect, being mediated by the subesophageal neurosecretory cells (Herlant-Meewis, 1957). Unfortunately, no reimplantations of the structures presumed to be involved in this positive control of egg laying have been attempted, and there is no experimental evidence for the endocrine control of any stages in oocyte development prior to egg laying.

C. Hirudinea

The leeches are hermaphrodite with a single pair of ovaries like the majority of oligochaetes. Neurosecretory cells are present in the brain, with a presumed neurohemal area on the posterior surface of the dorsal commissure (Hagadorn, 1966a; Hagadorn and Nishioka, 1961). The cerebral neurosecretory system is undoubtedly involved in gamete maturation, as has been demonstrated by studies of spermatogenesis (Hagadorn, 1966b).

III. MOLLUSCA

A. General Considerations

The great diversity in form and habitat shown by the molluscs is reflected in their reproductive mechanisms. Thus the cephalopods and scaphopods are exclusively gonochoristic and the lamellibranchs largely so, with only about 4% of the species hermaphrodite. The opisthobranch and pulmonate gastropods are hermaphrodite, but the prosobranchs are mostly gonochoristic (Fretter and Graham, 1964). The polyplacophoran amphineurans are gonochoristic and the aplacophora hermaphrodite.

The gonads are single in polyplacophoran amphineurans and paired in aplacophorans, single in gastropods and cephalopods, and single or paired in lamellibranchs. The gonads often take the form of branched, blind-ended tubules or acini.

In the lamellibranchs, oocyte development proceeds without associated follicular or nurse cells, although adjacent cells of the germinal epithelium may lengthen and partly cover the stalk by which the oocyte is connected to the basal membrane of the ovarial wall. In gastropods and polyplacophoran amphineurans, germinal epithelial cells form nurse cells, extending on the inner side of the germinal epithelium. When the oocytes begin to grow and extend into the gonadal cavity, the nurse cells form follicles around each oocyte. In the cephalopods, follicle cells form a membrana granulosa around the oocyte, which are, in turn, surrounded by a layer of connective tissue cells, the theca (Raven, 1958).

B. Endocrine System

1. Neurosecretion

The presence of neurosecretory cells in the nerve ganglia of molluscs has been known for many years (Scharrer, 1935, 1937). All the major groups contain neurosecretory cells, although in most instances their function is uncertain (Gabe, 1966).

Neurosecretory cells are present in the buccal ganglia of polyplacophoran amphineurans (Martoja, 1967); their absence from the cephalic part of the nervous system is considered primitive (Vicente and Gasquet, 1970).

Of twenty-five species of prosobranch gastropods examined by Gabe (1953a), neurosecretory cells are undoubtedly present in the cerebral ganglia of ten species, in the pleural ganglia of seventeen species, in the supraintestinal ganglia of sixteen species, and in the subintestinal and abdominal ganglia of only six and three species, respectively; neurosecretory cells are absent from the pedal and buccal ganglia. In the pulmonate gastropods, the cerebropleural and supraintestinal ganglia are major sites for neurosecretory cells (Gabe, 1954), although other ganglia may contain various numbers of the cells (Lever, 1957; Lever *et al.*, 1961). Opisthobranch gastropods have obvious neurosecretory cells in the cerebral ganglia, and the pleural ganglia in many species also contain many such cells (Gabe, 1953b). Clusters of large neurosecretory cells ("bag cells"), whose function is now known (Section III,C), are present in the abdominal ganglion of *Aplysia californica* (Coggeshall, 1967; Frazier *et al.,* 1967). The variety in distribution of neurosecretory cells among the nerve ganglia of opisthobranch gastropods is correlated with the anatomical variability of the nervous system in this group (Gabe, 1966). Similar variations in the number and distribution of neurosecretory cells occur in the lamellibranchs (Gabe, 1955, 1966). In the cephalopods, neurosecretory cells are found in the outer layers of the visceral lobes of the "brain" (Alexandrowicz, 1964, 1965; Berry and Cottrell, 1970).

Despite the large number of descriptions of neurosecretory cells in the molluscs, it is not certain that all cells designated "neurosecretory" are in fact so, since many neuronal inclusions can stain in the same way as neurosecretions with the reagents used (Gabe, 1966; Simpson *et al.,* 1966; Durchon, 1967b; Highnam and Hill, 1969). Until recently, the problem of molluscan neurosecretion was difficult to understand because of the apparent absence of neurohemal organs in the phylum (Gabe, 1966). Neurohemal areas in gastropods have now been described; they occur under the perineurium of nerves, along the intercerebral commissures and their connectives, and in the connective tissue sheath around the cerebral ganglia (Simpson *et al.,* 1966; Simpson, 1969; Wendelaar Bonga, 1970). In cephalopods, neurohemal areas are found in the inner layers of the vena cava, adjacent to the lumen of the vessel (Alexandrowicz, 1964, 1965; Martin, 1968; Berry and Cottrell, 1970). Neurosecretory axon terminals also occur in the connective tissue of blood vessels in the polyplacophoran amphineurans (Vicente and Gasquet, 1970).

2. Endocrine Glands

In cephalopods, the optic glands are paired organs situated upon the optic tracts connecting the brain with the optic lobes (Boycott and Young, 1956; Wells and Wells, 1959, 1969; Wells, 1960). The glands are possibly nervous in origin, arising from the regions at the sides of the vertical lobe of the brain (Boycott and Young, 1956). The optic glands are well vascularized and innervated, but without trace of neurosecretory axons.

In freshwater pulmonate snails, dorsal bodies are attached to the perineurium of the cerebral ganglia (Lever, 1957, 1958; Lever *et al.,* 1965; Joosse, 1964, 1972; Boer *et al.,* 1968). In all species, paired mediodorsal bodies occur; in some, there are additional paired laterodorsal bodies (Lever *et al.,* 1965; Joosse, 1972). It has been suggested that the dorsal bodies are innervated by neurosecretory axons from cells in the cerebral ganglia (Lever, 1958), although no direct contact between neurosecretory cells and dorsal bodies exists in *Lymnaea stagnalis, Ancylus fluviatilis, Australorbis glabratus,* and *Planorbarius corneus* (Boer *et al.,* 1968). In *Lymnaea stagnalis,* neurosecretory axons from cells in the cerebral ganglia occur within both the medio- and laterodorsal bodies. Because of their position near capillaries and blood spaces in the narrow strands of connective tissue which separate the cell groups of the dorsal bodies, it has been suggested they are merely part of the perineural neurohemal system (Wendelaar Bonga, 1970). The possibility that control by the dorsal body is affected by blood-borne neurosecretions, as in the vertebrate adenohypophysis, does not appear to have been considered.

In terrestrial pulmonates, structures comparable with the dorsal bodies are located in the thick connective tissue sheath around the cerebral ganglia and the cerebral commissure (Kuhlmann, 1966; Laryea, 1970). In prosobranch and opisthobranch gastropods the equivalent "organes juxtaganglionnaires" are attached to the anterodorsal regions of the cerebral ganglia (Martoja, 1965a,b; Vicente, 1969a,b,c). The "organe juxtacommissural" of polyplacophoran amphineurans, situated between the intercerebral commissures and an anterior blood space, is probably homologous with the organes juxtaganglionnaires of the prosobranchs and opisthobranchs (Vicente, 1970; Vicente and Gasquet, 1970).

The optic tentacles of stylommatophoran pulmonates are said to contain endocrine elements, probably neurosecretory (Lane, 1962, 1963; Sanchez, 1962), although this has been rejected (Bierbauer *et al.,* 1965; Rogers, 1969). The rhinophores of several species of opisthobranch gastropods are also said to contain endocrine cells (Vicente, 1969a,b,c). The endocrine status of these structures must at present be considered doubtful (Joosse, 1972), although there is experimental evidence that the tentacles are in some way involved in the control of reproductive development (Section III,C,2).

C. Endocrine Control of Ovarian Development

1. Amphineura

Neurosecretory cells and the organe juxtacommissural have been implicated in ovarian development in *Acanthochitona discrepans* and *Chiton olivaceus* (Vicente and Gasquet, 1970), but experimental evidence is lacking thus far.

2. Gastropoda

Until recently, gastropod reproductive endocrinology was concerned mainly with the establishment of correlations between histological cycles in neurosecretory cells and reproductive events (Highnam and Hill, 1969). However, at present, extirpations and reimplantations of suspected endocrine centers, together with the results of gonad cultures with and without hormone-producing tissues, provide a sound basis for gastropod endocrinology (Joosse, 1972).

   a. Differentiation of Gametes. Gamete development and sex reversal in hermaphrodite gastropods have attracted most attention. Hermaphrodite glands of *Helix aspersa* can be cultured for long periods in a suitable medium (Gomot and Guyard, 1964). Spermatogonia and spermatocytes survive for only a short time in the isolated gonad, the germinal epithelium producing oocytes and organized follicles (Gomot and Guyard, 1964; Guyard, 1969). When hemolymph from a snail in the male phase is added to the culture, spermatogenesis ensues (Gomot and Guyard, 1964). Correspondingly, cerebral ganglia from snails in the female phase stimulate oogenesis (Guyard, 1967). It is now known that isolated gonads autodifferentiate into ovaries (Guyard, 1969), other endocrine factors being responsible for testicular development. Similarly, isolated juvenile gonads of *Calyptraea sinensis* transform into ovaries after 20 days incubation without trace of spermatogenesis; gonads in the initial stages of spermatogenesis, or even in active spermatogenesis, transform into ovaries after 17 and 32 days incubation, respectively (Streiff, 1967). Isolated gonads of *Patella vulgata* also show ovarian development (Choquet, 1964, 1965), although it is possible that a mitogenic factor from the cerebral ganglia initially stimulates oogonial mitosis (Choquet, 1971).

   In general, this evidence suggests that female gametes arise by autodifferentiation and that spermatogenesis is stimulated by a factor in the hemolymph from animals in the male phase (Gomot and Guyard, 1964) presumably originating in the cerebral ganglia, since spermatogenesis occurs in hermaphrodite glands incubated with these ganglia (Choquet, 1965; Streiff, 1967; Guyard, 1969). Moreover, in protandric hermaphrodites, periods of spermatogenic repose are induced by a factor from the tentacles (Choquet, 1971), and incubation of gonads with tentacles also inhibits spermatogonial mitosis (Choquet, 1965, 1971). In *Arion californicus*, tentacle extirpation increases spermatogenesis, whereas the injection of tentacular homogenates suppresses precocious spermatogenesis (Gottfried and Dorfman, 1970a). The results of Choquet (1965, 1971) and Gottfried and Dorfman (1970a) are contrary to those of Pelluet and Lane (1961) and Pelluet (1964) in which the tentacles appear to have a stimulating effect upon spermatogenesis in several species of slugs. In *Ariolimax columbianus*, removal of the optic tentacles increases the galactogen content of the albumen gland, together with the incorporation of [^{14}C]glucose into galactogen; injection of tentacle homogenates reverses these effects (Meenakshi and Scheer, 1969). In *Helix aspersa,* tentacle removal leads to modifications of the hepatopancreas (Sanchez and Sablier, 1962). Perhaps the role of the tentacles is to affect metabolism rather than to control spermatogenesis directly, and tentacle removal may have differential effects according to the metabolic status of the animals.

b. Vitellogenesis. In cultures, the autodifferentiation of oocytes does not proceed beyond the growth phase; for vitellogenesis, a hormone from the cerebral ganglia is necessary (Guyard, 1967; Streiff, 1967; Choquet, 1971). However, in *Lymnaea stagnalis,* cauterization of the neurosecretory cells in the dorsal parts of the cerebral ganglia allows vitellogenesis to proceed, although shell growth is retarded. Removal of the dorsal bodies inhibits vitellogenesis, whereas their reimplantation is followed by renewed vitellogenesis (Joosse and Geraerts, 1969; Joosse, 1972). It is possible that the factor from the "cerebral ganglia," which also stimulates vitellogenesis in other species, actually has its source in the dorsal bodies or their equivalents.

In *Lymnaea stagnalis,* early oocytes and an accompanying follicle cell migrate from their origin in the germinal epithelial ring in each acinus of the ovotestis, to an area of the acinus which is apposed to the adjacent digestive gland (Joosse and Reitz, 1969). Here the oocyte grows and is covered with follicle cells, although one side of the oocyte lies against the acinar wall (and hence adjacent to the digestive gland) (Joosse and Reitz, 1969). It is possible that this arrangement of acinus and digestive gland plays an important part in the transfer of vitellogenic materials into the oocyte.

c. Ovulation/Spawning. In *Aplysia californica,* clusters of neurosecretory cells occur at the bases of the two pleurovisceral connectives where these join the parietovisceral ganglion (abdominal ganglion) (Coggeshall, 1967; Frazier *et al.,* 1967). Neurohemal areas associated with the neurosecretory cells are found within the connective tissue sheath of the abdominal ganglion and the distal parts of the pleurovisceral connectives (Frazier *et al.,* 1967).

Sea water extracts of the bag cells induce egg laying when injected into *Aplysia* (Kupfermann, 1967). Extracts of the abdominal ganglion (without bag cells) and of the distal parts of the pleurovisceral connectives induce egg laying (Strumwasser *et al.,* 1969), presumably because of their content of bag cell axon terminals within the neurohemal areas (Toevs and Brackenbury, 1969).

The *Aplysia* bag cells contain two specific proteins not found in any other neural tissue. One of these has exactly the same distribution within the bodies of the bag cells, the sheaths of the abdominal ganglion, and the pleurovisceral connectives as the active egg-laying factor (Toevs and

Brackenbury, 1969; Strumwasser *et al.,* 1969). Both the egg-laying factor and the bag cell-specific protein are destroyed by pronase and are heat denatured. It is suggested that the two are either identical or that the protein is a carrier for another active molecule like the neurophysin of vertebrates (Toevs and Brackenbury, 1969).

Each abdominal ganglion contains at least five times the threshold amount of the factor required for normal egg laying in *Aplysia* (Kupfermann, 1970). The bag cells are homogeneous and have remarkably similar properties. Normally they are electrically silent and do not respond to the stimulation of peripheral nerves, but prolonged repetitive spike activity, lasting up to 55 minutes, occurs when the pleurovisceral connectives are stimulated (Kupfermann and Kandel, 1970). All the bag cells in one cluster are invariably synchronously active after such stimulation, and there is also some synchrony with the bag cells of the opposite cluster. It is suggested that the electrophysiological properties of the cells are such that relatively prefixed amounts of hormone are released after stimulation (Kupfermann and Kandel, 1970).

Eggs are laid about 1 hour after injection of bag cell extract (Kupfermann, 1967; Strumwasser *et al.,* 1969), although they appear in the small hermaphrodite duct next to the ovotestis within a minute after injection of the extract (Coggeshall, 1970). During the remainder of the time spent within the genital tract, the eggs are packaged into a continuous cordon of gelatinous capsules. Coggeshall (1970) suggests that the bag cell hormone acts upon the muscle cells around the ovotestis causing their contraction, and forcing the ripe oocytes into the small hermaphrodite duct, and perhaps additionally causing dissolution of the small junctions holding the ripe oocytes to the follicle cells.

The number of bag cells in *Aplysia* increases about three times during maturation (Coggeshall, 1967), while the bag cell-specific protein increases ninefold (Toevs and Brackenbury, 1969). The bag cell secretion is present throughout the year (Strumwasser *et al.,* 1969; Toevs and Brackenbury, 1969), but bag cell extracts are maximally effective during the summer months (Strumwasser *et al.,* 1969; Kupfermann, 1970). It is suggested that, at other times of the year, a spawning inhibitor is present in the ovotestis (Strumwasser *et al.,* 1969; Toevs and Brackenbury, 1969) as in the gonads of starfish (Section VI,C). However, the ovotestis of *Aplysia* has a seasonal rhythm and its cyclic response to bag cell extract may reflect only the cyclic maturation of oocytes in the ovotestis (Coggeshall, 1970).

A rapid ovulation mechanism is also found in *Lymnaea stagnalis* (Joosse, 1972). It is possible that neuroendocrine processes involved in ovulation are universal in the gastropods.

d. Development of the Genital Tracts. In hermaphrodite gastropods, the proximal parts of the reproductive tract are themselves hermaphrodite, but the distal parts separate into male and female ducts, each with their own accessory glands. In terrestrial pulmonates, implantation of undifferentiated tracts into a host in the male phase induces development of the male parts of the implant; implantation into a female phase host leads to the development of the female duct and accessory glands (Laviolette, 1954; Runham and Hunter, 1970). The development of the genital tract is thus controlled by two hormones: one, present in early maturation, induces the development of the male parts of the tracts; the other, present later, controls the female parts of the tract (Runham and Hunter, 1970).

The gonad was originally thought to be the source of these hormones (Abeloos, 1943; Laviolette, 1954), although the gonad does not appear to contain obvious endocrine cells and injections of gonad extracts do not have any affect upon reproductive tract development. In *Ariolimax californicus* the cerebral ganglia produce the maturation hormones (Gottfried *et al.*, 1967). In *Calyptraea* and *Crepidula* species, as well as the gonochoristic *Littorina littorea,* the optic tentacles are responsible for the morphogenesis of the male genital tract and the pediopleural complex for its regression (Streiff *et al.*, 1970). In *Lymnaea stagnalis,* the dorsal bodies are involved in the growth and differentiation of the genital tract and, together with neurosecretory cells in the cerebral ganglia, control general metabolism (Joosse and Geraerts, 1969; Joosse, 1972). It is likely that hormones controlling the growth and differentiation of the genital tract may have various sources in different gastropod species (Joosse, 1972).

3. Lamellibranchiata

In *Mytilus edulis* and *Dreissena polymorpha,* histological cycles of secretion in neurosecretory cells in various nerve ganglia have been correlated with reproductive activity, although the effects of extirpation of such ganglia do not fully confirm such a function (Lubet, 1955, 1956; Antheunisse, 1963).

4. Cephalopoda

a. Oogenesis. The activity of the optic glands is regulated by an inhibitory nerve supply from the subpendunculate lobe of the brain (Wells and Wells, 1959, 1969). Isolated optic glands show a marked increase in size on the fourth day of incubation (Durchon and Richard, 1967). When ovaries of *Sepia officinalis* are incubated with optic glands, among other effects (see below), large numbers of mitotic figures appear within the germinal epithelium (Durchon and Richard, 1967). The optic gland hormone thus influences oogonial multiplication as well as the early stages of follicle cell development.

b. Vitellogenesis. Removal of the nervous inhibition of the optic glands of *Octopus* results in an increase in the size and activity of the glands, and is followed by enlargement of the ovary from one-five hundredth of the body weight to as much as one-fifth within 5 weeks of the procedure (Wells and Wells, 1959). In young females, the optic glands respond similarly to removal of the nervous inhibition, but the ovaries do not necessarily increase in size (Wells, 1960). The optic gland hormone stimulates vitellogenesis only if the oocytes are competent to respond. It has been shown by means of cultures of ovaries and optic glands that the oocytes of *Sepia officinalis* must be more than 0.3 mm in diameter before they respond to the optic gland hormone by increasing in size and initiating vitellogenesis (Durchon and Richard, 1967).

Vitellogenic proteins are synthesized within the ovaries, probably by the follicle cells, in *Octopus,* rather than being manufactured elsewhere and transported by the blood to the ovaries as in arthropods and vertebrates (O'Dor and Wells, 1973).

The optic glands presumably respond to photoperiod (Wells and Wells, 1959; Richard, 1971). Female *Octopus* die after laying their eggs and brooding them. There is no evidence for specific ovulation/spawning hormones in the cephalopods. The female reproductive tract with its associated glands enlarges and becomes functional under the independent control of the optic gland hormone (Wells, 1960), unlike the reproductive tract of the male, which regresses after castration (Taki, 1944).

IV. CRUSTACEA

A. General Considerations

The sexes are usually separate in the Crustacea, although many sedentary and parasitic forms are hermaphrodite. Protandric hermaphroditism is common in the prawns.

The ovaries of Crustacea are usually paired, and are tubular or saclike, particularly in the Malacostraca, to which most information about hormonal control mechanisms relates. The germinal epithelium is bandlike, extending along the lateral or ventral walls of the ovaries. The oocytes are enclosed in follicles formed from germinal epithelium cells, which are indistinguishable from those which transform into oocytes.

Unlike the majority of insects (Section V), the postmetamorphic Malacostraca continue to molt and grow. The juvenile forms are undifferentiated both morphologically and gonadally and sexual development takes place progressively. In some species, secondary sexual differentiation is not complete until well after gonadal maturity; in others, a "molt of puberty" marks the significant development of the secondary sex characters, and is preceded by gonadal differentiation.

Female reproduction alternates with molting, occurring during the intermolt periods, although molting egg-bearing individuals of *Carcinus maenas* and *Paratelphusa hydrodromous* have occasionally been observed (Cheung, 1966; Adiyodi, 1968a). Some decapod crustaceans molt throughout the year, or molt several times in a specific season, while others molt annually. The molting pattern can, therefore, impose perennial, seasonal, or annual reproductive cycles upon the species concerned. Copulation can occur only after a molt, when the female reproductive organs are soft enough to be deformed by the male intromittent organs (Carlisle, 1960). Since some decapods cease to molt when they have reached a certain size (Teissier, 1935; Carlisle and Knowles, 1959), the relationship between mating and female reproduction in these species is unclear.

B. The Endocrine System

The malacostracan endocrine system comprises four major endocrine centers: neurosecretory cells in the optic lobes with their associated neurohemal organs, the sinus glands; ectodermal epithelial endocrine glands, the Y organs, situated in either the antennary or the second maxillary segments (Gabe, 1953c, 1956); the androgenic glands in males, which are mesodermal derivatives associated with the vasa deferentia; and the ovaries in females.

The optic lobe neurosecretory cell centers are commonly called X organs, although the homologies of X organs in different groups are uncertain (Carlisle and Knowles, 1959; Gorbman and Bern, 1962). The X organ–sinus gland complex of crustaceans is analogous to the cerebral neurosecretory cell–corpora cardiaca system in insects (Section V,B) and the hypothalamo–hypophysial system in vertebrates. Some axon terminals in the sinus glands originate from neurosecretory cells in the brain, and other important groups of neurosecretory cells with associated neurohemal organs occur elsewhere in the nervous system (Carlisle and Knowles, 1959; Highnam and Hill, 1969). The neurosecretory hormones are peptides or small molecular weight proteins (Fingerman, 1966; Fingerman and Bartell, 1969; Bartell and Fingerman, 1969; Kleinholz, 1970; Terwilliger *et al.,* 1970; Berlind *et al.,* 1970; Berlind and Cooke, 1970).

The Y organs are analogous to the insect thoracic glands (Section V,B) and are presumably the source of the crustacean ecdysones which are essential for molting (Hampshire and Horn, 1966; Galbraith et al., 1968; Faux et al., 1969). The secretory activity of the Y organs is controlled by an inhibitory neurosecretory hormone from the X organ–sinus gland complex. The androgenic glands induce testis formation in genetic males, together with the sperm duct primordia, spermatogenesis, and the secondary sexual characters (Charniaux-Cotton, 1954, 1957, 1960a, 1962). The histo-chemistry and ultrastructure of the androgenic glands suggests that the hor-mone is a polypeptide or protein, although an associated or additional steroid hormone is not unlikely (Sarojini, 1963; Gilgan and Idler, 1967; Tcholakian and Eik-Nes, 1969).

The ovaries appear to be the source of hormones which induce both the permanent female secondary sexual characters and the temporary ones associated with incubation of the eggs (Charniaux-Cotton, 1960a; Balesdent, 1965; Reidenbach, 1967). The importance of other hormones in these inductions needs reassessment (Adiyodi and Adiyodi, 1970). The chemical nature of the ovarian hormones is unknown.

C. Endocrine Control of Egg Production

1. Primary Sex Determination

Androgenic glands are functional in males but degenerate in females (Charniaux-Cotton, 1954, 1956, 1957, 1960a, 1962; Charniaux-Cotton and Kleinholz, 1964; Charniaux-Cotton et al., 1966). Androgenic glands trans-planted into young females transform the ovaries into testes, and masculinize the females both with regard to secondary sexual characters and behavior. Ovaries implanted into normal males are similarly transformed into testes, but when implanted into males without androgenic glands remain as ovaries (Charniaux-Cotton, 1960a, 1962). The primordial gonad thus autodifferentiates into ovary in genetic females.

Early ovarian growth follows somatic growth, and is, therefore, associated with the production of ecdysones by the Y organs (Passano and Jyssum, 1963). In the shore crab, Carcinus maenas, bilateral extirpation of Y organs from females with vitellogenic oocytes prevents subsequent molt-ing, but has little effect upon the course of oocyte development. When the operation is performed in younger females, oogonial mitosis almost ceases and no oocytes develop (Arvy et al., 1954, 1956). Early oocyte growth may also be dependent upon ecdysones, because degenerated oocytes are found after bilateral Y organ extirpation in young females (Arvy et al., 1956). The early dependence of ovarial growth, including oogonial proliferation upon the ecdysones (Arvy et al., 1954, 1956; Démeusy, 1962), recalls the situation in the insects (Section V,B).

2. Vitellogenesis

In many species of decapod crustaceans, removal of the eyestalks leads to accelerated vitellogenesis, provided that the pubertal molt has already occurred (Adiyodi and Adiyodi, 1970). The implantation of sinus glands, extracts thereof, or extracts of the pars ganglionaris X organ will all inhibit ovarian development in females whose eyestalks have been removed (Panouse, 1943, 1947; Carlisle, 1953, 1959; Rangnekar and Deshmukh, 1968). There is little doubt that the X organ–sinus gland system in decapod crustaceans produces an "ovary-inhibiting hormone" (Panouse, 1943).

Removal of the eyestalks from prepubertal females causes molting rather than vitellogenesis (Passano, 1960; Passano and Jyssum, 1963; Bliss, 1966). This suggests that the result of the operation depends upon the developmental state of the animals. Molting in juveniles takes precedence over vitellogenesis, presumably because the ovaries have not reached the requisite degree of differentiation. In adults, vitellogenesis assumes greater importance than molting. However, eyestalk removal in postpubertal females at certain times of the year leads to molting rather than vitellogenesis (Gomez, 1965; Adiyodi, 1968a; Adiyodi and Adiyodi, 1970). Either the ovaries are periodically or cyclically responsive to removal of the ovary-inhibiting hormone, or some other factor intervenes in the control of vitellogenesis.

A gonad-stimulating principle has been suggested (Otsu and Hanaoka, 1951; Otsu, 1963). Implantation of thoracic ganglia into young females causes ovarian enlargement (Otsu, 1963; Gomez, 1965). Brain implantations have a similar effect (Gomez and Nayar, 1965) but other parts of the central nervous system do not (Gomez, 1965). Cycles of secretion of neurosecretory cells within the brain and thoracic ganglia have been correlated with vitellogenesis (Parameswaran, 1955; Matsumoto, 1958, 1962; Kurup, 1964; Adiyodi and Adiyodi, 1970). It is unfortunate that the site(s) of origin of the ovary-stimulating hormones precludes extirpation experiments. The control of vitellogenesis by alternately fluctuating levels of ovary-inhibiting and ovary-stimulating hormones, produced by neurosecretory centers in different parts of the central nervous system, suggests separate identities of the ovary-inhibiting and molt-inhibiting hormone (Adiyodi and Adiyodi, 1970). The problem has been extensively discussed (Carlisle, 1953; Passano, 1960; Charniaux-Cotton and Kleinholz, 1964) and the molt-inhibiting hormone must be considered essential for ovarian development in that it inhibits the Y organ and the processes leading to molting. It is suggested that the molt and ovary-inhibiting hormones act synergistically during postmolt phases, but that they become antagonistic during intermolt stages, during which reproductive development can take place (Adiyodi and Adiyodi, 1970).

In protandric hermaphrodites, eyestalk removal at about the time of sex reversal promotes ovarian development, whereas the injection of eyestalk extracts at that time inhibits ovarian growth and delays sex reversal (Carlisle, 1953). The inference that high levels of ovary-inhibiting hormone prevent sex reversal in protandric hermaphrodites (Carlisle, 1953) is not, at present, generally accepted. Androgenic gland hormone inhibits vitellogenesis even after eyestalk removal (Charniaux-Cotton, 1960a; Charniaux-Cotton and Kleinholz, 1964) and sex reversal is delayed if androgenic glands are implanted at the time of sex reversal (Charniaux-Cotton, 1960b). The androgenic glands atrophy under the influence of high titers of ovary (= gonad)-inhibiting hormone (Hoffman, 1968; Adiyodi and Adiyodi, 1970). Sex reversal in protandric hermaphrodites is thus ultimately controlled by the X organ–sinus gland system acting upon the androgenic glands. The ways in which the ovary-inhibiting and -stimulating hormones exert their effects upon vitellogenesis is unknown. There is evidence that a major lipoprotein constituent of yolk is synthesized extraovarially, probably in the hepatopancreas, and transferred to the vitellogenic oocytes via the hemolymph (Adiyodi, 1968b,c, 1969a,b,c). The ovary-inhibiting hormone prevents the mobilization of this and other yolk proteins (Adiyodi, 1968b). Female-specific proteins have been identified in a number of crustaceans (Barlow and Ridgway, 1969; Horn and Kerr, 1969; Kerr, 1969; Besse and Mocquard, 1968). Hepatopancreatic and hemolymph levels of phospholipids and carbohydrates also fluctuate in relation to the ovarian cycle (Adiyodi and Adiyodi, 1970; Dean and Vernberg, 1965; Telford, 1968; Lockwood, 1968), and hepatopancreatic RNA and enzyme syntheses appear to be controlled by eyestalk hormones (Fingerman *et al.,* 1967; Kurup, 1964; Kurup and Scheer, 1966; Wang and Scheer, 1963). Whether the extraovarial synthesis and mobilization of such materials is necessarily followed by vitellogenesis, or whether the ovary-inhibiting and -stimulating hormones act gonadotropically, or whether some other endocrine intervention is involved is uncertain.

3. Ovarian Hormones

In amphipods, the development of the female secondary sexual characters are controlled by hormones produced by the ovaries; oostegites are produced at the next molt when ovaries are transplanted into males deprived of their androgenic glands; ovigerous hairs are temporary sex characters, developing on the margins of the oostegites at the molt preceding the laying of eggs, and are produced only by vitellogenic ovaries (Charniaux-Cotton, 1956, 1957, 1960a,b, 1962). It is thus likely that the ovarian hormone controlling the permanent secondary sex characters is produced intermittently, perhaps from the follicle cells. Production of the latter hormone is thus probably influenced by the ovary-inhibiting and -stimulating hormones (Charniaux-Cotton, 1957, 1960a,b; Charniaux-Cotton and Kleinholz, 1964). In some species, certain permanent female secondary characters may be neutral, differentiating in females but inhibited by the androgenic gland in males (Reidenbach, 1967). In decapod crustaceans the control of secondary sexual characters by ovarian hormones is implied (Reinhard, 1950) but not yet proved (Adiyodi and Adiyodi, 1970). In any event, ovarian hormones have little to do with the control of oocyte development per se, unlike the relationship between androgenic glands and spermatogenesis in males.

4. Ovulation

As with the insects (Section V,D) ovulation and spawning are rarely distinguished as separate processes in the crustaceans. Increased neurosecretory cell activity associated with ovulation has been described by Matsumoto (1958), and an inhibitory control of spawning by eyestalk factors is suggested in *Menippe mercenaria* (Cheung, 1969). Environmental factors affect oviposition in *Orconectes virulis,* probably channeled through the neuroendocrine system (Aiken, 1969). The mechanism needs more thorough investigation before definite conclusions are drawn.

V. INSECTA

A. General Considerations

Insect ovaries are usually paired and consist of a variable number of tubules, the ovarioles. Each ovariole is enveloped in a simple epithelial sheath and a layer of connective tissue, continued into the terminal filament, which attaches the ovary to the body wall. The ovariole is divided into three regions: germarium, vitellarium, and ovariole stalk. The germarium contains the oogonia, enveloped and interspersed with mesodermal cells. At the base of the germarium are other mesodermal cells forming the prefollicular tissue. The vitellarium is the most extensive part of the ovariole, in which the oocyte grows and lays down yolk, surrounded by follicle cells derived from the germarial prefollicular tissue. Adjacent oocytes in the vitellarium are separated by plugs of modified follicle cells, the interfollicular tissue. The ovariole stalk connects each ovariole with a lateral oviduct, and contains a plug of follicular tissue which is ruptured during ovulation.

Ovarioles are of two main types: panoistic, which are without trophocytes (nurse cells), and meroistic, in which trophocytes are associated with each oocyte. Meroistic ovarioles may be polytrophic, in which the trophocytes accompany the oocyte into the vitellarium either contained within a common follicular epithelium (endofollicular) or in closely connected follicles (exofollicular). Alternatively, they may be telotrophic (acrotrophic), in which the trophocytes remain within the germarium but are connected with the oocyte in the vitellarium by strands of cytoplasm, the nutritive cords.

In panoistic ovarioles, all germ cells in the germarium transform into oocytes, whereas in meroistic ovarioles an oocyte and its associated trophocytes are descended from a common oogonium (King, 1970). The panoistic oocyte nucleus shows continuous and vigorous RNA synthesis, at least in its early stages (Zinsmeister and Davenport, 1971), and this, together with nucleolar RNA synthesis (Zalokar, 1960), suggests that the oocyte is involved in protein synthesis on its own account, although the follicle cells can assist by the synthesis or transfer of vitellogenic precursors (Lusis, 1963; Bell, 1970; Dufour *et al.,* 1971). Proteins derived from an extraovarial source are also incorporated as yolk (Engelmann, 1970). In meroistic oocytes, on the other hand, RNA and other materials are derived from the trophocytes both in telotrophic (Bonhag, 1955; Vanderberg, 1963) and in polytrophic types (Ramamurty, 1963, 1964; Bier, 1965; King, 1970) during the early stages of development. In many species, however, the bulk of the oocyte proteins comes from extraovarial and follicle cell sources (Telfer, 1954, 1960; Anderson and Telfer, 1969; de Loof and de Wilde, 1970a; Cruickshank, 1971).

The fully grown oocyte is surrounded by the vitelline membrane produced by the follicle cells alone (King and Koch, 1963; King, 1964) or in association with proteins synthesized in the oocyte cortex (Cruickshank, 1971). The follicle cells finally cover the oocyte with the more or less complex chorion (Beament, 1946). After ovulation, the follicle cells degenerate and may form a plug at the base of the ovariole called the corpus luteum. Corpora lutea are also formed from follicle cells which have invaded and destroyed developing oocytes (Phipps, 1949, 1960; King and Richards, 1968), although these can sometimes be distinguished from the corpora lutea of ovulation (Lusis, 1963). The insect corpus luteum is functionally and morphologically quite unlike that of vertebrate ovaries.

B. The Endocrine System

The insect endocrine system has four major components; groups of neurosecretory cells in the brain; the corpora cardiaca; the corpora allata;

and the thoracic glands or their equivalents. The cerebral neurosecretory cells and the corpora cardiaca form a neurosecretory system analogous to the hypothalamoneurohypophysial system of vertebrates. During embryogenesis, the corpora cardiaca develop as evaginations of the roof of the foregut in the same series as the ganglia and connectives of the dorsal sympathetic nervous system. The corpora allata and the thoracic glands are ectodermal epithelial endocrine glands derived from the first and second maxillary pouches. The endocrine system has a variety of arrangements in different insect species (Cazal, 1948).

The cerebral neurosecretory system produces a number of protein or peptide hormones (Ichikawa and Ishizaki, 1961; Gersch, 1961; Fraenkel *et al.*, 1966; Mills and Neilsen, 1967; Mordue and Goldsworthy, 1969). The thoracic glands, either directly or in association with some other tissue (Locke, 1969; Weir, 1970), secrete steroid hormones, the ecdysones (Karlson and Sekeris, 1966). The corpora allata produce one or more terpenoid compounds, the juvenile hormones (Röller *et al.*, 1967; Dahm *et al.*, 1968; Meyer *et al.*, 1968).

Growth and development of the body form are controlled by the synergistic action of the ecdysones and juvenile hormone (Wigglesworth, 1964; Highnam, 1970), the thoracic glands and corpora allata producing their secretions under the overall control of the cerebral neurosecretory system. Ovarian growth during the larval stages is controlled in a manner similar to that of the somatic tissues. In the pterygote insects, the thoracic glands degenerate during or shortly after the last molt, although exceptions are known (Cassier and Fain-Maurel, 1968, 1969a,b; Fain-Maurel and Cassier, 1968). Adult pterygotes consequently do not molt, and when oocyte development in the adult is hormonally controlled, the cerebral neurosecretory system and the corpora allata are the most obvious sources of the hormones concerned. Ecdysone, in fact, in some way inhibits oocyte development in the adult (Engelmann, 1959; Robbins *et al.*, 1968, 1970) in spite of its apparent developmental effects in the larval stages.

C. Endocrine Control of Oocyte Vitellogenesis

1. The Corpora Allata

In the bloodsucking bug, *Rhodnius prolixus,* allatectomy (removal of the corpus allatum) prevents vitellogenesis: the oocyte nucleus degenerates and the oocyte is invaded and resorbed by the follicle cells (Wigglesworth, 1936). The dependence of vitellogenesis upon the corpus allatum hormone has now been confirmed in a number of insect species representing most of the major orders (Engelmann, 1968, 1970). The exceptions are Lepidoptera with short-lived adults (Bounhiol, 1936; Bodenstein, 1938; Williams, 1952), although in some species total fecundity is reduced after allatectomy (Röller, 1962) and, in *Pieris brassicae,* vitellogenesis is controlled by the corpora allata (Karlinsky, 1963, 1967a,b,). In the Phasmidae, similarly, vitellogenesis proceeds in the absence of the corpora allata (Pflugfelder, 1937) although fecundity in some species may be reduced (Neugebauer, 1961). In the grasshopper, *Melanoplus sanguinipes,* a proportion of females still lay eggs after allatectomy (Dogra and Ewen, 1970), apparently as a consequence of regeneration from the allatal nerve.

The way in which the corpus allatum hormone promotes vitellogenesis is controversial. The hormone could affect the extraovarial synthesis of vitellogenic precursors which are then taken up from the hemolymph by the developing oocytes, or it could stimulate the oocytes directly, and the consequent demand for vitellogenic materials would induce their production elsewhere. The hormone could therefore be either "metabolic" or truly gonadotropic. The metabolic action of the corpus allatum hormone is well documented. In the grasshopper, *Melanoplus differentialis,* the hormone facilitates the production and mobilization of previtellogenic materials at an extraovarial site, probably the fat body (Pfeiffer, 1945). In the cockroach, *Periplaneta americana,* allatectomy results in a large increase in the fat content of the fat body, thought to be a consequence of an altered protein metabolism (Bodenstein, 1953), although in the locust, *Schistocerca gregaria,* similar effects of allatectomy upon the fat body are interpreted as being consequent upon reduced locomotor activity (Odhiambo, 1966a,b). Failure of protein synthesis following allatectomy occurs in the stick insect, *Dixippus morosus* (L'Hélias, 1957), and the procedure reduces muscle transaminase activity in *Periplaneta americana* (Wang and Dixon, 1960), as well as the production of other muscle enzymes in the Colorado beetle, *Leptinotarsa decemlineata* (de Kort, 1969). Protein metabolism is drastically altered by allatectomy in larvae of the wax moth, *Galleria mellonella* (Röller, 1962), and also reduces oxygen consumption in other insects (Thomsen, 1949; Thomsen and Hamburger, 1955; Sägesser, 1960).

Modified hemoglobin is found in the oocytes of a number of bloodsucking insects (Wigglesworth, 1943), and the suggestion that proteins in the hemolymph can be taken up directly by developing oocytes has been confirmed in saturniid moths (Telfer, 1954, 1960; Telfer and Rutberg, 1960). More significantly in the moth, *Hyalophora cecropia,* a female sex-specific protein is present in the hemolymph (Telfer and Williams, 1953), and although all hemolymph proteins are taken up by the oocytes to some extent, the sex-specific protein is taken up preferentially (Telfer, 1960). Female sex-specific proteins, or vitellogenins, have now been identified in other saturniids (Laufer, 1960); in cockroaches of different species (Engelmann, 1965, 1969, 1970; Engelmann and Penney, 1966; Adiyodi, 1967; Menon, 1963; Adiyodi and Nayar, 1966; Thomas and Nation, 1966; Nielsen and Mills, 1968; Scheurer, 1969; Bell, 1969, 1970; Prabhu and Hema, 1970; Bell and Barth, 1970); in locusts (Hill, 1962; Tobe and Loughton, 1967; Kulkarni and Mehrotra, 1970; Dufour *et al.*, 1971; Engelmann *et al.*, 1971); in the mosquito *Aedes* (Roth and Porter, 1964); in the fleshfly *Sarcophaga bullata* (Wilkens, 1967); in the housefly *Musca* (Bodnaryk and Morrison, 1968); and in the Colorado beetle, *Leptinotarsa decemlineata* (de Loof and de Wilde, 1970a,b). The vitellogenin(s) may be present in small amounts in the hemolymph, and can be detected with certainty only by immunological techniques (Hill, 1972). Without the use of such methods, the stated absence of vitellogenins in *Locusta migratoria* (Bentz, 1969; Bentz *et al.*, 1970) should be accepted cautiously.

It is now known that vitellogenins are produced only when the corpus allatum hormone is circulating (Coles, 1964, 1965a,b; Engelmann, 1965, 1966; Engelmann and Penney, 1966; Engelmann *et al.*, 1971) and that they are synthesized when other cerebral hormones are excluded (Engelmann, 1969; Bell, 1969, 1970; Bell and Barth, 1970, 1971; de Loof and de Wilde, 1970a,b). This provides the best evidence for an extraovarial metabolic action of the corpus allatum hormone. It is suggested that the hormone activates the fat body genome to produce specific mRNA's for vitellogenin synthesis (Minks, 1967; Engelmann, 1972), although an inducing effect by the ovaries cannot be entirely ruled out since transplantation of ovaries into male insects is followed a considerable time later by the appearance of vitellogenins in the hemolymph (Prabhu and Hema, 1970). It must be supposed that vitellogenins are not normally produced by fat bodies of male insects, in spite of the presence of corpus allatum hormone, because of genetic sex differences.

Despite this metabolic effect of the corpus allatum hormone, there is equally good evidence for a gonadotropic action. Implanted corpora allata can have a local effect upon vitellogenesis (Joly, 1945) which would not be expected if the only action of the hormone were extraovarial. Injections of the hormone can also act locally upon vitellogenesis (de Loof and de Wilde, 1970b). Rapid and complete oocyte resorption follows allatectomy in *Schistocerca gregaria* (Highnam *et al.*, 1963a), the time interval being too short for the complete run.down of a protein synthetic mechanism. Moreover, in *Schistocerca gregaria,* resorption will occur in some oocytes even in partially ovariectomized animals, and can be prevented by the addition of a corpus allatum hormone mimic (Highnam *et al.*, 1963b). Developing oocytes are more permeable to vital dyes (Iwanoff and Mestcherskaja, 1935), and the corpus allatum hormone could therefore act by altering oocyte membrane permeabilities allowing the uptake of previtellogenic materials from the hemolymph. The hormone could also affect the synthesis or transfer of vitellogenic materials by the follicle cells (Pfeiffer, 1945; Novak *et al.*, 1959; Dufour *et al.*, 1971; de Loof and Lagasse, 1970a,b).

Perhaps the obvious conclusion is that the corpus allatum hormone has a dual role: to promote vitellogenin synthesis and to facilitate the uptake of yolk precursors. Vitellogenins injected into allatectomized *Periplaneta americana* are not taken up by the oocytes unless corpus allatum hormone is also injected or corpora allata implanted (Bell, 1969). Starved cockroaches begin to resorb their oocytes after 10 days, and the concentration of vitellogenic blood proteins increases markedly; yolk deposition recommences when corpus allatum hormone or a mimic is injected (Bell, 1971). Finally, when the ovaries of *Eublaberus posticus* are transplanted into males, the greatest development of yolk spheres in the oocytes occurs only when vitellogenins and corpus allatum hormone are injected together rather than singly (Bell and Barth, 1971).

The exact correlation between vitellogenic protein production and its utilization implied by the duality of action of the corpus allatum hormone is clearly metabolically "efficient," and could be of crucial importance when external or internal conditions are such that resorption of all or a proportion of oocytes becomes necessary (Highnam and Lusis, 1962; Highnam *et al.*, 1963a,b; Highnam, 1972).

2. The Cerebral Neurosecretory System

In the blowfly, *Calliphora erythrocephala,* removal of the cerebral neurosecretory cells prevents oocyte development, the egg chambers remaining smaller than after allatectomy (Thomsen, 1952). Destruction or removal of the cerebral neurosecretory cells similarly prevents oocyte development in the locusts *Schistocerca gregaria* (Highnam, 1962a,b), *Schistocerca paranensis* (Strong, 1965a,b), *Locusta migratoria* (Girardie, 1964, 1966), and *Anacridium aegyptium* (Geldiay, 1965); in the grasshopper, *Gomphocerus rufus* (Loher, 1965, 1966a,b); in a number of mosquito species (Lea, 1963, 1967, 1972); in the beetle, *Tenebrio molitor* (Mordue, 1965a,b,c); and in the fleshfly, *Sarcophaga bullata* (Wilkens, 1967, 1968). In the hemipteran, *Oncopeltus fasciatus,* and the phasmids, *Carausius morosus* and *Clitumnus extradentatus,* destruction of the neurosecretory cells does not prevent oocyte development, but the overall fecundity of the females is reduced (Johansson, 1958; Dupont-Raabe, 1952, 1956).

Because of the reduced size of the egg chambers in *Calliphora erythrocephala* after removal of the cerebral neurosecretory cells, together with the resemblance between these flies and those kept without meat, it seemed possible that the operation interferes with protein metabolism (Thomsen, 1952). This suggestion is confirmed by the reduced midgut proteinase activity following the operation (Thomsen and Møller, 1959a,b, 1963). More directly, the hemolymph protein concentration is reduced following destruction of the cerebral neurosecretory cells in *Schistocerca gregaria* (Hill, 1962), *Gomphocerus rufus* (Loher, 1965, 1966a,b), and *Tenebrio molitor* (Mordue, 1965a,b,c). It has been suggested, therefore, that neurosecretory hormones control vitellogenic protein synthesis, and that the proteins are taken up from the hemolymph by the developing oocytes under the control of the gonadotropic corpus allatum hormone (Highnam, 1971).

The later evidence for vitellogenin synthesis controlled by the corpus allatum hormone, together with possible side effects of the operations for removal or destruction of the cerebral neurosecretory cells (Engelmann, 1960, 1965; Engelmann and Penney, 1966), casts doubt on the hypothesis that neurosecretory hormones control protein synthesis. However, in *Leucophaea maderae,* five out of six hemolymph protein fractions are stimulated by neuroendocrine factors, the sixth being stimulated only when the corpora allata are present and active (Scheurer, 1969). In *Leptinotarsa decemlineata,* both brain and corpus allatum hormones are necessary for the synthesis of vitellogenic blood proteins (de Loof, 1969; de Loof and de Wilde, 1970b). Cerebral neurosecretory cells and the corpus allatum are equally essential for oocyte development in mosquitoes (Lea, 1967, 1970, 1972).

The control mechanisms for oocyte development may differ in different insects (Engelmann, 1970). However, in those insects where both neurosecretory and corpus allatum hormones are necessary, it is likely that the neurosecretory factors can affect protein synthesis generally, while the corpus allatum hormone induces the production of specific vitellogenins as well as acting gonadotropically (Minks, 1967). Moreover, as pointed out by de Loof and de Wilde (1970a), almost all the hemolymph proteins can be detected in the oocytes, and although the vitellogenins make up the bulk of protein yolk, the trace amounts of the other proteins may be of great importance during subsequent embryogenesis.

In the fleshfly, *Sarcophaga bullata,* a hormone from the cerebral neurosecretory system is said to act gonadotropically, facilitating the uptake of proteins produced under the control of the corpus allatum hormone (Wilkens, 1967, 1969). If this is true, the roles of the two hormones are thus exactly opposite those postulated for blowflies and locusts.

The cerebral neurosecretory system also produces hormones involved in the control of water balance (Mordue, 1969, 1971; Vietinghoff, 1967; Highnam *et al.,* 1965; Cazal and Girardie, 1968; Berridge, 1966), of carbohydrate metabolism (Goldsworthy, 1969, 1970, 1971; Highnam and

Goldsworthy, 1972; Mordue and Goldsworthy, 1969; Mordue *et al.,* 1970; Steele, 1961, 1963), and of lipid mobilization (Mayer and Candy, 1969). It is likely that variations in the production or release of these factors can have profound, though secondary, effects upon oocyte vitellogenesis.

3. Interactions between the Cerebral Neurosecretory System and the Corpora Allata

There is good evidence that the activity of the corpora allata is controlled by a neurosecretory "allatotropic" factor from the cerebral neurosecretory cells (Highnam, 1964; de Wilde, 1964; Masner *et al.,* 1970), although, in some species at least, cerebral control of the glands is neural in nature (Engelmann, 1968, 1970). In *Locusta migratoria,* it is claimed that the corpora allata are stimulated and inhibited by neurosecretions originating in different types of cerebral neurosecretory cells (Girardie, 1966, 1967; Girardie and Girardie, 1967).

The corpus allatum hormone can have a stimulatory effect upon the production and/or release of factors from the cerebral neurosecretory system (Highnam *et al.,* 1963a; Nayar, 1962; Lea and Thomsen, 1962, 1969; Cassier, 1967; Thomsen and Lea, 1969). In *Calliphora erythrocephala,* the reciprocity between the corpora allata and the cerebral neurosecretory system could be via the intervention of hemolymph metabolites and selective feeding (Strangways-Dixon, 1962), although this is considered unlikely (Lea and Thomsen, 1969; Thomsen and Lea, 1969).

The difficulties inherent in interpreting the effects upon oocyte development of removal and implantation of the parts of the cerebral endocrine system, when feedback mechanisms between these parts are involved, may have given rise to much of the confusion which surrounds our views about the endocrine control of reproduction in insects. These difficulties are now being resolved with the current availability of pure preparations of corpus allatum hormone and of compounds which mimic its effect.

4. Cyclic Development of Oocytes

Many insects produce successive batches of eggs, the oocytes nearest the oviducts developing synchronously. It has been suggested that these "terminal" oocytes preferentially become vitellogenic because they are developmentally ahead of the penultimate and other oocytes within the ovariole, and compete more successfully for available vitellogenic materials and corpus allatum hormone (Highnam *et al.,* 1963a,b). Once laid, the places of the oocytes are assumed by the next oocytes in line and the cycle is repeated. In ovoviviparous insects, and those which carry an ootheca for a time before deposition, nervous stimuli generated by the egg cases inhibit the activity of the neuroendocrine system (Engelmann, 1957, 1959, 1960, 1964, 1970; Barth, 1968). The nature of the reproductive process itself thus imposes a cyclic rhythm of activity upon the neuroendocrine system, with concomitant gonotropic cycles.

In other insects, however, the cyclic activity of the cerebral neurosecretory system and the corpora allata must be controlled by a different mechanism. The balance between rhythmic feeding activity and the utilization of previtellogenic materials by the oocytes is one possibility (Highnam *et al.,* 1963a,b; Highnam and West, 1971; Hill *et al.,* 1966, 1968; Hill and Goldsworthy, 1968). A possible humoral inhibitory feedback mechanism to the neuroendocrine system (Nayar, 1956a,b, 1958a,b,c; Engelmann, 1957, 1964, 1965; Scharrer and von Harnack, 1961; Doane, 1961; Minks, 1967) has now been placed on a more solid footing (Adams *et al.,* 1968; Adams and Nelson, 1969; Adams, 1970; Clift, 1971), and should stimulate investigations for "oostatic hormones" in other insect species.

D. Ovulation and Oviposition

Ovulation and oviposition are not necessarily synchronous in insects, since many species may retain ovulated eggs in their oviducts. Consequently, although there is evidence that oviductal contractions are facilitated by hormones from the cerebral neurosecretory system (Enders, 1956; Nayar, 1958a,b; Highnam, 1963, 1965; Davey, 1967), direct hormonal intervention in ovulation has not been investigated. In *Hyalophora cecropia,* increased oviposition is due to release of an intrinsic corpus cardiacum hormone, triggered by the reception of spermatozoa by the female (Truman and Riddiford, 1971).

E. Environmental Control of Oocyte Development

1. Short-Term Stimuli

Many stimuli, arising from crowding, mating, feeding, locomotion, and pheromones, increase both the rate of production and the numbers of eggs in different insect species (Highnam, 1964, 1965; Engelmann, 1968, 1970; Barth, 1968). The stimuli act to increase the activities of the cerebral neurosecretory system and the corpora allata, with concomitant effects upon oocyte vitellogenesis. By these means, the chances of fertilization of the largest number of eggs, laid in optimal conditions for embryonic and postembryonic development, are greatly increased.

2. Long-Term Stimuli

Many female insects react to unfavorable environmental conditions by an arrest of reproductive activity. The adults may continue to feed (when the condition is called gonotropic dissociation), or feeding may be arrested and metabolism reduced, the insects entering a true diapause. During diapause, the corpora allata and the cerebral neurosecretory system are inactive until there is some change in the photoperiod, or the quantity and quality of food. The subject is too vast to be discussed in detail here, but excellent reviews are available (de Wilde, 1964, 1970).

VI. ECHINODERMATA

A. General Considerations

Echinoderms are usually dioecious although some species may be protandrous hermaphrodites, while a few are true self-fertilizing hermaphrodites. The primitively single gonad is retained only in the Holothuroidae. In the remaining groups, the gonads have a radially symmetrical arrangement corresponding to the body symmetry.

Histologically, the ovaries of all echinoderms are very similar, with a central lumen surrounded by the germinal epithelium which may invaginate to form ovarian folds (Mauzey, 1966). Connective tissue and muscle layers lie outside the germinal epithelium, and the whole ovary is surrounded by peritoneum. In the echinoids *Echinometra lucunter* and *Mespilia globulus,* the outermost layer is said to be the germinal epithelium, oogonia from which migrate through the muscle and connective tissue layers. After division in the ovarial lumen, the oogonia develop into oocytes which are then aligned along the ovarial wall to form the inner epithelial layer (Tennent and Ito, 1941). Developing oocytes later pass into the ovarial lumen.

Cells in the vicinity of growing oocytes may form rudimentary follicles in some echinoderms; they are poorly developed and transient in echinoids, but better developed and longer lasting in asteroids and holothuroids. The follicle cells assist in the nutrition, and sometimes resorption, of the oocytes (Bal, 1970). The presence or absence of follicles around the developed oocytes is of significance in the control of maturation and spawning in the asteroids and echinoids.

B. Endocrine System

Although neurosecretory hormones play an important part in the control of oocyte maturation and spawning (see below), discrete neurosecretory systems and endocrine glands are unknown in the echinoderms. Neurosecretory cells are associated with Lange's nerve (part of the hyponeural system) in the starfish *Asterias rubens* (von Hehn, 1970) and with the radial nerves in the ophiuroid *Ophiopholis aculeata* (Fontaine, 1962). Otherwise, although neurosecretion in the radial nerves is directly involved in reproductive processes, the cells which produce the factors are unknown.

C. Control of Oocyte Maturation and Spawning

Oogonial proliferation and vitellogenesis have been little studied in echinoderms. Most information relates to the control of the final stages of oocyte development, maturation and spawning, and is confined almost exclusively to asteroids. Water extracts of dried, lyophilized radial nerves injected into the coelomic cavity of *Asterias forbesi* induce spawning within 30 minutes (Chaet and McConnaughy, 1959; Chaet and Musick, 1960; Chaet, 1964; Kanatani, 1964). The gamete-shedding substance is a heat-labile, dialyzable polypeptide (Kanatani and Noumura, 1962; Chaet, 1966a), with a molecular weight of about 2200 (Kanatani, 1969a; Chaet, 1966a,b). The substance is not sex- or species-specific (Hartman and Chaet, 1962; Noumura and Kanatani, 1962; Chaet, 1964, 1966a), although exceptions are known. *Henricia leviuscula* and *Othelia tenuispina* do not respond to shedding substance prepared from other species (Chaet, 1966a), and these species produce fewer and larger eggs than the other species tested (Chaet, 1966a). Moreover, *Asterina pectinifera* nerve extract will induce spawning in *Asterias amurensis,* but the converse does not obtain (Noumura and Kanatani, 1962). It was suggested that the shedding substance induces contractions in the ovarial wall and that it required Ca^{2+} for this action (Chaet, 1966a,b). At spawning, germinal vesicle breakdown is induced, so that the shedding substance indirectly promotes oocyte maturation in addition to its spawning action (Chaet, 1964, 1966a).

The shedding substance is related to the presence of neurosecretory granules, 1 to 2 μm in diameter, in specific layers of the radial nerves (Chaet, 1966a). The concentration of shedding substance is the same along the full lengths of the radial nerves and in the nerve ring itself (Kanatani and Ohguri, 1966). The perikarya of the neurosecretory neurons presumably responsible for the production of the substance have not been identified, although some neuronal perikarya in Lange's nerve associated with the lateral portions of the radial nerves do contain neurosecretory granules (von Hehn, 1970).

Because of the proximity of the radial nerves to the oral surface, and the ease with which asteroids can absorb amino acids and other substances from seawater (see Ferguson, 1964), it was suggested that the gamete-shedding substance passes outside the body to be taken up by the peripheral tissues and then carried to the coelom where the ovarian effect is induced (Chaet, 1966a,b). In *Asterias amurensis,* although extracts of the tube feet and body wall have considerable gamete-shedding properties, so also does coelomic fluid when the starfish undergo spawning (Kanatani and Ohguri, 1966) and a seawater route of entry to the coelom is denied (Kanatani and Shirai, 1969). In the ophiuroid *Ophiopholis aculeata,* it has been suggested that neurosecretion may diffuse into the coelom from the radial nerves (Fontaine, 1962), and that the same may occur in asteroids. It is also possible that neurosecretion may pass along nerves to the genital ducts and the external epithelium of the gonads in *Asterina gibbosa* (Bruslé, 1969).

It is now known that the radial nerve shedding substance does not act directly to induce spawning and maturation in the oocytes. In *Asterias forbesi,* a small molecular weight substance can be extracted from ovaries which causes breakdown of the germinal vesicles in isolated immature oocytes maintained in normal or Ca^{2+}-free seawater (Schuetz and Biggers, 1967; Kanatani and Shirai, 1967, 1970). The factor increases in amount on the addition of radial nerve substance to ovaries (Schuetz and Biggers, 1967), and the increase is proportional to the amount of nerve substance (GSS) added (Schuetz, 1969). The ovarian factor is heat stable and nonproteinaceous (Schuetz, 1969; Kanatani, 1969a). The radial nerve factor is now called "gonad-stimulating substance" and the material it induces "meiosis-stimulating or -inducing substance" (MIS) (Kanatani, 1969a, 1972; Kanatani *et al.,* 1969), although, since meiosis is completed under the action of the compound, "maturation-stimulating substance" may be a more descriptive and exact term (Dodd, 1972).

When 20 kg of *Asterias amurensis* ovaries were incubated in 100 liters of GSS seawater, 8.5 mg pure MIS were extracted, and finally identified as 1-methyladenine (Kanatani *et al.,* 1969). Synthetic 1-methyladenine is very effective in inducing oocyte maturation and spawning in starfish (Kanatani, 1969b). When tested upon five asteroid species, pure 1-methyladenine induced spawning in each, but one echinoid and one holothuroid similarly injected did not react (Stevens, 1970). Concentrations as low as $1 \times 10^{-6} M$ are fully effective in inducing spawning in *Patiria miniata* (Stevens, 1970). MIS has two distinct effects: it increases the dissolution of cementing substances between the follicle cells and between the follicle cells and the oocyte surface (at least in some starfish: *Asterias amurensis* and *Asterias forbesi* strip off follicle cells when isolated in normal seawater), and it causes germinal vesicle breakdown in the oocyte (Kanatani and Shirai, 1967). In the five starfish species examined by Stevens (1970), a lower concentration of 1-methyladenine was required for oocyte maturation than for stripping off follicle cells ($1.3 \times 10^{-7} M$ as compared with $<1 \times 10^{-6}$

M). 1-Ethyladenine is also effective in inducing oocyte maturation and 1-methyladenosine acts on oocytes in the ovary, but not on isolated oocytes (Kanatani and Shirai, 1971). 1-Methyladenosine is split into 1-methyladenine ribose by the enzyme 1-methyladenosine ribohydrolase which is present in ovarial tissue (Kanatani and Shirai, 1971), or specifically the ovarian wall (Schuetz, 1970).

With a constant concentration of gonad-stimulating substance, the quantity of 1-methyladenine produced increases in proportion to the number of oocytes with follicles in the incubation medium, and follicle cells alone produce equivalent amounts of the compound (Hirai and Kanatani, 1971). The neurosecretory hormone must, therefore, act upon the follicle cells to produce 1-methyladenine, and the original suggestion that gonad-stimulating substance acts upon the ovarial wall to induce its contraction (Chaet, 1966a,b) is thus no longer valid. It is likely that freeing the oocytes from their follicles facilitates spawning under pressure from the expanded ovaries (Kanatani, 1969a).

Production of 1-methyladenine by follicle cells assumes greater significance when the site of action of the compound is considered. Microinjection of 1-methyladenine into the oocyte fails to induce germinal vesicle breakdown (Kanatani and Hiramoto, 1970), whereas enucleated oocytes produce fertilization membranes after insemination only when they are previously incubated with 1-methyladenine (Hirai *et al.,* 1971). These results suggest that cytoplasmic maturation at the surface of the oocyte is induced by 1-methyladenine without the participation of germinal vesicle material (Hirai *et al.,* 1971), and that germinal vesicle breakdown is consequent upon this cytoplasmic maturation.

The time at which spawning and oocyte maturation occurs could be determined by the release of gonad-stimulating substance into the coelom, perhaps induced by rising temperatures or dehydration (Kanatani and Ohguri, 1966). However, Chaet (1966a,b) has suggested that the action of gonad-stimulating substance is inhibited by another radial nerve neurosecretory factor which he has termed "shedhibin." Fluctuations in the relative proportions of "shedhibin" present would then determine the actual time of spawning due to gonad-stimulating substance (Chaet, 1966a). The situation is further complicated by the reported presence of antimitotic substances present in starfish ovaries early in the breeding season which inhibit the action of gonad-stimulating substance (Ikegami *et al.,* 1967). Heilbrunn *et al.* (1954) suggested a heparinlike compound as the mitotic inhibitor, but Ikegami *et al.* (1967) have identified L-glutamic acid as the compound involved. When a range of pure amino acids was similarly tested, only aspartic acid proved to inhibit the gonad-stimulating substance, and then only by about 10% of that of authentic L-glutamic acid (Ikegami *et al.,* 1967).

VII. COMPARATIVE ASPECTS

Ovarian extracts from invertebrates have long been known to elicit estrogenic responses in mammals (Hanström, 1939). In the scallop, *Pecten hericius,* estrogenic activity is equivalent to about 10 µg/kg wet tissue (Botticelli
et al., 1961). Estrogens (Donahue, 1948, 1952, 1957), and specifically estra-
diol-17β (Lisk, 1961), have been found in the eggs of the American lobster,
Homarus americanus; α-ketolic steroids occur in the ovaries of other crustaceans (Brodzicki, 1966). Estradiol-17β, together with progesterone, is
present in the ovaries of the starfish, *Pisaster ochraceus* (Botticelli *et al.,*
1960), and probably also in those of the sea urchin *Strongylocentrotus franciscanus* (Botticelli *et al.,* 1961).

Enzymes involved in steroid biosynthesis and metabolism have been
reported from the gonads and associated tissues of molluscs (Gottfried and
Dorfman, 1969, 1970a,b,c; Gottfried and Lusis, 1966; Gottfried *et al.,* 1967;
Idler *et al.,* 1969), crustaceans (Gilgan and Idler, 1967; Tcholakian and Eik-
Nes, 1969), insects (Lehoux *et al.,* 1968; Lehoux and Sandor, 1969; Dubé
and Lemonde, 1971), and sea urchins (Hathaway and Black, 1969),
although the conversions described may not necessarily occur in nature
(Lehoux and Sandor, 1970).

In male lobsters and crabs, steroidogenic enzymes are presumably
necessary for the normal endocrine function of the androgenic glands
(Gilgan and Idler, 1967; Tcholakian and Eik-Nes, 1969), and it is suggested
that estrogen may inhibit molting in the egg-bearing female lobster
(Donahue, 1955). In the slug, *Ariolimax californicus,* the optic tentacle
inhibition of spermatogenesis may operate by modifying the sequence of
steroidogenic conversions (Gottfried and Dorfman, 1970a,b,c). In the sea
urchin, *Strongylocentrotus franciscanus,* enzymes in the gut convert estra-
diol-17β to estradiol-3-sulfate at the rate of 6 to 7 µmole/gm protein/hour,
probably as a detoxification or excretory process, suggesting a high
turnover of steroid by the gonads (Creange and Szego, 1967). In *Schistocerca gregaria,* steroidogenic enzymes may also serve as detoxification
and excretory mechanisms (Dubé and Lemonde, 1971).

The significance of ovarian steroids in invertebrates is not understood. In
gonochoristic and hermaphrodite mollusks and crustaceans (and in
Lampyris noctiluca among the insects), oogonia autodifferentiate from the
germinal epithelium, whereas spermatogonia are induced by extragonadal
hormones—from the cerebral ganglia in gastropod molluscs and from the
androgenic glands in crustaceans. Sex reversal in protandric hermaphrodites in both gastropod molluscs and crustaceans is also controlled by
extragonadal hormones—a tentacular hormone in the gastropods, and a hormone from the eyestalk neurosecretory system in crustaceans, which

indirectly prevents spermatogenesis by inhibiting the secretory activity of the androgenic glands.

Oogonial proliferation is hormonally controlled in gastropod and cephalopod molluscs, crustaceans, insects, and possibly also in *Arenicola marina* in the annelids. The hormones originate in the cerebral ganglia (or perhaps the dorsal bodies) in gastropods, the optic glands in cephalopods, the Y organs in crustaceans, the thoracic glands in insects (both ultimately controlled by neurosecretory tropic hormones), and the cerebral ganglia in *Arenicola*. The arthropod hormones are steroids, while those of the gastropods, cephalopods, and *Arenicola* are probably peptides or proteins.

Oocyte vitellogenesis is hormonally controlled by the dorsal bodies in gastropod molluscs and by the, perhaps equivalent, optic glands in cephalopods. In nereid polychaetes, the reduction in titer of a cerebral ganglion hormone promotes vitellogenesis; in oligochaetes and leeches, vitellogenic hormones have not been described. Neurosecretory hormones in crustaceans inhibit and promote vitellogenesis, acting either directly or indirectly upon the ovaries. In insects, neurosecretory and corpus allatum hormones act together to promote vitellogenesis, the former acting indirectly, the latter both indirectly and directly. Vitellogenic hormones are unknown in starfish.

Oocyte maturation is controlled by a brain hormone in *Arenicola*, and by a combination of neurosecretory and ovarian (follicular) hormones in starfishes. In the latter group, both inhibitory and stimulatory neurosecretory hormones are present; whether the 1-methyladenine induced by the neurosecretory gonad-stimulating hormone is a hormone in the strict sense is doubtful. It now seems to have a much more localized effect than was originally thought, and should perhaps be considered as a step in the mode of action of the neurosecretory hormone. It would be interesting if the brain hormone in *Arenicola* proved to operate in a similar manner. In any event, a clear distinction should be made between meiosis-inducing and maturation-inducing substances in these organisms (Dodd, 1972).

Oocyte ovulation and/or spawning is controlled by neurosecretory hormones in *Aplysia* and perhaps other gastropods, by the presumably neurosecretory maturation-inducing hormone in *Arenicola*, perhaps by neurosecretory hormones in crustaceans and insects, and by the maturation-inducing substance in starfish. Ovulation and/or spawning is opposed by inhibitors in starfish and perhaps also in gastropods.

The development and differentiation of the reproductive tract does not seem to be controlled by ovarian hormones in any of the groups examined, except possibly in the crustaceans, but by the independent action of the extraovarian hormones influencing oocyte development. Thus, neurosecretory and dorsal body hormones affect the female parts of the reproductive tract in hermaphrodite gastropods, the optic gland hormone operates similarly in cephalopods (although the male tract is controlled via the testis), and the corpus allatum hormone independently affects accessory gland development in insects.

In the invertebrate animals described here, hormonal control of egg production appears to be concentrated upon different stages of oogenesis in the various groups, but this impression may merely reflect the interests of workers and the ease of manipulation of their experimental material. The diversity of invertebrate structure and habit would presuppose a corresponding variety of their endocrine mechanisms. The endocrine glands and hormone chemistry of a fish may profitably be compared with those of a mammal, but a similar exercise even between related phyla such as the Annelida and Mollusca, or the Crustacea and Insecta is impossible. The principle of an endocrine control of reproduction is common to many kinds of animal: the methods by which this is achieved are as varied as the animals themselves. Perhaps the only unifying theme is the almost universal use of neurosecretory mechanisms to promote reproductive synchrony with environmental events by which nervous and other stimuli can be transduced into hormonal messages. Only the cephalopod molluscs seem to be without a neurosecretory involvement in reproduction, possibly because the relationship between visual stimuli and optic gland activity is as intimate as it is.

REFERENCES

Abeloos, M. (1943). Effets de la castration chez un mollusque, *Limax maximus* L. *C. R. Acad. Sci. (Paris)* **216**, 90–91.

Aboim, A. N. (1945). Développement embryonnaire et post-embryonnaire des gonades normales et agamétiques de *Drosophila melanogaster. Rev. Suisse Zool.* **52**, 53–154.

Adams, T. S. (1970). Ovarian regulation of the corpus allatum in the housefly, *Musca domestica. J. Insect Physiol.* **16**, 349–360.

Adams, T. S., and Nelson, D. R. (1969). The effect of the corpus allatum and the ovaries on the amount of pupal and adult fatbody in the housefly, *Musca domestica. J. Insect Physiol.* **15**, 1729–1749.

Adams, T. S., Hintz, A. M., and Pomonis, J. G. (1968). Oöstatic hormone production in houseflies *Musca domestica* with developing ovaries. *J. Insect Physiol.* **14**, 983–993.

Adiyodi, K. G. (1967). The nature of haemolymph proteins in relation to the ovarian cycle in the viviparous cockroach, *Nauphoeta cinerea. J. Insect Physiol.* **13**, 1189–1195.

Adiyodi, K. G., and Adiyodi, R. G. (1970). Endocrine control of reproduction in decapod crustacea. *Biol. Rev. Cambridge Philos. Soc.* **45**, 121–165.

Adiyodi, K. G., and Nayar, K. K. (1966). Haemolymph proteins and reproduction in *Periplaneta americana. Curr. Sci.* **35**, 587–588.

Adiyodi, R. G. (1968a). On reproduction and moulting in the crab *Paratelphusa hydrodromous. Physiol. Zool.* **41**, 204–209.

Adiyodi, R. G. (1968b). Protein metabolism in relation to reproduction and moulting in the crab *Paratelphusa hydrodromous* (Herbst). Part I. Electrophoretic studies on the mode of utilisation of soluble proteins during vitellogenesis. *Indian J. Exp. Biol.* **6**, 144–147.

Adiyodi, R. G. (1968c). Protein metabolism in relation to reproduction and moulting in the crab *Paratelphusa hydrodromous* (Herbst). Part II. Fate of conjugated proteins during vitellogenesis. *Indian J. Exp. Biol.* **6**, 200–203.

Adiyodi, R. G. (1969a). On the storage and mobilisation of organic resources in the hepatopancreas of a crab (*Paratelphusa hydrodromous*). *Experientia* **25**, 43–44.

Adiyodi, R. G. (1969b). Protein metabolism in relation to reproduction and moulting in the crab [*Paratelphusa hydrodromous* (Herbst)]. III. RNA-activity and protein yolk biosynthesis during normal vitellogenesis and under conditions of acute inanition. *Indian J. Exp. Biol.* **7**, 13–16.

Adiyodi, R. G. (1969c). Protein metabolism in relation to reproduction and moulting in the crab *Paratelphusa hydrodromous* (Herbst). IV. Moulting. *Indian J. Exp. Biol.* **7**, 135–137.

Aiken, D. E. (1969). Ovarian maturation and egg-laying in the cray fish *Orconectes virilis:* influence of temperature and photoperiod. *Can. J. Zool.* **48**, 931–935.

Alexandrowicz, J. S. (1964). The neurosecretory system of the vena cava in Cephalopoda. I. *Eledone cirrosa. J. Mar. Biol. Assoc. U.K.* **44**, 111–132.

Alexandrowicz, J. S. (1965). The neurosecretory system of the vena cava in Cephalopoda. II. *Sepia officinalis* and *Octopus vulgaris. J. Mar. Biol. Assoc. U.K.* **45**, 209–228.

Anderson, L. M., and Telfer, W. H. (1969). A follicle cell contribution to the yolk spheres of moth oocytes. *Tissue & Cell* **1**, 633–644.

Antheunisse, L. J. (1963). Neurosecretory phenomena in the zebra mussel *Dreissena polymorpha* Pallas. *Arch. Neerl. Zool.* **15**, 237–314.

Arvy, L., Echalier, G., and Gabe, M. (1954). Modification de la gonade de *Carcinides maenas* L. après ablation bilaterale de l'organe Y. *C. R. Acad. Sci. (Paris)* **239**, 1853–1855.

Arvy, L., Echalier, G., and Gabe, M. (1956). Organe Y et gonade chez *Carcinides maenas* L. *Ann. Sci. Nat., Zool. Biol. Anim.* [11] **18**, 263–267.

Bal, A. K. (1970). Ultrastructural changes in the accessory-cells and the oocyte surface of the sea-urchin *Strongylocentrotus dröbachiensis* during vitellogenesis. *Z. Zellforsch. Mikrosk. Anat.* **111**, 1–14.

Balesdent, M. L. (1965). Recherches sur la sexualité et le déterminisme des caractères sexuels d'*Asellus aquaticus* Linné (Crustacé Isopode). *Bull. Acad. Soc. Lorraines Sci.* **5**, 1–231.

Barlow, J., and Ridgway, G. J. (1969). Changes in serum protein during the molt and reproductive cycles of the American lobster (*Homarus americanus*). *J. Fish. Res. Board Can.* **26**, 2101–2109.

Bartell, C. K., and Fingerman, M. (1969). Application of continuous particle electrophoresis for the separation of lipoprotein having melanin-dispersing activity from eyestalks of fiddler crabs (*Uca pugilator*). *Biol. Bull. (Woods Hole, Mass.)* **137**, 390–391.

Barth, R. H. (1968). The comparative physiology of reproductive processes in cockroaches. I. Mating behaviour and its endocrine control. *Adv. Reprod. Physiol.* **3**, 167–207.

Baskin, D. G. (1970). Studies on the infracerebral gland of the polychaete annelid, *Nereis limnicola*, in relation to reproduction, salinity and regeneration. *Gen. Comp. Endocrinol.* **15**, 352–360.

Baskin, D. G. (1971). A possible neuroendocrine system in polynoid polychaetes. *J. Morphol.* **133**, 93–104.

Baskin, D. G., and Golding, D. W. (1970). Experimental studies on the endocrinology and reproductive biology of the viviparous polychaete annelid, *Nereis limnicola* Johnson. *Biol. Bull.* (*Woods Hole, Mass.*) **139**, 461–475.

Beament, J. W. L. (1946). The formation and structure of the chorion of the egg in an Hemipteran, *Rhodnius prolixus. Q. J. Microsc. Sci.* **87**, 393–439.

Bell, W. J. (1969). Dual role of juvenile hormone in the control of yolk formation in *Periplaneta americana. J. Insect Physiol.* **15**, 1279–1290.

Bell, W. J. (1970). Demonstration and characterisation of two vitellogenic blood proteins in *Periplaneta americana:* an immunochemical analysis. *J. Insect Physiol.* **16**, 291–299.

Bell, W. J. (1971). Starvation induced oocyte resorption and yolk protein salvage in *Periplaneta americana. J. Insect Physiol.* **17**, 1099–1111.

Bell, W. J., and Barth, R. H. (1970). Quantitative effects of juvenile hormone on reproduction in the cockroach *Byrsotria fumigata. J. Insect Physiol.* **16**, 2303–2313.

Bell, W. J., and Barth, R. H. (1971). Initiation of yolk deposition by juvenile hormone. *Nature* (*London*), *New Biol.* **230**, 220–222.

Bentz, F. (1969). Variations protéiniques de l'ovaire de *Locusta migratoria* (L.) (Orthoptère) en premier phase de maturation. *C. R. Acad. Sci. Ser. D* **269**, 494–496.

Bentz, F., Girardie, A., and Cazal, M. (1970). Etude eléctrophorétique des variations de la proteinémie chez *Locusta migratoria* pendant la maturation sexuelle. *J. Insect Physiol.* **16**, 2257–2270.

Berlind, A., and Cooke, I. M. (1970). Release of a neurosecretory hormone as peptide by electrical stimulation of crab pericardial organs. *J. Exp. Biol.* **53**, 679–686.

Berlind, A., Cooke, I. M., and Goldstone, M. W. (1970). Do the monoamines in crab pericardial organs play a role in peptide neurosecretion? *J. Exp. Biol.* **53**, 669–677.

Berridge, M. J. (1966). The physiology of excretion in the cotton stainer *Dysdercus fasciatus* Signoret. IV. Hormonal control of excretion. *J. Exp. Biol.* **44**, 553–566.

Berry, C. F., and Cottrell, G. A. (1970). Neurosecretion in the vena cava of the cephalopod *Eledone cirrosa. Z. Zellforsch. Mikrosk. Anat.* **104**, 107–115.

Besse, G., and Mocquard, J. P. (1968). Etude par électrophorèse des quantités relatives des protéines de l'hémolymphe d'individus normaux et de femelles castrées chez deux Crustacés Oniscoides: *Porcellio dilatatus* Brandt, et *Ligia oceanica* L. *C. R. Acad. Sci. Ser. D* **267**, 2017–2019.

Bier, K. (1965). Zur Funktion der Nährzellen im meroistischen Insektenovar unter besonderer Berücksichtigung der oogenese adephager Coleopteren. *Zool. Jahrb., Abt. Allg. Zool. Physiol. Tiere* **71**, 371–384.

Bierbauer, J., Torok, L. J., and Teichmann, I. (1965). Cytologische und histochemische Untersuchungen am neurosekretorischen system der Augententakel von Pulmonaten. *Zool. Jahrb., Abt. Allg. Zool. Physiol. Tiere* **71**, 545–551.

Bierne, J. (1964). Maturation sexuelle anticipée par decapitation de la femelle chez l'Hétéronemerte *Lineus ruber* Müller. *C. R. Acad. Sci. (Paris)* **259**, 4841–4843.

Bierne, J. (1966). Localisation dans les ganglions cérébroides du centre régulateur de la maturation sexuelle chez la femelle de *Lineus ruber* Müller (Hétéronémertes). *C. R. Acad. Sci. Ser. D* **262**, 1572–1575.

Bierne, J. (1968). Facteur androgeñe et différenciation du sexe chez la Némerte *Lineus ruber* Müller. L'effet "free-martin" dans la parabiose hétérosexuée. *C. R. Acad. Sci. Ser. D* **267**, 1646–1648.

Bierne, J. (1970). Recherches sur la différentiation sexuelle au cours de l'ontogenèse et de la régénération chez la némertien *Lineus ruber* Müller. *Ann. Sci. Natl., Zool. Biol. Anim.* [12] **12**, 181–298.

Bliss, D. E. (1966). Physiological processes and neurosecretion as related to ecdysis and reproduction. Relation between reproduction and growth in decapod crustaceans. *Am. Zool.* **6**, 231–233.

Bobin, G., and Durchon, M. (1952). Etude histologique du cerveau de *Perinereis cultrifera*. Mise en évidence d'une complèx cérébro-vasculaire. *Arch. Anat. Microsc. Morphol. Exp.* **41**, 25–40.

Bodenstein, D. (1938). Untersuchungen zum Metamorphoseproblem. III. Über die Entwicklung der Ovarian im thoraxlosen Puppenabdomen. *Biol. Zentralbl.* **58**, 328–332.

Bodenstein, D. (1953). Studies on the humoral mechanisms in growth and metamorphosis of the cockroach *Periplaneta americana*. III. Humoral effects on metabolism. *J. Exp. Zool.* **124**, 413–416.

Bodnaryk, R. P., and Morrison, P. E. (1968). Immunochemical analysis of the origin of a sex specific accumulated blood protein in female houseflies. *J. Insect Physiol.* **14**, 1141–1146.

Boer, H. H., Slot, J. W., and van Andel, J. (1968). Electron microscopical and histochemical observations on the relation between medio-dorsal bodies and neurosecretory cells in the basommatophoran snails *Lymnaea stagnalis, Ancylus fluviatilis, Australorbis glabratus* and *Planorbarius corneus*. *Z. Zellforsch. Mikrosk. Anat.* **87**, 435–450.

Boilly-Marer, Y. (1969). Isolement de la substance responsable du déclenchement de la danse nuptiale et de l'émission des produits sexuels chez *Platynereis dumerilii* Aud. et M.Edn. (Annelide Polychète). Premiers resultats. *C. R. Acad. Sci. Ser. D* **268**, 93–96.

Bonhag, P. F. (1955). Histochemical studies on the ovarian nurse tissue and oocytes of the milkweed bug, *Oncopeltus fasciatus* (Dallas). I. Cytology, nucleic acids, and carbohydrates. *J. Morphol.* **96**, 381–439.

Botticelli, C. R., Hisaw, F. L., and Wotiz, H. H. (1960). Estradiol-17β and progesterone in ovaries of starfish (*Pisaster ochraceous*). *Proc. Soc. Exp. Biol. Med.* **103**, 875–877.

Botticelli, C. R., Hisaw, F. L., and Wotiz, H. H. (1961). Estrogens and progesterone in the sea urchin (*Strongylocentrotus franciscanus*) and pecten (*Pecten hericius*). *Proc. Soc. Exp. Biol. Med.* **106**, 887–889.

Bouchard-Madrelle, C. (1967). Influence de l'ablation d'une partie au de la totalité du complexe neural sur le fonctionnement des gonades de *Ciona intestinalis* (Tunicier, Ascidiacé). *C. R. Acad. Sci. Ser. D* **264**, 2055–2058.

Bounhiol, J. J. (1936). Métamorphose après ablation des corpora allata chez le ver a soie (*Bombyx mori* L.). *C. R. Acad. Sci. (Paris)* **203**, 388–389.

Boycott, B. B., and Young, J. Z. (1956). The subpendunculate body and nerve and other organs associated with the optic tract of cephalopods. *In* "Bertil Hanstrom: Zoological Papers in Honour of His Sixty-Fifth Birthday" (K. G. Wingstrand, ed.), pp. 76–105. Zoological Institute, Lund, Sweden.

Brien, P. (1963). Substance gamétique, substance sexuelle chez les hydres d'eau douce. *Gen. Comp. Endocrinol.* **3**, 689–690.

Brodzicki, S. (1966). Localization of lipids and α-ketolic steroids in the ovary of Crustacea. *Histochim. Cytochim. Lipides, Actes Symp. Int. Histol., 5th, 1963* Vol. 1, pp. 259–268.

Bruslé, J. (1969). Aspects ultrastructuraux de l'innervation des gonades chez l'étoile de mer *Asterina gibbosa* P. *Z. Zellforsch. Mikrosk. Anat.* **98**, 88–98.

Carlisle, D. B. (1953). Studies on *Lysmata seticaudata* Risso (Crustacea Decapoda). V. The ovarian inhibiting hormone and the hormonal inhibition of sex reversal. *Pubbl. Stn. Zool. Napoli* **24**, 355–372.

Carlisle, D. B. (1959). On the sexual biology of *Pandalus borealis*. II. The termination of the male phase. *J. Mar. Biol. Assoc. U.K.* **38**, 481–492.

Carlisle, D. B. (1960). Sexual differentiation in Crustacea Malacostraca. *Mem. Soc. Endocrinol.* **7**, 9–16.

Carlisle, D. B., and Knowles, F. G. W. (1959). "Endocrine Control in Crustaceans." Cambridge Univ. Press, London and New York.

Cassier, P. (1967). La reproduction des insectes et la régulation de l'activité des corps allates. *Annee Biol.* **6**, 596–670.

Cassier, P., and Fain-Maurel, M-A. (1968). Origine golgienne et évolution des corps vacuolaires impliqués dans la dégénérescence des glandes de mue chez *Locusta migratorioides* (R. et F.). *C. R. Acad. Sci. Ser. D* **266**, 1290–1293.

Cassier, P., and Fain-Maurel, M-A. (1969a). Étude infrastructurale des glandes de mue de *Locusta migratoria migratorioides* (R. et F.). III. Sur la persistance ou la dégénérescence des glandes ventrales chez les imagos solitaires. *Arch. Zool. Exp. Gen.* **110**, 203–224.

Cassier, P., and Fain-Maurel, M-A. (1969b). Étude infrastructurale des glandes de mue de *Locusta migratoria migratorioides* (R. et F.). IV. Evolution des glandes de mue au cours de la maturation sexuelle et des cycles ovariens chez les solitaires verts. *Arch. Zool. Exp. Gen.* **110**, 267–278.

Cazal, P. (1948). Les glandes endocrines rétro-cérébrales des insectes. *Bull. Biol. Fr. Belg., Suppl.* **32**, 1–227.

Cazal, M., and Girardie, A. (1968). Contrôle humoral de l'équilibre hydrique chez *Locusta migratoria migratorioides*. *J. Insect Physiol.* **11**, 655–668.

Chaet, A. B. (1964). A mechanism for obtaining mature gametes from starfish. *Biol. Bull.* (*Woods Hole, Mass.*) **126**, 8–13.

Chaet, A. B. (1966a). The gamete shedding substance of starfish: a physiological-biochemical study. *Am. Zool.* **6**, 263–271.

Chaet, A. B. (1966b). Neurochemical control of gamete release in starfish. *Biol. Bull.* (*Woods Hole, Mass.*) **130**, 43–58.

Chaet, A. B., and McConnaughy, R. A. (1959). Physiologic activity of nerve extracts. *Biol. Bull.* (*Woods Hole, Mass.*) **117**, 407–408.

Chaet, A. B., and Musick, R. S. (1960). A method for obtaining gametes from *Asterias forbesi*. *Biol. Bull.* (*Woods Hole, Mass.*) **119**, 292.

Charniaux-Cotton, H. (1954). Découverte chez un Crustacé Amphipode (*Orchestia gammarella*) d'une glande endocrine responsable de la différenciation des caractères sexuels primaires et secondaires males. *C. R. Acad. Sci. (Paris)* **239**, 780–782.

Charniaux-Cotton, H. (1956). Déterminisme endocrinien des caractères sexuels d'*Orchestia gammarella* (Crustacé Amphipode). *Ann. Sci. Nat., Zool. Biol. Anim.* [11] **18**, 305–310.

Charniaux-Cotton, H. (1957). Croissance, régénération et déterminisme endocrinien des caractères sexuels d'*Orchestia gammarella* (Pallas). Crustacé Amphipode. *Ann. Sci. Nat., Zool. Biol. Anim.* [11] **19**, 411–559.

Charniaux-Cotton, H. (1960a). Sex determination. *In* "The Physiology of Crustacea" (T. H. Waterman, ed.), Vol. 1, pp. 411–447. Academic Press, New York.

Charniaux-Cotton, H. (1960b). Physiologie de l'inversion sexuelle chez la Crevette à hermaphroditisme proterandrique funcionnel, *Lysmata seticaudata. C. R. Acad. Sci. (Paris)* **250**, 4046–4048.

Charniaux-Cotton, H. (1962). Androgenic gland of crustaceans. *Gen. Comp. Endocrinol., Suppl.* **1**, 241–247.

Charniaux-Cotton, H., and Kleinholz, L. H. (1964). Hormones in invertebrates other than insects. *In* "The Hormones" (G. Pincus, K. V. Thimann, and E. B. Astwood, eds.), Vol. 4, pp. 135–198. Academic Press, New York.

Charniaux-Cotton, H., Zerbils, C., and Meusy, J. J. (1966). Monographie de la glande androgène des Crustacés supérieurs. *Crustaceana* **10**, 113–136.

Cheung, T. S. (1966). The interrelations among three hormonal-controlled characters in the adult female shore crab, *Carcinus maenas* L. *Biol. Bull. (Woods Hole, Mass.)* **130**, 59–66.

Cheung, T. S. (1969). The environment and hormonal control of growth and reproduction in the adult female stone crab, *Menippe mercenaria* (Say). *Biol. Bull. (Woods Hole, Mass.)* **136**, 327–346.

Choquet, M. (1962). Effet inhibiteur de l'hormone cérébrale sur l'évolution des cellules sexuelles chez *Nereis pelagica* L. (A. P.). *C. R. Soc. Biol.* **156**, 1112–1114.

Choquet, M. (1964). Culture organotypique de gonades de *Patella vulgata* L. (Mollusque, Gastéropode, Prosobranche), *C. R. Acad. Sci. (Paris)* **258**, 1089–1091.

Choquet, M. (1965). Recherche en culture organotypique sur la spermatogénèse chez *Patella vulgata* L. Rôle des ganglions cérébroïdes et des tentacules. *C. R. Acad. Sci. (Paris)* **261**, 4521–4524.

Choquet, M. (1971). Etude du cycle biologique et de l'inversion du sexe chez *Patella vulgata* L. (Mollusque, Gastéropode, Prosobranche). *Gen. Comp. Endocrinol.* **16**, 59–73.

Clark, M. E. (1964). Biochemical studies on the coelomic fluid of *Nephtys hombergi* (Polychaeta: Nephtyidae), with observations on changes during different physiological states. *Biol. Bull. (Woods Hole, Mass.)* **127**, 63–84.

Clark, R. B. (1956). The blood vascular system of *Nephtys. Q. J. Microsc. Sci.* **97**, 235–249.

Clark, R. B. (1959). The neurosecretory system of the supra-oesophageal ganglion of *Nephtys* (Annelida, Polychaeta). *Zool. Jahrb., Abt. Allg. Zool. Physiol. Tiere* **68**, 395–424.

Clark, R. B. (1961). The origin and formation of the heteronereis. *Biol. Rev. Cambridge Philos. Soc.* **36**, 199–236.

Clark, R. B. (1962). The hormonal control of growth and reproduction in polychaetes, and its evolutionary implications. *Mem. Soc. Endocrinol.* **12**, 323–327.

Clark, R. B. (1965). Endocrinology and the reproductive biology of polychaetes. *Annu. Rev. Oceanogr. Mr. Biol.* **3**, 211–255.

Clark, R. B. (1966). The integrative action of a worm's brain. *Symp. Soc. Exp. Biol.* **20**, 345–379.

Clark, R. B. (1969). Endocrine influences in annelids. *Gen. Comp. Endocrinol., Suppl.* **2**, 572–581.

Clark, R. B. and Ruston, R. J. G. (1963). The influence of brain extirpation on oogenesis in the polychaete *Nereis diversicolor. Gen. Comp. Endocrinol.* **3**, 529–541.

Clift, A. D. (1971). Control of germarial activity and yolk deposition in nonterminal oocytes of *Lucilia cuprina. J. Insect Physiol.* **17**, 601–606.

Coggeshall, R. E. (1967). A light and electron microscope study of the abdominal ganglion of *Aplysia californica. J. Neurophysiol.* **30**, 1263–1287.

Coggeshall, R. E. (1970). A cytologic analysis of the bag cell control of egg laying in *Aplysia. J. Morphol.* **132**, 461–486.

Coles, G. C. (1964). Some effects of decapitation on metabolism in *Rhodnius prolixus* Stål. *Nature (London)* **203**, 323.

Coles, G. C. (1965a). Haemolymph proteins and yolk formation in *Rhodnius prolixus* Stål. *J. Exp. Biol.* **43**, 425–431.

Coles, G. C. (1965b). Studies on the hormonal control of metabolism in *Rhodnius prolixus* Stål. I. The adult female. *J. Insect Physiol.* **11**, 1325–1330.

Creange, J. E., and Szego, C. M. (1967). Sulphation as a metabolic pathway for oestradiol in the sea urchin *Strongylocentrotus franciscanus. Biochem. J.* **102**, 898–907.

Cruickshank, W. J. (1971). Follicle cell protein synthesis in moth oocytes (*Anagasta kichniella*). *J. Insect Physiol.* **17**, 217–232.

Dahm, K. H., Röller, H., and Trost, B. M. (1968). The juvenile hormone. IV. Stereochemistry of juvenile hormone and biological activity of some of its isomers and related compounds. *Life Sci.* **7**, 129–138.

Dales, R. P. (1950). The reproduction and larval development of *Nereis diversicolor* O.F. Müller. *J. Mar. Biol. Assoc. U.K.* **29**, 321–360.

Davey, K. G. (1967). Some consequences of copulation in *Rhodnius prolixus. J. Insect Physiol.* **13**, 1629–1636.

Davis, C. W. C. (1967). A comparative study of larval embryogenesis in the mosquito *Culex fatigans* Wiedemann (Diptera: Culicidae) and the sheep fly, *Lucilia sericata* Meigen (Diptera: Calliphoridae). I. Description of embryonic development. *Aust. J. Zool.* **15**, 547–579.

Dawson, A. B., and Hisaw, F. L. (1964). The occurrence of neurosecretory cells in the neural ganglia of Tunicates. *J. Morphol.* **114**, 411–423.

Dean, J. M., and Vernberg, F. J. (1965). Variations in the blood glucose level of Crustacea. *Comp. Biochem. Physiol.* **14**, 29–34.

de Kort, C. A. D. (1969). Hormones and the structural and biochemical properties of the flight muscles in the Colorado beetle. *Meded. Landbovwhogesch. Wageningen* **69**, 1–64.

Delavault, R. (1966). Determinism of sex. *In* "Physiology of Echinodermata" (R. A. Boolootian, ed.), pp. 615–638. Wiley, New York.

de Loof, A. (1969). Hormonal control of the synthesis of an important haemolymph protein in the Colorado beetle, *Leptinotarsa decemlineata* Say. *Gen. Comp. Endocrinol.* **13**, 518.

de Loof, A., and de Wilde, J. (1970a). The relation between haemolymph proteins and vitellogenesis in the Colorado beetle, *Leptinotarsa decemlineata. J. Insect Physiol.* **16**, 157–169.

de Loof, A., and de Wilde, J. (1970b). Hormonal control of synthesis of vitellogenic female protein in the Colorado beetle, *Leptinotarsa decemlineata. J. Insect Physiol.* **16**, 1455–1466.

de Loof, A., and Lagasse, A. (1970a). The ultrastructure of the follicle cells of the ovary of the Colorado beetle in relation to yolk formation. *J. Insect Physiol.* **16**, 211–220.

de Loof, A., and Lagasse, A. (1970b). Juvenile hormone and the ultrastructural properties of the fat body of the adult Colorado beetle, *Leptinotarsa decemlineata* Say. *Z. Zellforsch. Mikrosk. Anat.* **106**, 439–450.

Démeusy, N. (1962). Rôle de la glande de mue dans l'évolution ovarienne du crabe *Carcinus maenas* Linné. *Cah. Biol. Mar.* **3**, 37–56.

de Wilde, J. (1964). Reproduction—endocrine control. *In* "Physiology of Insecta" (M. Rockstein, ed.), Vol. 1, pp. 59–90. Academic Press, New York.

de Wilde, J. (1970). Hormones and insect diapause. *Mem. Soc. Endocrinol.* **18**, 487–514.

Dhainaut, A. (1966a). Présence de membranes annelées extra- et intra-nucléaires au cours de l'ovogénèse naturelle et expérimentale chez *Nereis pelagica* L. (Annelide Polychète). *C. R. Soc. Biol.* **160**, 749–752.

Dhainaut, A. (1966b). Influence de l'hormone cérébrale sur la consommation d'oxygène chez *Nereis diversicolor* O. F. Müller (Annelide Polychète). *C. R. Soc. Biol.* **160**, 1002–1004.

Dhainaut, A. (1967). Etude de la vitellogenèse chez *Nereis diversicolor* O. F. Müller (Annélide Polychète) par autoradiographie à haute résolution. *C. R. Acad. Sci. Ser. D* **265**, 434–436.

Dhainaut, A. (1968). Etude par autoradiographie à haute résolution de l'élaboration des mucopolysaccharides acides au cours de l'ovogénèse de *Nereis pelagica* L. (Annelide Polychète). *J. Microsc.* (Paris) **7**, 1075–1080.

Dhainaut, A. (1970a). Etude cytochimique et ultrastructurale de l'évolution ovocytaire de *Nereis pelagica* L. (Annélide Polychète). I. Ovogénèse naturelle. *Z. Zellforsch. Mikrosk. Anat.* **104**, 375–389.

Dhainaut, A. (1970b). Etude cytochimique et ultrastructurale de l'évolution ovocytaire de *Nereis pelagica* L. (Annélide Polychète). II. Evolution expérimentale en l'absence d'hormone cérébrale. *Z. Zellforsch. Mikrosk. Anat.* **104**, 390–404.

Dhainaut, A., and Porchet, M. (1967). Evolution ovocytaire en l'absence d'hormone cérébrale chez *P. cultrifera* G. (A. P.). *C. R. Acad. Sci. Ser. D* **264**, 2807–2810.

Dhainaut-Courtois, N. (1966a). Le complexe cérébrovasculaire de *Nereis pelagica* L. (Annélide Polychète). Données histologiques et ultrastructurales. *C. R. Acad. Sci. Ser. D* **262**, 2048–2051.

Dhainaut-Courtois, N. (1966b). Le complexe cérébrovasculaire de *Nereis pelagica* L. Origine des cellules infracérébrales et structure de la paroi du reseau vasculaire. *C. R. Soc. Biol.* **6**, 1232–1234.

Dhainaut-Courtois, N. (1968a). Contribution à l'étude du complexe cérébrovasculaire des Néréidiens. Cycle évolutif des cellules infracérébrales de

Nereis pelagica L. (Annélide Polychète); étude ultrastructurale. *Z. Zellforsch. Mikrosk. Anat.* **85**, 466–482.

Dhainaut-Courtois, N. (1968b). Etude histologique et ultrastructurale des cellules nerveuses du ganglion cérébrale de *Nereis pelagica* L. (Annélide Polychète). Comparaison entre les types cellulaires I-IV et ceux décrits antérieurement chez les Néréidae. *Gen. Comp. Endocrinol.* **11**, 414–443.

Doane, W. W. (1961). Developmental physiology of the mutant female sterile (2) adipose of *Drosophila melanogaster*. III. Corpus allatum-complex and ovarian transplantation. *J. Exp. Zool.* **146**, 275–298.

Dodd, J. M. (1955). The hormones of sex and reproduction and their effects in fish and lower chordates. *Mem. Soc. Endocrinol.* **4**, 166–184.

Dodd, J. M. (1972). In discussion to Joosse (1972).

Dogra, G. S., and Ewen, A. B. (1970). Egg laying by allatectomized females of the migratory grasshopper, *Melanoplus sanguinipes. J. Insect Physiol.* **16**, 461–469.

Donahue, J. K. (1948). Fluorimetric and biological determination of estrogens in the eggs of the american lobster (*Homarus americanus*). *Proc. Soc. Exp. Biol. Med.* **69**, 179–181.

Donahue, J. K. (1952). Studies on ecdysis in the american lobster (*Homarus americanus*). I. The lobster egg as a source of estrogenic hormone. *State Maine Dep. Sea Shore Fish., Res. Bull.* No. **8**.

Donahue, J. K. (1955). Studies on ecdysis in the american lobster (*Homarus americanus*). IV. Estrogenic hormone as a possible moult-inhibitor in the egg-bearing female. *State Maine Dep. Sea Shore Fish., Res. Bull.* No. **24**.

Donahue, J. K. (1957). Chromatographic identification of lobster egg estrogen. *State Maine Dep. Sea Shore Fish., Res. Bull.* No. **28**.

Dubé, J., and Lemonde, A. (1971). Caractérisation des stéroïde déshydrogénases de l'ovaire des *Schistocerca gregaria. J. Insect Physiol.* **17**, 13–27.

Dufour, D., Taskar, S. P., and Perron, J. M. (1971). Ontogenesis of a female specific protein from the locust, *Schistocerca gregaria. J. Insect Physiol.* **16**, 1369–1377.

Dupont-Raabe, M. (1952). Contribution à l'étude du rôle endocrine du cerveau et notamment de la pars intercerebralis chez les phasmides. *Arch. Zool. Exp. Gen.* **89**, 128–138.

Dupont-Raabe, M. (1956). Quelques données relatives aux phénomènes de neurosécrétion chez les phasmides. *Ann. Sci. Nat., Zool. Biol. Anim.* [11] **18**, 293–303.

Durchon, M. (1952). Recherches expérimentales sur deux aspects de la reproduction chez les Annélides Polychètes: l'épitoquie et la stolonisation. *Ann. Sci. Nat., Zool. Biol. Anim.* [11] **14**, 119–206.

Durchon, M. (1959). Contribution à l'étude de la stolonisation chez les Syllidiens (Annélides Polychètes). I. Syllinae. *Bull. Biol. Fr. Belg.* **93**, 155–219.

Durchon, M. (1962). Neurosecretion and hormonal control of reproduction in Annelida. *Gen. Comp. Endocrinol., Suppl.* **1**, 227–240.

Durchon, M. (1967a). "L'endocrinologie des Vers et des Mollusques." Masson, Paris.

Durchon, M. (1967b). Rôle du système nerveux dans la régénération chez les Annelides. *Bull. Soc. Zool. Fr.* **92**, 319–331.

Durchon, M., and Porchet, M. (1970). Dosage de l'activité endocrine cérébrale au cours du cycle génitale femelle chez *Nereis diversicolor* O. F. Müller (Annélide Polychète). *C. R. Acad. Sci. Ser. D* **270**, 1689–1691.

Durchon, M., and Richard, A. (1967). Etude en culture organotypique, du rôle endocrine de la gland optique dans la maturation ovarienne chez *Sepia officinalis* L. (Mollusque Cephalopode). *C. R. Acad. Sci. Ser. D* **264**, 1497–1500.

Durchon, M., and Wissocq, J. C. (1964). Contribution à l'étude de la stolonisation chez les Syllidiens (Annélides Polychètes). II. Autolytinae. *Ann. Sci. Nat., Zool. Biol. Anim.* [12] **6**, 160–208.

Enders, E. (1956). Die hormonal Steuerung rhythmischer Bewegungen von Insekton-Ovidukten. *Verh. Dtsch. Zool. Ges.* **19**, 113–116.

Engelmann, F. (1957). Die Steuerung der Ovarfunktion bei der ovoviviparen Schabe *Leucophaea maderae* (Fabr.). *J. Insect Physiol.* **1**, 257–278.

Engelmann, F. (1959). Über die Wirkung implantierter Prothoraxdrüsen im adulten Weibchen von *Leucophaea maderae*. *Z. Vergl. Physiol.* **41**, 456–470.

Engelmann, F. (1960). Mechanisms controlling reproduction in two viviparous cockroaches (Blattaria). *Ann. N.Y. Acad. Sci.* **89**, 516–536.

Engelmann, F. (1964). Inhibition of egg maturation in a pregnant viviparous cockroach. *Nature (London)* **202**, 724–725.

Engelmann, F. (1965). The mode of regulation of the corpus allatum in adult insects. *Arch. Anat. Microsc. Morphol. Exp.* **54**, 387–404.

Engelmann, F. (1966). Corpus allatum controlled protein biosynthesis in *Leucophaea maderae*. *In* "Symposium on Insect Endocrines, Brno" (V. J. A. Novak, ed.), pp. 110–116. Praha.

Engelmann, F. (1968). Endocrine control of reproduction in insects. *Annu. Rev. Entomol.* **13**, 1–26.

Engelmann, F. (1969). Female specific protein: biosynthesis controlled by corpus allatum in *Leucophaea maderae*. *Science* **165**, 407–409.

Engelmann, F. (1970). "The Physiology of Insect Reproduction." Pergamon, Oxford.

Engelmann, F. (1972). The effect of juvenile hormones on RNA and protein synthesis in adult insects. *Gen. Comp. Endocrinol., Suppl.* **3**, 168–173.

Engelmann, F., and Penney, D. (1966). Studies on the endocrine control of metabolism in *Leucophaea maderae*. I. The haemolymph proteins during egg maturation. *Gen. Comp. Endocrinol.* **7**, 314–325.

Engelmann, F., Hill, L., and Wilkens, J. L. (1971). Juvenile hormone control of female specific protein synthesis in *Leucophaea maderae, Schistocerca vaga* and *Sarcophaga bullata*. *J. Insect Physiol.* **17**, 2179–2191.

Fain-Maurel, M.-A., and Cassier, P. (1968). Etude infrastructurale des glandes de mue de *Locusta migratoria migratorioides* (R. et F.). I. Evolution cyclique au cours des stades larvaires et genèse du produit de sécrétion; l'ecdysone. *Arch. Zool. Exp. Gen.* **109**, 445–476.

Farner, D. S. (1962). In discussion to Clark (1962).

Faux, A., Horn, D. H. S., Middleton, E. J., Fales, H. M., and Lowe, M. E. (1969). Moulting hormones of a crab during ecdysis. *Chem. Commun.* No. **4**, pp. 175–176.

Ferguson, J. (1964). Nutrient transport in starfish. II. Uptake of nutrients by isolated organs. *Biol. Bull. (Woods Hole, Mass.)* **126**, 391–406.

Fingerman, M. (1966). Neurosecretory control of pigmentary effectors in crustaceans. *Am. Zool.* **6**, 169–179.

Fingerman, M., and Bartell, C. K. (1969). Gel filtration of chromatophorotropins from eyestalk extracts of the fiddler crab, *Uca pugilator*, and the prawn, *Palaemonetes vulgaris*. *Biol. Bull. (Woods Hole, Mass.)* **137**, 399.

Fingerman, M., Daniniczak, T., Miyawaki, M., Oguro, C., and Yamomoto, Y. (1967). Neuroendocrine control of the hepatopancreas in the crayfish *Procambarus clarkii*. *Physiol. Zool.* **40**, 23–30.

Fontaine, A. R. (1962). Neurosecretion in the ophiuroid *Ophiopholis aculeata*. *Science* **138**, 908–909.

Fraenkel, G., Hsiao, C., and Seligman, M. (1966). Properties of bursicon: an insect hormone that controls cuticular tanning. *Science* **151**, 91–93.

Frazier, W. T., Kandel, E. R., Kupfermann, I., Waziri, R., and Coggeshall, R. E. (1967). Morphological and functional properties of identified neurons in the abdominal ganglion of *Aplysia californica*. *J. Neurophysiol.* **30**, 1288–1351.

Fretter, V., and Graham, A. (1964). Reproduction. *In* "Physiology of Mollusca" (K. M. Wilbur and C. M. Yonge, eds.), Vol. 1, pp. 127–164. Academic Press, New York.

Gabe, M. (1953a). Particularités histologiques des cellules neurosécrétrices chez quelques Prosobranches Monotocardes. *C. R. Acad. Sci.* (*Paris*) **236**, 333–336.

Gabe, M. (1953b). Particularités histologiques des cellules neurosécrétrices chez quelques Gastéropodes Opisthobranches. *C. R. Acad. Sci.* (*Paris*) **236**, 2161–2163.

Gabe, M. (1953c). Sur l'existence chez quelques Crustacés Malacostracés, d'un organe comparable à la glande de mue des Insectes. *C. R. Acad. Sci.* (*Paris*) **237**, 1111–1113.

Gabe, M. (1954). La neurosécrétion chez les invertebrés. *Annee Biol.* **30**, 6–62.

Gabe, M. (1955). Particularités histologiques des cellules neurosécrétrices chez quelques Lamellibranches. *C. R. Acad. Sci.* (*Paris*) **240**, 1810–1812.

Gabe, M. (1956). Histologie comparée de la glande de mue (organe Y) des Crustacés Malacostracés. *Ann. Sci. Nat., Zool. Biol. Anim.* [11] **18**, 145–152.

Gabe, M. (1966). "Neurosecretion." Pergamon, New York.

Galbraith, M. N., Horn, D. H. S., Middleton, E. J., and Hackney, R. J. (1968). Structure of deoxycrustecdysone, a second crustacean moulting hormone. *Chem. Commun.* No. **2**, pp. 83–85.

Geigy, R. (1931). Action de l'ultra-violet sur le pôle germinal dans l'oeuf de *Drosophila melanogaster* (castration et mutabilité). *Rev. Suisse Zool.* **38**, 187–288.

Geldiay, S. (1965). Hormonal control of adult diapause in the Egyptian grasshopper, *Anacridium aegyptium* L. *Gen. Comp. Endocrinol.* **5**, 680.

Gersch, M. (1961). Insect metamorphosis and the activation hormone. *Am. Zool.* **1**, 53–57.

Gilgan, M. W., and Idler, D. R. (1967). The conversion of androstenedione to testosterone by some lobster (*Homarus americanus* Milne Edwards) tissues. *Gen. Comp. Endocrinol.* **9**, 319–324.

Girardie, A. (1964). Action de la pars intercérébralis sur le développement de *Locusta migratoria* L. *J. Insect Physiol.* **10**, 599–609.

Girardie, A. (1966). Contrôle de l'activité génitale chez *Locusta migratoria*. Mise en évidence d'un facteur gonadotrope et d'un facteur allatotrope dans la pars intercérébralis. *Bull. Soc. Zool. Fr.* **91**, 423–439.

Girardie, A. (1967). Contrôle neuro-hormonal de la métamorphose et de la pigmentation chez *Locusta migratoria cinerascens* (Orthoptére). *Bull. Biol. Fr. Belg.* **101**, 79–114.

Girardie, A., and Girardie, J. (1967). Étude histologique, histochimique et ultrastructurale de la pars intercerebralis chez *Locusta migratoria* L. (Orthoptere). *Z. Zellforsch. Mikrosk. Anat.* **78**, 54–75.

Golding, D. W. (1970). The infracerebral gland in *Nephtys*—a possible neuroendocrine complex. *Gen. Comp. Endocrinol.* **14**, 114–126.

Golding, D. W. (1972). Neuroendocrine structures and their role in reproduction in polychaetes. *Gen. Comp. Endocrinol., Suppl.* **3**, 580–590.

Golding, D. W. (1974). A survey of neuroendocrine phenomena in non-arthropod invertebrates. *Biol. Rev.* **49**, 161–224.

Golding, D. W., Baskin, D. G., and Bern, H. A. (1968). The infracerebral gland of *Nereis. J. Morphol.* **124**, 187–216.

Goldsworthy, G. J. (1969). Hyperglycaemic factors from the corpus cardiacum of *Locusta migratoria. J. Insect Physiol.* **15**, 2131–2140.

Goldsworthy, G. J. (1970). The action of hyperglycaemic factors from the corpus cardiacum of *Locusta migratoria* on glycogen phosphorylase. *Gen. Comp. Endocrinol.* **14**, 78–85.

Goldsworthy, G. J. (1971). The effects of removal of the cerebral neurosecretory cells on haemolymph and tissue carbohydrate in *Locusta migratoria migratorioides* R & F. *J. Endocrinol.* **50**, 237–240.

Gomez, R. (1965). Acceleration of development of gonads by implantation of brain in the crab *Paratelphusa hydrodromous. Naturwissenschaften* **9**, 216.

Gomez, R., and Nayar, K. K. (1965). Certain endocrine influences in the reproduction of the crab *Paratelphusa hydrodromous. Zool. Jahrb., Abt. Allg. Zool. Physiol. Tiere* **71**, 694–701.

Gomot, L., and Guyard, A. (1964). Evolution en culture in vitro de la glande hermaphrodite de jeunes escargots de l'espéce *Helix aspersa* Müll. *C. R. Acad. Sci. (Paris)* **258**, 2902–2905.

Gorbman, A., and Bern, H. A. (1962). "A Textbook of Comparative Endocrinology." Wiley, New York.

Gottfried, H., and Dorfman, R. I. (1969). The occurrence of *in vivo* cholesterol biosynthesis in an invertebrate, *Ariolimax californicus. Gen. Comp. Endocrinol., Suppl.* **2**, 590–593.

Gottfried, H., and Dorfman, R. I. (1970a). Steroids of invertebrates. IV. On the optic tentacle-gonadal axis in the control of the male-phase ovotestis in the slug (*Ariolimax californicus*). *Gen. Comp. Endocrinol.* **15**, 101–119.

Gottfried, H., and Dorfman, R. I. (1970b). Steroids of invertebrates. V. The *in vitro* biosynthesis of steroids by the male-phase ovotestis of the slug (*Ariolimax californicus*). *Gen. Comp. Endocrinol.* **15**, 120–138.

Gottfried, H., and Dorfman, R. I. (1970c). Steroids of invertebrates. VI. Effect of tentacular homogenates *in vitro* upon post-androstenedione metabolism in the male phase of *Ariolimax californicus* ovotestis. *Gen. Comp. Endocrinol.* **15**, 139–142.

Gottfried, H., and Lusis, O. (1966). Steroids of invertebrates: the *in vitro* production of 11-ketotestosterone and other steroids by the eggs of the slug *Arion ater rufus* (Linn.). *Nature (London)* **212**, 1488–1489.

Gottfried, H., Dorfman, R. I., and Wall, P. E. (1967). Steroids of invertebrates; production of estrogens by an accessory reproductive tissue of the slug *Arion ater rufus* (Linn.). *Nature (London)* **215**, 409–410.

Guyard, A. (1967). Féminisation de la glande hermaphrodite juvénile d'*Helix aspersa* Müll. associée *in vitro* au ganglion cérébroïde d'Escargot adulte ou de Paludine femelle. *C. R. Acad. Sci. Ser. D* **265**, 147–149.

Guyard, A. (1969). Autodifférenciation femelle de l'ébauche gonadique de l'escargot *Helix aspersa* Müll cultivée sur milieu anhormonal. *C. R. Acad. Sci. Ser.* D **268**, 966–969.

Hagadorn, I. R. (1966a). Neurosecretion in the Hirudinea and its possible role in reproduction. *Am. Zool.* **6**, 251–261.

Hagadorn, I. R. (1966b). The histochemistry of the neurosecretory system in *Hirudo medicinalis. Gen. Comp. Endocrinol.* **6**, 288–294.

Hagadorn, I. R., and Nishioka, R. S. (1961). Neurosecretion and granules in neurones of the brain of the leech. *Nature (London)* **191**, 1013–1014.

Hampshire, F., and Horn, D. H. S. (1966). Structure of crustecdysone, a crustacean moulting hormone. *Chem. Commun.* No. **2**, pp. 37–38.

Hanström, B. (1939). "Hormones in Invertebrates." Oxford Univ. Press, London and New York.

Hartman, H. B., and Chaet, A. B. (1962). Gamete shedding with radial nerve extracts. *Fed. Proc. Fed. Am. Soc. Exp. Biol.* **21**, 363.

Hathaway, R. R., and Black, R. E. (1969). Interconversions of estrogens and related developmental effects in sand dollar eggs. *Gen. Comp. Endocrinol.* **12**, 1–11.

Hathaway, D. S., and Selman, G. G. (1961). Certain aspects of cell lineage and morphogenesis studies in embryos of *Drosophila melanogaster* with ultra-violet micro-beam. *J. Embryol. Exp. Morphol.* **9**, 310–325.

Hauenschild, C. (1955). Photoperiodizität als Ursache des von Mondphase abhängigen Metamorphose-Rhythmus bei dem Polychaeten *Platynereis dumerilii. Z. Naturforsch., Teil B* **10**, 658–662.

Hauenschild, C. (1956). Neue experimentelle Untersuchungen zum Problem der Lunarperiodizität. *Naturwissenschaften* **43**, 361–363.

Hauenschild, C. (1959). Hemmender Einfluss der Proventrikelregion auf Stolonisation und Oocyten—Entwicklung bei dem Polychaeten *Autolytus prolifer. Z. Naturforsch., Teil B* **14**, 87–89.

Hauenschild, C. (1960). Lunar periodicity. *Cold Spring Harbor Symp. Quant. Biol.* **25**, 491–497.

Hauenschild, C. (1964). Postembryonale Entwicklungs steuerung durch ein Gehirn-Hormon bei *Platynereis dumerilii. Zool. Anz., Suppl.* **27**, 111–120.

Hauenschild, C. (1965). Hormone bei Nereiden und anderen niederen Wirbellosen. *Zool. Jahrb., Abt. Allg. Zool. Physiol. Tiere* **71**, 511–544.

Hauenschild, C. (1966). Die hormonale Einfluss des Gehirnsauf die Sexuelle Entwicklung bei dem Polychaeten *Platynereis dumerilii. Gen. Comp. Endocrinol.* **6**, 26–73.

Hauenschild, C., and Fischer, A. (1962). Neurosecretory control of development in *Platynereis dumerilii. Mem. Soc. Endocrinol.* **12**, 297–312.

Heilbrunn, L. V., Chaet, A. B., Dunn, A., and Wilson, W. L. (1954). Antimitotic substance from ovaries. *Biol. Bull. (Woods Hole, Mass.)* **106**, 158–168.

Herlant-Meewis, H. (1956). Croissance et neurosécrétion chez *Eisenia foetida* (Sav.). *Ann. Sci. Nat., Zool. Biol. Anim.* [11] **18**, 185–198.

Herlant-Meewis, H. (1957). Reproduction et neurosécrétion chez *Eisenia foetida* (Sav.). *Ann. Soc. R. Zool. Belg.* **87**, 151–183.

Highnam, K. C. (1962a). Neurosecretory control of ovarian development in *Schistocerca gregaria. Q. J. Microsc. Sci.* **103**, 57–72.

Highnam, K. C. (1962b). Neurosecretory control of ovarian development in the desert locust. *Mem. Soc. Endocrinol.* **12**, 379–390.

Highnam, K. C. (1963). Endocrine effects of crowding and isolation in insects. *Proc. Int. Congr. Zool., 16th, 1963* Vol. 3, pp. 6–7.

Highnam, K. C. (1964). Endocrine relationships in insect reproduction. *Symp. R. Entomol. Soc. London* **2**, 26–42.

Highnam, K. C. (1965). Some aspects of neurosecretion in arthropods. *Zool. Jahrb., Abt. Allg. Zool. Physiol. Tiere* **71**, 558–582.

Highnam, K. C. (1970). Mode of action of arthropod steroid and other hormones. *Adv. Steroid Biochem. Pharmacol.* **1**, 1–50.

Highnam, K. C. (1971). Estimates of neurosecretory activity during maturation in female locusts. *In* "Insect Endocrines" (V. J. A. Novak and K. Slama, eds.), pp. 81–90. Czech. Acad. Sci., Prague.

Highnam, K. C. (1972). Hormonal control of maturation in locusts. *In* "Proceedings of the International Study Conference on the Current and Future Problems of Acridology" (C. F. Hemming and T. H. C. Taylor, eds.), pp. 123–138. Centre for Overseas Pest Research, London.

Highnam, K. C., and Goldsworthy, G. J. (1972). Regenerated corpora cardiaca and hyperglycemic factor in *Locusta migratoria. Gen. Comp. Endocrinol.* **18**, 83–88.

Highnam, K. C., and Hill, L. (1969). "Comparative Endocrinology of the Invertebrates." Arnold, London.

Highnam, K. C., and Lusis, O. (1962). The influence of mature males on the neurosecretory control of ovarian development in the desert locust. *Q. J. Microsc. Sci.* **103**, 73–83.

Highnam, K. C., and West, M. W. (1971). The neuropilar neurosecretory reservoir of *Locusta migratoria migratorioides* R. & F. *Gen. Comp. Endocrinol.* **16**, 574–585.

Highnam, K. C., Lusis, O., and Hill, L. (1963a). The role of the corpora allata during oocyte growth in the desert locust, *Schistocerca gregaria* Forsk. *J. Insect Physiol.* **9**, 587–596.

Highnam, K. C., Lusis, O., and Hill, L. (1963b). Factors affecting oocyte resorption in the desert locust *Schistocerca gregaria* Forsk. *J. Insect Physiol.* **9**, 827–837.

Highnam, K. C., Hill, L., and Gingell, D. J. (1965). Neurosecretion and water balance in the male desert locust, *Schistocerca gregaria* (Forsk). *J. Zool.* **147**, 201–215.

Hill, L. (1962). Neurosecretory control of haemolymph protein concentration during ovarian development in the desert locust. *J. Insect Physiol.* **8**, 609–619.

Hill, L. (1972). Hormones and the control of metabolism in insects. *Gen. Comp. Endocrinol., Suppl. 3* 174–183.

Hill, L., and Goldsworthy, G. J. (1968). The utilization of reserves during starvation of larvae of the migratory locust. *Comp. Biochem. Physiol.* **36**, 61–71.

Hill, L., Mordue, W., and Highnam, K. C. (1966). The endocrine system, frontal ganglion and feeding during maturation in the female desert locust. *J. Insect Physiol.* **12**, 1197–1208.

Hill, L., Luntz, A. J., and Steele, P. A. (1968). The relationship between somatic growth, ovarian growth and feeding activity in the adult desert locust. *J. Insect Physiol.* **14**, 1–20.

Hirai, S., and Kanatani, H. (1971). Site of production of meiosis-inducing substance in ovary of starfish. *Exp. Cell Res.* **67**, 224–227.

Hirai, S., Kubota, J., and Kanatani, H. (1971). Induction of cytoplasmic maturation by 1-methyladenine in starfish oocytes after removal of the germinal vesicle. *Exp. Cell Res.* **68,** 137–143.

Hoffman, D. L. (1968). Seasonal eyestalk inhibition of the androgenic glands of a protandric shrimp. *Nature (London)* **218,** 170–172.

Horn, E. C., and Kerr, M. S. (1969). The haemolymph proteins of the blue crab, *Callinectes sapidus.* 1. Haemocyanins and certain other major constituents. *Comp. Biochem. Physiol.* **29,** 493–508.

Howie, D. I. D. (1959). The spawning of *Arenicola marina* (L.). I. The breeding season. *J. Mar. Biol. Assoc. U.K.* **38,** 395–406.

Howie, D. I. D. (1961a). The spawning of *Arenicola marina* (L.). II. Spawning under experimental conditions. *J. Mar. Biol. Assoc. U.K.* **41,** 127–144.

Howie, D. I. D. (1961b). The spawning of *Arenicola marina* (L.). III. Maturation and shedding of the ova. *J. Mar. Biol. Assoc. U.K.* **41,** 771–783.

Howie, D. I. D. (1963). Experimental evidence for the humoral stimulation of ripening of the gametes and spawning in the polychaete *Arenicola marina* (L). *Gen. Comp. Endocrinol.* **3,** 660–668.

Howie, D. I. D. (1966). Further data relating to the maturation hormone and its site of secretion in *Arenicola marina* L. *Gen. Comp. Endocrinol.* **6,** 347–361.

Howie, D. I. D., and McLeneghan, C. M. (1965). Evidence for a feedback mechanism influencing spermatogonial division in the lugworm (*Arenicola marina* L). *Gen. Comp. Endocrinol.* **5,** 40–44.

Hubl, H. (1953). Die inkreterischen Zellelemente in Gehirn der *Lumbriciden. Arch. Entwicklungsmech. Org.* **146,** 421–432.

Hubl, H. (1956). Uber die Beziehungen der Neurosekretion zum Regenerationsdeschehen bei *Lumbriciden* nebst Beschreibung eines neuartigen neurosekretorischen Zelltyps im Unterschlundganglion. *Arch. Entwicklungsmech. Org.* **149,** 73–87.

Ichikawa, M., and Ishizaki, H. (1961). Brain hormone of the silkworm *Bombyx mori. Nature (London)* **181,** 933–934.

Idler, D. R., Sangalang, G. B., and Kanazawa, A. (1969). Steroid desmolase in gonads of a marine invertebrate, *Placopecten magellanicus* Gmelin. *Gen. Comp. Endocrinol.* **12,** 222–230.

Ikegami, S., Tamura, S., and Kanatani, H. (1967). Starfish gonad: action and chemical identification of spawning inhibitor. *Science* **158,** 1052–1053.

Iwanoff, P. P., and Mestcherskaja, K. A. (1935). Die physiologischen Besonderheiten der geschlechtlich unreifen Insekenovarien und die Zyklischen Veranderungen ihrer Eigenschaften. *Zool. Jahrb., Abt. Allg. Zool. Physiol. Tiere* **55,** 281–384.

Johansson, A. S. (1958). Relation of nutrition to endocrine-reproductive functions in the milkweed bug *Oncopeltus fasciatus* (Dallas) (Heteroptera: Lygaeidae). *Nytt. Mag. Zool.* **7,** 1–132.

Joly, P. (1945). La fonction ovarïenne et son contrôle humoral chez les Dytiscides. *Arch. Zool. Exp. Gen.* **84,** 49–164.

Joosse, J. (1964). Dorsal bodies and dorsal neurosecretory cells of the cerebral ganglia of *Lymnaea stagnalis* L. *Arch. Neerl. Zool.* **16,** 1–103.

Joosse, J. (1972). Endocrinology of reproduction in molluscs. *Gen. Comp. Endocrinol., Suppl.* **3,** 591–601.

Joosse, J., and Geraerts, W. J. (1969). On the influence of the dorsal bodies and the adjacent neurosecretory cells on the reproduction and metabolism of *Lymnaea stagnalis. Gen. Comp. Endocrinol.* **13,** 511.

Joosse, J., and Reitz, D. (1969). Functional anatomical aspects of the ovotestis of *Lymnaea stagnalis. Malacologia* **9**, 101–109.

Kanatani, H. (1964). Starfish spawning: action of gamete-shedding substance obtained from radial nerves. *Science* **146**, 1177–1179.

Kanatani, H. (1969a). Mechanism of starfish spawning: action of neural substance on the isolated ovary. *Gen. Comp. Endocrinol. Suppl.* **2**, 582–589.

Kanatani, H. (1969b). Induction of spawning and oocyte maturation by 1-methyladenine in starfishes. *Exp. Cell Res.* **57**, 333–337.

Kanatani, H. (1972). On the meiosis-inducing substance produced in the starfish gonad by neural substance. *Gen. Comp. Endocrinol., Suppl.* **3**, 571–579.

Kanatani, H., and Hiramoto, Y. (1970). Site of action of 1-methyladenine in inducing oocyte maturation in starfish. *Exp. Cell Res.* **61**, 280–284.

Kanatani, H., and Noumura, T. (1962). On the nature of active principles responsible for gamete shedding in the radial nerves of starfishes. *J. Fac. Sci., Univ. Tokyo* **9**, 403–416.

Kanatani, H., and Ohguri, M. (1966). Mechanism of starfish spawning. I. Distribution of active substance responsible for maturation of oocytes and shedding of gametes. *Biol. Bull. (Woods Hole, Mass.)* **131**, 104–114.

Kanatani, H., and Shirai, H. (1967). *In vitro* production of meiosis-inducing substance by nerve extract in ovary of starfish. *Nature (London)* **216**, 284–286.

Kanatani, H., and Shirai, H. (1969). Mechanism of starfish spawning. II. Some aspects of action of a neural substance obtained from radial nerve. *Biol. Bull. (Woods Hole, Mass.)* **137**, 297–311.

Kanatani, H., and Shirai, H. (1970). Mechanism of starfish spawning. III. Properties and action of meiosis-inducing substance produced in gonad under influence of gonad stimulating substances. *Dev., Growth & Differ.* **12**, 119–140.

Kanatani, H., and Shirai, H. (1971). Chemical structural requirements for induction of oocyte maturation and spawning in starfishes. *Dev., Growth & Differ.* **13**, 53–64.

Kanatani, H., Shirai, H., Nakanishi, K., and Kurokawa, T. (1969). Isolation and identification of meiosis-inducing substance in starfish *Asterias amurensis. Nature (London)* **221**, 273–274.

Karlinsky, A. (1963). Effets de l'ablation des corpora allata imaginaux sur le développement ovarïen de *Pieris brassicae* (Lepidoptère). *C. R. Acad. Sci. (Paris)* **256**, 4101–4103.

Karlinsky, A. (1967a). Corpora allata et vitellogénèse chez les Lepidoptères. *Gen. Comp. Endocrinol.* **9**, 511–512.

Karlinsky, A. (1967b). Influence des corpora allata sur le fonctionnement ovarïen en milieu mâle de *Pieris brassicae* L. (Lepidoptère). *C. R. Acad. Sci. Ser. D* **265**, 2040–2042.

Karlson, P., and Sekeris, C. (1966). Ecdysone, an insect steroid hormone, and its mode of action. *Recent Prog. Horm. Res.* **22**, 473–502.

Kerr, M. S. (1969). The haemolymph proteins of the blue crab, *Callinectes sapidus.* II. A lipoprotein serologically identical to oocyte lipovitellin. *Dev. Biol.* **20**, 1–17.

King, P. E., and Richards, J. G. (1968). Accessory nuclei and annulate lamellae in Hymenopteran oocytes. *Nature (London)* **218**, 488.

King, R. C. (1964). Early stages of insect oogenesis. *In* "Insect Reproduction" (K. C. Highnam, ed.), pp. 13–25. Royal Entomological Society, London.

King, R. C. (1970). "Ovarian Development in *Drosophila melanogaster.*" Academic Press, New York.

King, R. C., and Koch, E. A. (1963). Studies on ovarian follicle cells of *Drosophila. Q. J. Microsc. Sci.* **104**, 297–320.

Kleinholz, L. H. (1970). A progress report on the separation and purification of crustacean neurosecretory pigmentary-effector hormones. *Gen. Comp. Endocrinol.* **14**, 578–588.

Korn, H. (1959). Vergleich-embryologische Untersuchungen an *Harmothoë* Kinberg 1857 (Polychaeta, Annelida). *Z. Wiss. Zool.* **161**, B, 345–443.

Kuhlmann, D. (1966). Der Dorsalkörper der Stylommatophoren (Gastropoda). *Z. Wiss. Zool.* **175.B**, 173–231.

Kulkarni, A. P., and Mehrotra, K. N. (1970). Amino acid nitrogen and proteins in the haemolymph of adult desert locusts, *Schistocerca gregaria. J. Insect Physiol.* **16**, 2181–2199.

Kupfermann, I. (1967). Stimulation of egg laying: possible neuro-endocrine function of bag cells of abdominal ganglion of *Aplysia californica. Nature (London)* **216**, 814–815.

Kupfermann, I. (1970). Stimulation of egg laying by extracts of neuroendocrine cells (bag cells) of abdominal ganglion of *Aplysia. J. Neurophysiol.* **33**, 877–881.

Kupfermann, I., and Kandel, E. R. (1970). Electrophysiological properties and functional interconnections of two symmetrical neurosecretory clusters (bag cells) in abdominal ganglion of *Aplysia. J. Neurophysiol.* **33**, 865–876.

Kurup, N. G. (1964). The incretory organs of the eyestalk and brain of the porcelain crab, *Petrolisthes cinctipes* Randall (Reptantia-Anomura). *Gen. Comp. Endocrinol.* **4**, 99–112.

Kurup, N. G., and Scheer, B. T. (1966). Control of protein synthesis in an anomuran crustacean. *Comp. Biochem. Physiol.* **18**, 971–973.

Lane, N. J. (1962). Neurosecretory cells in the optic tentacles of certain pulmonates. *Q. J. Microsc. Sci.* **103**, 211–226.

Lane, N. J. (1963). Neurosecretion in the tentacles of pulmonate gastropods. *J. Endocrinol.* **26**, xix–xx.

Laryea, A. A. (1970). Quoted by Runham and Hunter (1970).

Laufer, H. (1960). Blood proteins in insect development. *Ann. N.Y. Acad. Sci.* **89**, 490–515.

Laviolette, P. (1954). Rôle de la gonade dans le déterminisme glandulaire du tractus génital chez quelques gastéropodes arionidae et limnaidae. *Bull. Biol. Fr. Belg.* **88**, 310–332.

Lea, A. O. (1963). Some relationships between environment, corpora allata, and egg maturation in Aedine mosquitoes. *J. Insect Physiol.* **9**, 793–809.

Lea, A. O. (1967). The medial neurosecretory cells and egg maturation in mosquitoes. *J. Insect. Physiol.* **13**, 419–429.

Lea, A. O. (1970). Endocrinology of egg maturation in autogenous and anautogenous *Aedes taeniorhynchus. J. Insect Physiol.* **16**, 1689–1696.

Lea, A. O. (1972). The corpus cardiacum of the mosquito: its ultrastructure and role as the regulator of egg maturation. *Gen. Comp. Endocrinol., Suppl.* **3**, 602–608.

Lea, A. O., and Thomsen, E. (1962). Cycles in the synthetic activity of the medial neurosecretory cells of *Calliphora erythrocephala* and their regulation. *Mem. Soc. Endocrinol.* **12**, 345–347.

Lea, A. O., and Thomsen, E. (1969). Size independent secretion by the corpus allatum of *Calliphora erythrocephala. J. Insect Physiol.* **15**, 477–482.

L'Hélias, C. (1957). Action du complexe rétrocérébrale sur le métabolisme chez le phasme *Carausius morosus. Bull. Biol., Suppl.* **44**, 1–96.

Lehoux, J. G., and Sandor, T. (1969). Conversion of testosterone to 4-androstene-3,17-dione by house cricket (*Gryllus domesticus*) male gonad preparations *in vitro. Endocrinology* **84**, 652–657.

Lehoux, J. G., and Sandor, T. (1970). The occurrence of steroids and steroid metabolizing enzyme systems in invertebrates. A review. *Steroids* **16**, 141–171.

Lehoux, J. G., Sandor, T., Lanthier, A., and Lusis, O. (1968). Metabolism of exogenous progesterone by insect tissue preparations *in vitro. Gen. Comp. Endocrinol.* **11**, 481–488.

Lever, J. (1957). Some remarks on neurosecretory phenomena in *Ferrissia* sp. (Gastropoda Pulmonata). *Proc. K. Ned. Akad. Wet.* **60**, 510–522.

Lever, J. (1958). On the relation between the medio-dorsal bodies and the cerebral ganglia in some pulmonates. *Arch. Neerl. Zool.* **13**, 194–201.

Lever, J., Kok, M., Meuleman, E. A., and Joosse, J. (1961). On the location of Gomori-positive neurosecretory cells in the central ganglia of *Lymnaea stagnalis. Proc. K. Ned. Akad. Wet.* **64**, 640–647.

Lever, J., de Vries, C. M., and Jager, J. C. (1965). On the anatomy of the central nervous system and the location of neurosecretory cells in *Australorbis glabratus. Malacologia* **2**, 219–230.

Lisk, R. P. (1961). Estradiol-17β in the eggs of the American lobster *Homarus americanus. Can. J. Biochem. Physiol.* **39**, 659–663.

Locke, M. (1969). The ultrastructure of the oenocytes in the molt/intermolt cycle of an insect. *Tissue & Cell* **1**, 103–154.

Lockwood, A. P. M. (1968). "Aspects of the Physiology of Crustacea." Oliver & Boyd, Edinburgh.

Loher, W. (1965). Hormonale Kontrolle der Oocytenentwicklung bei der Heuschrecke *Gomphocerus rufus* L. *Zool. Jahrb., Abt. Allg. Zool. Physiol. Tiere* **71**, 677–684.

Loher, W. (1966a). Nervöse und hormonale Kontrolle des Sexualverhaltens beim Weibchen der *Heuschrecke Gomphocerus rufus* L. *Verh. Dtsch. Zool. Ges. Jena,* pp. 386–391.

Loher, W. (1966b). Die Steuerung sexueller Verhaltens-weisen und der Oocytenentwicklung bei *Gomphocerus rufus* L. *Z. Vergl. Physiol.* **53**, 277–316.

Lubet, P. (1955). Le déterminisime de la ponte chez les Lamellibranches (*Mytilus edulis* L.). Intervention des ganglions nerveux. *C. R. Acad. Sci. (Paris)* **241**, 254–256.

Lubet, P. (1956). Effets de l'ablation des centres nerveux sur l'émission des gamètes chez *Mytilus edulis* L. et *Chlanys varia* L. (Mollusques Lamellibranches). *Ann. Sci. Nat., Zool. Biol. Anim.* [11] **18**, 175–183.

Lusis, O. (1963). The histology and histochemistry of development and resorption in the terminal oocytes of the desert locust, *Schistocerca gregaria. Q. J. Microsc. Sci.* **104**, 57–68.

Martin, R. (1968). Fine structure of the neurosecretory system of the vena cava in *Octopus. Brain Res.* **8**, 201–205.

Martoja, M. (1965a). Existence d'un organe juxtaganglionnaire chez *Aplysia punctata* Cuv. (Gasteropode Opisthobranche). *C. R. Acad. Sci. (Paris)* **260**, 4615–4617.

Martoja, M. (1965b). Données relatives à l'organe juxtaganglionnaire des Prosobranches Diotocardes. *C. R. Acad. Sci. (Paris)* **261**, 3195–3196.

Martoja, M. (1967). Sur l'existence de cellules neurosécrétrices chez quelques Mollusques Polyplacophores. *C. R. Acad. Sci. Ser. D* **264**, 1461–1463.

Masner, P., Huot, L., Corrivault, G.-W., and Prudhomme, J. C. (1970). Effects of reserpine on the function of the gonads and its neuroendocrine regulation in tenebrioid beetles. *J. Insect Physiol.* **16**, 2327–2344.

Matsumoto, K. (1958). Morphological studies on the neurosecretion in crabs. *Biol. J. Okayama Univ.* **4**, 103–176.

Matsumoto, K. (1962). Experimental studies on the neurosecretory activities of the thoracic ganglion of a crab, *Hemigrapsus. Gen. Comp. Endocrinol.* **2**, 4–11.

Mauzey, K. P. (1966). Feeding behaviour and reproductive cycles in *Pisaster ochraceus. Biol. Bull. (Woods Hole, Mass.)* **131**, 127–144.

Mayer, R. J., and Candy, D. J. (1969). Control of haemolymph lipid concentration during locust flight: an adipokinetic-hormone from the corpora cardiaca. *J. Insect Physiol.* **15**, 611–620.

Meenakshi, V. R., and Scheer, B. T. (1969). Regulation of galactogen synthesis in the slug *Ariolimax columbianus. Comp. Biochem. Physiol.* **29**, 841–845.

Menon, M. (1963). Endocrine influences on protein and fat in the haemolymph of the cockroach, *Periplaneta americana. Proc. Int. Congr. Zool., 16th, 1963* Vol. 1, p. 297.

Meyer, A. S., Schneiderman, H. A., Hanzmann, E., and Ko, J. H. (1968). The two juvenile hormones from the cecropia silkmoth. *Proc. Natl. Acad. Sci. U.S.A.* **60**, 853–860.

Mills, R. R., and Nielsen, D. J. (1967). Hormonal control of tanning in the American cockroach. V. Some properties of the purified hormone. *J. Insect Physiol.* **13**, 273–280.

Minks, A. K. (1967). Biochemical aspects of juvenile hormone action in the adult *Locusta migratoria. Arch. Neer. Zool.* **17**, 175–258.

Mordue, W. (1965a). Studies on oocyte production and associated histological changes in the neuro-endocrine system in *Tenebrio molitor* L. *J. Insect Physiol.* **11**, 493–503.

Mordue, W. (1965b). The neuro-endocrine control of oocyte development in *Tenebrio molitor* L. *J. Insect Physiol.* **11**, 505–511.

Mordue, W. (1965c). Neuro-endocrine factors in the control of oocyte production in *Tenebrio molitor* L. *J. Insect Physiol.* **11**, 617–629.

Mordue, W. (1969). Hormonal control of Malphigian tube and rectal function in the desert locust, *Schistocerca gregaria. J. Insect Physiol.* **15**, 273–285.

Mordue, W. (1971). The hormonal control of excretion and water balance in locusts. *In* "Insect Endocrines" (V. J. A. Novak and K. Slama, eds.) pp. 79–83. Czech. Acad. Sci., Prague.

Mordue, W., and Goldsworthy, G. J. (1969). The physiological effects of corpus cardiacum extracts in locusts. *Gen. Comp. Endocrinol.* **12**, 360–369.

Mordue, W., Highnam, K. C., Hill, L., and Luntz, A. J. (1970). Environmental effects upon endocrine mediated processes in locusts. *Mem. Soc. Endocrinol.* **18**, 111–136.

Naisse, J. (1963). Détermination sexuelle chez *Lampyris noctiluca* L. (Insecte Coléoptère Malacoderme). *C. R. Acad. Sci. (Paris)* **256**, 799–800.

Naisse, J. (1965). Contrôle endocrinïen de la différenciation sexuelle chez les insectes. *Arch. Anat. Microsc. Morphol. Exp.* **54**, 417–428.

Nayar, K. K. (1956a). Studies on the neurosecretory system of *Iphita limbata* Stal (Hemiptera). III. The endocrine glands and the neurosecretory pathways in the adult. *Z. Zellforsch. Mikrosk. Anat.* **44**, 697–705.

Nayar, K. K. (1956b). Studies on the neurosecretory system of *Iphita limbata* Stal (Pyrrhocoridae: Hemiptera). IV. Observations on the structure and functions of the corpora cardiaca of the adult insect. *Proc. Natl. Inst. Sci. India, Part B* **22**, 171–184.

Nayar, K. K. (1958a). Neurosecretory system and egg laying in the insect *Iphita limbata* Stal. *J. Biol. Sci.* **1**, 26–29.

Nayar, K. K. (1958b). Studies on the neurosecretory system of *Iphita limbata* Stal. V. Probable endocrine basis of oviposition in the female insect. *Proc. Indian Acad. Sci., Sect. B* **47**, 233–251.

Nayar, K. K. (1958c). Probable endocrine mechanism controlling oviposition in the insect *Iphita limbata* Stal. *Int. Symp. Neurosekret., 2nd, 1958* pp. 102–104.

Nayar, K. K. (1962). Effects of injecting juvenile hormone extracts on the neurosecretory system of adult male cockroaches (*Periplaneta americana*). *Mem. Soc. Endocrinol.* **12**, 371–378.

Neugebauer, W. (1961). Wirkungen der Extirpation und Transplantation der Corpora allata auf den Sauerstoffverbrauch, die Eibildung und den Fettkörper von *Carausius* (*Dixippus*) *morosus* Br. et Redt. *Arch. Entwicklungsmech. Org.* **153**, 314–352.

Nielsen, D. J., and Mills, R. R. (1968). Changes in electrophoretic properties of haemolymph and terminal oocyte proteins during vitellogenesis in the American cockroach. *J. Insect Physiol.* **14**, 163–170.

Noumura, T., and Kanatani, H. (1962). Induction of spawning by radial nerve extracts in some starfishes. *J. Fac. Sci., Univ. Tokyo Sect. 4* **9**, 397–402.

Novak, V. J. A., Slama, K., and Wenig, K. (1959). Effect of implantation of corpus allatum on the oxygen consumption in *Pyrrhocoris apterus*. *Acta Symp. Evol. Insects, 1959* pp. 147–151.

Odhiambo, T. R. (1966a). The metabolic effects of the corpus allatum hormone in the male desert locust. I. Lipid metabolism. *J. Exp. Biol.* **45**, 45–50.

Odhiambo, T. R. (1966b). The metabolic effects of the corpus allatum hormone in the male desert locust. II. Spontaneous locomotor activity. *J. Exp. Biol.* **45**, 51–63.

O'Dor, R. K., and Wells, M. J. (1973). Yolk protein synthesis in the ovary of *Octopus vulgaris* and its control by the optic gland gonadotropin. *J. Exp. Biol.* **59**, 665–674.

Otsu, T. (1963). Bihormonal control of sexual cycle in freshwater crab, *Potaman dehaemi*. *Embryologica* **8**, 1–20.

Otsu, T., and Hanaoka, K. I. (1951). Relation between body weight and precocious differentiation of ova in eyestalkless crab (in Japanese with English summary). *Bull. Yamagata Univ., Nat. Sci.* **1**, 269–274.

Panouse, J. B. (1943). Influence de l'ablation de pédoncle oculaire sur la croissance de l'ovarie chez la crevette *Leander serratus*. *C. R. Acad. Sci.* (*Paris*) **217**, 553–555.

Panouse, J. B. (1947). La glande du sinus et la maturation des produits génitaux chez les crevettes. *Bull. Biol. Fr. Belg., Suppl.* **33**, 160–163.

Parameswaran, R. (1955). Neurosecretory cells in *Paratelphusa hydradromous* (Herbst). *Curr. Sci.* **24**, 23–24.

Passano, L. M. (1960). Molting and its control. *In* "The Physiology of Crustacea" (T. H. Waterman, ed.), Vol. 1, pp. 473–536. Academic Press, New York.

Passano, L. M., and Jyssum, S. (1963). The role of the Y-organ in crab proecdysis and limb regeneration. *Comp. Biochem. Physiol.* **9**, 195–213.

Pelluet, D. (1964). On the hormonal control of cell differentiation in the ovotestis of slugs (Gastropoda: Pulmonata). *Can. J. Zool.* **42**, 196–199.

Pelluet, D., and Lane, N. J. (1961). The relation between neurosecretion and cell differentiation in the ovotestis of slugs (Gastropoda: Pulmonata). *Can. J. Zool.* **39**, 789–805.

Pfeiffer, I. G. (1945). Effect of the corpora allata on the metabolism of adult female grasshopper. *J. Exp. Zool.* **99**, 183–233.

Pflugfelder, O. (1937). Bau, Entwicklung und Funktion der Corpora allata und cardiaca von *Dixippus morosus* Br. *Z. Wiss. Zool.* **149**, B 477–512.

Phipps, J. (1949). The structure and maturation of the ovaries in British Acrididae (Orthoptera). *Trans. R. Entomol. Soc. London* **100**, 233–247.

Phipps, J. (1960). Ovulation and oocyte resorption in Acridoidea (Orthoptera). *Proc. R. Entomol. Soc. London* **A41**, 78–86.

Porchet, M. (1970). Relations entre le cycle hormonal cérébral et l'évolution ovocytaire chez *Perinereis cultrifera* Grube (annelide polychete). *Gen. Comp. Endocrinol.* **15**, 220–231.

Poulson, D. F., and Waterhouse, D. F. (1960). Experimental studies on pole cells and midgut differentiation in Diptera. *Aust. J. Biol. Sci.* **13**, 541–567.

Prabhu, V. K. K., and Hema, P. (1970). Effect of implantation of ovaries in the male cockroach, *Periplaneta americana. J. Insect Physiol.* **16**, 147–156.

Ramamurty, P. S. (1963). Über die Herkunft der Ribonukleisäure in den wachsenden Eizellen der Skorpions fliege *Panorpa communis* (Insecta, Mecoptera). *Naturwissenschaften* **50**, 383–384.

Ramamurty, P. S. (1964). On the contribution of the follicle epithelium to the deposition of yolk in the oocyte of *Panorpa communis* (Mecoptera). *Exp. Cell Res.* **33**, 601–605.

Rangnekar, P. V., and Deshmukh, R. D. (1968). Effect of eyestalk removal on the ovarian growth of the marine crab, *Scylla serrata* (Forskal). *J. Anim. Morphol. Physiol.* **15**, 116–126.

Raven, C. P. (1958). "Morphogenesis." Pergamon, Oxford.

Reidenbach, J. M. (1967). Démonstration expérimentale de la neutralité des caractères sexuels externes de la femelle chez le Crustacé Isopode *Idotea baltica blasteri* Audouin. *C. R. Acad. Sci. Ser. D* **265**, 1321–1323.

Reinhard, E. G. (1950). An analysis of the effects of a saculinid parasite on the external morphology of *Callinectes sapidus* Rathbun. *Biol. Bull. (Woods Hole, Mass.)* **98**, 277–288.

Richard, A. (1971). Action qualitative de la lumière dans le déterminisme du cycle sexuel chez le céphalopode *Sepia officinalis* L. *C. R. Acad. Sci. Ser. D* **272**, 106–109.

Robbins, W. E., Kaplanis, J. N., Thompson, M. J., Shortino, T. J., Cohen, C. F., and Joyner, S. C. (1968). Ecdysones and analogs: effects on development and reproduction of insects. *Science* **161**, 1158–1160.

Robbins, W. E., Kaplanis, J. N., Thompson, M. J., Shortino, T. J., and Joyner, S. C. (1970). Ecdysones and synthetic analogues: molting hormone activity and inhibitive effects on insect growth, metamorphosis and reproduction. *Steroids* **16**, 105–125.

Rogers, D. C. (1969). Fine structure of the epineural connective tissue sheath of the suboesophageal ganglion in *Helix aspersa. Z. Zellforsch. Mikrosk. Anat.* **102,** 99 112.

Röller, H. (1962). Über den Einfluss der Corpora allata auf den Stoffwechsel der Wachsmotte. *Naturwissenschaften* **49,** 524 525.

Röller, H., Dahm, K. H., Sweely, C. C., and Trost, B. M. (1967). The structure of the juvenile hormone. *Angew. Chem.* **6,** 179 180.

Roth, T. F., and Porter, K. R. (1964). Yolk protein uptake in the oocyte of the mosquito, *Aedes aegyptii* L. *J. Cell Biol.* **20,** 313 332.

Runham, N. W., and Hunter, P. J. (1970). "Terrestrial Slugs." Hutchinson, London.

Sägesser, H. (1960). Über die Wirkung der Corpora allata auf den Sauerstoff verbrauch bei der Schaber, *Leucophaea maderae* (F.). *J. Insect Physiol.* **5,** 264 285.

Sanchez, S. (1962). Histophysiologie neuro-hormonale chez quelques mollusques gastéropodes. I. Complexes neuroendocriniens. *Bull. Soc. Zool. Fr.* **87,** 309 319.

Sanchez, S. and Sablier, H. (1962). Histophysiologie neuro-hormonale chez quelques mollusques gastéropodes. II. Corrélations hormonales. *Bull. Soc. Zool. Fr.* **87,** 319 330.

Sarojini, S. (1963). Comparison of the effects of androgenic hormone and testosterone propionate on the female Ocypod crab. *Curr. Sci.* **32,** 411 412.

Scharrer, B. (1935). Über das Hanström'sche Organ X bei Opisthobranchiern. *Pubbl. Stn. Zool. Napoli* **15,** 132 142.

Scharrer, B. (1937). Über sekretorisch tätige Nervenzellen bei wirbellosen Tieren. *Naturwissenschaften* **25,** 131 138.

Scharrer, B., and von Harnack, M. (1961). Histophysiological studies on the corpora allata of *Leucophaea maderae.* III. The effect of castration. *Biol. Bull. (Woods Hole, Mass.)* **121,** 193 208.

Scheurer, R. (1969). Endocrine control of protein synthesis during oocyte maturation in the cockroach *Leucophaea maderae. J. Insect Physiol.* **15,** 1411 1419.

Schmid, L. A. (1947). Induced neurosecretion in *Lumbricus terrestris. J. Exp. Zool.* **104,** 365 377.

Schroeder, P. C. (1971). Studies on oogenesis in the polychaete annelid *Nereis grubei* (Kinberg). II. Oocyte growth rates in intact and hormone-deficient animals. *Gen. Comp. Endocrinol.* **16,** 312 322.

Schuetz, A. W. (1969). Chemical properties and physiological actions of a starfish radial nerve factor and ovarian factor. *Gen. Comp. Endocrinol.* **12,** 209 221.

Schuetz, A. W. (1970). Effects of 1-methyladenosine on isolated gonads and oocytes of the starfish; evidence for an ovarian wall nucleosidase. *Biol. Bull. (Woods Hole, Mass.)* **139,** 435.

Schuetz, A. W., and Biggers, J. D. (1967). Regulation of germinal vesicle breakdown in starfish oocytes. *Exp. Cell Res.* **46,** 624 628.

Sengel, P., and Georges, D. (1966). Effets de l'éclairement et de l'ablation du complexe neural sur la ponte de *Ciona intestinalis. C. R. Acad. Sci. Ser. D* **263,** 1876 1879.

Sengel, P., and Kiery, M. (1962). Rôle du complexe formé par la glande neurale, le ganglion nerveux et l'organe vibratile sur la différenciation sexuelle des gonades de *Molgula manhattensis* (Tunicier Ascidiace). *Bull. Soc. Zool. Fr.* **87,** 615 628.

Simpson, L. (1969). Morphological studies of possible neuroendocrine structures in *Helisoma tenue* (Gastropoda: Pulmonata). *Z. Zellforsch. Mikrosk. Anat.* **102**, 570–593.

Simpson, L., Bern, H. A., and Nishioka, R. S. (1966). Survey of evidence for neurosecretion in Gastropod molluscs. *Am. Zool.* **6**, 123–138.

Smith, R. I. (1950). Embryonic development in the viviparous nereid polychaete, *Neanthes lighti* Hartman. *J. Morphol.* **87**, 417–465.

Steele, J. E. (1961). Occurrence of a hyperglycaemic factor in the corpus cardiacum of an insect. *Nature (London)* **192**, 680–681.

Steele, J. E. (1963). The site of action of insect hyperglycaemic hormone. *Gen. Comp. Endocrinol.* **3**, 46–52.

Stevens, M. (1970). Procedures for induction of spawning and meiotic maturation of starfish oocytes by treatment with 1-methyladenine. *Exp. Cell Res.* **59**, 482–484.

Strangways-Dixon, J. (1962). The relationship between nutrition, hormones and reproduction in the blowfly *Calliphora erythrocephala* Meig. III. The corpus allatum in relation to nutrition, the ovaries, innervation and the corpus cardiacum. *J. Exp. Biol.* **39**, 293–306.

Streiff, W. (1967). Etude endocrinologique de cycle sexuel chez un Mollusque hermaphrodite protandre *Calyptraea sinensis* L. III. Mise en évidence par culture *in vitro* de facteurs hormonaux conditionnant l'évolution de la gonade. *Ann. Endocrinol.* **28**, 641–656.

Streiff, W., Le Breton, J., and Silberzahn, N. (1970). Non spécificité des facteurs hormonaux responsable de la morphogenèse et du cycle du tractus génital mâle chez les Mollusques Prosobranches. *Ann. Endocrinol.* **31**, 548–556.

Strong, L. (1965a). The relationships between the brain, corpora allata, and oocyte growth in the Central American locust, *Schistocerca* sp. I. The cerebral neurosecretory system, the corpora allata, and oocyte growth. *J. Insect Physiol.* **11**, 135–146.

Strong, L. (1965b). The relationships between the brain, corpora allata and oocyte growth in the Central American locust, *Schistocerca* sp. II. The innervation of the corpora allata, the lateral neurosecretory system, and oocyte growth. *J. Insect Physiol.* **11**, 271–280.

Strumwasser, F., Jacklet, J. W., and Alvarez, R. B. (1969). A seasonal rhythm in the neural extract induction of behavioural egg-laying in *Aplysia*. *Comp. Biochem. Physiol.* **29**, 197–206.

Taki, I. (1944). Studies on *Octopus*. 2. The sex and genital organ. *Jpn. J. Malacol.* **19**, 267–310. [In Japanese: observations summarized in *J. Fac. Fish. Anim. Husb., Hiroshima Univ.* **5**, 345–417 (1964).]

Tcholakian, R. K., and Eik-Nes, K. B. (1969). Conversion of progesterone to 11-deoxycorticosterone by the androgenic gland of the blue crab (*Callinectes sapidus* Rathbun). *Gen. Comp. Endocrinol.* **12**, 171–173.

Teissier, G. (1935). Croissance des variants sexuels chez *Maia squinada* L. *Trav. Stn. Biol. Roscoff* **13**, 93–130.

Telfer, W. H. (1954). Immunological studies of insect metamorphosis. II. The role of sex limited blood proteins in egg formation by the *Cecropia* silkworm. *J. Gen. Physiol.* **37**, 539–558.

Telfer, W. H. (1960). The selective accumulation of blood proteins by the oocytes of saturniid moths. *Biol. Bull. (Woods Hole, Mass.)* **118**, 338–351.

Telfer, W. H., and Rutberg, L. D. (1960). The effect of blood protein depletion on the growth of the oocytes in the Cecropia moth. *Biol. Bull. (Woods Hole, Mass.)* **118**, 352–366.

Telfer, W. H., and Williams, C. M. (1953). Immunological studies of insect metamorphosis. I. Qualitative and quantitative description of the blood antigens of the *Cecropia* silkworm. *J. Gen. Physiol.* **36**, 389–413.

Telford, M. (1968). The identification and measurement of sugars in the blood of three species of Atlantic crabs. *Biol. Bull. (Woods Hole, Mass.)* **135**, 574–584.

Tennent, D. H., and Ito, T. (1941). A study of the oogenesis of *Mespilia globulus* (Linne). *J. Morphol.* **69**, 347–404.

Terwilliger, R. C., Terwilliger, N. B., Clay, G. A., and Belamarich, F. A. (1970). The subcellular localization of a cardioexcitatory peptide in the pericardial organs of the crab, *Cancer borealis. Gen. Comp. Endocrinol.* **15**, 70–79.

Thomas, K. K., and Nation, J. L. (1966). Control of a sex limited haemolymph protein by corpora allata during ovarian development in *Periplaneta americana. Biol. Bull. (Woods Hole, Mass.)* **130**, 254–263.

Thomsen, E. (1949). Influence of the corpus allatum on the oxygen consumption of adult *Calliphora erythrocephala* Meig. *J. Exp. Biol.* **26**, 137–149.

Thomsen, E. (1952). Functional significance of the neurosecretory brain cells and the corpus cardiacum in the female blowfly, *Calliphora erythrocephala* Meig. *J. Exp. Biol.* **29**, 137–172.

Thomsen, E., and Hamburger, K. (1955). Oxygen consumption of castrated females of the blowfly, *Calliphora erythrocephala* Meig. *J. Exp. Biol.* **32**, 692–699.

Thomsen, E., and Lea, A. O. (1969). Control of the medial neurosecretory cells by the corpus allatum in *Calliphora erythrocephala. Gen. Comp. Endocrinol.* **12**, 51–57.

Thomsen, E., and Møller, I. (1959a). Neurosecretion and intestinal proteinase in an insect, *Calliphora erythrocephala* Meig. *Nature (London)* **183**, 1401–1402.

Thomsen, E., and Møller, I. (1959b). Further studies on the function of the neurosecretory cells of the adult *Calliphora* female. *Acta Symp. Evol. Insects 1959* pp. 121–126.

Thomsen, E., and Møller, I. (1963). Influence of neurosecretory cells and of corpus allatum on intestinal protease activity in the adult *Calliphora erythrocephala* Meig. *J. Exp. Biol.* **40**, 301–321.

Tobe, S. S., and Loughton, B. G. (1967). The development of blood proteins in the African migratory locust. *Can. J. Zool.* **45**, 975–984.

Toevs, L. A., and Brackenbury, R. W. (1969). Bag cell-specific proteins and the humoral control of egg laying in *Aplysia californica. Comp. Biochem. Physiol.* **29**, 207–216.

Truman, J. W., and Riddiford, L. M. (1971). Role of the corpora cardiaca in the behavior of saturniid moths. II. Oviposition. *Biol. Bull. (Woods Hole, Mass.)* **140**, 8–14.

Vanderberg, J. P. (1963). Synthesis and transfer of DNA, RNA, and protein during vitellogenesis in *Rhodnius prolixus* (Hemiptera). *Biol. Bull. (Woods Hole, Mass.)* **125**, 556–575.

Vicente, N. (1969a). Étude histologique et histochimique du système nerveux central, des rhinophores et de la gonade chez les gastéropodes opisthobranches. *Tethys* **1**, 833–874.

Vicente, N. (1969b). Corrélations neuroendocrines chez *Aplysia rosea* ayant subi l'ablation de divers ganglions nerveux. *Tethys* **1**, 875–900.

Vicente, N. (1969c). Contribution à l'étude des gastéropodes opisthobranches du Golfe de Marseille. II. Histophysiologie du système nerveux. Etude des phénomènes neurosécrétoires. *Recl. Trav. St. Mar. Endocrine* **62**, 13–121.

Vicente, N. (1970). Observations sur l'ultrastructure d'un organe juxtacommissural dans le système nerveux du Chiton (Mollusque Polyplacophore). *C. R. Soc. Biol.* **164**, 601–607.

Vicente, N., and Gasquet, M. (1970). Etude du système nerveux et de la neurosécrétion chez quelques mollusques polyplacophores. *Tethys* **2**, 515–546.

Vietinghoff, V. (1967). Neurohormonal control of renal function in *Carausius morosus*. *Gen. Comp. Endocrinol.* **9**, 503.

von Hehn, G. (1970). Über den Feinbau des hyponeuralen Nervensystems des Seesternes (*Asterias rubens* L.). *Z. Zellforsch. Mikrosk. Anat.* **105**, 137–154.

Wang, D. H., and Scheer, B. T. (1963). UDPG-glycogen transglucosylase and a natural inhibitor in crustacean tissue. *Comp. Biochem. Physiol.* **9**, 263–274.

Wang, S., and Dixon, S. E. (1960). Studies on the transaminase activity of muscle tissue from allatectomised cockroaches, *Periplaneta americana*. *Can. J. Zool.* **38**, 275–283.

Weir, S. B. (1970). Control of moulting in an insect. *Nature (London)* **228**, 580–581.

Wells, M. J. (1960). Optic glands and the ovary of *Octopus*. *Symp. Zool. Soc. London* **2**, 87–107.

Wells, M. J., and Wells, J. (1959). Hormonal control of sexual maturity in *Octopus*. *J. Exp. Biol.* **36**, 1–33.

Wells, M. J., and Wells, J. (1969). Pituitary analogue in the octopus. *Nature (London)* **222**, 293–294.

Wendelaar Bonga, S. E. (1970). Ultrastructure and histochemistry of neurosecretory cells and neurohaemal areas in the pond snail *Lymnaea stagnalis* (L.). *Z. Zellforsch. Mikrosk. Anat.* **108**, 190–224.

Wigglesworth, V. B. (1936). The function of the corpus allatum in the growth and reproduction of *Rhodnius prolixus* (Hemiptera). *Q. J. Microsc. Sci.* **79**, 91–121.

Wigglesworth, V. B. (1943). The fate of haemoglobin in *Rhodnius prolixus* (Hemiptera) and other blood-sucking arthropods. *Proc. R. Soc. London, Ser. B* **131**, 313–339.

Wigglesworth, V. B. (1964). The hormonal regulation of growth and reproduction in insects. *Adv. Insect Physiol.* **2**, 247–336.

Wilkens, J. L. (1967). The control of egg maturation in *Sarcophaga bullata* (Diptera). *Am. Zool.* **7**, 723–724.

Wilkens, J. L. (1968). The endocrine and nutritional control of egg maturation in the fleshfly *Sarcophaga bullata*. *J. Insect Physiol.* **14**, 927–943.

Wilkens, J. L. (1969). The endocrine control of protein metabolism as related to reproduction in the fleshfly *Sarcophaga bullata*. *J. Insect Physiol.* **15**, 1015–1024.

Williams, C. M. (1952). Physiology of insect diapause. IV. The brain and pro-thoracic glands as an endocrine system in the cecropia silkworm. *Biol. Bull. (Woods Hole, Mass.)* **103**, 120–138.

Zalokar, M. (1960). Sites of ribonucleic acid and protein synthesis in *Drosophila*. *Exp. Cell Res.* **19**, 184–186.

Zinsmeister, P. P., and Davenport, R. (1971). RNA and protein synthesis in the cockroach ovary. *J. Insect Physiol.* **17**, 29–34.

2

The Development of Estrogen-Sensitive Tissues of the Genital Tract and the Mammary Gland

A. Raynaud

I. INTRODUCTION

With the progress of research in recent years, embryology and endocrinology become increasingly linked together. Alongside the fundamental contribution of embryology to the knowledge of the formation and morphogenesis of the endocrine glands, physiological embryology shows how the internal secretions play a role in the growth and modeling of the embryo, particularly the sexual modeling. The embryologist and the endocrinologist have to study, to varying degrees, the response of various

63

embryonic or young tissues to the same hormonal agent (see Willier, 1955; Jost, 1969; Jost and Picon, 1970). Thus, one problem the reproductive physiologist faces is the relationship between the embryonic origin of an epithelium of the genital tract and its response to sex hormones, in fetal life or after birth.

In the adult animal, estrogenic ovarian hormones have very specific effects, frequently reversible, on target organs such as vaginal epithelium, urethra, or certain prostatic lobes; this sensitivity is apparent in the fetus. In the first edition of this book, the author devoted a chapter (Raynaud, 1962) to the study of the histogenesis of urogenital and mammary tissues sensitive to estrogens in several species and to the interpretation of the various epithelial reactions in the different parts of the genital tract under the influence of the estrogenic hormones. Subsequent studies, by descriptive embryologists and histochemists on the embryonic genital tract, have made it necessary to alter some of these interpretations. The present chapter is a review of the new results.

II. HISTORICAL BACKGROUND

A brief summary of the main effects of estrogens on the genital tract in males and females is given below, but the reader is referred to Raynaud (1962) for more detailed information.

In intact or castrated male rodents, estrogens cause: (a) hypertrophy of some of the prostatic lobes in which the normal lining epithelium consisting of a single layer of cuboidal cells is replaced by a squamous, epidermoid epithelium which desquamates abundantly (mouse: Lacassagne, 1933; De Jongh, 1933, 1935; Burrows, 1934, 1935a,b; Burrows and Kennaway, 1934; rat: Korenchevsky and Dennison, 1934, 1935; Pfeiffer, 1936); (b) transformation of the single-layered secretory epithelium in the extremities of the ejaculatory ducts, some of the prostatic lobes, and sometimes in the urethra into a keratinized, epidermoid epithelium (Lacassagne, 1933, 1935b, 1936; Burrows, 1934, 1935a,b, 1937; Weller *et al.*, 1936).

These observations were soon extended to other species of mammals; thus, it was found that the prostatic utricle of monkeys was particularly sensitive to estrogens, its epithelium being transformed (either completely or only the basal part of the organ, according to the species) into a thick, cornified epithelium (Courrier and Gros, 1935; Parkes and Zuckerman, 1935; Courrier, 1936; Zuckerman and Parkes, 1936a,b,c; Zuckerman and Sandys, 1939). The same type of reaction was later observed in the guinea pig (Courrier and Cohen-Solal, 1936) and in the cat (Gros, 1939). In the dog, the whole prostate reacts to estrogens, the epithelial lining becoming squamous. Urethral keratinization has been observed in monkeys (Van Wagenen, 1935; Courrier and Gros, 1938; Zuckerman, 1936, 1940; Starkey and Leathem, 1939), in cats (Raynaud, 1937; Courrier and Gros, 1938), and in the ground squirrel, *Citellus* (Wells, 1936; Wells and Overholser, 1940). Thus, in the presence of estrogen, certain epithelia of the male genital tract undergo metaplasia and are transformed into stratified epithelium of the squamous type, sometimes with very marked keratinization; this reaction recalls the changes undergone by the vaginal epithelium at estrus and has often been termed a "vaginal-type reaction."

In most species, the vagina of intact or ovarietomized females becomes keratinized after treatment with estrogens; with prolonged treatment or increased dosage, areas of stratified, squamous epithelium appear in the cervix, in the body of the uterus, and even in the posterior part of the uterine horns (Lacassagne, 1935a,c).

The development in males receiving estrogens of epithelia of epidermoid type, often greatly keratinized, is difficult to explain. By analogy, the modified epithelia were first considered as being derived from the caudal parts of the Müllerian ducts. This "Müllerian hypothesis" was formulated by Lacassagne (1933) and Burrows and Kennaway (1934) and was developed in this form by Weller *et al.* (1936) to account for the transformations of certain lobes of the mouse prostate in response to estrogens: "perhaps the reason for these specific reactions of the vagina and prostate to the same hormone may be traced to a similarity in their embryological origin . . . Insofar as the most marked effect of estrin involves only one part of the prostate, that region where remnants of the Müllerian ducts would be sought, it is possible that these ducts contribute to the formation of the prostate." However, the Müllerian hypothesis had to be abandoned when it was found that the epithelium of other regions of the genital tract, such as the ejaculatory ducts and urethra, may show this epidermoid metaplasia. Zuckerman (1936, 1940) then proposed the now classical concept which bears his name: "Epithelial metaplasia and stratification in the reproductive tract in response to oestrogenic stimulation, whatever be its histochemical basis, may in general be regarded anatomically as a primary response of tissue in whose development oestrogen-sensitive sinus epithelium has either played a direct or an indirect part." The formation of a stratified, squamous epithelium in regions other than the urogenital sinus itself was explained by the invasion of these regions by cells derived from the urogenital sinus (e.g., extremities of the Wolffian canals and the prostatic utricle).

Zuckerman's hypothesis stimulated a considerable amount of research. To verify the hypothesis it was necessary to follow the development of the embryonic urogenital sinus and to determine the relations between its epithelium and the neighboring rudiments in a wide variety of species. Raynaud's (1962) appraisal of the literature on the histogenesis of the urogenital sinus indicated that most observations, for example, those on the mouse fetus, were in agreement with Zuckerman's hypothesis, although there were a few doubtful details in the mole and hedgehog, which at first sight do not appear to accord with the concept. Since 1962, many authors have made detailed studies of the development of the genital apparatus, particularly the vagina.

III. ORIGIN OF THE EPITHELIAL CONSTITUENTS OF THE VAGINA

Various theories have been proposed to explain the formation of the vagina. The vaginal duct can be derived from two main components, the caudal part of the Müllerian ducts and the urogenital sinus, which have come into contact in the embryo. The proportion of these two components varies with the species. In some species, the contribution of the caudal parts of the Wolffian ducts to the formation of the vagina has been described. A general account of the differences between representatives of the principal families of placental and marsupial mammals is found in Raynaud (1969). Detailed studies have been reported for the fetus of the rat (Bengmark and Forsberg, 1959; Forsberg and Olivecrona, 1964, 1965), mouse (Forsberg, 1963; Juillard and Delost, 1963a,b, 1964a; Forsberg and Olivecrona, 1964, 1965), hamster (Forsberg, 1960), rabbit (Forsberg, 1961), dog and cow (Forsberg, 1963), sheep (Bulmer, 1964), and man (Matèjka, 1959, 1963; Forsberg, 1963).

A. Rodents

Forsberg, Juillard, and Delost have conducted most of the work on formation of the rodent vagina mainly with the use of mouse embryo. When the fetus is 14 days of age, the Müllerian ducts extend into two short expansions on the dorsal surface of the urogenital sinus. The expansions reunite to form first: an epithelial bridge and then the sinus cord which includes the caudal parts of the Müllerian ducts. The caudal extremities of the Wolffian ducts insert into similar sinus pouches. The linings of the Wolffian ducts usually degenerate within the sinus expansions but in some 17- to 18-day-old fetuses, the lining can be recognized and followed as far as the interior of the fold of epithelium of the urogenital sinus that projects between the caudal parts of the Müllerian ducts (see Raynaud, 1942, Figs. 7 and 9). The sinus diverticula extend in a cranial direction and are recognizable in the 18-day fetus and in the neonate (see Raynaud, 1942, Figs. 13 and 14). Thus, at 18 days, this region of the female genital tract consists, from head to tail, of: the caudal extremities of the Müllerian ducts, a cord of cells from the cranio-dorsal part of the sinus (sinus cord), and the urogenital sinus, which divides frontally into a ventral urethra and a dorsal cord forming part of the vagina. Vaginal rudiment formation is completed during the course of the nineteenth day of intrauterine life and the first days of postnatal life.

Forsberg (1963) has shown that the caudal parts of the Müllerian ducts constitute the cranial part of the vagina. The Müllerian vagina is distinguished from the cervix uteri by the rudiments of the vaginal fornices, and, at term, it represents three-tenths of the total length of the vagina. It is followed by the short sinus cord. The rest of the vagina is derived from the urogenital sinus. After birth, the sinus vagina shortens and at the age of 4 days the Müllerian vagina forms six-tenths of the vaginal rudiment and the sinus vagina only four-tenths (Forsberg, 1963). Juillard and Delost (1963a,b) have confirmed these points (Fig. 1).

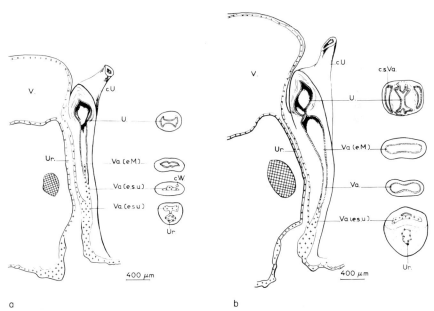

Fig. 1. Structure of the mouse vagina; (a) at birth; (b) 4 days after birth. On the left, the genital and urinary ducts are represented in sagittal section; on the right, transverse sections of the vagina at different levels. c.s.Va., vaginal fornices; c.U., uterine horn; c.W., Wolffian duct; e.s.u., epithelium of urogenital sinus; U., uterus; Ur., urethra; V., bladder; Va., vagina; Va. (e.M.), Müllerian part of the vagina; Va. (e.s.u.), sinus part of the vagina. (From Juillard and Delost, 1963a.)

Having established the topography of the genital ducts we can now consider the structural modifications shown by the epithelia of the two parts of the vaginal rudiment. In the fetus and in the neonate the cells of the urogenital sinus contain abundant PAS-positive material, part of which is glycogen (Raynaud, 1940, 1942, 1962). At the level where the Müllerian ducts come in contact with the urogenital sinus, the content of PAS-positive material is much less in the Müllerian cells than in the sinus cells, and this, of course, simplifies the task of distinguishing these epithelia. After birth, the structure of the epithelium becomes progressively modified in the caudal extremities of the Müllerian ducts; at the 4-day stage the Müllerian epithelium contains two layers, a basal layer formed of one or two rows of small cells with clear cytoplasm, and a superficial layer, nearest the lumen, formed of tall cylindrical cells. In some ways, the cells of the basal zone resemble those of the urogenital sinus. The superficial cells undergo reduction of size and number, and degeneration granules can be seen. Finally, this thin layer of cells is rejected and the epithelium of the Müllerian vagina takes on the appearance of a stratified epithelium continuous with that of the sinus vagina. The origin of the small basal cells with clear cytoplasm of the Müllerian vagina is uncertain. They may be derived from the urogenital sinus itself, or they may be formed from the pseudostratified cylindrical cells of the superficial layer. Forsberg has tried for many years (1963–1970) to solve this question, using the methods listed below.

1. Distribution of Mitosis

The use of colchicine shows that the mitotic rate is considerably higher in the Müllerian epithelium than in the sinus epithelium; in the Müllerian vagina the rate is lower in the basal zone than in the superficial zone. Moreover, an autoradiographic study with tritiated thymidine showed that the labeling of the cell was weak in the epithelium of the urogenital sinus, but appreciable in both layers of the Müllerian vagina. The label was approximately equal in the basal and superficial zones (Forsberg, 1965a). These results indicate that the basal layer of small cells must be formed from the cylindrical cells of the Müllerian vagina.

2. Enzyme Distributions

The results of studies of alkaline and acid phosphatases and of PAS-positive substances are in agreement with the study of distribution of mitotic figures (Forsberg and Olivecrona, 1964, 1965; Forsberg, 1967).

3. Experimental Studies

These have involved grafting of the Müllerian region after separation from the rest of the vaginal rudiment. The superior part of the Müllerian vagina of a newborn female (Fig. 2, mv) was detached cranially before its junction with the sinus cord (Fig. 2, sc) in such a way as to avoid contamination of the Müllerian graft with sinus cells. The isolated Müllerian graft was grafted into the thigh muscles of a newborn mouse; whole vaginal grafts served as controls. The grafts developed well and epithelial differentiation continued for days 6 to 14 of the experiment. Microscopic examination showed that the isolated Müllerian epithelium contained, as in the controls, a superficial and a basal zone of cells, the latter subsequently giving rise to the whole of the basal epithelial layer of the anterior part of the vagina. Forsberg (1965b) therefore concluded that the vaginal epithelium of the mouse has a double origin, the posterior part coming from the urogenital sinus and the anterior part from the Müllerian epithelium; but an influence of the sinus epithelium on the differentiation of the Müllerian epithelium was not excluded. A second series of experiments involved the same type of graft, but the rudiments of the Müllerian vagina were taken from 14.5-day fetuses (Forsberg and Norell, 1966). Microscopic examination of the remaining genital apparatus of the donor fetus showed that the posterior part of the Müllerian ducts was present and attached to the urogenital

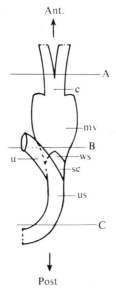

Ant.

Post

Fig. 2. Diagrammatic representation of the vaginal rudiment in the mouse at birth. The part of the anlage between the lines A and B was grafted as "Müllerian vagina" and the part between A and C was the control as "whole vaginal anlage." c, cervix; mv, Müllerian vagina; sc, cord of sinus epithelium, the cranial part of the sinus vagina anlage; u, urethra; us, urogenital sinus; ws, wedge of sinus epithelium protruding into the lumen of the Müllerian vagina. (From Forsberg, 1965b.)

sinus. Forsberg believed that this was sufficient proof that the removed Müllerian rudiment was free of cells from the sinus epithelium. The grafts of Müllerian vagina were studied after 12 days and were found to have differentiated normally into a basal and a superficial layer. These results confirmed previous conclusions that the anterior part of the mouse vaginal epithelium is derived from Müllerian epithelium. According to Forsberg and Norell (1966), these results also suggest that it is rather unlikely that the differentiation of the Müllerian epithelium is induced by the sinus epithelium; it is more reasonable to consider the relation of mesenchyme to epithelium as an important factor in the process of differentiation of the Müllerian vaginal epithelium.

An immunological study on the same subject was reported by Forsberg and Nord (1969). Rabbits were injected with uterine epithelium from fertile mice to obtain a specific antibody, and the distribution of antigen was investigated in fetal, neonatal, and young mice. A reaction was found in the Müllerian epithelium of the neonatal and young mice, but none was present in either the sinus or Wolffian epithelia of the fetuses. Specific fluorescence was observed in the narrow superficial zone of the whole of the Müllerian epithelium of neonates, but there was none in the basal layers of the vaginal Müllerian epithelium.

In the rat and the hamster, formation of the vagina appears to be similar to that found in the mouse (Bengmark and Forsberg, 1959; Forsberg, 1960, 1963), but there are differences in the content of certain enzymes and of PAS-positive material in the epithelia.

In summary, the vaginal rudiment in the mouse, rat, and hamster contains: (a) a cranial part formed by the posterior, fused parts of the Müllerian ducts; (b) a short intermediate part, the sinus cord, formed by a craniodorsal expansion of the urogenital sinus; and (c) a posterior part, the sinus vagina, formed by a dorsal part of the urogenital sinus. The proportion of the Müllerian and sinus components is different in the three species (Fig. 3). The Müllerian epithelium persists and differentiates, in the posterior parts of the Müllerian ducts, to form the epithelium of the Müllerian vagina; it is not replaced by the epithelium of the urogenital sinus, except for a short segment of its caudal part.

B. Rabbit

In the rabbit there is no frontal division of the urogenital sinus, but the Müllerian epithelium near the sinus degenerates being replaced by an epithelium derived from the Wolffian ducts rather than the urogenital sinus (Forsberg, 1961, 1962, 1963). This conclusion was based on a study of the histology and the distribution of β-glycuronidase and acid and alkaline

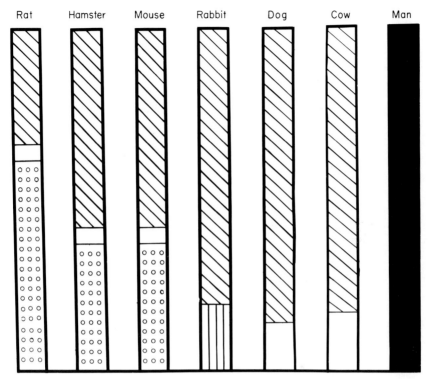

Fig. 3. Diagrammatic representation of the relative proportions of the sinus, Müllerian, and Wolffian epithelia of different species based on study of the latest stage showing a clear distinction of the epithelia. Small circles, sinus epithelium; unshaded, sinus
epithelium which grows cranially; diagonal bars, Müllerian epithelium; vertical bars, Wolffian epithelium; black, uncertain origin. (From Forsberg, 1963.)

phosphatase activities in the epithelia. Thus, the caudal part of the rabbit
vagina (about one-sixth to one-seventh of the total vaginal anlage) would be
Wolffian in origin, and the remainder, Müllerian, with the sinus forming the
vestibule of the vagina (Fig. 3). This interpretation differs from that given
previously by Jost (1947) and Carr (1953) who described the development
from the urogenital sinus of a segment of sinus epithelium which became
intercalated between the caudal extremities of the Müllerian ducts and the
urogenital sinus.

C. Dog

There is no longitudinal partitioning of the urogenital sinus in the dog. At
birth, sinus epithelium represents about one-tenth of the total length of the

vaginal rudiment (Fig. 3) and the rest is Müllerian in origin (Forsberg, 1963).

D. Cow

In the cow, the greater part of the epithelial lining of the vagina is composed of Müllerian epithelium; only the caudal part of the vagina is derived from the urogenital sinus (Fig. 3). Between the 70- and the 150-mm stage, degeneration of the caudal parts of the Müllerian ducts takes place and there is a displacement in the cranial direction of the boundary (zone of fixation of the genital ducts) between the sinus and Müllerian epithelia. The sinus epithelium also replaces the Wolffian epithelium in the caudal parts of the Wolffian ducts (Forsberg, 1963).

E. Sheep

In earlier work, Bulmer (1952) had shown that the epithelium of the superior vaginal segment in the sheep was Müllerian in origin. Later (1964), he studied the glycogen content of the Müllerian and sinus epithelia at different stages. From the 58-mm stage, the stratified pavement epithelium of the urogenital sinus contains much glycogen and little cytoplasmic protein; the prismatic Müllerian epithelium has a relatively low glycogen content. At midgestation, the Müllerian epithelium in the caudal part of the superior vagina is transformed into stratified polygonal epithelium with an accumulation of glycogen. This epithelium shows a certain resemblance to that of the urogenital sinus. From these observations, Bulmer (1964) concluded that the epithelium of Müllerian origin may be in two forms: a stratified pavement type and a prismatic type, according to its position in the genital tract and the influences to which it is submitted. Bulmer considered that certain influences may act on the Müllerian vaginal epithelium to stimulate its transformation into a stratified squamous epithelium; these influences might come from the epithelium of the urogenital sinus, but there would be no replacement of Müllerian cells by sinus cells. Bulmer also stressed the difference between the sheep and the mouse in the amount of PAS-positive material in the Müllerian cells, citing the facts mentioned by Raynaud in 1962. However, these facts refer to different stages in the development of the fetal and neonatal mouse; only the caudal part of the Müllerian ducts is considered in the neonate. Moreover, the comparison of glycogen content in cells of Müllerian origin in a sheep fetus 15 to 20 cm long and in fetal or neonatal mice is hardly justified. Bulmer (1964) noted that, in the younger fetal stages, the prismatic Müllerian epithelium contains little glycogen, and this contrasts with the high content in the sinus cells. This corresponds with the situation described in the mouse (Raynaud, 1942) and the differences between the vaginal epithelia of the sheep and the mouse may not therefore be as great as Bulmer (1964) has suggested.

F. Man

The formation of the human vagina presents some difficult problems for the embryologist, and there is still no general agreement on the interpretation of certain histological aspects. Forsberg (1963) gives a review of the principal work on human vaginal morphogenesis; here only a few special points will be mentioned. According to Matèjka (1959), the epithelium is derived entirely from the urogenital sinus. The Müllerian epithelium does not degenerate to make room for the sinus epithelium, but ceases growth at the sinus boundary where it forms the endocervix. A thin layer of sinus epithelium arises from the dorsal part of the sinus pocket where the genital ducts insert, and this grows cranially to form the rudiment of the vaginal epithelium. Fluhmann (1960) considers, based on histological and endocrinological grounds, that the squamous and the cylindrical epithelia of the cervix uteri are derived from the epithelium of the urogenital sinus.

Forsberg (1963) has used 41 female fetuses with crown-rump lengths of 60 to 230 mm in an extensive morphological study. This was supplemented by a histochemical study of 37 fetuses to see if the parts of the vaginal rudiment could be distinguished by their content of PAS-positive material and enzyme activities. Two principal findings emerged. First in the region where the uterovaginal canal is in contact with the urogenital sinus, the dorsal lining of the sinus, between the points of insertion of the Wolffian ducts, is formed of a special epithelium, termed "differentiated epithelium," which differs in appearance from the sinus epithelium. The histochemical reactions suggested to Forsberg that, in the region of the Müllerian tubercle, the epithelium of the median walls of the Wolffian duct is incorporated into the dorsal lining of the urogenital sinus, thus giving rise to the differentiated epithelium. Second, in the caudal part of the uterovaginal canal at the point of contact with the differentiated epithelium, the epithelium is at first pseudostratified and later becomes stratified squamous in type, the transformation progressing cranially. A solid epithelial layer, the vaginal plate, appears between the urogenital sinus and the uterovaginal canal and this increases as the stratified, squamous Müllerian epithelium degenerates, finally giving rise to the vaginal epithelium. The origin of the vaginal plate is not definitely known. Some authors consider that it is composed of Müllerian epithelium, others of sinus epithelium. According to Forsberg, the epithelium of this plate is paler than the sinus epithelium, but a comparison of enzyme activity in the epithelia of the urogenital sinus, of the Wolffian ducts, and of the vaginal plate shows a greater similarity between the activity of several enzymes (arylsulfatase, acid phosphatase, esterase, and leucine aminopeptidase) in the epithelia of the vaginal plate and the Wolffian ducts than in the epithelium of the urogenital sinus. These facts favor derivation of the vaginal plate from the epithelium of the Wolffian ducts. However, comparable contents of β-glycuronidase and DPN-diaphorase are found in the epithelia of the vaginal plate and the urogenital sinus. Thus, if there is to be a choice between a Müllerian or Wolffian origin of the vaginal epithelium in man, Forsberg (1963) considers the Wolffian to be the more probable and concludes: "All findings point to the fact that the epithelium over the Müllerian tubercle, from which region the cranially growing vaginal plate originates, is a Wolffian derivative. Man differs markedly from all other species studied, because the Müllerian epithelium is not included in the vaginal epithelium in the adult, but definitely undergoes a transformation into stratified squamous epithelium in the fetus."

G. Conclusions

There seems, therefore, to be considerable variation in the origin of the components of the vagina in the various species so far studied, and the participation of the different epithelial components are shown diagramatically in Fig. 3. The concept that the Wolffian epithelium is involved in the formation of the vagina in the rabbit and man gives rise to several problems. The Wolffian epithelium is known to degenerate if androgenic hormones are lacking and, in the fetuses of both sexes and in the adult female, it does not keratinize under the influence of estrogens. The possibility that the Wolffian epithelium may be influenced by the sinus to undergo transformation should to be verified by further investigation.

IV. THE EFFECTS OF ESTROGENS

Many papers have described the effects and mechanism of action of steroid hormones on the vagina of the neonatal rat and mouse and of immature or adult animals. The administration of estradiol or of testosterone to the neonate (or even to the fetus) can produce constant estrus, the nature of which depends on the dose administered and the sensitivity of the animals injected. However vaginal keratinization thus produced may be independent of the ovary, or may depend on ovarian secretions. In the latter instance, it is probable that the steroid administered affects the hypothalamo–hypophysial axis, stimulating the continuous liberation of small quantities of gonadotropins which cause secretion of ovarian hormone (Gardner, 1959a,b; Takasugi et al., 1962; Takasugi, 1963, 1966; Takewaki, 1964; Kimura et al., 1967a,b,c; Kimura and Nandi, 1967).

In vitro experiments, in which vaginal epithelium is grown in the absence or presence of estradiol or testosterone and then transplanted into 4-day-old females, suggest that the mechanism of induction of constant estrus by estrogens and androgens is different. The estrogens have a direct effect on vaginal epithelium in culture, causing irreversible modifications independent of the ovary; the effects of testosterone are indirect (Kimura et al., 1967c). During these studies of the effect of estradiol on the neonate, it was found that hyperplastic lesions developed in the vaginal epithelium, together with hypertrophy of the uterine epithelium and various changes in the genital and mammary apparatus (Dunn and Green, 1963; Takasugi and Bern, 1964; Kimura and Nandi, 1967; Kimura et al., 1967a).

Recent work has followed three main lines. (1) The places of action and the primary mechanisms of action of estrogens on the receptor cell have been investigated (see Volume III, Chapter 2), particularly with respect to the effects on the nuclear synthesis of RNA (Mueller et al., 1958; Mueller, 1965; Means and Hamilton, 1966) and the binding with cytoplasmic and nuclear proteins (Talwar et al., 1968; Baulieu et al., 1967; Jensen et al., 1969; Talwar, 1970; Baulieu and Rochefort, 1970; Teng and Hamilton, 1970). (2) The effect of estrogens on mitosis has been extensively studied (see Volume III, Chapter 3; Tice, 1961; Perrotta, 1962; Martin and Claringbold, 1960; Martin and Finn, 1968; Galand et al., 1967; Finn et al., 1969; Forsberg, 1970). (3) The synergism of estrogens and progesterone in the differentiation of the uterus and its secretions has been investigated. Takasugi (1963) has indicated that the continuous estrus produced by estrogen treatment of the neonate must be the result of changes provoked by the hormone at the cell level in the vagina.

In this section, consideration will be given only to the effects of estrogens on the development of different types of vaginal epithelium.

A. Forsberg's Research

When 1 μg estradiol/day is administered to mice from birth (Forsberg, 1966), no changes in the vaginal epithelium are seen for the first 2 to 3 days. At 4 days of age, the sinus epithelium starts to stratify and at 5 days it has become squamous and stratified (Fig. 4a). In the Müllerian vagina, the

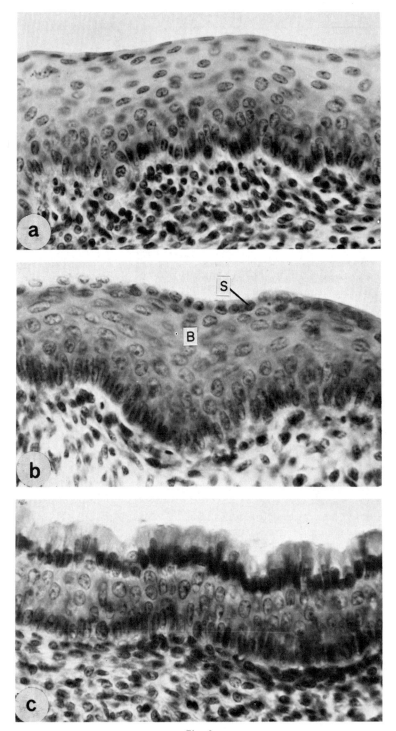

Fig. 4

basal zone which is formed from columnar Müllerian epithelium (see Section III,A) thickens at 4 days, and at 5 days, in the posterior part of the Müllerian vagina, it is differentiated into squamous, stratified epithelium while the superficial zone becomes flattened (Fig. 4b). In the cranial part of the Müllerian vagina, the superficial zone remains as tall, columnar cells, and the basal zone is formed of several layers of cells with swollen ovoid nuclei (Fig. 4c). At a more advanced age, the epithelium of the Müllerian part of the vagina tends to become entirely squamous and stratified, but at 21 days the cranial part still has a superficial layer of tall cells containing mucus. The Müllerian vaginal epithelium thus becomes squamous and stratified like that lining the sinus vagina.

This histological study was complemented by an autoradiographic investigation in which tritiated thymidine was administered at 2 or 6 hours after injection of estradiol. Two hours after thymidine injection, the number of labeled cells in the pseudostratified epithelium of the Müllerian vagina was higher than that in the sinus epithelium. Forty-eight hours after injection, when the Müllerian epithelium is differentiated into two zones, numerous labeled cells appear in the superficial and basal zones. Some labeled cells are present in the epithelium of the sinus vagina. Forsberg concludes (1966) that the basal cells of the Müllerian epithelium are derived from the columnar cells and are, therefore, Müllerian in origin. If the sinus epithelium had given rise to the basal zone present in most of the Müllerian vagina, a considerable difference in the labeling of the cells of the basal and superficial zones would have been expected because of the isotope dilution.

It is not possible to postulate from these results a growth of the sinus epithelium in a cranial direction after estradiol treatment. The hormone simply accelerated the normal process of differentiation of the basal Müllerian layer and in treated animals the zone was much better developed than in control animals of the same age. The experiment establishes that the Müllerian epithelium is capable of responding to estrogens by changing into a squamous, stratified epithelium; this capacity is not restricted to the epithelium of the urogenital sinus. In young mice treated with estradiol and normal young mice, differentiation of the basal Müllerian zone starts in the zone adjacent to the urogenital sinus and then extends in a cranial direction. Forsberg has suggested (1965a,b,c, 1969) that, as for the sheep fetus (Bulmer, 1964), the urogenital sinus can exercise an influence on the Mül-

Fig. 4. Histological sections of the vaginal epithelium of a 5-day-old female mouse given daily subcutaneous injections of 1 μg estradiol. (a) The sinus vagina showing a stratified squamous epithelium; (b) the caudal part of the Müllerian vagina where the basal zone (B) is well differentiated into a stratified, squamous epithelium, and the superficial zone (S) is a layer of low cuboidal cells; (c) the cranial part of the Müllerian vagina where the basal zone is well differentiated, hyperplasic, and covered by a superficial zone of columnar cells. × 477. (From Forsberg, 1966.)

lerian epithelium. This concept has been modified as a result of the grafting experiments (see Section IV,B) and mesenchyme–epithelium interactions are postulated.

B. Juillard and Delost's Research

Juillard and Delost (1964a,b, 1965b, 1966) and Juillard (1966, 1969) have investigated the origin of keratinizing epithelia of the mouse vagina in a number of experiments. In one series, a single subcutaneous injection of 0.5 or 1 mg estradiol dipropionate was administered to neonatal mice of the Swiss strain, which were killed after 2 to 15 days. The sinus epithelium was keratinized 4 days after the injection and keratinization became more intense by 6 days. At 6 days, in the posterior and middle part of the Müllerian vagina, and beneath the columnar epithelium which had not yet been shed into the lumen, there was an epidermoid epithelium which appeared to grow from the urogenital sinus toward the uterus. In the superior part as far as the cervix, there were distinct regions of swollen, slightly stained cells (Figs. 5a and b) and these regions ultimately (at 8 or 11 days) gave rise to an epidermoid epithelium which became keratinized and was shed into the lumen as the columnar cell layer. When estrogens were administered to 3-day-old mice, the effects were comparable. After 1 day there was a layer of small basal cells that was continuous with the sinus epithelium and beneath the columnar epithelium of the uterus there were one or two layers of small transparent cells which appeared to have grown from the caudal region (Fig. 6).

The results of Juillard and Delost (1964b, 1965a,b) therefore support the concept of cranial development of the sinus epithelium beneath the atrophying Müllerian epithelium. However, the discontinuity between the regions of small translucent cells needs explanation and Juillard and Delost suggest that they are derived from isolated groups of sinus cells which have infiltrated the epithelium, in the fetus or the neonate, and have retained the capacity to keratinize under the influence of estrogen. Thus, these authors consider that the sinus epithelium is able to proliferate and grow cranially to invade the vagina and the uterus.

In a second series of experiments, Juillard and Delost (1966) and Juillard (1966) tried to find evidence of an inductive function of the urogenital sinus on the embryonic Müllerian duct by grafting together a very young Müllerian duct and a urogenital sinus. Segments of Müllerian ducts were taken before (13-day-old fetuses) and after (14 days and older) they had reached the urogenital sinus. These segments were then grafted alone or in contact with portions of the urogenital sinus into the anterior chamber of the eye of adult female mice treated with estrogens. In hosts killed 10–12 days after

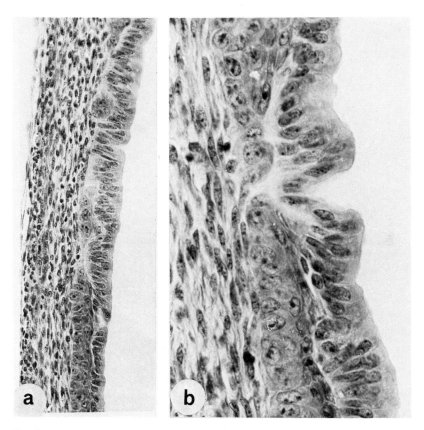

Fig. 5. Sections of the superior part of the vagina of a 6-day-old mouse treated with estradiol at birth, showing the discontinuous areas of small cells beneath the pyramidal Müllerian epithelium. (a) ×320; (b) ×900. (From Juillard and Delost, 1964b.)

grafting, 13-day stage Müllerian ducts grafted in contact with sinus epithelium retained their characteristic epithelium of columnar cells (Fig. 7a). A similar result was obtained with a 14-day stage in which attachment to the sinus had already occurred (Fig. 7b and c). When Müllerian ducts were removed from the neonate and grafted under the same conditions, a basal zone of several layers of transparent cells appeared beneath the layer of columnar cells.

Comparable results have been obtained from *in vitro* experiments (Juillard, 1966). Therefore, Juillard (1966) concluded that the inferior part of the Müllerian duct, when removed from a 13-day fetus before its attachment to the sinus, never becomes keratinized in a graft or in culture, even when it is in the presence of several pieces of sinus and remains in close contact with the sinus epithelium or mesenchyme. However, when similar

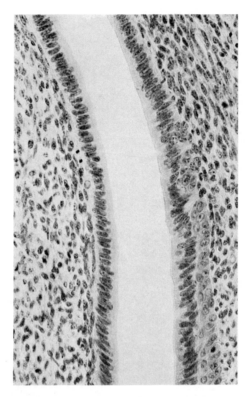

Fig. 6. Longitudinal section of the uterus of a 6-day-old mouse which had received 0.5 mg estradiol dipropionate at 3 days of age. A discontinuous epithelial layer of small, clear cells can be seen, which appear to progress cranially beneath the layer of cylindrical cells of the Müllerian uterine epithelium. ×455. (From Juillard and Delost, 1965b.)

Fig. 7. Grafts of urogenital sinus (s.u.g.) and Müllerian duct (c.M.) of young mouse embryos into the anterior chamber of the eye of adult female mice treated with estradiol. (a) Müllerian duct and two portions of urogenital sinus taken from a 12-day 13-hour embryo and grafted side by side for 11 days. The sinus epithelium has become thickened and keratinized but the Müllerian epithelium shows the normal appearance for a uterine horn. ×425. (From Juillard, 1966.) (b) Junctional zone of the Müllerian vagina (Va.m.) and urogenital sinus taken from a 14-day embryo grafted for 14 days. The Müllerian epithelium retains its typical structure while the sinus has become keratinized. ×425. (From Juillard, 1966.) (c) Section of the Müllerian duct taken above the level of the junction with the urogenital sinus in a 14-day 12-hour embryo grafted for 12 days in contact with a sinus from an embryo of the same age. The keratinized urogenital sinus has not modified the structure of the Müllerian epithelium. ×425. (From Juillard, 1966.)

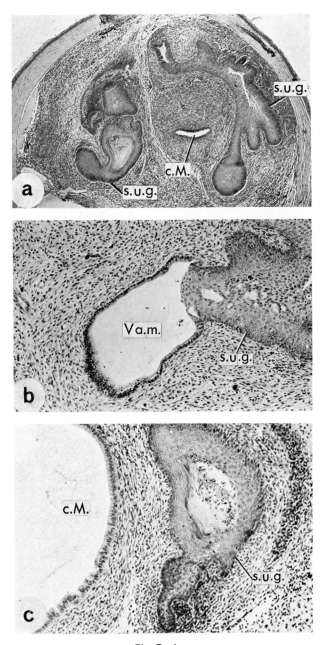

Fig. 7a, b, c

grafts are made from the uterus or from the Müllerian vagina of an 18-day fetus or of a neonate, a layer of small transparent cells appears beneath the layer of prismatic Müllerian cells (Juillard, 1966). Thus, it seems that the potential of the rudiments has been modified between days 13 and 18 of fetal life. Juillard and Delost (1966) consider that, from the end of day 14, sinus cells spread beneath the Müllerian epithelium and the keratinization that subsequently occurs in the Müllerian derivatives is believed to be due to their presence and further development.

The fact that the urogenital sinus does not influence the 13-day Müllerian duct graft, does not necessarily invalidate the induction theory of Forsberg (see Section IV,A), because in Juillard's grafts the embryonic Müllerian duct was not completely divested of the surrounding mesenchymatous layers. In these conditions, therefore, it is possible that the urogenital sinus could not exert the same influence on the Müllerian epithelium in the anterior chamber of the eye as it does *in vivo* where the caudal extremities of the Müllerian ducts have penetrated the mesenchyme. Experiments are now being conducted with grafts of Müllerian ducts from which the greater part of the mesenchyme has been removed and the results are awaited with interest. At the present time, Forsberg's results on counts of mitotic figures, autoradiographic, histochemical, and immunological studies, favor the concept that the urogenital sinus influences the caudal parts of the Müllerian ducts that are in contact with it.

C. Other Research

Several authors (Perrotta, 1962; Peckham and Kiekhofer, 1962; Husbands and Walker, 1963; Peckham *et al.*, 1963) have investigated whether the vaginal epithelium of the rat and mouse has two distinct cell populations, one responding to estrogens by keratinization and the other synthesizing glycogen and mucins, or only one type of cell which is differentiated according to the hormonal environment at the time of formation. The results of these investigations are inconclusive. Forsberg and Olivecrona (1965) state that the fate of the superficial cells is definitely fixed and cannot be changed by the action of estradiol. These cells produce mucin and do not keratinize. It is the cells destined to form the basal zone from the superficial zone which would synthesize glycogen and keratin in the normal animal, while those remaining in the superficial zone would synthesize glycogen and mucin. After administration of sufficiently high doses of estradiol, the capacity of the basal zone cells in the Müllerian vagina to synthesize mucin would be inhibited and these cells would be able to form keratin. Estradiol could thus change the determination of the basal layer cells of the mouse vaginal epithelium. Takasugi (1963) has shown that estradiol could have a permanent effect on the vaginal epithelium at the cell level. However, estradiol could also be acting simultaneously on the sinus epithelium cells to inhibit their latent ability to synthesize mucins and allow them to produce keratin only (Forsberg and Olivecrona, 1965). Juillard (1969) has shown that the simultaneous administration of estradiol and testosterone or estradiol and progesterone caused mucification of the epithelium of the Müllerian vagina, but not the sinus epithelium. When estradiol is administered to the neonatal mouse, proliferation of the columnar Müllerian epithelium of the cervix uteri is inhibited (Forsberg, 1970). The same effect has been found *in vitro,* but the uterine epithelium was not affected (Forsberg and Lannerstad, 1970).

V. INFLUENCE OF ESTROGENS ON THE MAMMARY RUDIMENTS

It is well established that, in the mouse, estrogens are able to alter the development of the mammary rudiments and to induce in them a great variety of malformations (Raynaud, 1947; Raynaud and Raynaud, 1956). These observations have been confirmed in the rat (Delost *et al.,* 1962, 1963; Jean and Delost, 1965; Jean, 1970). In general, estrogens (estradiol, dihydrostilbestrol) injected early in fetal life inhibit the development of the mammary anlage (*sensu stricto*) and exaggerated development of the ectodermal parts of the rudiment. The ectoderm covering the mammary area (nipple ectoderm) becomes thickened at the surface, or in the underlying layers, sometimes penetrating deeper into the mesenchyme, and forms the epithelial lining of a cutaneous pocket (coelomast) in which superficial keratinized layers accumulate. The ectodermal cells of the epithelial stalk connecting the mammary rudiment to the surface epithelium proliferate under the influence of estrogens and form massive epithelial invaginations which penetrate into the dermis. Some of the cells forming the mammary rudiment are inhibited in their development or are destroyed (amastia or micromastia). It is possible that estrogens may also modify the differentiating capacity of some of the rudiment cells and reduce them to a simple ectodermal differentiation (Raynaud, 1971).

VI. ESTROGENS AND THE UROGENITAL TRACT IN REPTILES

The urodaeum of reptiles corresponds to the urogenital sinus of mammals. In 1938, Dantschakoff drew attention to the fact that in embryos of both sexes the epithelium of the cloaca and its derivatives show the Allen–Doisy reaction in that they respond to estrogens. Histological and histochemical investigations (Raynaud and Raynaud, 1963; Raynaud et al., 1968) have shown the similarities that exist between the urodaeal cells and those of the mammalian urogenital sinus; both contain an abundance of PAS-positive material which consists partly of glycogen. Like the lining of the rodent vagina, the epithelial lining of the urodaeum shows seasonal modifications under the influence of ovarian secretions. At the time of mating, the period corresponding to estrus, the urodaeum is lined with stratified epithelium that has a thick Malpighian layer, with a superficial stratified pavement epithelium showing desquamation. The superficial cells do not become keratinized but patches of epithelial cells, two to ten layers thick, are shed into the urodaeal lumen.

We have studied the localization of the epithelial proliferation in the genital ducts of the slowworm (Anguis fragilis L.) at the time of mating (Raynaud and Raynaud, 1963; Raynaud et al., 1968); proliferation was found in: (a) the whole of the urodaeum, but more in the dorsal and dorsolateral than in the ventral epithelium; (b) the caudal part of the Wolffian ducts; and (c) the caudal part of the ureters. In contrast, the epithelium in the caudal parts of the Müllerian ducts, the "vaginal segment" (Giersberg, 1923, Jacobi, 1936), does not undergo true stratification. Even at the culmination of estrus this epithelium is different from that of the urodaeum; it forms folds but consists of alternating ciliated and mucous cells.

Because of the long period between birth and first estrus, we used young animals to study the responses of the genital ducts to estrogens. In this respect, reptiles are of particular interest because the dilated caudal extremities of the Müllerian ducts become juxtaposed to the dorsal lining of the urodaeum but do not open into this cloacal chamber until several months after birth. Thus, in the slowworm, the Müllerian ducts reach the urodaeum at the 120-mg stage and, at this time, numerous Müllerian cells degenerate close to the urodaeal lining. The Müllerian ducts remain in contact with, but do not open into, the genital urodaeal pouch for the remainder of fetal life (from the 120- to the 500-mg stage). It is only a year after birth that the ducts open into the cloaca, and blind ducts have been found by the author in slowworms weighing 3 or 4 gm in May and June.

In young females (weighing 2.5–4.69 gm) treated with doses of 0.2 to 1.0 mg estradiol benzoate and killed 8–16 days after the first injection, histological examination showed that the Müllerian ducts were not yet open into the urodaeum. In animals which had been treated for 10 days or more, the urodaeal epithelium in the genital pouch had become thickened and stratified; it was composed of 7 to 15 layers of cells surmounted by a layer of mucous cells in the process of being shed into the lumen. The Müllerian epithelium of the "vaginal segment" showed no stratification comparable with that seen in the urodaeal epithelium; it was thickened but the cells retained a pseudostratified arrangement. At the very end of the vaginal segment, the cells had a swollen nucleus with a large nucleolus; two or three nuclei were sometimes superimposed in the epithelium while the adjoining epithelium was single layered. In the connective tissue which surrounds the Müllerian ducts, similar cells appeared singly or in groups and were not found elsewhere in the connective tissue. It was not possible to determine their origin; they could be Müllerian cells which have been isolated at a time when the Müllerian ducts reached the urodaeum or they could be derived from the urodaeal epithelium. In all these animals, the Müllerian ducts ended in the wall of the urodaeal genital pouch, without opening into the urodaeum. The caudal parts of the ureters and Wolffian ducts were hypertrophied and the walls consisted of four or five layers of small cells surmounted by a layer of mucous cells.

Comparable modifications were observed in the treated slowworms killed at a more advanced age. For example, at 28 to 37 days after the first injection, the urodaeal epithelium reacted strongly to estrogens; it became stratified, 100–150 μm along the lining of the genital pouch, and consisted of ten to twenty layers of cells, the most superficial of which had been shed into the lumen. In the vaginal segment, the epithelium had remained relatively thin (10–30 μm) with, occasionally, a multiplication and an irregular, pseudostratified arrangement of the cells, but no true stratification. Figure 8a shows the striking contrast between the structure of the vaginal and urodaeal epithelium. In a female which had received a total dose of 1.48 mg estradiol benzoate and was killed 29 days after the first injection, a short region of stratified epithelium (connected to that of the genital pouch) was present in a secondary fold of the vagina. This fold was separated from the main body of the vagina and appeared to be degenerating; no comparable modification was visible in the main vagina. In another female, killed 37 days after the first injection (total dose of estradiol benzoate injected was 1.72 mg), there was no stratified epithelium in the vagina. For these reasons it is probable that the modifications which have been observed in one animal represent an upward movement of a region of urodaeal epithelium from the genital pouch into the lumen of a diverticulum of the principal vagina.

In the terminal parts of the Wolffian ducts and ureters (principal and secondary ureter) the epithelium also reacted to estrogens; several basal layers with a stratified arrangement surmounted by a superficial layer of large mucous cells were formed (Fig. 8b and c). In the Wolffian ducts this structure was only found in a short terminal segment; above this, the epithelium of the Wolffian duct was similar to the normal type but showed

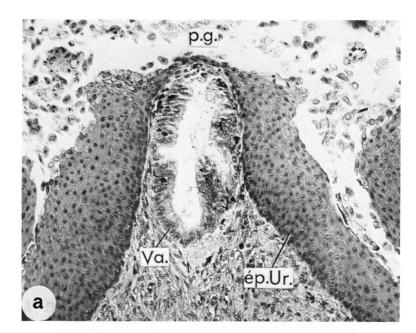

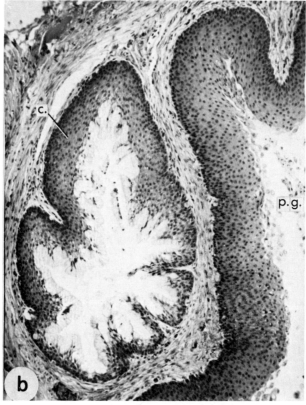

Fig. 8a and b

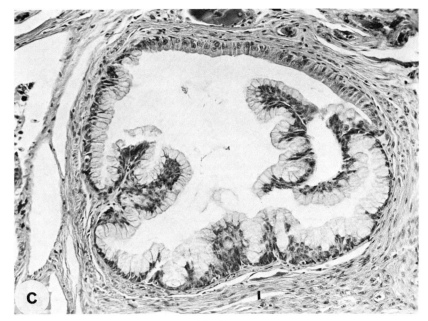

Fig. 8c

Fig. 8. Histological sections of the genital tract of slowworms treated with estradiol. (a) The dorsal lining of the urodaeal pouch (p.g.) at the level of the termination of the right vagina (Va.) from a 13.7-cm, 8.9-gm slowworm treated with 1.72 mg estradiol benzoate (in five injections) and killed 22 days after the first injection. The section shows the scarcely modified structure of the vaginal epithelium about 200 μm above the point where it opens into the urodaeum. The epithelium of the vagina has remained thin but the genital pouch epithelium (ep. Ur.) is thickened and desquamating ($\times$170). (b) The caudal part of the common Wolffian duct and ureter (c.) and the dorsal wall of the genital urodaeal pouch (p.g.) of a 13.7-cm slowworm treated with 1.72 mg estradiol. The epithelium of the genital pouch is stratified and the superficial layers are sloughing; the duct epithelium is reacting similarly but is still surmounted by a layer of mucous cells ($\times$112). (c) Section across the ureter close to its entry into the kidney, of a 5.7-gm slowworm treated with 1.48 mg estradiol benzoate in four injections and killed 38 days after the first injection. For almost the whole length of the ureter there are two layers along the ventral part, a basal layer of small, stratified cells, and a superficial layer of large mucous cells. This type of epithelium, shown in the lower half of the section, gives way in the dorsal half of the duct to another type formed of a single layer of cells ($\times$180).

numerous pyknotic cells. Complete degeneration was found in a female weighing 8.9 gm and treated with a total dose of 1.7 mg estradiol benzoate (given in five injections) and killed after 28 days. In the ureters, however, there was a considerable length where the epithelium consisted of stratified basal layers cells surmounted by a superficial layer of large cells rich in mucus. The thickness of the basal stratified layer diminished progressively with distance from the urodaeum, and close to the insertion of the ureters

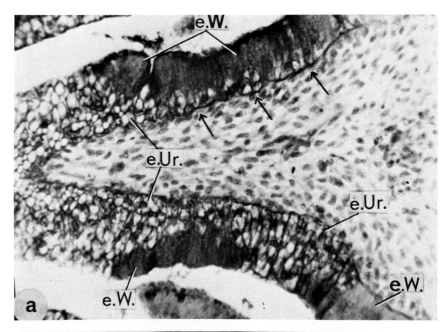

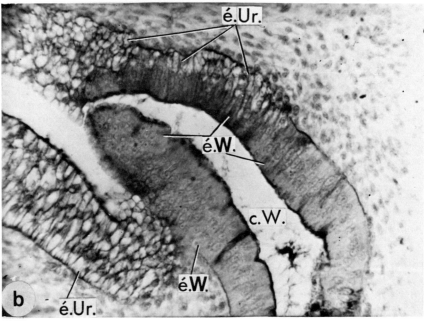

Fig. 9a and b

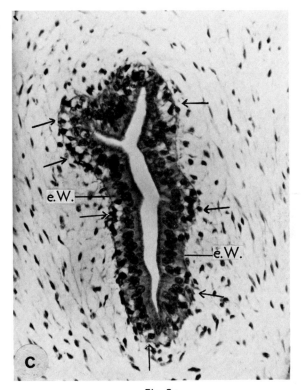

Fig. 9c

Fig. 9. Sections of the genital tract of embryos of the reticulated python (*Python reticulatus*). (a and b) Section at level where the Wolffian ducts (c.W.) open into the urodaeum in a 22-day embryo python. The juxtaposition of the epithelial cells of the urodaeum (e.Ur.) and of cells of the Wolffian duct epithelium (e.W.) can be clearly seen and groups of urodaeal cells beneath the Wolffian cells are arrowed. (PAS-hemalum; ×380). (c) Caudal extremity of the common ureter and Wolffian ducts (c.W.) in 54-day embryo python. Groups of small, clear cells (arrowed) beneath the Wolffian epithelium (e.W.) appear similar to urodaeal cells. ×380.

into the kidney this type of epithelium gave place to another type that was a single layer of cells with spherical nuclei (Fig. 8c).

These studies of normal adult females at estrus and young slowworms treated with estradiol show that:

a. The epithelium of the urodaeal wall reacts to estrogens by proliferation and thickening, and by stratification with desquamation of the superficial layers.

b. In normal females and, in the conditions of our experiments, in young animals treated with estradiol, the epithelium bounding the vagina does not change at estrus into stratified squamous epithelium. Instead, it remains pseudostratified with ciliated and glandular cells.

c. The epithelial lining of the caudal parts of the ureters and the Wolffian ducts reacts to estradiol in the same way as the urodaeal epithelium; it is transformed into a thick stratified epithelium.

The transformation of the epithelium of the ureters and Wolffian ducts is hard to explain, but embryological studies on the slowworm have shown that this epithelium is morphologically similar to that of the urodaeum. This is particularly clear in the region where the Wolffian duct joins and penetrates the dorsal diverticulum of the urodaeum; here, the urodaeal epithelium appears to replace the caudal parts of the epithelium of the ureters and Wolffian ducts, and the two epithelia can be seen to be superimposed in places (see Raynaud *et al.*, 1968, Figs. 40 and 41). A similar dual structure (Fig. 9a and b) has been found by the author in a 22-day embryo (17 cm) of the reticulated python (*Python reticulatus*): the superficial layer is formed of PAS-positive columnar cells which are identical with those lining the Wolffian duct. The basal layers are made of small cells which have a large, clear cytoplasmic area and are similar to those of the urodaeal epithelium. It seems probable that the Wolffian epithelium penetrates the dorsal diverticulum of the urodaeum, or that this diverticulum grows around the caudal part of the Wolffian duct and the Wolffian and urodaeal epithelia overlap. Then the basal membrane of the Wolffian epithelium disappears and the small urodaeal cells thus become subjacent to, and eventually mixed with, the cells of the Wolffian duct (Fig. 9a and b). The juxtaposition of urodaeal and Wolffian cells is apparent for about 180 μm along that part of the Wolffian duct situated behind the junction with the ureter, but is not found in the ureter. However, in older, 40-day embryos, urodaeal cells are present along 375 μm of the Wolffian duct and are apparent in the ureter. At 54 days (Fig. 9c), urodaeal cells are spread along 600 μm of the Wolffian ducts and 300 μm of the ureter. During the course of fetal development, therefore, there is an increasing length of the caudal segment of the Wolffian duct and of the ureter in which Wolffian and urodaeal cells are juxtaposed.

In summary, the lining of the reptile urogenital tract, particularly the urodaeum and the caudal parts of the Wolffian ducts and ureters, which may be lined by urodaeal epithelium, forms a stratified desquamating epithelium under the influence of estrogens. The vaginal Müllerian epithelia of normal females or of young animals treated with estradiol do not react in this way.

The evidence so far available for reptiles can be interpreted as conforming to Zuckerman's hypothesis.

VII. DISCUSSION

Two main conclusions can be reached. First, in the males of several species, estrogens stimulate stratification of the epithelium, often proceeding as far as epidermoid metaplasia, in the prostatic lobes, in the lower part or in the whole of the prostatic utricle, in the ejaculatory ducts, and in the urethra. The reacting epithelium is derived from the urogenital sinus and the majority of observations accord with Zuckerman's hypothesis (1936, 1940). Investigations on reptiles of both sexes treated with estrogens give results that can be interpreted in the same way. However there are variations in the sensitivity to estradiol of the primordia from one species to another and even exceptions; they have been discussed by different authors, particularly by Zuckerman (1940), Thorbord (1948), and Raynaud (1962).

(a) Within a single species there are variations in sensitivity to estradiol of different prostatic primordia, e.g., in the mouse, the cranial lobes (coagulating glands) are the more sensitive and may become completely keratinized, unlike that of the other lobes. This may be explained by a different determination during ontogeny of the epithelia of different lobes, or by differing sensitivities of the various regions of the urogenital sinus from which the accessory glands are derived, the craniodorsal part of the sinus showing maximum sensitivity to estrogens. (b) There are variations in sensitivity to estradiol of different primordia between one species and another. Thus, the keratinization which appears so rapidly in the coagulating glands of the mouse does not appear in other species, or only does so after prolonged treatment. The secretory capacities of the epithelia of the accessory glands must vary qualitatively and quantitatively from one species to another. The determination of these epithelia has been different and their capacities to react to the same stimulus (e.g., estrogen) may also be different. In the urethra of the cat, ferret, dog, and monkey, but not in the rat and guinea pig, the epithelium proliferates and readily becomes keratinized under the influence of estrogens. These specific variations may be interpreted as due to differences in determination of the epithelia, and do not invalidate Zuckerman's hypothesis. (c) In a few cases, stratification and sometimes keratinization occurs in the epithelia of organs which are not derived from the urogenital sinus. The classical example is that of the seminal vesicles of the hedgehog (*Erinaceus europaeus* L.), which has been discussed previously

(Mombaerts, 1943; Raynaud, 1962). Embryological studies have shown that the vesicular gland primordia are first formed from the Wolffian duct but then become separated from it and insert into the urogenital sinus. It seems certain, therefore, that the special reaction of these glands is linked with this histogenetic pattern.

Stratification and keratinization may occur in organs distant from the urogenital sinus, usually as pathological responses to inflammatory reactions, infections, and irritation of epithelia by chemicals. Such cases have no physiological bearing on the subject of the present discussion.

Second, in females, the epithelium of the sinus vagina and that of derivatives of the urogenital sinus react to estrogens by stratification and keratinization. There are variations in detail from one species to another, and these are linked to the capacities of the cells of the sinus itself, as in the rabbit, or to secondary embryonic factors. These responses conform to Zuckerman's hypothesis.

Until quite recently the reaction of the Müllerian epithelium to estradiol was considered to be of the glandular type and fundamentally different from the reaction of the epithelium of the urogenital sinus to the same hormone. However, the work of Forsberg has produced a number of important results which favor the concept that in the caudal extremities of the Müllerian ducts the epithelium may become stratified and keratinized under the influence of estrogens. Bulmer's findings (1964) on the sheep fetus corroborate those of Forsberg. This reaction of the Müllerian epithelium to estrogens does not really contradict Zuckerman's general concept, for it is not the whole of the Müllerian duct epithelium which reacts, but only that at the caudal extremities of the ducts. This epithelium is influenced by the urogenital sinus and it is possible that the sinus mesenchyme is an important factor which modifies the reactive capacity of the normal Müllerian epithelium and gives it the ability to react to estrogens by stratification and keratinization. This suggestion implies that the pattern of response of an epithelium to a given agent, such as a hormone, is the consequence of the determination, during early embryonic life, of the functional capacities of the epithelium. As was emphasized in the first edition (Raynaud, 1962), this problem cannot be resolved solely by the techniques usually employed in causal embryology. We still know little about the determination of the genital rudiments. However, Didier (1970, 1971) has recently shown that the Wolffian duct induces the formation of the Müllerian duct in the chick embryo. Furthermore, the mesenchyme is known to play an important role in the differentiation of a number of organ rudiments (Grobstein, 1967, 1968; Wessells, 1968). Recently, Propper and Gomot (1967) and Propper (1968), confirming the suppositions of Balinsky (1950), have shown that it is the mesenchyme subjacent to the mammary crest which determines the formation of this crest. Kratochwill (1969) has established that the mesenchyme subjacent to the submaxillary salivary gland rudiment was capable of directing the growth and ramification of the epithelial mammary rudiment along lines very similar to those characteristic of salivary glands; "Mammary epithelium in salivary mesenchyme undergoes a morphogenesis clearly different from its usual one and reminiscent of that of the salivary gland. This response involves directive, organ-specific factors supplied by the mesenchyme." Further work on the influence of mesenchyme on the differentiation of epithelia of various organs (e.g., lung, kidney, stomach) has established that each mesenchyme can produce two types of inductive factor: (a) a morphogenetic, nonspecific factor which stimulates the morphogenesis of epithelial cells, along lines directed by their genetic makeup; and (b) an inductive, organ-specific factor which specifically influences the differentiation of the epithelium which it reaches. This latter factor would be weakly diffusible and would act only on the epithelium which is directly adjacent to it. David (1967, 1968, 1969, 1971) has shown that the gastric epithelium of a 10-day rabbit fetus may assume the multistratified type and become keratinized when placed in contact with mesenchyme of the esophagus.

There is, as yet, little detailed work on the mesenchyme subjacent to the genital tract epithelia, although it has been emphasized (Raynaud, 1962, pp. 220–221) that research should be orientated in this direction. However, in the light of the evidence from other studies of organ rudiments, it seems possible that the mesenchyme of the urogenital sinus could influence the evolution of the Müllerian epithelium toward a type of stratified epithelium capable of becoming keratinized. However, the presumed inductive factor could not act equally on all the epithelia which are in contact with the sinus. Thus, in mammals, the caudal extremities of the Wolffian ducts in proximity to the sinus retain their characteristic epithelium; it is only in the terminal part of the ejaculatory ducts, where the sinus epithelium is present from the outset, that the effect of estrogens is to produce stratification and keratinization. A study of reptile embryos has yielded analogous facts (see Section VI).

As stressed by Bulmer (1964), the possibility that the Müllerian epithelium may be changed under the inductive influence of the urogenital sinus into a stratified epithelium, does not invalidate the general principles of Zuckerman's hypotheses; it merely serves to enlarge the terms of the hypothesis and to give it a more general value. The concept can, therefore, now be presented in the following terms:

"The metaplasia and stratification of a genital tract epithelium in response to oestrogenic stimulation, whatever be its histochemical basis, may be considered as representing the primary reaction of a tissue in which the epithelium of the urogenital sinus has been present from the outset, or has penetrated secondarily, or in which development has taken place under the directing morphogenetic influence of the urogenital sinus."

This definition fits most of the facts now known.

REFERENCES

Balinsky, B. I. (1950). On the developmental processes in mammary glands and other epidermal structures. *Trans. R. Soc. Edinburgh* **62,** 1-31.

Baulieu, E. E., and Rochefort, H. (1970). Mécanisme d'action des hormones sexuelles. *In* "Mécanismes d'action intracellulaires des hormones" (R. Vokaer, ed.), pp. 53-69. Masson, Paris.

Baulieu, E. E., Alberga, A., and Jung, I. (1967). Récepteurs hormonaux. Liaison de divers oestrogènes à des protéines utérines. Une méthode de dosage de l'oestradiol. *C. R. Acad. Sci. Ser. D* **265,** 501-504.

Bengmark, S., and Forsberg, J.-G. (1959). On the development of the vagina. *Acta Anat.* **37,** 106-125.

Bulmer, D. (1952). Observations on the development of the lower end of the vagina in the sheep. *J. Anat.* **86,** 233-245.

Bulmer, D. (1964). The epithelia of the developing female genital tract in the sheep. *Acta Anat.* **57,** 349-366.

Burrows, H. (1934). Action of oestrin on the coagulating glands and on certain vestigial structures in the mouse (*Mus musculus*). *Nature (London)* **134,** 570.

Burrows, H. (1935a). Pathological conditions induced by oestrogenic compounds in the coagulating gland and prostate of the mouse. *Am. J. Cancer* **23,** 490-512.

Burrows, H. (1935b). The localisation of response to oestrogenic compounds in the organs of male mice. *J. Pathol. Bacteriol.* **41,** 423-429.

Burrows, H. (1937). Changes induced by oestrone in the bulbo-urethral (Cowper's) gland of the male mouse. *J. Pathol. Bacteriol.* **44,** 699-701.

Burrows, H., and Kennaway, N. M. (1934). On some effects produced by applying oestrin to the skin of mice. *Am. J. Cancer* **20,** 48-57.

Carr, E. B. (1953). The development of the rabbit vagina. *J. Anat.* **87,** 424-431.

Courrier, R. (1936). Notes d'endocrinologie sexuelle chez le Singe d'Algérie. *Bruxelles-Med.* **16,** 674-682.

Courrier, R., and Cohen-Solal, G. (1936). L'utricule prostatique chez le Cobaye soumis à la folliculinisation. *C. R. Soc. Biol.* **121,** 903-905.

Courrier, R., and Gros, G. (1935). Action de la folliculine chez le Singe mâle impubère. Modifications des annexes. *C. R. Soc. Biol.* **118,** 686-689.

Courrier, R., and Gros, G. (1938). Sur l'action de la folliculine chez le Chat mâle. *C. R. Soc. Biol.* **129,** 8-10.

Dantschakoff, V. (1938). Sur la validité du test d'Allen et Doisy dans l'embryon de Lézard des deux sexes. *C. R. Soc. Biol.* **128,** 895-899.

David, D. (1967). L'influence de divers mésenchymes sur la différenciation de l'épithélium gastrique du foetus de Lapin, en culture *in vitro*. *C. R. Acad. Sci. Ser. D* **264,** 1062-1065.

David, D. (1968). Influences de la nature du mésenchyme sur la différenciation de l'épithélium gastrique de foetus de Lapin âgés de 13 jours. *C. R. Soc. Biol.* **162,** 1142-1144.

David, D. (1969). Action comparée de mésenchymes hétérologues et d'extraits d'organes embryonnaires sur la morphogenèse gastrique en culture *in vitro. Ann. Embryol. Exp. Morphol.* **2,** 419–432.

David, D. (1971). "Etude expérimentale de l'organogenèse de l'estomac, chez le foetus de Lapin," Sér. E, No. 138. Thèse de Sciences, Faculté des Sciences de l'Université de Clermont-Ferrand.

de Jongh, S. E. (1933). Menformon und Schwangerschaft. *Acta Brevia Neerl. Physiol., Pharmacol. Microbiol.* **3,** 68–69.

de Jongh, S. E. (1935). Der Einfluss von Geschlechtshormonen auf die Prostata und ihre Umgebung bei der Maus. *Acta Brevia Neerl. Physiol., Pharmacol. Microbiol.* **5,** 28.

Delost, P., Jean, Ch., and Jean, Cl. (1962). Production expérimentale de malformations mammaires chez le foetus de Rat par injection d'oestradiol à la mère au 14 ème jour de la gestation. *C. R. Soc. Biol.* **156,** 2048–2052.

Delost, P., Jean, Ch., and Jean, Cl. (1963). Malformations de la glande mammaire et du mamelon chez le foetus produites par l'oestradiol injecté à la Ratte gestante. *J. Physiol. (Paris)* **55,** 237–238.

Didier, E. (1970). Sur le rôle du canal de Wolff dans la formation de l'ostium müllérien chez l'embryon de poulet. *C. R. Acad. Sci. Ser. D* **270,** 2188–2190.

Didier, E. (1971). Le canal de Wolff induit la formation de l'ostium müllérien: démonstration expérimentale chez l'embryon de Poulet. *J. Embryol. Exp. Morphol.* **25,** 115–129.

Dunn, T. B., and Green, A. W. (1963). Cyst of the epididymis, cancer of the cervix, granular cell myoblastoma and other lesions after estrogen injection in newborn mice. *J. Natl. Cancer Inst.* **31,** 425–455.

Finn, C. A., Martin, L., and Carter, J. (1969). A refractory period following oestrogenic stimulation of cell division in the mouse uterus. *J. Endocrinol.* **44,** 121–126.

Fluhmann, C. F. (1960). The developmental anatomy of the cervix uteri. *Obstet. Gynecol.* **15,** 62–69.

Forsberg, J.-G. (1960). On the development of the hamster vagina. *Acta Anat.* **41,** 16–37.

Forsberg, J.-G. (1961). On the development of the rabbit vagina. *Z. Anat. Entwicklungsgesch.* **122,** 459–481.

Forsberg, J.-G. (1962). The periodic acid-Schiff reaction in the vagina and cervical epithelia of fetal, immature and adult rats. *J. Histochem. Cytochem.* **10,** 29–35.

Forsberg, J.-G. (1963). "Derivation and differentiation of the vaginal epithelium." Håkan Ohlssons Boktryckeri, Lund.

Forsberg, J.-G. (1965a). Mitotic rate and autoradiographic studies on the derivation and differentiation of the epithelium in the mouse vaginal anlage. *Acta Anat.* **62,** 266–282.

Forsberg, J.-G. (1965b). An experimental approach to the problem of the derivation of the vaginal epithelium. *J. Embryol. Exp. Morphol.* **14,** 213–222.

Forsberg, J.-G. (1965c). Origin of vaginal epithelium. *Obstet. Gynecol.* **25,** 787–791.

Forsberg, J.-G. (1966). The effect of estradiol-17β on the epithelium in the mouse vaginal anlage. *Acta Anat.* **63,** 71–88.

Forsberg, J.-G. (1967). Histochemical studies on the changing enzyme pattern at the lumen formation in the solid sinus vagina in the mouse. *Z. Anat. Entwicklungsgesch.* **126,** 117–126.

Forsberg, J.-G. (1969). The development of a typical epithelium in the mouse uterine cervix and vaginal fornix after neonatal oestradiol treatment. *Br. J. Exp. Pathol.* **50,** 187–195.

Forsberg, J.-G. (1970). An estradiol mitotic rate inhibiting effect in the Müllerian epithelium in neonatal mice. *J. Exp. Zool.* **175,** 369–374.

Forsberg, J.-G., and Lannerstad, B. (1970). The *in vitro* response of the mouse Müllerian epithelium to estradiol. *Acta Embryol. Morphol. Exp.* **1,** 45–51.

Forsberg, J.-G., and Nord, L. (1969). Immunofluorescence studies on the distribution of mouse uterine epithelial antigen in foetal, immature and adult mice. *J. Embryol. Exp. Morphol.* **21,** 85–95.

Forsberg, J.-G., and Norell, K. (1966). Differentiation of the epithelium in early grafts of the mouse Müllerian vaginal region. *Experientia* **22,** 402–406.

Forsberg, J.-G., and Olivecrona, H. (1964). The activity of acid and alkaline phosphatase during development of the vaginal anlage in rat and mouse. *Z. Zellforsch. Mikrosk. Anat.* **63,** 362–373.

Forsberg, J.-G., and Olivecrona, H. (1965). Further studies on the differentiation of the epithelium in the mouse vaginal anlage. *Z. Zellforsch. Mikrosk. Anat.* **66,** 867–877.

Galand, P., Rodesch, F., Leroy, F., and Chrétien, J. (1967). Radioautographic evaluation of the estrogen dependent proliferative pool in the stem cell compartment of the mouse uterine and vaginal epithelia. *Exp. Cell Res.* **48,** 595–604.

Gardner, W. V. (1959a). Experimental induction of uterine, cervical and vaginal cancer in mice. *Cancer Res.* **19,** 170–176.

Gardner, W. V. (1959b). Sensitivity of the vagina to estrogen: genetic and transmitted differences. *Ann. N.Y. Acad. Sci.* **83,** 145–159.

Giersberg, H. (1923). Untersuchungen über Physiologie und Histologie des Eileiters der Reptilien und Vögel; nebst einen Beitrage zur Fasergenese. *Z. Wiss. Zool.* **120,** 1–92.

Grobstein, C. (1967). Mechanisms of organogenetic tissue interaction. *Natl. Cancer Inst., Monogr.* **26,** 279–299.

Grobstein, C. (1968). Developmental significance of interface materials in epitheliomesenchymal interaction. *In* "Epithelial-Mesenchymal Interactions" (R. Fleischmajer, ed.), pp. 173–176. Williams & Wilkins, Baltimore, Maryland.

Gros, G. (1939). Sur l'action du diethylstilboestrol chez le Chat. *C. R. Soc. Biol.* **131,** 172–174.

Husbands, M. E., and Walker, B. E. (1963). Differentiation of vaginal epithelium in mice given estrogen and thymidine-H^3. *Anat. Rec.* **147,** 187–198.

Jacobi, L. (1936). Ovoviviparie bei einheimischen Eidechsen, Vergleichende Untersuchungen an den Eiern und am Oviduct von *Lacerta agilis, Lacerta vivipara* und *Anguis fragilis. Z. Wiss. Zool. Abt. A* **148,** 401–464.

Jean, Ch. (1970). "Malformations mammaires induites chez la Souris par une action oestrogène prénatale." Sér E, No. **123** Thèse de Sciences, Université de Clermont Ferrand.

Jean, Ch., and Delost, P. (1965). Ostrogènes et malformations congénitales expérimentales de la morphogénèse mammaire. *C. R. Soc. Biol.* **159,** 2357–2362.

Jensen, E. V., Suzuki, T., Numata, M., Smith, S., and de Sombre, E. R. (1969). Estrogen binding substances of target tissues. *Steroids* **13,** 417–427.

Jost, A. (1947). Recherches sur la différenciation sexuelle de l'embryon de Lapin. *Arch. Anat. Microsc. Morphol. Exp.* **36,** 151–200 and 242–270.

Jost, A. (1969). The extent of foetal endocrine autonomy. *Foetal Auton., Ciba Found. Symp., 1968* pp. 79–89.

Jost, A., and Picon, L. (1970). Hormonal control of fetal development. *Adv. Metab. Disord.* **4,** 123–184.

Juillard, M.-T. (1966). Sur l'origine de la kératinisation du vagin et de l'utérus chez la Souris: arguments apportés par les greffes et les cultures. *C. R. Soc. Biol.* **160,** 1860–1864.

Juillard, M.-T. (1969). Greffes intra-oculaires des ébauches müllériennes et sinusaires du canal utéro-vaginal de Souris. Action des hormones sexuelles. *C. R. Soc. Biol.* **163,** 2401.

Juillard, M.-T., and Delost, P. (1963a). Etude morphologique du développement post-natal du vagin et de l'utérus chez la Souris. *C. R. Soc. Biol.* **157,** 2203–2207.

Juillard, M.-T., and Delost, P. (1963b). Les transformations du sinus urogénital de la naissance a l'âge adulte chez la Souris femelle. *Gen. Comp. Endocrinol.* **3,** 710 (Abstr. 57).

Juillard, M.-T., and Delost, P. (1964a). Sur la formation du vagin de la Souris. *J. Physiol.* (*Paris*) **56,** 382.

Juillard, M.-T., and Delost, P. (1964b). Transformations provoquées par l'oestradiol dans la structure du vagin de la Souris nouveau-née. *C. R. Soc. Biol.* **158,** 1497–1501.

Juillard, M.-T., and Delost, P. (1965a). Sur l'origine embryologique du vagin et de l'utérus chez la Souris. *Gen. Comp. Endocrinol.* **5,** 690 (Abstr. 54).

Juillard, M.-T., and Delost, P. (1965b). Les effets de l'oestradiol au cours de l'ontogenèse sur la différenciation du vagin et de l'utérus, chez la Souris. *C. R. Soc. Biol.* **159,** 1541–1544.

Juillard, M.-T., and Delost, P. (1966). A propos de l'influence inductrice du sinus dans la morphogenèse du vagin; évolution des ébauches müllériennes et sinusaires après greffes simultanées dans l'oeil. *J. Physiol.* (*Paris*) **58,** 540–541.

Kimura, T., and Nandi, S. (1967). Nature of induced persistent vaginal cornification in mice. IV. Changes in the vaginal epithelium of old mice treated neonatally with estradiol or testosterone. *J. Natl. Cancer Inst.* **39,** 75–93.

Kimura, T., Basu, S. L., and Nandi, S. (1967a). Nature of induced persistent vaginal cornification in mice. I. Effect of neonatal treatment with various doses of steroids. *J. Exp. Zool.* **165,** 71–88.

Kimura, T., Basu, S. L., and Nandi, S. (1967b). Nature of induced persistent vaginal cornification in mice. II. Effect of estradiol and testosterone on vaginal epithelium of mice of different ages. *J. Exp. Zool.* **165,** 211–222.

Kimura, T., Basu, S. L., and Nandi, S. (1967c). Nature of induced persistent vaginal cornification in mice. III. Effects of estradiol and testosterone on vaginal epithelium *in vitro. J. Exp. Zool.* **165,** 497–504.

Korenchevsky, V., and Dennison, M. (1934). The effects of oestrone on normal and castrated male rats. *Biochem. J.* **28,** 1474–1485.

Korenchevsky, V., and Dennison, M. (1935). Histological changes in the organs of rats injected with oestrone alone or simultaneously with oestrone and testicular hormone. *J. Pathol. Bacteriol.* **41,** 323–337.

Kratochwill, K. (1969). Organ specificity in mesenchymal induction demonstrated in the embryonic development of the mammary gland of the mouse. *Dev. Biol.* **20,** 46–71.

Lacassagne, A. (1933). Métaplasie épidermoïde de la prostate provoquée chez la

Souris par des injections répétées de fortes doses de folliculine. *C. R. Soc. Biol.* **113,** 590–592.

Lacassagne, A. (1935a). Modifications progressives de la structure du conduit tubo-utérin, chez des Lapines soumises à partir de la naissance, à des injections répétées d'oestrone (folliculine). *C. R. Soc. Biol.* **120,** 685–689.

Lacassagne, A. (1935b). Modifications de l'épithelium vésical chez la Souris atteinte de rétention urinaire à la suite d'injections d'oestrone. *C. R. Soc. Biol.* **120,** 833–835.

Lacassagne, A. (1935c). Modifications progressives de l'utérus de la Souris sous l'action prolongée de l'oestrone. *C. R. Soc. Biol.* **120,** 1156–1158.

Lacassagne, A. (1936). Histophysiological consequences of prolonged treatment with oestrone, begun at birth. *In* "Certain Biological Problems Relating to Cancer, Hormones and Radiation," Lecture series, pp. 1–78. Int. Cancer Res. Found., Philadelphia, Pennsylvania.

Martin, L., and Claringbold, P. J. (1960). The mitogenic action of oestrogens in the vaginal epithelium of the ovariectomized mouse. *J. Endocrinol.* **20,** 173–186.

Martin, L., and Finn, C. A. (1968). Hormonal regulation of cell division in epithelial and connective tissues of the mouse uterus. *J. Endocrinol.* **41,** 363–371.

Matèjka, M. (1959). Die Morphogenese der menschlichen Vagina und ihre Gestzmässigkeiten. *Anat. Anz.* **106,** 20–37.

Matèjka, M. (1963). Studien über die Histogenese der Plattenepithel-Zylinder- epithelgrenze an der Portio vaginalis uteri des Menschen, unter Berücksichtigung der Bedeutung phylogenetischen Faktoren für die Ontogenese. *Anat. Anz.* **112,** 426–446.

Means, A. R., and Hamilton, T. H. (1966). Early estrogen action: concomitant stimulations within two minutes of nuclear RNA synthesis and uptake of RNA precursor by the uterus. *Proc. Natl. Acad. Sci. U.S.A.* **56,** 1594–1598.

Mombaerts, J. (1943). Le sinus urogénital et les glandes sexuelles annexes du hérisson (*Erinaceus europaeus, L.*). Recherches anatomiques, histologiques, embryologiques et expérimentales. *Arch. Biol.* **55,** 393–554.

Mueller, G. C. (1965). The role of RNA and protein synthesis in oestrogen action. *In* "Mechanisms of Hormone Action" (P. Karlson, ed.), pp. 228–245. Academic Press, New York.

Mueller, G. C., Herranen, A. M., and Jervell, K. F. (1958). Studies on the mechanism of action of estrogens. *Recent Progr. Horm. Res.* **14,** 95–139.

Parkes, A. S., and Zuckerman, S. (1935). Experimental hyperplasia of the prostate. *Lancet* **1,** 925–937.

Peckham, B., and Kiekhofer, W. (1962). Cellular behavior in the vaginal epithelium of estrogen-treated rats. *Am. J. Obstet. Gynecol.* **83,** 1021–1026.

Peckham, B., Ladinsky, J., and Kiekhofer, W. (1963). Autoradiographic investigation of estrogen response mechanisms in rat vaginal epithelium. *Am. J. Obstet. Gynecol.* **87,** 710–714.

Perrotta, C. (1962). Initiation of cell proliferation in the vaginal part and uterine epithelia of the mouse. *Am. J. Anat.* **111,** 195–204.

Pfeiffer, C. A. (1936). Effects of ovarian transplants upon the development and maintenance of the seminal vesicle and prostate gland of the albino rat. *Anat. Rec.* **65,** 213–237.

Propper, A. (1968). Relations épidermo-mésodermiques dans la différenciation de l'ébauche mammaire d'embryon de Lapin. *Ann. Embryol. Morphog.* **1,** 151–160.

Propper, A., and Gomot, L. (1967). Interactions tissulaires au cours de l'organogenèse de la glande mammaire chez l'embryon de Lapin. *C. R. Acad. Sci. Ser. D* **22**, 2573–2575.

Raynaud, A. (1937). Modification apportées précocement dans la structure de la vessie et de l'urètre du Chat par des injections de dihydrofolliculine. *C. R. Soc. Biol.* **126**, 215–218.

Raynaud, A. (1940). Répartition topographique du glycogène dans le tractus urogénital de la Souris nouveau-née. *C. R.Soc. Biol.* **134**, 574–577.

Raynaud, A. (1942). Recherches embryologiques et histologiques sur la différenciation sexuelle normale de la Souris. *Bull. Biol. Fr. Belg. Suppl.* **29**, 1–114.

Raynaud, A. (1947). Effet des injections d'hormones sexuelles à la Souris gravide, sur le développement des ébauches de la glande mammaire des embryons. II. Action de fortes doses de substances oestrogènes. *Ann. Endocrinol.* **8**, 318–329.

Raynaud, A. (1962). The histogenesis of urogenital and mammary tissues sensitive to oestrogens. *In* "The Ovary" (S. Zuckerman, ed.), 1st ed., pp. 179–230. Academic Press, New York.

Raynaud, A. (1969). Les organes génitaux des Mammifères. *In* "Traité de Zoologie" (P.-P. Grassé, ed.), Vol. 16, pp. 149–636. Masson, Paris.

Raynaud, A. (1971). Foetal development of the mammary gland and hormonal effects on its morphogenesis. *In* "Lactation" (I. R. Falconer, ed.), pp. 3–29. Butterworth, London.

Raynaud, A., and Raynaud, J. (1956). La production expérimentale de malformations mammaires chez le foetus de Souris, par l'action des hormones sexuelles. *Ann. Inst. Pasteur Paris* **90**, 39–91 and 187–219.

Raynaud, A., Pieau, C., and Raynaud, J. (1968). Contribution à l'étude de la formation du cloaque chez l'Orvet (*Anguis fragilis* L.). *Mem. Mus. Natl. Hist. Nat., Paris, Ser. A* **52**, 1–64.

Raynaud, J., and Raynaud, A. (1963). Données histologiques et histochimiques relatives à l'épithélium de la paroi de l'urodaeum de l'embryon d'Orvet (*Anguis fragilis L.*). *C. R. Soc. Biol.* **157**, 8–13.

Starkey, W. F., and Leathem, J. H. (1939). Action of estrone on sexual organs of immature male cats. *Anat. Rec.* **75**, 85–89.

Takasugi, N. (1963). Vaginal cornification in persistent-estrous mice. *Endocrinology* **72**, 607–619.

Takasugi, N. (1966). Persistent changes in vaginal epithelium in mice induced by short-term treatment with estrogens beginning at different early postnatal ages. *Proc. Jpn. Acad.* **42**, 151–155.

Takasugi, N., and Bern, A. H. (1964). Tissue changes in mice with persistent vaginal cornification induced by early postnatal treatment with estrogen. *J. Natl. Cancer. Inst.* **33**, 855–865.

Takasugi, N., Bern, A. H., and DeOme, K. B. (1962). Persistent vaginal cornification in mice. *Science* **138**, 438–439.

Takewaki, K. (1964). Persistent changes in uterus and vagina in rats given injections of estrogen for the first thirty postnatal days. *Proc. Jpn. Acad.* **40**, 42–47.

Talwar, G. P. (1970). Mechanism of action of estrogens. *Gen. Comp. Endocrinol., Suppl.* **2**, 123–134.

Talwar, G. P., Sopori, M. L., Biswas, D. K., and Segal, S. J. (1968). Nature and characteristics of the binding of oestradiol-17β to a uterine macromolecular fraction. *Biochem. J.* **107**, 765.

Teng, C. S., and Hamilton, T. H. (1970). Regulation by estrogen of organ-specific synthesis of a nuclear acidic protein. *Biochem. Biophys. Res. Commun.* **40,** 1231–1238.

Thorbord, J. V. (1948). On the influence of oestrogenic hormones on the male accessory genital system. *Acta Endocrinol. (Copenhagen), Suppl.* **2,** 1–214.

Tice, L. W. (1961). Some effects of oestrogen on uterine epithelial mitosis in the mouse. *Anat. Rec.* **189,** 515–524.

van Wagenen, G. (1935). The effects of oestrin on the urogenital tract of the male monkey. *Anat. Rec.* **63,** 387–403.

Weller, D., Overholser, M. D., and Nelson, W. O. (1936). The effect of estrin on the prostate gland of the albino rat and mouse. *Anat. Rec.* **65,** 149–163.

Wells, L. J. (1936). Effect of oestrin on accessory reproduction organs of the male ground squirrel (*Citellus tridecemlineatus*). *Anat. Rec.* **78,** 43–51.

Wells, L. J., and Overholser, M. D. (1940). Effects of estrogen on the genital tract and urethra of anestrous female ground squirrels. *Anat. Rec.* **78,** 43–51.

Wessells, N. K. (1968). Problems in the analysis of determination mitosis and differentiation. *In* "Epithelial-Mesenchymal Interactions" (R. Fleischmajer, ed.), pp. 132–151. Williams & Wilkins, Baltimore, Maryland.

Willier, B. H. (1955). Ontogeny of endocrine correlation. *In* "Analysis of Development" (B. H. Willier, P. A. Weiss, and V. Hamburger, eds.), Sect. X, pp. 574–619. Saunders, Philadelphia, Pennsylvania.

Zuckerman, S. (1936). An embryological interpretation of changes induced by oestrogens in the male reproductive tract. *Lancet* **i,** 135–136.

Zuckerman, S. (1940). The histogenesis of tissues sensitive to oestrogens. *Biol. Rev. Cambridge Philos. Soc.* **15,** 231–271.

Zuckerman, S., and Parkes, A. S. (1936a). Effect of sex hormones on the prostate of monkeys. *Lancet* **i,** 242–247.

Zuckerman, S., and Parkes, A. S. (1936b). Morphological and functional homologies of the male and female reproductive systems. *Br. Med. J.* **2,** 864–865.

Zuckerman, S., and Parkes, A. S. (1936c). The effect of oestrone on the prostate and uterus masculinus of various species of primate. *J. Anat.* **70,** 323–330.

Zuckerman, S., and Sandys, O. C. (1939). Further observations on the effects of sex hormones on the prostate and seminal vesicles of monkeys. *J. Anat.* **73,** 597–616.

3

Transplantation of the Ovary

P. L. Krohn

I. INTRODUCTION

It is necessary to introduce the detailed discussion of the particular problems of ovarian transplantation with a short statement of the general principles on which successful transplantation of any tissue is thought to depend, and to define a number of relevant words, misuse of which has led to much confusion in the past.

Grafts or transplants are of several types:

1. *Autograft.* Tissue transferred from one part of a body to another part of the same body.

2. *Isograft.* Tissue transferred from one member of a species to another member which is virtually genetically identical with the donor, i.e., exchanges between identical twins or between two members of any one of the numerous highly inbred strains of mice or of fewer inbred strains of guinea pigs, or between hybrids of two inbred strains. Many so-called

101

inbred strains, particularly of rats, are insufficiently homogeneous genetically to fulfil this requirement but the situation is improving. There is no more reason now for complicating the interpretation of transplantation experiments in rats, than in mice, by the use of heterogeneous animals.

3. *Allograft* (synonym: homograft). Tissue transferred from one member of a randomly bred species to another, or of one inbred strain to another inbred strain.

4. *Xenograft* (synonym: heterograft). Tissue transferred from a member of one species to a member of another species.

Grafts may be either (a) *orthotopic,* when they are placed in a position in the host equivalent to that which they occupied in the donor, or (b) *heterotopic.* They may also be *isochronic,* when host and donor are of the same age or *heterochronic* when host and donor are of different ages. A distinction has also been made by Dempster (1955) between "implants," which are the common form of grafts and which depend for their continued viability on the development of a new circulation between the graft bed and the graft, and "transplants," in which direct anastomoses between host and donor venous and arterial systems are brought about surgically at the time of grafting. This latter form is becoming widely applied to ovarian grafts in larger species, particularly in domestic animals (see Section III,C).

A large body of work has established the validity of rules governing the successful transplantation of tissues (Snell, 1958), which are based on the general belief that the host responds to a foreign graft by developing a form of immunity reaction (Medawar, 1958) to genetically controlled antigenic substances released by the graft (histocompatibility antigens).

There are exceptions to the general rules which depend either on previous treatment of the host (e.g., irradiation: Dempster *et al.,* 1950; cortisone: Billingham *et al.,* 1951), or on special properties of the site of grafting (e.g., anterior chamber of the eye, brain: Medawar, 1948), which render the passage of antigenic material to the reactive centers of the host more difficult or impossible, or interfere with the normal development of the immune response. More recently, the concept of "immunological tolerance," the introduction of antilymphocytic sera and the use of other immunosuppressant drugs have dramatically widened the opportunities for mitigating the ordinary responses to a foreign graft. It will be seen later that the reported indifference of endocrine tissues to the laws of transplantation is more apparent than real. Nevertheless, there remains a body of discrepant information which implies that the ovary possesses unique properties distinguishing it from the other endocrine organs.

Successful transplantation is one of the steps classically required to demonstrate the reality of the presumed endocrine nature of a tissue. It was first used for the testis, as by Hunter in the eighteenth century and especially by Berthold in 1849. Transplantation of the ovary seems to have been first carried out by Knauer (1896) in rabbits. Within a few years Foà (1900, 1901) made the important discovery that the development of an immature ovary is accelerated if it is transplanted into an adult animal. Yet most early workers seem to have accepted the idea that ovarian allografts were unsuccessful (for reviews see Schultz, 1910; Tschernischoff, 1914).

II. ASSESSMENT OF FUNCTION OF A GRAFTED OVARY

Complete function of an ovarian graft requires the production of off-spring from a graft in a host whose own ovaries have been totally removed. This, in turn, implies that the graft has been transplanted orthotopically. Such a stringent test is often not necessary and evidence of the normal production of hormones by the graft is usually sufficient. Transplantation after preservation at low temperatures is possible though usually the return to activity is delayed (see Krohn, 1962a, for references).

The simpler and most commonly used tests of function generally demonstrate only the production of estrogen. Thus, the estrous cycle can readily be studied in spayed rodents by taking vaginal smears. Such records provide additional information about the physiological condition of the graft which can be derived from (i) the interval between grafting and the first postoperative estrous period, (ii) the duration of the positive response, and (iii) the character of the response, i.e., whether normal cycles, or periods of constant cornification or pseudopregnancy occur. Restoration of the normal cycles of running activity has also occasionally been used in rats (Wang *et al.,* 1925). Chapekar *et al.* (1966) have used the estrogenic output of a graft to assess the function of ovaries after they have been cultured *in vitro.*

In rodents, the capacity to secrete progesterone is shown by the fact that grafts can maintain the development of deciduomata in the uterus (Kullander, 1956; Bartosik and Szarowski, 1970) and can also maintain a pregnancy which was begun by transferring fertilized ova into pseudo-pregnant spayed hosts which then received grafts (Boot, 1956). In primates and man, the return of menstrual cycles can be accepted as satisfactory evidence of ovarian function (Mandl and Zuckerman, 1949; van Wagenen and Gardner, 1953).

The capacity of ovarian grafts to form steroids has been tested in tissue slice preparations by Rice and Segaloff (1966), and the formation of progesterone, 20α-hydroxy-4-pregnen-3-one, androstenedione, and cortico-steroid by ovarian grafts in the spleen has been measured by direct estimation of the content of portal and vena caval blood (Leavitt and Carlson, 1969; see also the work on ovarian grafts in sheep and pigs, Section III,C). Grafted ovaries may also secrete considerable quantities of androgen, especially if the graft's environment is cooler than normal. They can also assume some of the functions of the adrenal cortex (see Krohn, 1962a, for references).

The behavior of ovarian grafts in the male is more difficult to assess and usually depends on histological examination at the end of the experiment. However, it is possible to transplant pieces of vagina as well (Martins, 1932; Gardner and Argyris, 1956) and to form artificial pockets from which smears can subsequently be taken.

It has been customary to transplant whole ovaries, often cut up into a few pieces. Falck (1959), however, has made microdissections of the ovaries before transplantation so that he could determine which particular component of the ovary was responsible for secreting estrogen.

III. SITES FOR TRANSPLANTATION

Ovaries have been transplanted to a wide variety of sites. Only those that have provided information of special interest will now be discussed.

A. Orthotopic Transplantation

Early experiments designed to study the autonomy of the germ plasm in foreign environments and to distinguish between the effects due to the expression of deficient genes and those due to the environment are described by Krohn (1962a). Orthotopic grafts have also been used by Krohn (1958, 1965) and by Jones and Krohn (1960) to distinguish between the contribution which aging of the ovary itself and of the general body environment makes to the decline in fertility as mice get older (see Section VI,A). The surgical technique adopted has been more radical and has achieved a freedom from the complication of regenerating host ovarian tissue which troubled earlier workers. With experience, this technique (described in Krohn, 1962b) has proved to be satisfactorily free of the other difficulties, such as, the formation of capsular adhesions, failure of vascularization, and the loss of irreplaceable oocytes. As many as seventeen litters and seventy-nine offspring have been obtained from an individual graft-bearing mouse. An alternative method (Parrott and Parkes, 1957) is to sterilize the animal by X irradiation before implanting the grafts alongside the host's own shrivelled ovaries.

Encouraged by the successes of organ transplants in clinical practice, more workers are attempting orthotopic ovarian transplantation in a wider range of animals with some success (dogs: Eraslan *et al.*, 1966; O'Leary *et al.*, 1969; pigs: Binns *et al.*, 1969). Yonemoto *et al.* (1969) have attempted allotransplantation of the uterus and ovaries with vascular anastomoses in dogs using immunosuppressive therapy to control the immune reaction against the grafts, but they were unable to report more success than the presence of viable uterine tissue in the few survivors. Betteridge (1970) has tackled the more difficult technical problem of vascular anastomosis in rabbits. Despite being cut off from a circulating blood supply for 1 to $1\frac{3}{4}$ hours and being subject to the immune response, the grafted ovaries were functional for up to 6 days.

B. Sites Permitting Direct Observation of Grafts

The anterior chamber of the eye has the great advantage that it allows repeated direct observations of grafts to be made. It also appears to be an immunologically privileged site (Ward *et al.*, 1953). Schochet (1920) first used it for transplanting ovaries and saw ova lying free in the anterior chamber 8 weeks after grafting. Such grafts in rabbit ovulate and form corpora lutea in response to mating or injection of chorionic gonadotropin (Allen and Priest, 1932; Podleschka and Dworzak, 1933, 1934). This response has been used in a modified Friedman test (Allen and Priest, 1932; Colonna, 1954). An extensive series of papers by Browning and his colleagues (Browning *et al.*, 1965; Crisp and Browning, 1968; Browning and Guzman, 1967, 1968) has reported detailed observations on the behavior of intraocular grafts, especially of the luteal tissue, in rats and mice in a wide range of hormonal environments.

It is usually considered that transplants are not only isolated from their normal autonomic innervation but that ovulation occurs in them without involvement of the nervous system. Jacobowitz and Laties (1970) have found in cats that ovarian slices transplanted to the anterior chamber became attached to the iris by vascular bridges and that adrenergic fibers (demonstrated by catecholamine fluorescence) reappeared to form the normal pattern of innervation. However, Gilbert and Wood-Gush (1969), studying the neural component of ovulation in hens, found that the normal vigorous parasympathetic and sympathetic innervation, associated not only with blood vessels but also with smooth muscle fibers, had virtually disappeared in transplants.

Dierschke and Wallace Williams (1970) have directly observed grafts of the ewe's ovary relocated in the vagina. The hamster's cheek pouch has been used for the same purpose (Caldwell *et al.,* 1966).

C. Transplantation by Vascular Anastomosis

Ovarian transplants in large animals, e.g., domestic farm animals, cannot be expected to survive adequately unless proper vascular anastomoses are prepared. Dempster (1954) successfully transplanted the ovary in greyhounds as long ago as 1954 but only recently have advances in vascular surgical techniques been applied, particularly by Goding, first, to the transplantation of the adrenal gland, and, later, to the ovary in sheep and pigs. The technique is fully described by Goding *et al.* (1967b) and McCracken and Baird (1969). The operation is carried out in two stages, 2 or preferably 3 months apart. First, a skin loop is prepared in the neck containing the left carotid artery and jugular vein. Later, the left ovary is transferred, a patch of the aorta and uterine veins being used for the anastomosis with the carotid and jugular vessels. Successful preparations are indicated by patent angiograms, behavioral estrus, progesterone secretion in the venous draining blood, and increased hormonal output in response to gonadotropin.

The technique has been used for two main purposes. First, it achieves anatomical separation of the ovary and uterus and the interruption of the normal pattern of innervation. Thereby it becomes possible to test the theory that control of the life span of the corpus luteum depends on the formation of a "luteolysin" in the uterus which acts locally and directly on the ovary. It also avoids the criticism that local surgical mutilation can produce abnormal luteal function. Goding *et al.* (1967a) showed that the cycle was lengthened by removal of the ovary in this way from any direct uterine influence and the corpus luteum persisted after the transfer. Harrison *et al.* (1968) subsequently transferred ovary and uterus together into the neck in the sheep and found normal estrous cycle lengths and levels of progesterone in the draining blood comparable with those coming from abdominal ovaries. It is not sufficient to transplant the uterus separately to the neighborhood of a previously transplanted ovary, as was discovered by Binns *et al.* (1969) who did not achieve normal activity in two sheep using this method. The normal connecting local pathway, as well as anatomical proximity, was required. Separation of the normal anatomical relationships of ovary and uterus to study the source of luteolysin has also been used by Bland and Donovan (1968) in guinea pigs. When the ovaries were transferred to the kidneys (by subcapsular insertion) the estrous cycle was prolonged. Transfer of the ovary to the uterus was followed by shortened cycles.

Second, it facilitates the taking of samples of blood from the ovarian vessels without concomitant surgical stress. Baird *et al.* (1968) report results from five ewes from whom blood was removed twice weekly for 4 to 9 weeks and McCracken *et al.* (1969) present comprehensive results of the effect of treatment with sheep gonadotropins on the steroid output of the transplanted ovary. Luteinizing hormone produces a dramatic hyperemia and large rises in steroid output. Follicle-stimulating hormone and prolactin were without effect. Further information is provided by Baird *et al.* (1971). Four ewes with transplanted ovaries were given infusions of ovine LH or HCG. The greatest increase in progesterone output occurred 10 minutes after the start of the infusion but after about 30 minutes output began to fall and continued to fall despite continued infusion. The response was not sensitive to changes in the rate of infusion. The pattern of blood flow through the ovary did not parallel the release of progesterone. Maximum blood flow was not attained until after the end of the last infusion.

D. Organs Draining into the Portal Circulation

The importance of the area drained by the portal circulation as a site for transplantation rests on the insight into hypophysial–ovarian relationships which it provides (Lipschütz, 1957). The liver cells are believed to inactivate estrogens secreted by the graft before they can act systemically (see Krohn, 1962a, for references to earlier work). Inactivation of estrogen is not as complete as was at first thought, even in rats in which the process is more nearly complete than in other species (mouse: Bernstorf, 1951; guinea pig: Wenner and Hofman, 1950; rabbit: Peckham and Greene, 1952; ferret: Donovan, 1969; hamster: Takewaki, 1955). Ber (1968a) believes that implantation into the omentum near the large curvature of the stomach is an improvement on the spleen model, but here, too, there is some stimulation of the vagina and uterus in the rat. Estrogen inactivation is very incomplete in monkeys. All the ordinary signs of hormonal stimulation in rhesus monkeys, such as coloring and swelling of the sexual skin and menstrual cycles, persist even when the only ovarian tissue in the body is one-half of one ovary transplanted to the spleen (van Wagenen and Gardner, 1950, 1953). The liver is more or less inactive in immature rats but at about the age of puberty its capacity to metabolize ovarian hormones is developing rapidly and is strongly developed by 41 days of age (Donovan and O'Keefe, 1966). Any progesterone that is secreted by a graft seems to be much less affected by passage through the liver. Kullander (1956) produced deciduomata in spayed rats bearing intrasplenic grafts. The hypophysial–gonadal system responds much more sluggishly to a fall in gonadal hormone levels induced by intrasplenic transplants in the ferret than in the rat (Donovan, 1969) and there were no essential differences between grafts in the spleen and kidney.

The most important late change in intrasplenic grafts is the development of tumors (Biskind and Biskind, 1949) resembling those which appear after irradiation of the ovary. The unrestrained proliferation of granulosa cells usually begins about 12 weeks after grafting when most of the follicles have disappeared, but it may be 12 months or more before tumors appear in guinea pigs, animals in which proliferation of luteal elements is more notable; metastases are rare. Tumors develop in ovaries from androgenized rats just as in normal ovarian transplants to the spleen (Kovacs, 1965).

Tumor formation does not occur in an intrasplenic graft as long as normal ovarian tissue remains in its usual position. On the contrary, the graft is quite inactive (Biskind and Kordan, 1949; Peckham and Randak, 1952). However, it is capable of rapid development when the remaining ovary is removed, even as long as 465 days after the original grafting (Biskind *et al.,* 1950), and soon develops the customary proliferations. Injections of gonadotropin will accelerate the development of tumors (Biskind *et al.,* 1953) and, conversely, growth is suppressed by injections of antigonadotropin (Ely, 1955). Treatment with estriol will also prevent the development of tumors but does not affect those which have already formed (Fels and Zorro, 1969).

These basic experiments have been largely responsible for the development of ideas about the potential of hormones to promote tumor formation. Transplantation of the ovary to the spleen so that its secretions are inactivated by the liver frees the anterior lobe of the pituitary from restraint. The additional gonadotropin secreted then induces proliferative changes in the graft which first require the continued presence of gonadotropin for maintenance (hormone-dependent stage) but which ultimately become autonomous and malignant. The stage in the life of a tumor when it still requires a stimulus for growth, and the nature and mode of action of the stimulus present important problems to research workers studying cancer.

IV. THE DEVELOPMENT OF AN OVARIAN AUTOGRAFT

Subcutaneous ovarian grafts often reconstitute a new bursa around themselves which is lined by cells derived from a proliferating germinal epithelium. Some (Butcher, 1947; Gaillard, 1953) believe that the success of a transplant largely depends on this reformation of the bursa. Equally often, however, the germinal epithelium is totally lost and the surface of the graft adheres firmly to surrounding connective tissue without damage to the graft. Whether the epithelium is present or absent seems to make no difference to the number of oocytes in the graft, a finding which lends support to those who doubt that the epithelium really deserves to be called "germinal" or that the continued oogenesis which this title implies is at all probable (see also Volume I, Chapter 2).

The series of histological changes which appear in ovarian autografts has been described by Harris and Eakin (1949). Twenty-four hours after grafting, the transplants are still without vascular connections and show a large central zone of necrosis around which is a thin rim of surviving tissue. Within 3 days the boundary between living and dead areas is more distinct and ingrowing vessels are already linking up with existing vessels in the graft, reestablishing a circulation. The central area of necrosis is removed by macrophage activity and within a few weeks has almost disappeared leaving behind only a small connective tissue scar. Many of the oocytes are lost from ischemia, even in the best conditions of grafting (Jones and Krohn, 1960). Fortunately, most of the small oocytes are to be found in that rim of cortical tissue which is most likely to survive. Follicular growth and the formation of corpora lutea take place from this viable fringe. Both Deanesly (1956) and Parkes (1956) have described and illustrated the range of normal structures which may be found in a graft a year or more after grafting. Follicular and luteal cysts are common and probably represent the results of a distorted hypophysial–ovarian balance. Corpora lutea with entrapped ova are common in ovarian grafts, particularly when the grafts are subcutaneous and without any capsule-covered ovarian bursa. However, the inadequate ovulatory response is likely to be the consequence of an unsatisfactory general reorganization of the grafted tissue. Even when a normal hormonal environment can be inferred from the normal responses of a contralateral intact ovary, the grafted adult ovary shows numerous accessory corpora lutea. Grafts carried out in the neonatal period, with a long period of reorganization available before examination, do not show this deficiency (Welschen, 1971). Abnormal cycles may result if insufficient ovarian tissue is transplanted. Mammary stimulation, uterine hypertrophy, and an adrenal degeneration syndrome develop some months after grafting according to Browning and White (1967). The frequency of this result is inversely proportional to the amount of grafted tissue until, with only $\frac{1}{32}$ of the ovary being used, there is no response.

It is generally agreed that normal autografts begin to secrete hormone again about 6 to 12 days after grafting (Long and Evans, 1922; Harris and Eakin, 1949; Parkes, 1956). Ben-Or (1965) found in mice that there was an inverse relationship between the age of the donor (over the range 3–18 days) and the time lag before the adult hosts showed estrous changes in vaginal smears. With newborn donors the latent interval was 11 days or more; for 18-day-old donors it was only 4 or 5 days. Autografts, once they have "taken" satisfactorily, can apparently produce estrogen for as long as a normal ovary, though the rhythm and regularity of the estrous cycle may be disturbed (Muhlbock and Boot, 1956). However, animals with orthotopic grafts are often less fertile than normal mice and become sterile earlier than is usual for the particular strain of mouse used (Jones and Krohn, 1960).

V. OVARIAN ALLOGRAFTS

A. Development

The general cellular response of the host to the grafting of foreign ovarian tissue does not differ from that which characterizes its response to other tissues such as skin (Medawar, 1944), kidney (Dempster, 1953), or vagina (Krohn, 1955). The first sign of any reaction occurs, according to Harris and Eakin (1949), about 5 to 7 days after grafting when lymphocytes begin to congregate around and within the graft. All organized ovarian structure soon disappears, the last being the luteal elements, and in 3 weeks or so the lymphocytes are being replaced by connective tissue elements. The dead tissue is removed by fatty phagocytic cells which may sometimes be mistaken for individual luteal cells.

If the vigor of the allograft reaction is subdued for any reason, healthy looking small oocytes and occasional corpora lutea may be found lying in a fibrous stroma, surrounded by loose infiltrations of lymphocytes. An ovary which is not fully compatible with the host may, therefore, survive the attack of a niggling, only partly effective, allograft reaction, in much the same way that an incompletely acceptable skin graft can survive, though only at the expense of the development of eczematous patches, scaliness, and small scabs on its surface epithelium.

Harris and Eakin (1949) state that allografts took longer to become established than autografts. Intrastrain grafts took 13–14 days to produce vaginal cornification compared with 8 days for autografts and 24 days for full allografts. Allografts also took longer to become established in Parkes' (1956) experiments, but the difference between the behavior of autografts and allografts was not statistically significant.

B. Factors Affecting Survival

In the early days of the development of ovarian physiology, chemists had not yet succeeded in making available potent preparations of female sex hormones. Pioneers, such as Lipschütz and Steinach, had, therefore, to rely on the method of transplanting the ovary to supply the host with what were presumed to be hormones. As far as one can tell from their papers, they paid no regard to any questions of genetic compatibility or incompatibility between donor and host. Nevertheless, they seem to have grafted gonads of either sex into hosts of either sex without any difficulty. Yet, at the same time, other workers seem to have been agreed that allografts would not survive (see Krohn, 1962a). Such differences of opinion have remained until the present day when the immunological consequences of grafting have become of more general interest.

Of the various factors which might abate the vigor of a normal allograft reaction, Halsted's law that the "physiological need of the organism" determines the fate of a graft is the oldest. Halsted (1909) believed that endocrine tissues could be grafted successfully only when the host animal had a total lack of the particular hormones secreted by the graft. This statement rests on experiments with parathyroid autografts and was not applied by Halsted to allografts. Indeed, he did not believe that they ever succeeded. Later workers were responsible for extending the application of his law to cover allografts, and increased endocrinological knowledge also allowed them to define "physiological need" in terms of an increased production of pituitary tropic hormones. In so doing these later workers appear to have forgotten that the parathyroid grafts had no apparent need for any tropic hormone. Halsted's law is now so generally applied to the behavior of allografts of endocrine organs that, regardless of the doubtful justification for the hypothesis, one must discuss the following problems: 1. Are allografts more likely to succeed in conditions where extra gonadotropin is available? 2. If this is so, by what mechanism is the result achieved?

Numerous experiments seem to show that grafts of any sort of ovary are more successful in spayed than in intact animals. However, it is not possible to decide whether a graft is producing estrogen or not when the host's ovaries are still present, nor is size alone much of a criterion that a graft is alive or dead. Indeed, even autografts seem to find it difficult or impossible to make use of circulating gonadotropin if the other ovary remains intact (Podleschka and Dworzak, 1933; Lane and Markee, 1941; Welschen, 1971). Such grafts lie dormant, but are not dead, until removal of the remaining ovary allows them in some way to respond to the gonadotropins which had been denied them earlier. This dormancy is very well demonstrated by the experiments of Stevens (1955). Using orthotopic transplantation and a color-marker gene technique to allow the source of offspring to be identified with certainty, he showed that practically none of the mice born to mothers with one isografted and one normal ovary came from the graft. If, however, he removed the normal ovary as long as 6–7 months after grafting, litters derived from the graft were soon born.

Two other points argue against the importance of increased gonadotropin. First, as Parkes (1956) points out, it is usual to remove the host's ovaries and insert the grafts during the same operation. In these circumstances there is no opportunity for the initial "taking" of the graft to be improved by additional gonadotropin. Second, unilateral spaying is followed by at least sufficient production of additional gonadotropin to cause compensatory hypertrophy of the other ovary. It does not, however, affect the survival of an allograft at all (Ingram and Krohn, 1956), and injected gonadotropin does not increase the chance that a graft will survive much longer than normal (Ferguson and Kirschbaum, 1954).

The allograft reaction against a transplant only takes place when sensitized cells (of the lymphocytic and plasma cell series) come in contact with the graft. This contact can be prevented by enclosing the graft in a chamber whose walls are made of Millipore filter material. The diameter of the pores allows sufficient passage of tissue fluids to nourish the graft, but they are too small for cells to penetrate. Using this technique, Sturgis and Castellanos (1958) have shown that protection can be afforded to ovarian allografts in rats. Their results were much less satisfactory in monkeys and human patients. Subsequently, Sturgis and Castellanos (1968) modified their technique by replacing the Millipore filter chamber with one fashioned from corneal tissue but had no greater success. Shaffer and Hulka (1969) have analyzed the use of corneal chambers to diminish transplantation reactions in rabbits. They found that the chambers did confer some protection and attributed it to a process of graft adaptation rather than to the induction of a form of tolerance. Other methods to combat a presumed immunological reaction against ovarian grafts have been tried; for example, blockade of the reticuloendothelial system (Arnold, 1927; Miller and Lampton, 1941), cortisone treatment (Ingram and Krohn, 1956), enhancement (Kalliss, 1958; Parkes, 1957, 1958), and normal serum treatment (Glaser and Nelken, 1969). The most successful has been the induction of immunological tolerance (Krohn, 1958).

There remains *prima facie* evidence that a degree of genetic incompatibility which precludes acceptance of a skin homograft will, nevertheless, allow an ovarian allograft to survive. The possibility that endocrine organs as a group are characterized by special immunological privileges provides a field of study which is of great interest. Several explanations of the observations are possible.

i. It is known that the survival of an allograft varies according to the amount of tissue grafted. The effective antigenic stimulus from an ovarian graft may be even smaller than would be thought at first because the bulk of the graft is destroyed before a new blood supply is available. Certainly the amount of ovarian tissue grafted is usually much less than the amount of an ordinary skin graft.

ii. The subcutaneous fatty tissue, usually the site of an ovarian graft, may be at least partly "privileged."

iii. Histocompatibility antigens produced by the ovary are either fewer in number or less capable of provoking a reaction than are those from the skin.

iv. A proliferating graft can in some way inactivate or absorb the "antibodies" that are produced by the host.

v. An hormonal imbalance caused by the stress of operation and the interference with normal endocrine relations increases the amount of adrenal hormones which are known to modify the normal development of the allograft reaction (Ingram and Krohn, 1956).

The various explanations are open to investigation to varying degrees. The evidence is against the first suggestion. The third possibility has been studied more closely. The number of genes controlling the formation of histocompatibility antigens in two strains of mice has been estimated for skin transplants by Barnes and Krohn (1957) and compared with an estimate for ovarian transplants (Hicken and Krohn, 1960). Assuming that the responsible genes segregate in the F_2 population, the proportion of surviving grafts from parent strain to F_2 hybrid host should be $(\frac{3}{4})^n$ where n is the number of genes involved. The method estimates the number of gene differences and does not take into account the number of genes which the two strains may have in common. The results indicated a difference between the two types of graft (fifteen or more genes for skin and not more than eleven for ovary) but this apparent confirmation of a diminished immunogenicity of ovarian grafts was again questioned when further experiments showed that animals which had accepted ovarian grafts subsequently accepted a skin graft from the same donor source. It was recognized that this finding might imply that an ovarian allograft could somehow condition the subsequent response of the host to another graft but it seemed more probable at the time that the explanation lay in random differences in the segregation of genes in the two separate F_2 populations. Linder's work showed this explanation to be inadequate.

Linder's first paper (1962a) was concerned with a combination of mouse strains which gave a weak but not absolutely uniform histoincompatibility reaction. A skin graft was given either to a normal host mouse or to one which had previously been spayed and had then received an ovarian graft from the same donor of the skin. Whenever the skin graft on the untreated host survived for more than 29 days, the skin graft on the host carrying the ovarian graft survived indefinitely. Linder then used (1962b,c) two genetically homogeneous strains (A × C57BL as donors to A × C57L hosts). This is a combination in which skin grafts survive for about 28 days compared to the 8–10 days found when histocompatibility differences are maximal. Spayed mice were given ovarian grafts and tested for their response to skin grafts from the same donor strain at intervals before or after the ovarian transplantation. If the two grafts were more or less contemporary (+5 to −5 days), the response to the skin graft was normal. If the skin transfer was delayed after the ovarian graft (but < 70 days) an accelerated rejection of the skin was found, indicating that the mouse had been sensitized by the ovarian graft. When the interval was 70 days the response was again neutral, but skin transferred 90 days after the first graft survived indefinitely provided that an adequately sized ovarian graft persisted. The skin graft was maintained only as long as the ovarian graft remained and broke down soon after the removal of the ovary.

The behavior of ovarian grafts in resisting rejection and inhibiting the normal response to subsequent skin grafts is specific to the ovary and does not represent a property of all endocrine grafts. It is not shown after grafts of the testis (or of submaxillary gland) and requires the absence of the host's ovaries and the development of cysts in the grafted tissue. In general, it seems to resemble the conditioning effect of transplants of liver in pigs which also modify the subsequent response to skin grafts (Calne, 1969).

These phenomena are not yet fully understood but they suggest that ovarian transplantation may have at least as much to contribute to the problem of transplantation immunology as it has to reproductive physiology. Yet in studying the apparent inconsistencies and unexpected responses in ovarian findings, other more simple explanations should not be overlooked. There are important differences in the ways transplants of skin and ovary are commonly performed. Skin is usually transferred orthotopically to a site where revascularization is vigorous and quick. Consequently, ischemic damage is only slight and drainage of antigenic material to lymph nodes is rapid and substantial. Ovaries, however, are usually transferred to avascular sites in fatty connective tissue where conditions are altogether different. Indeed, preliminary experiments (P. L. Krohn, unpublished observations) suggest that small pieces of mouse skin may survive when they are treated like ovarian grafts and embedded in fatty tissue in the axilla. Parrott (1960) too, has shown, in the rat strain used by Billingham and Parkes

(1955), that orthotopic transplants of ovary did not survive when subcutaneous transplants did. She also found that surviving subcutaneous grafts were soon destroyed if they were moved to the orthotopic position. Altogether, it seems that the development of a connective tissue capsule around the graft slows down the passage of antigen into the host and impedes the attack of host cells on the graft and may provide a situation comparable with that found when a Millipore filter or corneal epithelium is used to cover grafts.

Antigenic relationships between ovarian grafts and grafts of skin and other tissues are also discussed by Goldstein *et al.* (1961) and by Barnes (1969) and Barnes and Crosier (1969).

Kornblum and Silvers (1968) also used the intrasplenic site to determine whether ovarian grafts and grafts of other tissues can induce "tolerance" of a subsequent test skin graft. Ovarian and testicular grafts were able to induce good tolerance over the single *H-3* histocompatibility barrier, slight tolerance when the *H-1* barrier was involved, but when both *H-1* and *H-3* incompatibilities were concerned the ovarian graft was ineffective. Surprisingly, the testis grafts, which were as effective as ovary grafts, did not survive histologically unlike the ovarian grafts. These results emphasize the point that only when compatibility barriers are slight are positive effects of ovarian grafting to be expected. Indeed, when incompatibility is extreme— as between the C57 and A strains of mouse—neither ovarian nor skin grafts have survived in F_2 generation hosts (Krohn, 1965). Transplantation of the ovary can also be used to indicate the degree of phylogenetic relationship among, for example, species of *Drosophila* (Kambysellis, 1968). The "index of oogenesis" is greatly dependent on the genetic relationship of host and donor species and survival to metamorphosis of larvae derived from grafted ovaries was only found in closely related species. Reciprocal transplants between species of *Drosophila* have also been carried out by Durr Kinsey (1967).

VI. THE USE OF OVARIAN GRAFTS

A. Factors Controlling Puberty and Menopause

The transfer of young ovaries into an adult environment is an obvious experiment to perform for anyone wishing to investigate the factors which determine the age of onset of puberty. The experiment was first carried out by Foà (1900, 1901) whose successful demonstration that the immature ovary precociously developed adult functions in its new milieu is an important landmark. Long and Evans (1922) and Lipschütz (1925) subsequently confirmed Foà's finding. Long and Evans (1922) failed in the reciprocal experiment of grafting adult ovaries into immature animals, probably because of immunological reactions, but Lipschütz (1925) was able to show that mature ovaries did not induce precocious development when transplanted into an immature host. Numerous other workers (see Engle, 1929; Goodman, 1934; Pfeiffer, 1934; May, 1940; Dunham *et al.*, 1941) have added further details, such as the interval before estrous cycles begin, their duration, and the histological appearances of grafts.

In addition, there is some evidence that the removal and immediate replacement of immature ovaries as autografts may accelerate the onset of puberty (Greep and Chester Jones, 1950, following Hohlweg and Dohrn, 1932, and Mandel, 1933). It was thought that the ovary, even when only 25 days old, already influenced and restrained the gonadotropic activity of the pituitary. Mandl and Zuckerman (1951), however, could not confirm these observations and proposed that any surgical procedure or the administration of an anesthetic is in itself sufficient stress to elicit a response from the anterior lobe of the pituitary.

Evidently the young ovary is capable of responding to gonadotropic stimuli coming either from a mature environment into which it has been transplanted or from pituitary tissue implanted into the intact young animal. Yet it is known that the young pituitary not only contains gonadotropin but, according to Harris and Jacobsohn (1952), can maintain normal cyclic behavior when grafted to an adult hypophysectomized host. There is, therefore, some unidentified further control which determines the age at which the pituitary releases gonadotropin. It is usual to locate such a factor or center in the hypothalamus.

With increasing age most animals become less fertile. Their litters become smaller and the intervals between them longer until finally they become sterile (Krohn, 1964). Failure of the ovary has often been regarded as the main reason and it is not surprising, therefore, that many attempts to rejuvenate aging females have been made by replacing old ovaries with young ones (Pettinari, 1928). Such experiments should show whether overall failure is to be attributed to the ovary or to some other component in the reproductive system. Conversely, the transfer of old ovaries into young animals should indicate whether or not it is possible to rejuvenate old ovaries by exposing them to a youthful environment. Krohn (1962b, 1966) has brought together the results of experiments of this type using inbred strains of mice and the technique of orthotopic ovarian transplantation mentioned earlier (Section III,A). The old hosts were mice whose entire reproductive history was known and who had reached the stage when conceptions are followed by abortion or resorption of the fetuses rather than by normal delivery. In technically satisfactory preparations, thirteen mice conceived and implanted blastocysts in the uterus, but only two gave birth to litters—the largest being of three young. A group killed at midterm showed three normal embryos, two possibly normal embryos, and forty-three placental moles or degenerating embryos compared to a total of 104 corpora lutea that looked normal, suggesting that control by the old pituitary was satisfactory. Nevertheless, many blastocysts (estimated by the difference between total corpora lutea and conceptuses recovered) must have disappeared entirely.

Ascheim (1964) has provided evidence from heterotopic transplants in rats that hypothalamic control of the function of the anterior pituitary is defective in old rats, and Harman and Talbert (1970) believe that, in C57 strain mice, there is histological evidence for changes in the corpora lutea of old animals. The simplest explanation for the failure to bring pregnancies to term is that the quality of the uterine environment deteriorates.

The converse experiment of transferring ovaries from mice which have reached the period when pregnancies are not completed to young hosts is more difficult technically, because of the disparity between the size of the graft and the ovarian bursa into which it must be placed, and because of the small number of oocytes in the aged graft which will be further depleted by the inevitable initial ischemic necrosis. Nevertheless, the experiment is rewarding (Krohn, 1966). Six successful preparations had from two to seven pregnancies and most of the litters contained four or more and up to nine young. Such results would certainly not have been achieved from the old ovary in its normal situation and strongly favor the view that the old egg is not to blame for reproductive failure. This view is confirmed by Talbert and Krohn's (1966) work on egg transfer into young and old recipients but does not fit with Blaha's (1964) work using the hamster. Blaha (1970) has extended his work by showing that perirenal grafts of young ovarian tissue can improve the reproductive performance of spayed old hamsters (tested by transfer of morulae) while young females receiving grafts from old hamsters showed no implantations.

Heterochronic transplants have also been performed in rats by Ber (1968b,c,d) over a limited age range. Young donors gave better grafts, as judged by the weight of the grafts, than older donors but the results were not affected by the age of the host. Growth capacity is intrinsic to the ovary; perhaps the younger grafts are more satisfactory because they contain more follicular tissue that can be stimulated to grow.

Transplantation can be used as a technique for studying the behavior of the ovary in an environment of a different age and to distinguish between the genetic and environmental components of any given parameter. Thus, morphological differences between strains, differences in reproductive lifespan, or differences in rate of loss of oocytes are all subjects in which analysis is possible by means of the transfer of the one or other or both strains into the common and compatible environment of the F_1 hybrid formed by mating the two strains.

Bittner and Huseby (1950) were first to show that morphological differences between A and C3H strain mouse ovaries were lost when the ovaries were grafted into C3H × A hybrids but they retained their differing capacity to induce mammary carcinoma (see Krohn, 1962a, for a discussion of the use of ovarian transplants in the study of cancer).

Orthotopic transplantation makes it possible to study the total reproductive performance of two strains in a common environment (Krohn, 1965). It is known, for example, that CBA mice become infertile earlier than A strain mice. If one ovary of each type is transplanted to a CBA × A host which is mated to an A male, the CBA ovary will give rise to CBA × A (i.e., agouti) offspring and the A ovary to A × A (i.e., white) offspring. During the first pregnancies, the proportion of agouti:white is roughly 50:50 but gradually alters until finally only all-white litters are obtained. In neither case is the fertility increased by the hybrid environment. The increased fertility of hybrid mice is due to the greater ovulation rate of the hybrid ovary in response to the pituitary stimulus. Hybrid ovaries transplanted in hybrid mice have litters almost twice as large as either of the parent strains (Table I).

It is also possible to study strain differences by transplanting ovaries from one strain of mouse into the environment of another which has been rendered immunologically tolerant to the donor (Krohn, 1958; Jones and Krohn, 1962). Finally, eggs from the ovaries of old hypophysectomized mice, which may persist long after they would ordinarily have disappeared,

TABLE I

Fertility of Mice with Orthotopic Ovarian Transplants [a]

Strain combination	Proportion fertile	Total no. of litters	Litters/ fertile mouse	Total no. of offspring	Total offspring/ fertile mouse
A–A	4/6	23	5.75	86	21.5
A–CBA × A	5/6	24	4.8	69	13.8
CBA–CBA	6/6	43	7.2	136	22.7
CBA–CBA × A	6/7	38	6.3	156	26.0
CBA × A–CBA × A	7/7	89	12.7	369	52.7

[a] From Krohn (1965).

can also be shown to be healthy, despite their chronological age, by orthotopic transfer of the ovary of the old hypophysectomized animal into normal young hosts (Jones and Krohn, 1961).

B. Sex Differences in Function of the Pituitary

The histological appearance and function of the ovarian grafts vary according to the sex of the host. Transplanted into a spayed female a graft behaves entirely normally; it contains follicles and corpora lutea which produce hormones in the normal pattern so that regular estrous cycles are maintained. When transplanted into a castrated male the ovaries rarely, if ever, ovulate or form corpora lutea (Goodman, 1934; Deanesly, 1938). Follicular development is intensified and cystic follicles or large follicles containing blood cells are frequently found. A ring of vaginal epithelium transplanted at the same time is constantly cornified and shows none of the usual cyclical changes. The grafted ovary will, however, respond normally to injected gonadotropin. The usual explanation for these observations is that the male pituitary secretes gonadotropin at a constant rate and not in the cyclic fashion characteristic of the female. According to van der Schoot and Zeilmaker (1970), however, regular cycles were observed for 5 to 9 weeks in smears taken from subcutaneous vaginal grafts in neonatally castrated rats given ovarian grafts. The cycles then abruptly ceased to be replaced by persistent cornification.

Pfeiffer's (1936, 1937) analysis of this situation suggested that the direction of pituitary function has not been determined at the time of birth. It can be made to develop the male type of secretory pattern by transplanting testicular tissue into the newborn female. The secretion of luteinizing hormone is then suppressed, the ovary fails to ovulate, and the vagina remains constantly cornified. A single injection of androgen soon after birth can bring about this change in function of the "sexual centers" more conveniently than neonatal gonadectomy or grafting (see Barraclough, 1961). According to Harris and Jacobsohn (1952), however, the male pituitary is able to maintain normal reproductive cycles if transplanted under appropriate conditions to hypophysectomized female rats. If this is so, the change, thought by Pfeiffer to occur in pituitary function, must be ascribed to some other center controlling the pituitary, and may be found in the hypothalamus. Martinez and Bittner (1956) have provided further evidence that the pituitary itself does not decide in which way ovarian grafts are to behave. They found that male and female pituitaries induce follicular activity and constant cornification in castrated male mice which received ovarian and vaginal grafts.

Harris (1964, 1970) has studied the effect in detail using ovarian grafts, usually in the anterior chamber of the eye where they can be observed directly, and, in males, vaginal grafts from which to assess the nature and cyclicity of hormones produced by the ovarian grafts. Large corpora lutea were formed in the ovarian grafts to fifty-eight of sixty-one male rats castrated within 24 hours of birth, in only seven of thirty-three rats castrated between 24 and 72 hours after birth, and in none of thirty-three castrated later. In this third group, the grafts showed signs of acyclic stimulation without ovulation or the formation of corpora lutea. A sham operation did not confer the ability to form corpora lutea. Male rats bearing vaginal grafts as well showed normal 5-day cycles of cornification similar to those found in normal females. They also gave the normal female pattern of running activity.

The exchange of ovaries between treated and untreated animals demonstrates clearly that the functional and histological changes in the ovaries of androgen-treated rats are not due to the effects of treatment directly on the ovary (Harris, 1964). This has also been shown to be true for mice (E. C. Jones and P. L. Krohn, unpublished observations). Orthotopic transplants of ovaries from treated to normal mice are normally fertile but normal ovaries transferred to the treated mice assume the characteristic morphology of treated ovaries and the hosts remain infertile.

The inhibition of LH formation in females does not seem to be complete according to Kovacs (1966) and other papers suggest that the situation in rats is not as simple as was originally thought (Wagner, 1968; Ladosky *et al.*, 1969). It is not affected by adrenalectomy (Adams Smith, 1966).

Dominguez *et al.* (1968) have extended the study of ovarian grafts in male hosts to guinea pigs. Newborn guinea pigs were castrated and grafted with young normal ovaries when 60 days old. All showed polycystic grafts and no corpora lutea. Differentiation of gonadotropic control by the central nervous system, therefore, takes place in this species before birth, and Brown-Grant (1969) has shown that in guinea pigs treatment with androgen needs to be carried out around days 33 to 37 of pregnancy to be successful.

Pieces of ovary have been used as a source of hormone, instead of pellets or crystals, in experiments on the control of the hypothalamo–hypophysial tract. Thus, Zeilmaker (1970) has implanted ovarian tissue stereotaxically into the hypothalamus and into the anterior lobe of the pituitary of rats. Only in the latter position was the operation followed by a series of pseudopregnancies, indicating a local effect of ovarian hormone causing release of prolactin.

REFERENCES

Adams Smith, W. N. (1966). Absence of adrenal influence on ovarian graft activity in male rats castrated at birth. *J. Physiol. (London)* **186**, 714–718.

Allen, E., and Priest, F. O. (1932). Physiological responses of ectopic ovarian and endometrial tissue. *Surg., Gynecol. Obstet.* **55**, 553–558.

Arnold, W. (1927). Ovarialtransplantation und Retikulo-endothelialsystem. *Klin. Wochenschr.* **1**, 551–552.

Ascheim, P. (1964). Résultats fournis par la greffe hétérochrone des ovaires dans l'étude de la régulation hypothalamo-hypophyso-ovarienne de la ratte senile. *Gerontologia* **10**, 65–75.

Baird, D. T., Goding, J. R., Ichikawa, Y., and McCracken, J. A. (1968). The secretion of steroids from the autotransplanted ovary in the ewe spontaneously and in response to systemic gonadotrophins. *J. Endocrinol.* **42**, 283–299.

Baird, D. T., Collett, R. A., and Land, R. B. (1971). The pattern of progesterone secretion by the autotransplanted ovary of the ewe in response to ovine luteinizing hormone and HCG. *J. Reprod. Fertil.* **25**, 308–309.

Barnes, A. D. (1969). The immunogenicity of brain, ovary and skin allografts in mice. *Transplantation* **8**, 379–382.

Barnes, A. D., and Crosier, C. (1969). Effect of antilymphocytic serum on the survival of ovary allografts in the mouse. *Nature (London)* **223**, 1059–1060.

Barnes, A. D., and Krohn, P. L. (1957). The estimation of the number of histocompatibility genes controlling the successful transplantation of normal skin in mice. *Proc. R. Soc. London, Ser. B* **146**, 505–526.

Barraclough, C. A. (1961). Production of anovulatory, sterile rats by single injections of testosterone propionate. *Endocrinology* **68**, 62–67.

Bartosik, D., and Szarowski, D. H. (1970). Support of deciduoma formation and growth by the transplanted ovary of the rat. *J. Reprod. Fertil.* **23**, 505–507.

Ben-Or, S. J. (1965). Morphological and functional development of the ovary of the mouse. II. The development of the ovary in transplantation conditions in adult spayed hosts. *J. Embryol. Exp. Morphol.* **14**, 111–118.

Ber, A. (1968a). Development of ovarian autografts near the stomach of adult castrated rats. *Endokrinologie* **53**, 48–61.

Ber, A. (1968b). The age influence on the development of ovarian autografts near the stomach of castrated rats. *Endokrinologie* **53**, 62–71.

Ber, A. (1968c). Development of single homochronic and heterochronic ovarian homografts near the stomach of castrated female rats. *Endokrinologie* **53**, 230–236.

Ber, A. (1968d). Simultaneous development of homochronic and heterochronic ovarian homografts near the stomach of castrated female rats. *Endokrinologie* **53**, 237–241.

Bernstorf, E. C. (1951). Incomplete hepatic inactivation of hormone produced by intrasplenically grafted ovary in the mouse. *Endocrinology* **49**, 302–309.

Betteridge, K. J. (1970). Homotransplantation of ovaries with vascular anastomoses in rabbits: response of transplants to human chorionic gonadotrophin. *J. Endocrinol.* **47**, 451–461.

Billingham, R. E., and Parkes, A. S. (1955). Studies on the survival of homografts of skin and ovarian tissue in rats. *Proc. R. Soc. London, Ser. B* **143**, 550–560.

Billingham, R. E., Krohn, P. L., and Medawar, P. B. (1951). Effect of cortisone on survival of skin homografts in rabbits. *Br. Med. J.* **1**, 1157–1163.

Binns, R. M., Harrison, F. A., Heap, R. B., and Linzell, J. L. (1969). Reproductive behaviour in the sheep and pig after transplantation of the ovary. *J. Reprod. Fertil.* **20**, 356–357.

Biskind, G. R., and Biskind, M. S. (1949). Experimental ovarian tumors in rats. *Am. J. Clin. Pathol.* **19**, 501–521.

Biskind, G. R., and Kordan, B. (1949). Effect of pregnancy on rat ovary transplanted to spleen. *Proc. Soc. Exp. Biol. Med.* **71**, 67–68.

Biskind, G. R., Kordan, B., and Biskind, M. S. (1950). Ovary transplanted to spleen in rats; effect of unilateral castration, pregnancy and subsequent castration. *Cancer Res.* **10**, 309–318.

Biskind, G. R., Bernstein, D. E., and Gospe, S. M. (1953). Effect of exogenous gonadotrophins on the development of experimental ovarian tumors in rats. *Cancer Res.* **13**, 216–220.

Bittner, J. J., and Huseby, R. A. (1950). Some inherited hormonal factors influencing mammary carcinogenesis in virgin mice. *In* "A Symposium on Steroid Hormones" (E. S. Gordon, ed.), pp. 361–373. Univ. of Wisconsin Press, Madison.

Blaha, G. C. (1964). Effects on age of the donor and recipient on the development of transferred golden hamster ova. *Anat. Rec.* **150**, 413–416.

Blaha, G. C. (1970). The influence of ovarian grafts from young donors on the development of transferred ova in aged golden hamsters. *Fertil. Steril.* **21**, 268–273.

Bland, K. P., and Donovan, B. T. (1968). The effect of autotransplantation of the ovaries to the kidneys or uterus on the oestrous cycle of the guinea-pig. *J. Endocrinol.* **41**, 95–103.

Boot, L. M. (1956). The hormonal function of transplanted ovaries in mice. *Acta Physiol. Pharmacol. Neerl.* **4**, 565–566.

Brown-Grant, K. (1969). Quoted by Harris (1970).

Browning, H. C., and Guzman, R. (1967). Intraocular ovarian isografts in female rats and their response to gonadotropins and to pituitary isografts. *Endocrinology* **81**, 1311–1318.

Browning, H. C., and Guzman, R. (1968). Intraocular ovarian isografts in male rats and their response to luteinizing hormone. *Endocrinology* **82**, 245–252.

Browning, H. C., and White, W. D. (1967). Relation between size of ovarian isografts and abnormal reproductive cycles in the mouse. *Anat. Rec.* **157**, 155–162.

Browning, H. C., Brown, A. J., Crisp, T. M., and Gibbs, E. W. (1965). Response of ovarian isografts to purified FSH, LH, LTH in partially and completely hypophysectomized mice. *Tex. Rep. Biol. Med.* **23**, 715–728.

Butcher, E. C. (1947). The peri-ovarian space and the development of the ovary in the rat. *Anat Rec.* **98**, 547–555.

Caldwell, B. V., Pawling, R. S., and Wright, P. A. (1966). Reestablishment of ovarian periodicity after transplantation to the Syrian Hamster cheek pouch. *Proc. Soc. Exp. Biol. Med.* **123**, 551–553.

Calne, R. Y. (1969). Liver transplantation. *Transplant. Rev.* **2**, 69–89.

Chapekar, T. N., Nayak, G. V., and Ranadive, K. J. (1966). Studies on the functional activity of organotypically cultured mouse ovary. *J. Embryol. Exp. Morphol.* **15**, 133–141.

Colonna, M. (1954). Use of follicle-stimulating hormone gonadotrophins to obtain functioning ovarian grafts in the anterior chamber of the eye. *Experientia* **10**, 340–341.

Crisp, T. M., and Browning, H. C. (1968). The fine structure of corpora lutea in ovarian transplants of mice following luteotrophic stimulation. *Am. J. Anat.* **122**, 169–192.

Deanesly, R. (1938). Androgenic activity of ovarian grafts in castrated male rats. *Proc. R. Soc. London, Ser. B* **126**, 122–135.

Deanesly, R. (1956). Cyclic function in ovarian grafts. *J. Endocrinol.* **13**, 211–220.

Dempster, W. J. (1953). Kidney homotransplantation. *Br. J. Surg.* **40**, 447–465.

Dempster, W. J. (1954). A technique for the study of the behaviour of the autotransplanted kidney, adrenals and ovary of the dog. *J. Physiol. (London)* **124**, 15–16.

Dempster, W. J. (1955). The transplanted adrenal gland. *Br. J. Surg.* **42**, 540–552.

Dempster, W. J., Lennox, B., and Boag, I. W. (1950). Prolongation of survival of skin homotransplants in the rabbit by irradiation of the host. *Br. J. Exp. Pathol.* **31**, 670–679.

Dierschke, D. J., and Wallace Williams, E. I. (1970). Intravaginal relocation of an ovary in the ewe: surgical technique and initial observations. *Biol. Reprod.* **2**, 71–77.

Dominguez, R., Carlevaro, E., and Buno, W. (1968). Evolution of ovarian grafts in male guinea-pigs castrated on the first day of life. *Experientia* **24**, 459–460.

Donovan, B. T. (1969). The functional capacity of ovarian tissue transplanted to the spleen or kidney in the ferret. *J. Endocrinol.* **45**, 91–97.

Donovan, B. T., and O'Keefe, M. C. (1966). The liver and the feedback action of ovarian hormones in the immature rat. *J. Endocrinol.* **34**, 469–478.

Dunham, L. J., Watts, R. M., and Adair, F. L. (1941). Development of newborn rat ovaries implanted in anterior chamber of adult rat's eyes. *Arch. Pathol.* **32**, 910–927.

Durr Kinsey, J. (1967). Interspecific ovary transplantation in *Drosophila. Transplantation* **4**, 509–512.

Ely, C. A. (1955). The effect of antigonadotropic serum on ovarian implants to the spleen of castrate mice. *Anat. Rec.* **121**, 289.

Engle, E. T. (1929). Pituitary-gonadal mechanism and heterosexual ovarian grafts. *Am. J. Anat.* **44**, 121–139.

Eraslan, S., Hamernik, R. J., and Hardy, J. D. (1966). Replantation of uterus and ovaries in dogs with successful pregnancy. *Arch. Surg. (Chicago)* **92**, 9–12.

Falck, B. (1959). Site of production of oestrogen in rat ovary as studied in microtransplants. *Acta Physiol. Scand.* **47**, Suppl. 163, 1–101.

Fels, E., and Zorro, A. R. (1969). Action de l'oestriol sur la greffe d'ovaire intrasplenique. *C. R. Soc. Biol.* **162**, 1843–1844.

Ferguson, L., and Kirschbaum, A. (1954). Effect of gonadotropic hormone on cross-strain grafting of ovaries between inbred strains of mice. *Anat. Rec.* **118**, 298.

Foà, C. (1900). La greffe des ovaires en relation avec quelques questions de biologie générale. *Arch. Ital. Biol.* **34**, 43–73.

Foà, C. (1901). Sur la greffe des ovaires. *Arch. Ital. Biol.* **35**, 364–372.

Gaillard, P. J. (1953). Endocrine glands. *Transplant. Bull.* **1**, 8–10.

Gardner, W. U., and Argyris, B. F. (1956). Differences in vaginal sensitivity of mice of inbred strains; localisation in the vagina. *Endocrinology* **60**, 532–546.

Gilbert, A. B., and Wood-Gush, D. G. M. (1969). Innervation of ovarian transplants in the domestic hen. *J. Reprod. Fertil.* **18**, 550–551.

Glaser, M., and Nelken, D. (1969). Effect of normal serum on survival of rat ovary allografts. *Transplantation* **8**, 753–754.

Goding, J. R., Harrison, F. A., Heap, R. B., and Linzell, J. L. (1967a). Ovarian activity in the ewe after autotransplantation of the ovary or uterus to the neck. *J. Physiol. (London)* **191**, 129P–130P.

Goding, J. R., McCracken, J. A., and Baird, D. T. (1967b). The study of ovarian function in the ewe by means of a vascular autotransplantation technique. *J. Endocrinol.* **39**, 37–52.

Goldstein, D. P., Marshall, D. C., and Sturgis, S. H. (1961). Antigenic relationships between skin and free ovarian homografts. *Surg. Forum* **12**, 108–110.

Goodman, L. (1934). Observations on transplanted immature ovaries in eyes of adult male and female rats. *Anat. Rec.* **59**, 223–251.

Greep, R. O., and Chester Jones, I. (1950). Steroid control of pituitary function. *Recent Progr. Horm. Res.* **5**, 197–261.

Halsted, W. S. (1909). Auto and isotransplantation in dogs of the parathyroid glandules. *J. Exp. Med.* **11**, 175–199.

Harman, S. M., and Talbert, G. B. (1970). The effect of maternal age on ovulation, corpora lutea of pregnancy and implantation failure in mice. *J. Reprod. Fertil.* **23**, 33–39.

Harris, G. W. (1964). Sex hormones, brain development and brain function. *Endocrinology* **75**, 627–648.

Harris, G. W. (1970). Hormonal differentiation of the developing central nervous system with respect to patterns of endocrine function. *Philos. Trans. R. Soc. London, Ser. B* **259**, 165–177.

Harris, G. W., and Jacobsohn, D. (1952). Functional grafts of anterior pituitary gland. *Proc. R. Soc. London, Ser. B* **139**, 263–276.

Harris, M., and Eakin, R. M. (1949). Survival of transplanted ovaries in rats. *J. Exp. Zool.* **112**, 131–164.

Harrison, F. A., Heap, R. B., and Linzell, J. L. (1968). Ovarian function in the sheep after autotransplantation of the ovary and uterus to the neck. *J. Endocrinol.* **40**, xiii.

Hicken, P., and Krohn, P. L. (1960). The histocompatibility requirements of ovarian grafts in mice. *Proc. R. Soc. London, Ser. B* **151**, 419–433.

Hohlweg, W., and Dohrn, M. (1932). Uber die Beziehungen zwischen Hypophysenvorderlappen und Keimdrüsen. *Klin. Wochenschr.* **11**, 233–235.

Ingram, D. L., and Krohn, P. L. (1956). Factors influencing the survival of ovarian homografts in rats. *J. Endocrinol.* **14**, 110–120.

Jacobowitz, D., and Laties, A. M. (1970). Adrenergic reinnervation of the cat ovary transplanted to the anterior chamber of the eye. *Endocrinology* **86**, 921–924.

Jones, E. C., and Krohn, P. L. (1960). Orthotopic ovarian transplantation in mice. *J. Endocrinol.* **20**, 135–146.

Jones, E. C., and Krohn, P. L. (1961). The effect of hypophysectomy on age changes in the ovaries of mice. *J. Endocrinol.* **21**, 497–508.

Jones, E. C., and Krohn, P. L. (1962). Effect of the maternal environment on strain specific differences in the ovaries of new-born mice. *Nature (London)* **195**, 1064–1066.

Kalliss, N. (1958). Immunological enhancement of tumor homografts in mice. *Cancer Res.* **18**, 992–1003.

Kambysellis, M. P. (1968). Interspecific transplantation as a tool for indicating phylogenetic relationships. *Proc. Natl. Acad. Sci. U.S.A.* **59**, 1166–1172.

Knauer, E. (1896). Einige Versuche über Ovarientransplantation bei Kaninchen. *Zentralbl. Gynaekol.* **20**, 524–528.

Kornblum, J., and Silvers, W. K. (1968). Modification of the homograft response following intrasplenic exposure to ovarian and testicular tissue. *Transplantation* **6**, 783–786.

Kovacs, K. (1965). Production of tumors in the ovary of androgenized rats after autotransplantation into the spleen. *Endocrinology* **77**, 754–755.

Kovacs, K. (1966). The development of corpora lutea in androgenized female rats with the ovary transplanted into the spleen. *Acta Anat.* **63**, 167–178.

Krohn, P. L. (1955). The behavior of autografts and homografts of vaginal tissue in rabbits. *J. Anat.* **89**, 269–282.

Krohn, P. L. (1958). Litters from C_3H and CBA ovaries orthotopically transplanted into tolerant A strain mice. *Nature (London)* **181**, 1671–1672.

Krohn, P. L. (1962a). Transplantation of the ovary. *In* "The Ovary" (S. Zuckerman, ed.), 1st ed., Vol. 2, pp. 435–462. Academic Press, New York.

Krohn, P. L. (1962b). Review lectures on senescence. II. Heterochronic transplantation in the study of ageing. *Proc. R. Soc. London, Ser. B* **157**, 128–147.

Krohn, P. L. (1964). The reproductive lifespan. *Proc. Int. Congr. Anim. Reprod. Artif. Insem., 5th, 1963* **2**, pp. 23–25.

Krohn, P. L. (1965). Transplantation of endocrine organs. *Br. Med. Bull.* **21**, 157–162.

Krohn, P. L. (1966). Transplantation and ageing. *In* "Topics in the Biology of Ageing" (P. L. Krohn, ed.), pp. 125–148. Wiley, New York.

Kullander, S. (1956). Experimental deciduoma in spayed rats with ovarian tissue autografted into the spleen. *Acta Endocrinol. Copenhagen* **21**, 204–210.

Ladosky, W., Noronha, J. G. L., and Jozini, I. F. (1969). Luteinization of ovarian grafts in rats related to the treatment with testosterone in the neonatal period. *J. Endocrinol.* **43**, 253–258.

Lane, C. E., and Markee, J. E. (1941). Responses of ovarian intraocular transplants to gonadotrophins. *Growth* **5**, 61–67.

Leavitt, W. W., and Carlson, I. H. (1969). Progesterone, 20α-hydroxy-Δ⁴-pregnen-3-one, androstenedione and corticosterone content of portal vein and vena caval blood in rats bearing ovarian grafts to the spleen. *J. Reprod. Fertil.* **18**, 172.

Linder, O. E. A. (1962a). Further studies on the state of unresponsiveness against skin homografts, induced in adult mice of certain genotypes by a previous ovarian homograft. *Immunology* **5**, 195–202.

Linder, O. E. A. (1962b). Modification of the homograft response after pre-treatment with ovarian grafts. *Ann. N.Y. Acad. Sci.* **99**, 680–688

Linder, O. E. A. (1962c). "Experimental studies on the modification of the homograft response." Ph.D. Thesis, Karolinska Inst., Stockholm.

Lipschütz, A. (1925). Influence de l'âge du porteur sur la fonction endocrine de la greffe ovarienne. *C. R. Soc. Biol.* **93**, 1066–1068.

Lipschütz, A. (1957). "Steroid Homeostasis, Hypophysis and Tumorigenesis." Heffer, Cambridge, England.

Long, J. A., and Evans, H. M. (1922). The oestrous cycle in the rat and its associated phenomena. *Mem. Univ. Calif.* **6**, 1–148.

McCracken, J. A., and Baird, D. T. (1969). The study of ovarian function by means of transplantation of the ovary in the ewe. *In* "The Gonads" (K. W. McKern, ed.), pp. 175–210. Appleton, New York.

McCracken, J. A., Uno, A., Goding, J. R., Ichikawa, Y., and Baird, D. T. (1969).

The in-vivo effects of sheep pituitary gonadotrophins on the secretion of steroids by the autotransplanted ovary of the ewe. *J. Endocrinol.* **45**, 425–440.

Mandel, J. J. (1933). Influence of ovarian grafts upon immature castrate rats. *Proc. Soc. Exp. Biol. Med.* **30**, 1415–1417.

Mandl, A. M., and Zuckerman, S. (1949). Ovarian autografts in monkeys. *J. Anat.* **83**, 315–324.

Mandl, A. M., and Zuckerman, S. (1951). The time of vaginal opening in rats after ovarian autotransplantation. *J. Endocrinol.* **7**, 335–338.

Martinez, C., and Bittner, J. J. (1956). A non-hypophyseal sex difference in estrous behavior of mice bearing pituitary grafts. *Proc. Soc. Exp. Biol. Med.* **91**, 506–509.

Martins, T. (1932). Greffes de vagin sur des rats mâles, et observation des cycles au moyen des frottis. *C. R. Soc. Biol.* **109**, 134–135.

May, R. N. (1940). La greffe bréphoplastique intraoculaire de l'ovaire chez la ratte. *Bull. Histol. Techn. Microsc.* **17**, 51–55.

Medawar, P. B. (1944). The behavior and fate of skin autografts and skin homografts in rabbits. *J. Anat.* **78**, 176–199.

Medawar, P. B. (1948). Immunity to homologous grafted skin. III. The fate of skin homografts transplanted to the brain, to subcutaneous tissue and to the anterior chamber of the eye. *Br. J. Exp. Pathol.* **29**, 58–69.

Medawar, P. B. (1958). The Croonian Lecture. The homograft reaction. *Proc. R. Soc. London, Ser. B* **148**, 145–166.

Miller, A. J., and Lampton, A. K. (1941). Influence of reticuloendothelial system on internal secretory activity of transplanted ovaries in the rat. *J. Urol.* **45**, 863–868.

Mühlbock, O., and Boot, L. M. (1956). La fonction hormonale d'ovaires, testicules et hypophyses transplantes chez des souches de souris génétiquement pures. *Ann. Endocrinol.* **17**, 338–343.

O'Leary, J. A., Feldman, M., and Gaennslen, D. M. (1969). Uterine and tubal transplantation. *Fertil. Steril.* **20**, 757–780.

Parkes, A. S. (1956). Survival time of ovarian homografts in two strains of rats. *J. Endocrinol.* **13**, 201–210.

Parkes, A. S. (1957). Attenuation of host reaction to interstrain ovarian homografts. *J. Endocrinol.* **14**, xxxvii–xxxviii.

Parkes, A. S. (1958). Enhancement of the survival of interstrain ovarian homografts. *Transplant. Bull.* **5**, 45–47.

Parrott, D. M. V. (1960). The effect of site of transplantation on host reaction to ovarian homografts. *Immunology* **3**, 244–253.

Parrott, D. M. V., and Parkes, A. S. (1957). Restoration of fertility in the X-irradiated mouse by orthotopic grafting of ovarian tissue. *J. Endocrinol.* **14**, xxxvi–xxxvii.

Peckham, B. M., and Greene, R. R. (1952). Experimentally produced granulosa cell tumors in rabbits. *Cancer Res.* **12**, 654–656.

Peckham, B. M., and Randak, E. F. (1952). "Atrophy" of ovarian grafts in unilaterally castrated rats. *Q. Bull. Northwest. Univ. Med. Sch.* **26**, 163–165.

Pettinari, V. (1928). "Greffe ovarienne et action endocrine de l'ovaire." Doin, Paris.

Pfeiffer, C. A. (1934). Functional capacities of ovaries of newborn after transplantation into adult ovariectomized rats. *Proc. Soc. Exp. Biol. Med.* **31**, 479–481.

Pfeiffer, C. A. (1936). Sexual differences of the hypophyses and their determination by the gonads. *Am. J. Anat.* **58**, 195–225.

Pfeiffer, C. A. (1937). Alterations in the percentage of cell types in the hypophysis by gonad transplantation in the rat. *Endocrinology* **21**, 812–820.

Podleschka, K., and Dworzak, H. (1933). Über Autotransplantationen von Ovarien in die vordere Augenkammer des Kaninchens. *Zentralbl. Gynaekol.* **57**, 2114–2122.

Podleschka, K., and Dworzak, H. (1934). Über die Funktion autoplastisch in die Augenvorderkammer verpflanzter Kaninchenovarien. *Arch. Gynaekol.* **155**, 381–407.

Rice, B. F., and Segaloff, A. (1966). Steroid hormone formation by the rat ovary. *Acta Endocrinol. (Copenhagen)* **51**, 131–139.

Schochet, S. S. (1920). The physiology of ovulation. *Surg., Gynecol. Obstet.* **31**, 148–149.

Schultz, W. (1910). Verpflanzungen der Eierstöcke auf fremde Species, Varietäten und Männchen. *Arch. Entwicklungsnech. Org.* **29**, 79–108.

Shaffer, C. F., and Hulka, J. F. (1969). Ovarian transplantation. Graft-host interactions in corneal encapsulated homografts. *Am. J. Obstet. Gynecol.* **103**, 78–85.

Snell, G. D. (1958). Transplantable tumours. *In* "The Physiopathology of Cancer" (F. Homburger, ed.), 2nd ed., pp. 293–345. Harper (Hoeber), New York.

Stevens, L. C. (1955). Survival of ovarian grafts in castrate and unilaterally ovariectomised female mice. *Transplant. Bull.* **2**, 45–46.

Sturgis, S. H., and Castellanos, H. (1958). Ovarian homografts in the primate: experience with Millipore filter chambers. *Am. J. Obstet. Gynecol.* **76**, 1132–1147.

Sturgis, S. H., and Castellanos, H. (1968). Ovarian homografts. *In* "The Ovary" (H. C. Mack, ed.), pp. 166–187. Thomas, Springfield, Illinois.

Takewaki, K. (1955). Behaviour of hamster ovary transplanted into spleen or wall of stomach. *J. Fac. Sci., Univ. Tokyo,* **7**, 465.

Talbert, G. B., and Krohn, P. L. (1966). Effect of maternal age on viability of ova and uterine support of pregnancy in mice. *J. Reprod. Fertil.* **11**, 399–406.

Tschernischoff, A. (1914). Die Eierstocksüberpflanzung speziell bei Saugetieren. *Beitr. Pathol. Anat. Allg. Pathol.* **59**, 162–206.

van der Schoot, P., and Zeilmaker, G. H. (1970). The function of ovarian grafts in neonatally castrated rats. *J. Endocrinol.* **48**, lxii.

van Wagenen, G., and Gardner, W. U. (1950). Functional intrasplenic ovarian transplants in monkeys. *Endocrinology* **46**, 265–272.

van Wagenen, G., and Gardner, W. U. (1953). Absence of hepatic inactivation of estrogen in monkey. *Yale J. Biol. Med.* **25**, 477–483.

Wagner, J. W. (1968). Luteinization of ovarian transplants in gonadectomized, pregnant mare's serum-primed immature male rats. *Endocrinology* **83**, 479–484.

Wang, G. H., Richter, C. P., and Guttmacher, A. P. (1925). Activity studies on male castrated rats with ovarian transplants and correlation of activity with the histology of the grafts. *Am. J. Physiol.* **73**, 581–599.

Ward, J. E., Gardner, H. L., and Newton, B. L. (1953). Anterior ocular ovarian grafts in the rabbit. *Am. J. Obstet. Gynecol.* **66**, 1200–1206.

Welschen, R. (1971). Corpora lutea atretica in ovarian grafts. *J. Endocrinol.* **49**, 693–694.

Wenner, R., and Hofman, K. (1950). Experimentelle Ergebnisse bei Ovarimplantationen in die Milz juveniler Meerschweinchen. *Gynaecologia* **139**, 382–387.

Yonemoto, R. H., DuSold, W. D., and Deliman, R. M. (1969). Homotransplanta-
    tion of uterus and ovaries in dogs. A preliminary report. *Am. J. Obstet.*
    Gynecol. **104,** 1143–1151.
Zeilmaker, G. H. (1970). Effects of implantation of ovarian and vaginal tissue in the
    hypothalamus and pituitary gland on prolactin secretion in the rat. *J.*
    Endocrinol. **48,** lxv–lvi.

4

Tumors of the Ovary

Georgiana M. Bonser and J. W. Jull

Ovarian tumors are as complex in histogenesis, structure, and function as is the ovary itself. The aim of the present chapter is to simplify the subject as far as is compatible with the vast literature which relates to it.

I. SPONTANEOUS TUMORS

A. In Man

1. Structure

For morbid anatomical descriptions the reader is referred to the accounts by Willis (1967) on general pathology and by Rewell (1960) on gynecological pathology. Although the classification is outdated, the atlas of Barzilai (1943) gives an outstanding pictorial survey. The classification adopted here (from Willis) is as follows.

129

a. Benign Fibropapillary Growths and Serous Cystadenomas. These are common tumors, varying greatly in structure from warty growths springing from the surface of the ovary to large cysts containing intracystic papillary projections and filled with serous fluid. The wall of the cyst may burst and the peritoneum become studded with papillomas. The papillae may have solid cores of cells resembling ovarian cortical stroma and thus form a solid tumor or fibroadenoma. There is no clear line of demarcation between serous cystadenomas and carcinomas.

b. Simple and Papillary Mucinous Cystadenomas. These are the most common of the large ovarian tumors. They are multiloculated and lined by tall mucin-bearing cells which form complex papillae. The cysts may contain mucinous fluid and blood. Serous areas may be detected with a core of ovarian stroma. A rare complication is that known as pseudomyxoma peritonei when the peritoneum is covered with large masses of myxomatous material resembling the inspissated mucin of a cystadenoma. The primary tumor may or may not be malignant.

c. Carcinomas. These are the second most common tumor of the ovary (see Table IV). They are highly malignant, usually occurring in women between 40 and 60 years of age. They may be solid or cystic, or both. They are composed of irregular papillae covered with serous, or less commonly, mucin-secreting, columnar epithelium of various degrees of anaplasia, but solid areas and patches of squamous metaplasia may be seen. These tumors spread locally and also metastasize. The majority arise from the more active papilliferous cystadenomas, the transition from the latter being clearly demonstrable. Rewell (1960) points out that some patients who have had apparently innocent cystadenomas removed may later develop adenocarcinoma.

d. Granulosa Cell, Theca Cell, and Luteal Tumors. It is rational to classify these types together as they are fundamentally tumors of the formative ovarian stroma, and may contain all three elements. They are often estrogenic, even in childhood, and may attain a large size and metastasize. The granulosa cell type has a folliculoid epithelial structure with small cysts; the theca cell type is composed of spindle cells containing lipid or resembling fibrocytes. Both elements may undergo luteinization.

e. Brenner Tumors. These are uncommon, rarely bilateral, usually solid and small or moderate in size, but may occasionally be very large. They are composed of solid or cystic epithelial clumps set in plentiful theca cell or fibrous tissue. They are usually unaccompanied by hormonal disturbances.

f. Arrhenoblastomas. These rare masculinizing tumors form a heterogeneous group. Salm (1967) showed clearly how some arise in ovarian hilar cells, the homologs of testicular interstitial cells, and some are mixed granulosa–hilus cell tumors. Others are pure granulosa–theca cell tumors. Dunnihoo *et al.* (1966) reviewed fifty hilar cell tumors. Reinke crystals were present in twenty-three, absent in nineteen, and no statement was made concerning their presence in the remainder. The authors considered that this was evidence that the tumors arose in hilus cells, as Reinke crystals are a characteristic of normal interstitial cells of the testis. Dunnihoo *et al.* proposed that the term hilar cell tumor should be reserved for those in which such crystals can be demonstrated. Female sex chromatin may also be present in such tumors. It is probable that some of these tumors arise in testes in genetic males of female phenotype, all traces of the testis having been obliterated by the tumor. Very rarely, tumors with this character occur in ectopic adrenal tissue adjacent to the ovary. The view of Teilum (1958) and others that Sertoli cell tumors produce masculinizing hormones is erroneous as this is a function of interstitial cells.

g. Dysgerminomas. These have a structure similar to that of testicular seminomas and occur predominantly in young subjects or in hermaphrodites or pseudohermaphrodites, but also in normal fertile females (see Volume II, Chapter 5). Of 117 tumors analyzed by Asadourian and Taylor (1969), twelve were admixed with carcinomas of other types. Dysgerminomas are solid tumors, which may be large, composed of large rounded epithelial cells, and infiltrated with lymphocytes. They are essentially malignant and capable of metastasis. Tumors classed by Scully (1953) and many other authors as "gonadoblastomas" are discussed below (Section I,A,2).

h. Teratomas. These are common tumors, usually symptomless, which form benign "dermoid cysts" that are lined by squamous epithelium. They may be characterized by hairs, sweat glands, and abundant sebaceous glands, separately or all in the same nodule. Rarely, the squamous lining may become malignant or the whole tumor may be solid and composed of undifferentiated tissues.

Other rare tumors may occur, e.g., the endometrial mixed mesodermal tumor containing epithelial, rhabdomyoblastic or chondrosarcomatous components (Willis, 1967; Anderson *et al.,* 1967). Apart from the cystadenoma–carcinoma series and benign tumors, all other types of tumor are rare.

2. Histogenesis

According to Willis (1967), ovarian tumors arise from gonadal tissues themselves and not from included heterotopic tissues. He also suggests that the tumors are interrelated and that some structural combinations are more frequent than others. Willis further believes that there is a continuous production during adult life of the differentiated components of the normal ovary (see Volume I, Chapter 4) from the multipotent ovarian stroma, and that many tumors originate from these tissues and exhibit divergent differentiation and maturational changes like those of the normal ovary. Mucous or squamous metaplasia occurring in the normal ovary or in the tumor is a fourth factor in determining the structure of ovarian neoplasms.

It had been hoped that a study of the histogenesis of experimentally induced ovarian tumors in the mouse might throw some light on the origin of human tumors. This has not proved possible for three reasons: the effect of the strain of mouse used has not been sufficiently defined, the ovarian response varies according to the type of carcinogenic stimulus used (chemical, hormonal or irradiation), and the induced tumors in the mouse are either granulosa–theca cell types or are tubular adenomas, which represent only a small proportion in a human tumor series. In the mouse, granulosa cell tumors are derived either from ovarian stroma following treatment with the chemical 9,10-dimethyl-1,2-benzanthracene (DMBA) or whole body irradiation, or from cystic hemorrhagic follicles in intrasplenic grafts. The tubular adenomas, which are strain-limited, are formed from germinal epithelium in chemically treated and irradiated mice but from follicles in intrasplenic grafts. The value of the experimental work lies in showing that a variety of agents have a tumor-inducing effect and their mode of action is not uniform. The development of tumors in intrasplenic grafts shows that hormonal hyperstimulation combined with cellular nutritional defect and changed environment lead to tumor induction. Doubtless similar factors determine the occurrence of human tumors.

The following are some of the views held at the present time in regard to the histogenesis of human ovarian tumors. The serous cystadenomas are derived from the germinal epithelium and the immature cortical stroma associated with it. Mucinous cystadenomas are either metaplastic variants of serous tumors or are derived from cystic follicles which have already undergone mucous metaplasia (Gardner, 1938; Rewell, 1960). Rarely, they arise from Brenner tumors. The view that some mucinous cystadenomas represent a one-sided development of a teratoma is supported by Fox *et al.* (1964) and Ruffolo and Withers (1966) but is questioned by Azzopardi and Hou (1965). Fox *et al.* (1964) used the presence of argyrophil and argentaffin cells in pseudomucinous cysts, especially those associated with dermoid cysts, as an argument that the lining of the cysts is similar to that of the large intestine and that some, at least, should therefore be regarded as teratomas, though others may have a metaplastic origin. Azzopardi and Hou (1965) found intestinal metaplasia in an endocervical carcinoma, but were cautious about postulating a teratomatous origin for tumors of nonendodermal derivation such as pseudomucinous cystadenomas. Other evidence in support of a monophyletic origin of pseudomucinous tumors is lacking.

Carcinomas arise most frequently in serous cystadenomas, overlapping of the two types being frequently seen. A decision as to whether the tumor is a benign cystadenoma or a malignant carcinoma on microscopical examination alone is frequently impossible as one type merges into the other. Some carcinomas arise in foci of ovarian endometriosis (Gray and Barnes, 1966) and a clear cell variety, formerly regarded as mesonephroma because of the presence of glomerular-like structures, has the same origin (Czernobilsky *et al.*, 1970) as does the rare mixed mesodermal ovarian tumor.

According to Willis (1967), granulosa and theca cell tumors and luteomas arise from the formative ovarian stroma. Granulosa cell tumors always contain thecal cells (Rewell, 1958). Hughesdon (1958) studied in great detail one small theca cell and one thecogranulosa cell tumor and showed that the former arose from and within a hyperplastic ovarian cortex, whereas the latter arose by metaplasia of the hilar tongue of cortical stroma. Some authors suggest an origin of granulosa cell tumors from follicular granulosa cells but this is rejected by Burrows (1943) and others.

The origin of Brenner tumors is the subject of much discussion. A suggested origin in Walthard nests has been largely discarded since Teoh's thorough investigations (1953a,b, 1959). The resemblance of the epithelial element to granulosa cells and of the stromal element to modified ovarian stroma suggests an origin similar to that of granulosa cell tumors and, as in the case of mouse granulosa cell tumors, there may be more than one mode of origin.

Since some masculinizing tumors belong to the granulosa–theca cell group, they have similar histogenesis. The variety of structure which they exhibit is due to the various amounts of lipid which the cells contain. Lipid cell tumors resembling adrenocortical tissue are luteinized thecomas and may be derived from medullary mesenchyme (Fox, 1968).

Dysgerminomas are regarded as of germ cell origin. Hughesdon (1959) supported this by demonstrating that the dysgerminoma structure resembles that of the embryonic ovary. Asadourian and Taylor (1969) showed that the DNA content of the tumor cells was twice that present in lymphocytes and that none of the tumor cells had DNA values suggesting a haploid chromosome content. Thus the origin of the tumor is from primary oocytes, since these are arrested in prophase and, therefore, have twice as much DNA as do nuclei of resting diploid cells.

Primordial germ cells do not have sex chromatin bodies; Asadourian and Taylor (1969) found female sex chromatin bodies in the somatic cells of all twenty-two patients evaluated, but Theiss *et al.* (1960) examined twenty dysgerminomas from normal females and all had the female type of nuclear chromatin. In Asadourian and Taylor's (1969) series of ninety-eight patients, none was hermaphrodite and these authors state that the "prevalent concept that ovarian dysgerminomas frequently occur in women with gonadal dysgenesis stems from Meyer's original reports. Subsequent studies have failed to demonstrate any relationship between dysgerminomas and gonadal dysgenesis." Meyer (1931) actually reviewed twenty-seven cases of dysgerminomas associated with hermaphroditism and twenty-one without. Teter (1969) has described three dysgerminomas in twelve patients with isolated gonadal dysgenesis and a negative sex chromatin pattern, and Andrews (1971) found a dysgerminoma in a patient with a "streak" gonad on the opposite side and a 44,XY chromosome pattern. Scott and Bain (1965) referred to fifteen reported cases of dysgerminomas in phenotypic females of which fourteen showed an XY cell strain in at least one tissue analyzed. If the thesis is correct that dysgerminomas arise from primary oocytes they would not be expected to arise in streak ovaries lacking such cells, though they might arise in ovaries deficient in numbers of oocytes.

In 1953, Scully described two dysgerminomas, which he called gonadoblastomas, occurring in girls aged 18 and 19 years, respectively, and causing masculinization. The neoplasms were composed of cells of three types; dysgerminoma cells, cells of sex cord type (Sertoli or granulosa), the latter often surrounding the former coronally, and cells of mesenchyme type (Leydig or theca), which were thought to be the hormone-producing element. Calcified eosiniphilic bodies were prominent. In 1970, Scully reviewed seventy-four similar cases among which thirteen tumors were coincident with ovarian tumors of other types. Some of the patients were young intersexes of three types: nonvirilized or virilized phenotypic females or males, nearly all those examined carrying a Y chromosome or being mosaics. Scully noted that the tumors, most of which were sex-chromatin negative, could arise in streak ovaries or testes and questioned whether these lesions could be regarded as neoplasms, since mitotic figures have been found only in the germ cell element and only the dysgerminoma cells seem to metastasize. As the germ cells seemed to give rise to other types of tumor, such as embryonal carcinoma and choriocarcinoma, Scully concluded that gonadoblastoma should be regarded as a form of germ cell malignancy *in situ*. Fox (1968) questioned the neoplastic nature of gonadoblastoma and suggested that it might be an hamartoma. Teter (1969) divided these tumors into four classes and called them gonocytoma I to IV, stressing the relationship to gonadal dysgenesis and the Y chromosome or mosaicism. Hughesdon and Kumarasamy (1970) examined three tumors in great detail and found no sex chromatin in the germ cells in dysgerminoma or gonadoblastoma areas. They concluded that gonadoblastoma is a primary germ cell tumor which has provoked a Sertoli or granulosa cell response rather than a leukocytic one. This occurs rarely in normal testis, somewhat less rarely in normal ovary, where it develops in an antecedent stromal reaction, and most commonly and typically in dysgenetic gonads. If these premises are correct and a gonadoblastoma is a form of dysgerminoma, it would be preferable to devise a term which would indicate its essential feature, i.e., the Sertoli or granulosa cell response. A microscopical decision on the type of gonad is often impossible because of obliteration by the tumor, but considerable literature is accumulating on this subject. Further studies may clarify the relationship between karyotype and gonadal lesions of various kinds, particularly with respect to their occurrence in ovaries or testes.

Theories of the origin of teratomas have been numerous and are discussed fully by Willis (1967). The idea of there being a "suppressed fetus attached to an otherwise normal individual" has long been discarded in favor of Willis's own belief that "teratomas are tumors arising from foci of plastic pluripotential embryonic tissue which escaped from the influence of the primary organizer during early embryonic development, this escape being in some way related to disturbances emanating from the invaginated organizing tissues of the primitive streak...." Rashad *et al.* (1966) reviewed the previous literature and estimated the sex chromatin content of the squamous cells of 187 benign cystic teratomas. They found that 23.8% of nuclei were chromatin positive and that six patients had a normal female XX chromosome complement. In the male, however, there was an occasional discrepancy between the nuclear sex of the host and that of the teratoma, thus casting doubt on a simple neoplastic origin of the tumors and suggesting a possible origin from germ cells. The authors considered this probable but not proved.

3. Hormone Production

Serous and mucinous cystadenomas and carcinomas do not produce hormonal effects; dysgerminomas only do so when associated with gonadoblastoma in phenotypic females, and signs of masculinization may then be present. According to Pedowitz *et al.* (1955), some dysgerminomas and teratomas cause precocious puberty in children and elevated gonadotropin levels may give a positive pregnancy test.

Many granulosa cell tumors produce no hormonal effects but a proportion cause hyperestrinism in adults and precocious puberty in children; the symptoms disappear after removal of the tumor. Bruk *et al.* (1960) studied hormone excretion in two girls, aged 2 and 3 years, who exhibited precocious puberty because of granulosa cell tumors. The total urinary estrogen was at the level of midcycle in the adult and fell after operation, but the amount detected in the tumor tissue was very small. Urinary FSH was within the adult range but subsided after operation. The ketosteroids were normal.

Some granulosa cell tumors are, however, androgenic, especially if luteinized. Marsh *et al.* (1962) described a girl aged 26 months; elevated urinary estrogens and a granulosa cell tumor which had undergone luteinization were found. The tumor tissue had a high capacity to convert [^{14}C]testosterone to radioactive estrogen *in vitro,* and the luteinized granulosa cells were believed to be the source of the hormone. Granulosa cell tumors always contain many thecal cells, however, and the estrogen could be produced by these or by the tumor cells, or both.

There is a probable association of granulosa cell tumor with cystic endometrial hyperplasia and carcinoma, more easily demonstrated after the menopause, thus supporting the positive role of estrogen in the induction of uterine carcinoma (see Volume II, Chapter 5). Luteinized granulosa cell tumors may be associated with diffuse uterine myogenesis or fibromyomata (Willis, 1967). Mammary enlargement and secretion have been noted but there is no direct association with mammary carcinoma.

Theca cell tumors may be masculinizing. Older patients with Brenner tumors occasionally exhibit endometrial hyperplasia and uterine bleeding, or more rarely masculinization. Tumors arising in adrenal rests adjacent to the ovary may have adrenal cortical function manifested by high 17-ketosteroid excretion.

Teratomas rarely produce hormones, the best known cases being those where functioning thyroid tissue induces thyrotoxicosis (*struma ovarii*). Hemolytic anemia has occasionally been observed in children with cystic teratomas (Allibone and Collins, 1951), the anemia remitting after surgical removal of the tumor. Fox (1968) states that the Coombs test can be positive or negative and that antibodies are produced to the fluid contents of the cyst, in which are split proteins or unstable lipoproteins which share a common antigen with erythrocytes.

Fox also cites cases of hypoglycemia associated with ovarian fibromas or fibrosarcomas, although no insulin-like substance has been detected. Associated Meig's syndrome or the multiple basal cell carcinoma syndrome has also been described. Some cases with ovarian carcinoma have shown hypercalcemia but there were no secondary deposits in bone and no parathyroid-like substance was detected. The carcinoid syndrome, associated with localized primary ovarian growths has also been recorded, the serotonin passing directly into the circulation rather than via the liver (Fox, 1968). Sometimes, a systemic condition can be seen in a selected tissue of a teratoma as, for example, an asthmatic reaction in the bronchial tissue of a teratoma (Thomson, 1945).

Tumors not themselves hormonally active, e.g., primary ovarian cancer or secondary gastric cancer (Krukenberg tumors), may induce hormone production by ovarian cells. Pfleiderer *et al.* (1968) showed by histochemical methods that the hormone-producing cells in a patient were of thecal origin, and postulated that the cause was compression of the ovarian stroma by the tumor. Scott *et al.* (1967) very thoroughly investigated the urinary hormone excretion patterns of a masculinized patient with secondary gastric cancer in the ovaries, and, after considering several theories of origin of the excess of 17-ketosteroids and 17-ketogenic steroids, accepted that the cause was stimulation of the ovarian stroma by the tumor. They proposed that tumors which had bossed or scattered surfaces, i.e., those with large surface areas, stimulated the stroma more than smooth, round tumors and thus had a quantitative stimulatory effect.

Kupperman (1969) described various metabolic pathways and tests which can be applied to demonstrate the mode of steroidogenesis by the ovary, adrenal, testis, and placenta. He concluded that there is no evidence to suggest abnormal steroidogenesis by ovarian neoplasms; Goldzieher (1968) does not entirely agree. Since hypersecretion of androgenic or estrogenic steroids may be ovarian or adrenal in origin, differentiation must be made, and he recommends the use of metopirone and human chorionic gonadotropin stimulatory tests, as well as suppression tests, to delineate the role of the ovary. These are tests usually outside the scope of hospital laboratories.

4. Genetic Factors

From time to time, reports appear of the occurrence of ovarian tumors in siblings, first cousins, or in two or more generations of a family. Too little information is available to regard these observations as other than fortuitous. The decisive role of strain in the experimental induction of ovarian tumors in the mouse (Section II,B,3) does, however, suggest that an hereditary factor may occasionally operate in man.

Chromosomal aberrations in the cells of ovarian tumors have been studied either in smears from solid tumors or in ascitic fluid. In a mucinous cystadenoma, Benedict *et al.* (1969) described three aneuploid metaphases containing approximately 60, 150, and 300 chromosomes and a long acrocentric marker chromosome similar to that previously reported in a malignant tumor, and concluded that at least one area of the tumor was undergoing malignant change, although there was no histological evidence

of invasion. Similarly, Gerli *et al.* (1968) concluded that a clonal karyo-
typical evolution toward aneuploidy had occurred in a tumor not yet
malignant. Fracarro *et al.* (1968) related the histological appearance of a
papilliferous cystadenoma to the aberrant chromosomal numbers in the
tumor compared with those of blood and suggested that the case illustrated
the cytogenetic basis for the start of evolution toward malignancy.
Aneuploidy, hypodiploidy, and structural alterations of chromosomes,
mainly to gigantism, were demonstrated in four ovarian carcinomas by Tortora (1967). However, Bowey and Spriggs (1967) point out that control
observations are almost entirely lacking and that technical difficulties have
not yet been overcome. When studying normal human endometrial cells
entering mitosis *in vivo* they found a modal chromosome number of 46;
cells with more than 46 chromosomes were not encountered, but there was a
large scatter of values to the left due to chromosome loss. It may be concluded that these methods cannot at this stage be used as a diagnostic test in
man.

The nuclear DNA content has also been compared with the histological
appearances of fourteen pseudomucinous tumors (Weiss *et al.*, 1969). The
frankly benign tumors had a diploid tetraploid and the malignant ones an
aneuploid distribution, while the intermediate group had one or the other.
Weiss *et al.* regard these results as relatively decisive but point out that
application to clinical practice would necessitate prospective assessment of
patients whose lesions fall into these two nuclear groups. Atkin (1971)
measured modal DNA values and sometimes chromosome numbers in
sixty-one cases of ovarian carcinoma. The prognosis was better for patients
with low ploidy (near diploid) than for those with high ploidy tumors. There
were differences in ploidy in different areas of some tumors, suggesting
regional development of polyploidy of a single cell line. When examined, the
karyotype of the host was of normal female type.

The histogenesis of dysgerminoma and gonadoblastoma has already been
described (Section I,A,2). It now remains to discuss the relationship of the
host's chromosomal pattern to neoplasia. Teter (1969) performed
laparotomy on fifty-five patients with gonadal dysgenesis, female
phenotype, and high urinary gonadotropin who were referred to him with
amenorrhea and retardation of sexual development. Table I summarizes his
findings. Tumors were of several types, but only occurred when a Y chromosome was present. Teter therefore recommends that operation should be
reserved for patients shown to carry a Y chromosome. Taylor *et al.* (1966)
described forty-one dysgenetic gonads and stated that gonadoblastomas
arose in patients with 46,XY chromosomes and rudimentary testes.
Andrews (1971) also noted a relationship between streak gonads, the Y
chromosome, and gonadal tumors. She pointed out that two sex chro-

TABLE I

Gonadal Abnormalities in Patients with Gonadal Dysgenesis and Female Phenotype[a]

Group	Diagnosis	No. of patients	Sex chromatin pattern	Karyotype	Tumors (no.)
1	Turner's syndrome	21		45X/46XX or 45X	None
2	Turner's syndrome with masculinization	5		45X/46XY	Gonadoblastoma (1) Interstitial hilus cell tumor (1)
3	Isolated (pure) gonadal dysgenesis	14	+	46XX	None
4	Isolated gonadal dysgenesis	12	−	46XY	Dysgerminoma (3) Gonadoblastoma (1) Brenner tumor (1)

[a] After Teter (1969).

mosomes are necessary for the proper development of a functional ovary or testis. Where there is complete or partial absence of one sex chromosome germ cell degeneration follows, the most common pattern associated with streak gonads being 44,XO. Such gonads would not be expected to develop tumors.

The possibility of an autoimmune state as a cause of ovarian defect is raised by Mawdesley-Thomas and Cooke (1967). A young rat was found at postmortem to have no ovaries and lymphocytic infiltration of the thyroid gland. Thyroiditis occurs in Turner's syndrome in man. Turner's syndrome also occurs in the mouse, but the animals are phenotypically normal and breed. Chromosome studies were not done, but Mawdesley-Thomas and Cooke thought that the association of a possible genetic defect and autoimmune disease was "interesting." Scott and Bain (1965) described four girls with dysgerminomas of the ovary who died of uremia due to chronic glomerulonephritis. An association may have existed in these cases between the autoimmune disease and the XY karyotype. Marshall and Dayan (1964) supported this hypothesis. Similarly, Gon et al. (1965) proposed an immune reaction in an 18-year-old girl with dysgerminoma and blood dyscrasia which remitted following removal of the tumor. This is a field in which there will be further development.

B. In Other Species

Ovarian tumors have been described in many mammals and occasionally in amphibians and reptiles. No doubt such tumors occur in other vertebrates and await description. There has been increased interest in animal tumors in the past decade and many of the papers quoted in Table II contain descriptions of tumors in organs other than ovaries. It is impossible to survey completely the published material here and the following is a selective account only. No attempt has been made to estimate the age of occurrence of animal tumors, as information is scattered and inaccurate and few life-span studies are available.

1. Incidence

It is difficult to estimate the incidence of ovarian tumors in animals because the populations in which they arise are usually unknown. Surveys of slaughterhouse material have the disadvantage that many animals are killed at a young age before the majority of tumors could be expected to appear. Surveys of animals in Zoological Gardens are handicapped by lack of knowledge of age in captivity. Colonies of laboratory animals provide useful populations but are largely confined to rodents, dogs, or monkeys. Much information has been obtained from material collected by veterinarians from necropsies or biopsies of domestic pets but this is highly selected information, usually of a pathological nature, which does not permit a study of incidence.

Anderson and Sandison (1969) organized a survey by meat inspectors of cattle, sheep, and pigs slaughtered in 100 abattoirs in Great Britain from October 1965 to September 1966 (Table III). Eleven and four ovarian tumors were found in 1.3×10^6 cattle and 3.7×10^6 pigs, whereas none was found in 4.5×10^6 sheep. The average age of the tumor-bearing cattle was at least 8 years and that of pigs $2\frac{1}{2}$ years. Anderson and Sandison (1969) also surveyed nine literature reports of tumors in cattle, sheep, and pigs. They found seventy ovarian tumors (2.7%) among 2594 cattle tumors of all kinds, one among 249 sheep tumors, and three among 332 pig tumors. Harvey *et al.* (1940) saw four granulosa cell tumors in calves. From these surveys and that of Brandly and Migaki (1963), it may be concluded that ovarian tumors are rare in cattle and probably even rarer in sheep and pigs.

Ovarian tumors have been described in dogs, horses, and cats, in that order of frequency. The practice of spaying domestic cats may partly account for the low recorded incidence in these animals. Few ovarian tumors have been found in monkeys and guinea pigs, again probably due to the early age at slaughter. Dow (1960) and Willis (1967) used an unselected series of 400 and 204 necropsies, respectively, in which to assess the frequency of canine tumors. Dow found twenty-five ovarian tumors (6.25%), whereas Willis found three (1.5%). Cotchin (1961), in a series of 4187 canine tumors collected at operation or as postmortem specimens, found

TABLE II

Summary of Tumors Described in Animals

(continued)

Species	Cyst-adenoma	Car-cinoma	GCT series	Dysgerm-inoma	Tera-toma	Other	Reference
A. Mammalia							
Eutheria							
Carnivora							
Dog	1			11	2		Dehner et al. (1970)
	30[a]	3	26[b]	11	2		Norris et al. (1970)
				2			Buergelt (1968)
	20	5	30	8		6	Cotchin (1961)
	10	1	13			1	Dow (1960)
		1	2				Federer (1958)
		1[c]					Frost (1963)
		1					Owen and Hall (1962)
				1			Taylor and Dorn (1967)
		1	1				Zaldivar (1967)
		2[c]	1				Willis (1967)
		1					Stünzi and Stünzi (1950)
Cat	1	1		2	1		Dehner et al. (1970)
			5		2		Norris et al. (1969a)
			1			1[d]	Federer (1958)
			1[c]			2	Chesterman and Pomerance (1965)
Ferret							
Artiodactyla							
Cow		7	3[e]			1	Anderson and Sandison (1969)
			1				Kanagawa et al. (1964)
		7	14				Misdorp (1967)
		3	26[b]		3	2	Norris et al. (1969b)
			18				Roberts (1953)
			1				Short et al. (1963)

TABLE II (*continued*)

Species	Cyst-adenoma	Car-cinoma	GCT series	Dysgerm-inoma	Tera-toma	Other	Reference
Cow (continued)	1		2				Yamauchi (1963)
					3		Dehner et al. (1970)
							Studer (1967)
		1	17				McEntee and Zepp (1953)
		4				1	Trotter (1911)
		5				1	Davis et al. (1933)
		1	6				Monlux et al. (1956)
			11				Plummer (1956)
			3				Cotchin (1960)
			16				Brandly and Migaki (1963)
	8		27	4	1	10	Leopold (1967)
			1				Sastry and Tweihaus (1964)
						9	Anderson and Sandison (1969)
Pig	11	2	1	1	2		Nelson et al. (1967)
		1	1				Brandly and Migaki (1963)
			2				Lombard and Havet (1962)
				1	1		Lombard and Havet (1962)
Sheep							
Perissodactyla							
Horse	10	2	9		3	10	Cordes (1969)
	1		10				Leopold (1967)
							Krediet (1931)
Peruvian wild ass		1	6		2		Norris et al. (1968)
			1				Norris et al. (1968)
Rodentia							
Springhaas				1			Dehner et al. (1970)
Ground squirrel					1		Dehner et al. (1970)
Rat	1	1	1			1	Carter (1968)
Mouse (various)	1	2	13			1	Carter (1968)

Mouse (noninbred)	1	51		2	4	Horn and Stewart (1952–1953)
Hamster		5				Fortner (1957)
Guinea pig		1		2		Jain et al. (1970)
						Willis (1962)
Agouti	1					Weir (1971)
Cetacea						
Blue whale	1	1				Rewell and Willis (1962)
Fin whale		2				Rewell and Willis (1962)
Primates						
Rhesus monkey	4	2		2	3	Martin et al. (1970)[f]
Squirrel monkey		1[c]				Rewell (1954)
Marsupialia						
Koala	12					Finckh and Bolliger (1963)
B. Aves						
Fowl	3	3		3		Awadhiya and Jain (1967)
C. Amphibia						
Leopard frog		1[c]				Lucké (1934)
D. Reptilia						
Python		1				Lucké (1934)

[a] Includes some intermediate in grade.
[b] Includes Sertoli cell.
[c] Bilateral.
[d] Leiomyoma in lion.
[e] One in a teratoma.
[f] With thecoma and luteoma.

TABLE III

Incidence of Ovarian Tumors in Cattle, Sheep, and Pigs[a]

Species	Total no. surveyed ($\times 10^6$)	Total no. of tumors	Ovarian tumors (no.)	Age (years)
Cattle	1.3	302	Fibroma (1)	10
			Granulosa cell (2)	5, Adult
			Granulosa cell in teratoma (1)	3
			Serous cystadenocarcinoma (3)	6, 10, 21
			Mucinous cystadenocarcinoma (3)	$1\frac{1}{2}$, 8, 8
			Adenoacanthoma (1)	Old
Sheep	4.5	107	None	
Pigs	3.7	139	Leiomyoma (1)	$1\frac{1}{2}$
			Granulosa cell (1)	3
			Serous cystadenocarcinoma (2)	$\frac{1}{2}$, 5

[a] After Anderson and Sandison (1969).

fifty-one ovarian tumors (1.2%). Clearly, no accurate estimation can be made in the absence of knowledge of the populations from which the tumor-bearing animals were drawn.

2. Structure

Ovarian tumors in animals can be classified in a scheme similar to that adopted for man. All types have been found in animals but the Brenner tumor has been described only in the dog (Bloom, 1954, quoted by Cotchin, 1961) and fowl (Campbell, 1951). In Table IV the tumors occurring in four species with greatest incidence are compared with those in man. From the

TABLE IV

Classification of Primary Ovarian Tumors in Four Domestic Species and Man

Species	Cystadenoma	Carcinoma	Granulosa cell series	Dysgerminoma	Teratoma	Other	Total
Dog	61	16	73 (37)	33	4	7	194
Ox	9	46	146 (64)	4	7	15	227
Horse	11	3	26 (47)	0	5	10	55
Pig	11	3	4 (22)	2	2	9	31
	92 (18.1)[a]	68 (13.4)	249 (49.1)	39 (7.7)	18 (3.6)	41 (8.0)	507
Man[b,c]	252 (52.3)	88 (18.3)	13 (2.7)	4 (0.8)	67 (13.9)	58 (12.0)	482

[a] Figures in parentheses are percentages.
[b] Two series examined in Liverpool and Madras.
[c] From Rewell (1960).

many descriptions and illustrations in the literature it is clear that animal tumors (for example, the induced mouse tumor depicted in Fig. 9) resemble closely those of man in structure and behavior. The striking feature is the large number of granulosa cell tumors in animals, the highest incidence being in the cow. These tumors are usually unilateral and benign, but they may spread into the pelvis or metastasize to distant organs. In the cow, Norris *et al.* (1969b) described nine metastasizing tumors out of thirteen, and in the cat three out of five. On the whole, granulosa cell tumors of animals are thought to be more irregular in structure and more aggressive than those in man, but their behavior varies in different species. Cotchin (1961) and Norris *et al.* (1970) noted a tendency for granulosa cell tumors to resemble a Sertoli cell tumor, one commonly found in the canine testis. Willis (1967), in describing three canine granulosa cell tumors, stated that "they show papillary and granulosa-cell structures resembling those of human tumours" and "more than any human material they suggest kinship of these two kinds of growths." The intermingling of serous papillary and granulosa cell tumors was also noted by Dow (1960) and Campbell (1951).

Teratomas comprise the most infrequent tumor (Table IV) and are usually benign, being composed of a variety of well-differentiated tissues. Willis (1967) noted peritoneal deposits of neuroglial tissue in a teratoma of a guinea pig and remarked that this occurs only rarely in man. Campbell (1951) emphasized the great histological variations present in the twelve unusual fowl ovarian tumors which he studied. He thought that separation of types could not be made on morphological grounds and suggested that endocrine activity might be a better criterion, as Burrows (1943) had done in relation to human tumors.

3. Histogenesis

Little is known of this aspect of animal tumors. Cotchin (1961) proposed that adenomas (including papillomas) and adenocarcinomas are derived from surface epithelium or epithelial cords in the underlying cortex. He supported this view by quoting the work of Jabara (1959, 1962) and O'Shea and Jabara (1967) who treated bitches with diethylstilbestrol and observed papillary adenomas or adenocarcinomas within 19 months, a very short induction period. The origin of the tumors was traced to surface epithelium or subsurface epithelial structures, but we are not convinced of the malignant nature of these changes. Cotchin (1961) also thought that granulosa cell tumors might arise from epithelial cords derived from surface epithelium but admitted an alternative origin from specialized ovarian cortical stroma, as believed by Norris *et al.* (1970). Most authors regard dysgerminomas as arising from germ cells.

4. Hormone Production

As in man, hormonal effects are most usually due to an excess of estrogen or androgen. The species described as affected are dogs, cats, cows, horses, and chickens.

Cotchin (1961) collected information about the clinical features accompanying sixty-three ovarian tumors in the bitch (Table V). Of thirty animals with granulosa cell tumors, nineteen had pyometra and three cystic endometrial hyperplasia (73%), whereas among thirty-three bitches with other types of tumor only fifteen (45%) had these conditions that could be regarded as being due to excessive estrogenic activity. Cotchin states "the common occurrence of pyometra and other disturbances of the genital system in the ageing bitch make it difficult to know whether the association of ovarian tumours is merely coincidental; but it seems likely, in view of experience in humans, that the observed symptoms were in some cases, at any rate, due to the functional activity of the tumour, and that the first sign of the development of such a tumour may be 'pyometra' or some related condition."

Dow (1960) thought that canine granulosa cell tumors were less often estrogenic than human ones. Jones and Gilmore (1963) pointed out that the testicular Sertoli cell tumor produces estrogens and believed that its ovarian counterpart does the same, but did not think the evidence adequate to establish the point. Norris et al. (1969b) studied twenty-six bovine granulosa and Sertoli cell tumors and thought that the latter were more active estrogenically, though less likely to metastasize. The same authors also observed cystic endometrial hyperplasia in five cats bearing granulosa cell tumors and in one with a lipid cell tumor. This animal was virilized, the symptoms disappearing after operation, but showed prolonged estrus for several weeks preoperatively. They suggested that the tumor may have produced more

TABLE V

Feminizing Disturbances in Bitches with Ovarian Tumors[a]

Feature	Adenoma	Adenocarcinoma	Granulosa cell tumor	Seminoma
Pyometra	4	2	19	2
Cystic endometrial hyperplasia	2	1	3	1
Vaginal hemorrhage	0	1	0	1
Persistent attraction to male dogs	0	0	1	0
Other symptoms	9	0	3	2
Symptoms not reported	5	1	4	2

[a] From Cotchin (1961).

TABLE VI

Steroid Concentrations in Samples of Bovine Cyst Fluid[a]

Steroid	Concentration (mg steroid/100 ml fluid)		
	Sample (i)	Sample (ii)	Sample (iii)
Progesterone	4.3	2.1	9.0
20β-Hydroxypregn-4-en-3-one	16.8	9.8	6.5
17α-Hydroxyprogesterone	<2.0	<2.0	0.8
Androstenedione	<2.0	<2.0	<0.5
Estrone	<0.5	<0.8	<0.2
Estradiol-17β	0.75	1.0	0.4

[a] From Short *et al.* (1963).

than one hormone from luteinized stromal or thecal cells. Bilateral thecomas in a ferret were associated with marked glandular hyperplasia of the endometrium (Chesterman and Pomerance, 1965). Among seven horses bearing granulosa cell tumors, none had clinical endocrine abnormality, but endometrial specimens were not available (Norris *et al.,* 1968). According to Cordes (1969), mares with granulosa cell tumors are infertile and have prolonged periods of estrus, but after extirpation of the tumors the estrous cycle and fertility apparently return to normal and most mares subsequently foal. Krediet (1931) described a virilized mare bearing a granulosa cell tumor.

Short *et al.* (1963) described a Friesian heifer aged 2¾ years, which had shown clinical signs of virilism for 7 months but in which the udder was enlarged and the teats distended with milk. At necropsy a typical cystic granulosa cell tumor, 15 cm in diameter, was present in the left ovary. Three samples of cyst fluid were submitted to biochemical examination (Table VI). The tumor seemed to be producing mainly progesterone, 20β-hydroxypregn-4-en-3-one and a little estradiol-17β (progesterone:estrogen ratio, 6:1). With the exception of 20β-hydroxypregn-4-en-3-one, all the steroid concentrations found in the cyst fluid were considerably lower than those found in normal or cystic bovine follicular fluid. The fact that no androgens could be detected casts considerable doubt on the original clinical diagnosis of virilism and it was concluded that the animal was nymphomaniacal rather than virilized. The authors believed that estradiol-17β was arising either from the normal theca interna cells or was being produced by the neoplastic granulosa cells themselves. Roberts (1953) made a biological assay of the fluid from a cystic granulosa cell tumor in a 15-month-old nymphomaniac cow but failed to find estrogenic hormones. It is clear that further biochemical studies would help to elucidate the mechanisms of abnormal hormonal states and it is obviously unwise to accept a clinical diagnosis of hormonal excess unless it is supported by biochemical determinations.

Fortner (1957) described five thecomas in Syrian hamsters, no mention being made of a hormonal effect in the hosts. However, he suggested hormonal imbalance as an etiological factor on the grounds that ovarian luteinization was a prominent feature.

5. Genetic Factors

There is little to incriminate genetic factors in the etiology of spontaneous animal ovarian tumors. McEntee and Zepp (1953), in describing seventeen granulosa cell tumors in the cow, found three from related animals in one herd, two being sired by the same bull that was the grandsire of the third. We have found no information leading to the idea that breed within a species is an etiological factor. The effect of strain in the experimental induction of ovarian tumors in mice is discussed in Section II.

II. INDUCED TUMORS

Despite its complexity, the ovary has considerable advantages over many other tissues in elucidating the changes which lead to tumor formation. The boundaries of the organ are well defined, the relationship of different cell types has been the subject of considerable study through many disciplines on the normal physiological plane, and there are recognized routes of progressive cellular differentiation and atrophy.

The variety of ovarian tumors induced by experimental procedures is limited to cell types which are closely related and often grouped together as "granulosa cell tumors" (see Section I,A,1). This terminology may sometimes be appropriate on morphological grounds but is misleading as to etiology. An idea of the histological appearances which are encountered can be obtained from Figs. 6–11.

Tumorigenesis in the ovary by irradiation and chemicals (Sections II,A,B) will be considered separately. Following this, the common effects on ovarian function will be outlined in Sections II,C,D. Work on the development of tumors in intrasplenic grafts of ovaries will be described in Section II,E and the recent studies on ovarian tumorigenesis following chronic progestin treatment are described in Section II,F.

The evolution of neoplasia in the ovary can only be appreciated in relation to the normal processes of differentiation which occur in the course of fulfilling the function of the organ, i.e., the production and shedding of the mature ovum. The histogenesis of induced ovarian tumors, and those which occur spontaneously in ovaries congenitally deficient in oocytes, is outlined in Section II,G.

A. Induction by Radiation

The effects of radiation on the ovaries of many species are considered in detail in Volume III, Chapter 1. It is, therefore, only necessary here to outline briefly the major events in our present knowledge of this form of tumorigenesis so that it may be evaluated relative to other forms of tumor induction in the ovary.

In 1932, Drips and Ford observed hyperplastic ovarian follicular cysts following irradiation of immature rats and it seems, in retrospect, that these were probably ovarian tumors of the granulosa cell type. The first large-scale induction of ovarian tumors following whole body irradiation was reported by Furth and Furth (1936). The histological appearance of the tumors, hormone secretory activity, and relationship to concurrent breast cancer were described by Furth and Butterworth (1936), who also made the fundamental distinction between "granulosa cell tumors" and "luteomata" arising from the follicular apparatus, and "tubular adenomata" which were not hormone secreting and were derived from the germinal epithelium. Confirmation of this early work has come subsequently from several sources (Geist *et al.*, 1939; Furth and Sobel, 1947; Lick *et al.*, 1949; Kaplan, 1950; Deringer *et al.*, 1954/55; Guthrie, 1958).

It appears that only the ovaries need to be irradiated (Lick *et al.*, 1949). Ovarian tumors occurred in all of twenty-four mice given whole body irradiation and in four mice where the radiation was localized to the ovaries. Tumors also developed if one ovary was irradiated and the other removed, but no tumors were found in fifteen mice in which the untreated ovary was left *in situ*. In a more extensive study (Kaplan, 1950), no tumors arose in nonirradiated ovaries transplanted intramuscularly into thirty-one mice which were ovariectomized and subjected to whole body irradiation before grafting. When irradiated ovaries were grafted into twenty-eight ovariectomized, irradiated mice and into twenty-six ovariectomized but nonirradiated mice, 61 and 73%, respectively, developed tumors. However, none developed in irradiated ovaries grafted into nonirradiated mice with their own ovaries intact. A single dose of 25 rads is sufficient to induce ovarian tumors, but higher doses are more effective (Furth, 1949). Pregnancies can occur among mice exposed to high doses of irradiation.

Ovarian tumors induced by radiation are transplantable to genetically related mice (Bali and Furth, 1949). Five morphological types of tumor were transplanted into F_1 Ak/Rf hybrid mice and the granulosa cell tumors and the luteomas showed evidence of estrogen and progestin secretion, respectively. Estrogenic activity and the development of hypervolemia in the host were marked characteristics of the 406 granulosa cell tumors in the following proportions: only estrogenic, 15.5%; only hypervolemic, 20.2%; estrogenic and hypervolemic, 23.1%; neither estrogenic nor hypervolemic, 36.7%.

Vaginal keratinization has been used by a number of workers as an indication of changes in estrogen secretion after irradiation. Geist *et al.* (1939) correlated an absence of estrus for 4 to 8 weeks with the period of follicular destruction immediately after treatment. Estrus recurred subsequently in all the mice, but the duration varied from 1 to 63 days. Only 37% were in estrus at the time, 6 months to 1 year after irradiation, when epithelial proliferation and downgrowth occurred, but renewed estrous activity occurred during the final senescent phase of the tumor.

B. Induction by DMBA

1. Mode of Exposure and Dosage

The compound 7,12-dimethylbenz(a)anthracene (DMBA, formula I, Fig. 1, previously known as 9,10-dimethyl-1,2-benzanthracene) is a potent carcinogen for many tissues. Ovarian tumor induction by this compound was first reported as an incidental observation by Engelbreth-Holm and Lefevre (1941) in two of sixty-six mice, but Marchant *et al.* (1954) and Howell *et al.* (1954) demonstrated that a high incidence of granulosa cell tumors follows adequate exposure to the chemical applied to the skin in oil. Since then many workers have shown similar incidences in intact mice by a variety of routes of administration including intragastric (Biancifiori *et al.*, 1961; Jull *et al.*, 1966; Krarup, 1967; Kuwara, 1967), intraperitoneal (Krarup, 1967; Kuwara, 1967), intravenous (Kuwara, 1967; Jull, 1969a), subcutaneous (Shisa and Nishizuka, 1968), or directly to the surface of the ovary (Krarup, 1969).

In the experiments of Marchant (1959a,b, 1960a,b, 1961a,b), Jull *et al.* (1966), Jull (1969a,b), and Jull and Russell (1970), ovaries were exposed to the carcinogen in a donor mouse before transplantation to an ovariectomized recipient in which tumors subsequently developed (Fig. 2). Since the time between administering the chemical and transplantation was often not more than 24 hours after stomach tubing or 1 hour after intravenous injection (in one case as little as 10 minutes), the possibility of anything but an "initiating action" of DMBA on the ovary was ruled out. The amount of

I

II.

III

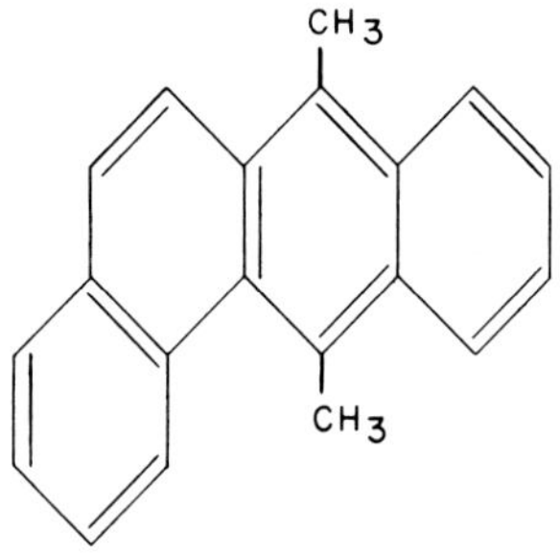

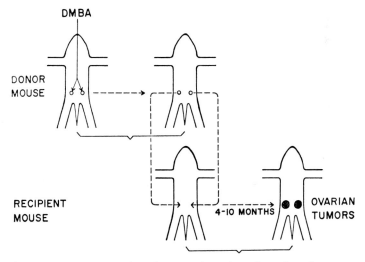

Fig. 1. Chemical formula of (I) 7,12-dimethylbenz(a)anthracene (DMBA); (II) 7-hydroxymethyl-12-methylbenz(a)anthracene; (III) 12-hydroxymethyl-7-methylbenz(a)-anthracene.

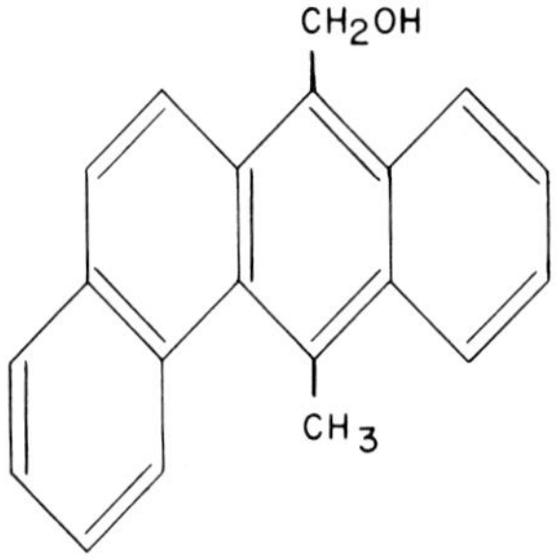

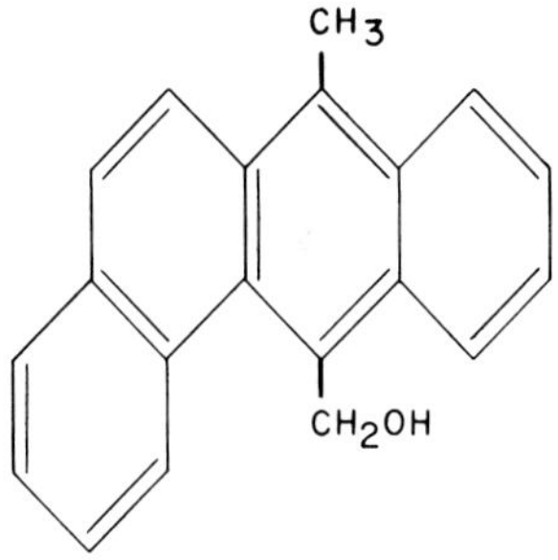

Fig. 2. Experimental procedure for transplantation of ovaries after exposure to a carcinogen.

DMBA transplanted with the graft to the otherwise untreated ovariec-tomized recipient mouse would be very small and insufficient to exert any systemic effect. A similar conclusion can be reached from the work of Krarup (1969) who applied a small amount of DMBA directly to the ovary at operation. The specificity of action of DMBA on the ovary is established beyond doubt, however, by the fact that exposure of ovaries to a solution of the carcinogen in organ culture medium for times ranging from 4 hours to 4 days is sufficient to induce neoplastic changes after their subsequent trans-plantation to ovariectomized recipient mice (Jull *et al.*, 1968).

The importance of a critical concentration of DMBA for the attainment of an appreciable tumor incidence has been demonstrated using organ cul-tures (Jull *et al.*, 1968), e.g., 0.5 μg DMBA/ml was without effect, but exposure of the ovary to 1.0 μg/ml or more gave significant numbers of tumors after reimplantation. It seems likely that the capacity of the treated mouse to metabolize DMBA will influence the effectiveness of small doses, which may be converted into inactive products of DMBA in animals pre-viously exposed to enzyme induction by related compounds (Jull, 1969a). Two of the major metabolites of DMBA are shown in Fig. 1. Neither of these has been adequately tested for carcinogenic activity in the intact mouse, but tests on ovaries exposed to them *in vitro* have revealed signifi-cant tumorigenic activity only with DMBA and its 12-hydroxylated metabolite (formula III, Fig. 1). Even this does not constitute conclusive evidence that these structures per se are active, as they may be converted to some other form inside the ovary. Alternatively, active molecules might not be absorbed into the cells.

2. Immediate Action of DMBA on the Ovary

The idea that DMBA in the mouse ovary causes an initial change, which can be developed to give actively growing macroscopic tumors with proper hormonal stimulation (Marchant, 1961a), is essentially an application of the concept of "initiation" and "promotion" put forward by Berenblum and Shubik (1947) in relation to the skin. A few efforts have been made to delineate the cellular processes involved. Jull and Jellinck (1968) showed that, despite the highly specific action of DMBA, localization of [^{14}C]DMBA in the mouse ovary was not particularly marked and was in fact much less than in some tissues, such as the adrenals and spleen (Table VII), in which it produces no chronic pathological changes. The rate of disappearance of radioactivity from the ovary was as great as from other tissues, being almost complete by 24 hours. Paradoxically, uptake of [^{14}C]DMBA by the rat ovary, on which it has no tumorigenic action, is rela-tively as great as in the mouse. It was calculated that, in the case of ovaries

TABLE VII

Content of Radioactivity (cpm/mg) in Tissues of CNZ Mice at Intervals after Intravenous Injection of a Suspension of 71 μg [^{14}C]DMBA in 2% Gelatin[a]

Tissue	Time after intravenous injection							
	15 minutes		1 hour		3 hours		24 hours	
	Total cpm/mg	Percentage ether soluble	Total cpm/mg	Percentage ether soluble	Total cpm/mg	Percentage ether soluble	Total cpm/mg	Percentage ether soluble
Right ovary	278	59	103	61	28	32	20	65
Left ovary	197	55	141	60	50	52	12	33
Right adrenal	727	73	691	66	133	60	22	50
Left adrenal	959	64	65	63	210	57	24	38
Liver	323	20	290	9	108	18	10	45
Spleen	450	25	470	16	74	15	21	19
Lungs	193	54	76	58	32	47	8	13
Right kidney	187	40	99	22	52	21	5	20
Left kidney	191	41	82	18	53	21	5	20
Blood	98	55	22	55	10	40	2	75

[a] From Jull and Jellinck (1968).

exposed to [¹⁴C]DMBA *in vitro*, the amount of radioactivity present at the time of reimplantation into an ovariectomized mouse was equivalent to only 3 ng DMBA. It is virtually impossible that such an amount could exert any action subsequent to its absorption, metabolism, and recirculation to the ovarian implant in the host.

DMBA has been commonly reported to cause a depression of DNA synthesis in cells. It was postulated that treatment with this compound might, therefore, affect the proliferating DNA-synthesizing granulosa cells of the ovary in a similar fashion (Pedersen and Krarup, 1969). By observing tritiated thymidine uptake of granulosa cells at different periods after DMBA treatment, Pedersen and Krarup showed that an early effect of the carcinogen was to increase the proportion of labeled granulosa cells as compared to controls. The duration of DNA synthesis (S phase) was not affected, but it was calculated that the doubling time of granulosa cells was reduced and hence the growth rate of follicles increased. It was concluded from these and related data that this effect of the carcinogen was indirect.

Using a system in which DMBA was injected intravenously into mice an hour before transplantation of the ovaries (Jull, 1969a), it was possible to show that pretreatment with various metabolic inhibitors did not affect the initiation of ovarian tumors. Among the compounds tested, actinomycin D and aflatoxin B1 are known to interact with DNA to inhibit RNA synthesis, ethionine and cycloheximide inhibit protein synthesis at different points, and metopirone, 2-methyl-1,2-bis(3-pyridyl)-1-propanone, has effects on steroid biosynthesis in addition to blocking the destructive action of DMBA on the rat adrenal. Interference at a number of points in the function of ovarian cells was therefore not incompatible with the immediate effects of DMBA in tumorigenesis. Despite the difficulties of interpreting such experiments, it seems probable that no new protein synthesis is necessary for DMBA to affect the mouse ovary. If a protein receptor is necessary, it must be present at the time of exposure to the carcinogen and have a life exceeding 4 hours, the time before DMBA that the inhibitors were injected.

In early experiments using IF strain mice in the transplantation system outlined in Fig. 2, there appeared to be some evidence that 11-deoxycorticosterone and 17-hydroxyprogesterone injected in high concentrations inhibited the initial tumorigenic action of DMBA *in vivo* (Jull *et al.*, 1966). Later and more extensive work with a wide range of steroid hormones and hybrid C3H × NZY mice has not confirmed this (J. W. Jull, unpublished). In view of the reduced tumor incidence after reimplantation of ovaries exposed *in vitro* to medium containing DMBA and either 11-deoxycorticosterone, 17-hydroxyprogesterone or 4-pregnen-3β,20β-diol (Jull *et al.*, 1968), however, there remains a possibility that under some conditions the induction of granulosa cell tumors by DMBA may be modified by some steroids.

3. Effects of the Hormonal Environment

Marchant (1961a) showed that DMBA administered to hypophysectomized mice can effect a change in the ovaries which results in granulosa cell tumors after their transfer to ovariectomized mice with intact pituitaries. However, no tumors arose in ovaries from intact mice exposed to DMBA after transplantation to hypophysectomized mice. The survival of both intact and hypophysectomized mice bearing DMBA-treated grafts was similar, so that the decisive influence of pituitary secretion in the emergence of these tumors was proved.

Marchant (1960a) initially demonstrated that the presence of ovarian tissue not exposed to DMBA completely inhibited the development of granulosa cell tumors. Extension of this work (Jull, 1969b) has shown that DMBA-treated ovaries remain atrophic when grafted subcutaneously into intact female mice. Subsequent removal of the host's own ovaries after intervals of up to 22 weeks resulted in tumors in the carcinogen-modified grafts. Ovaries from untreated mice, grafted into intact females in the same way, also remain atrophic until the host ovaries are removed, and then they grow and differentiate normally.

Genetic constitution is one of the factors that determines the hormone environment in which ovarian grafts develop. Thus, ovaries of mouse strain C57 are relatively resistant to tumor induction by DMBA (Marchant, 1957), but yield a significantly higher incidence when grafted into ovariectomized hybrids of C57 mice with the susceptible IF strain. In similar experiments (J. W. Jull, unpublished), no tumors occurred in ovaries of Balb/C strain mice exposed *in vivo* to DMBA and then grafted subcutaneously into ovariectomized Balb/C recipients. However, when the ovariectomized host was a hybrid of Balb/C with the highly susceptible NZY strain, transplantation was followed by a significant incidence of tumors in the implanted DMBA-treated Balb/C ovaries.

It is accepted that the pituitary gonadotropins of male and female mice are quantitatively different, although it is probable that their FSH and LH components are qualitatively the same. The difference is emphasized by the fact that DMBA-treated ovaries from the unsusceptible Balb/C and C57 strains will, nevertheless, give rise to granulosa cell tumors when transplanted into castrated males. Information regarding the pituitary secretion of gonadotropins in castrated male mice of these strains, demonstrating the essential differences in composition from that of ovariectomized females, would be invaluable in determining the basic hormonal stimulus for this type of neoplasia. With the advent of radioimmunoassay techniques for measuring protein hormones, this approach now awaits development.

4. Other Chemicals

Although DMBA is the most potent chemical initiator of ovarian tumor-igenesis, other polycyclic hydrocarbons have also been found to be active. For example, 3-methylcholanthrene (Biancifiori *et al.*, 1961) has moderate activity and a few ovarian tumors have followed administration of benz(a)pyrene (Mody, 1960). 7,8,12-Trimethylbenz(a) anthracene is comparable in its activity to DMBA (Uematsu and Huggins, 1968).

C. Effects of Irradiation and DMBA on Ovarian Follicles

Most workers using radiation and DMBA-induced ovarian tumors have noted that the most pronounced early effect of either carcinogenic agent is a depletion in the number of follicles and their eventual complete disappearance. Pedersen *et al.* (1969), Krarup (1969), and Pedersen and Krarup (1969) have measured the effects of radiation and DMBA on oocyte survival. Ovaries were painted directly with a solution of DMBA in oil or the compound was administered intraperitoneally or by stomach tube. Oocytes were classified by size: (i) small, 20 μm or less in diameter, and (ii) growing and large, greater than 20 μm in diameter. Counts of the numbers of oocytes were made at intervals after exposure to DMBA or radiation at the age of 21 days. Untreated controls and others whose ovaries were exposed to oil only were counted at the same time intervals. The number of small oocytes in normal mice decreased from initial levels of 3000–4000 at 22 days of age to a few hundred at 12 months (Fig. 3). The corresponding numbers of large and growing oocytes initially decreased with age to the time of puberty, then there was a progressive increase to the age of 4 to 5 months, followed by a gradual decline in older mice. After DMBA treatment, the number of small oocytes remained the same as in the controls for at least 2 weeks. There was then a rapid elimination of small oocytes followed by a more gradual rate of depletion resulting in complete disappearance within 2 or 3 months. After 21 days of age, the numbers of growing and large oocytes in DMBA-treated mice at first decreased at the same rate as in the controls, but continued to decline slowly until complete exhaustion occurred, instead of increasing in numbers at puberty. Krarup (1970b) has correlated the number of subsequent pregnancies with the time of DMBA treatment and shown that the latent period before the onset of

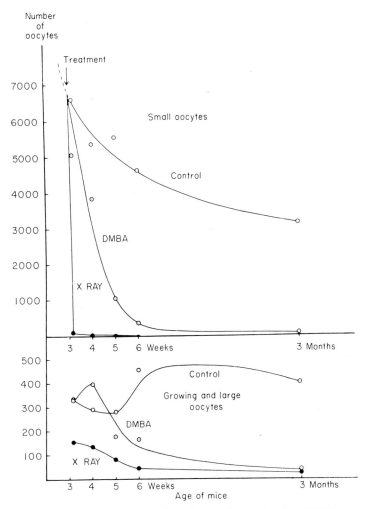

Fig. 3. Number of oocytes in control ovaries and in ovaries after DMBA treatment or X irradiation.

neoplasia in the ovaries is proportional to the time lapse since the last litter was born, that is, the time at which the last oocytes disappeared.

Figure 3 illustrateş that the destruction of small oocytes was even more rapid in irradiated animals than in those treated with DMBA. There was a decline in the number of small oocytes in mice treated at 21 days from the control figures of 6800 to less than 100, 24 hours later. Loss of the remaining small oocytes occurred within 3 weeks. The number of growing and large oocytes was reduced by one-half in the corresponding period. Peters

(1969) has succinctly reviewed the extensive literature on the effects of irradiation on the oocyte population of mouse ovaries at different ages. She has considered the evidence under the headings of the incidence of total sterility, the fertility span, the average number of litters, litter size, and total reproductive capacity. The hormonal changes consequent upon premature aging of the ovary which follows radiation damage constitute the mechanism for the functional and morphological changes which ensue. Furth (1949) showed the reduction in the number of pregnancies which results from these changes in oocyte population.

D. Ovarian Response to Gonadotropins

When the ovary only is irradiated, the changes in ovarian growth and function which occur without further treatment are due to alterations in its response to endogenous gonadotropin. Even when radiation is to the whole body there is no evidence that the ovarian changes are secondary to an initial action on pituitary gonadotropin function. When DMBA is injected into intact female rats (Moon, 1964; Stern *et al.*, 1965) or mice (Jull and Phillips, 1969), however, there is evidence of immediate changes in the function of the pituitary, which is compatible with the concept that the chemical interferes with the secretion of LH for a few days, and consequently there is a release of LTH. It also seems likely that the weight loss, which is common to mice and rats after administration of polycyclic hydrocarbons (Haddow *et al.*, 1937; Flaks, 1948; Hipkin, 1966; Jull and Phillips, 1969), is due to inhibition of LTH secretion by the pituitary.

Irradiation (Essenberg, 1949; VanEck-Vermande and Freud, 1949; Mandl and Zuckerman, 1956) and DMBA (Jull and Phillips, 1969) cause a reduction in the response of the mouse ovary to exogenous gonadotropins. In the experiments with DMBA there was a decrease in the ovarian weight in most strains of mice within 3 days after DMBA administration; rat ovaries were not similarly affected. The weight gain in response to chorionic or pregnant mares' serum gonadotropin (PMSG) was eliminated by DMBA in immature mice and the chemical caused a reduction of this response in most mature mice. Although immature ovaries of Sprague-Dawley (SD) rats responded less to PMSG after DMBA treatment, the ovaries of mature SD and immature and mature Fischer rats responded as well after DMBA treatment as did untreated controls. Unlike mice, the unimpaired response in rats correlates with the facts that rat follicles are not destroyed by DMBA, nor are ovarian tumors a sequel to its administration in this species.

E. Tumorigenesis in Intrasplenic Ovarian Implants

Biskind and Biskind (1944) were the first to show that granulosa cell tumors develop in intrasplenic grafts of ovaries in castrated rats. Subsequent studies confirmed these results in rats (Biskind and Biskind, 1948; Biskind *et al.*, 1950; Peckham and Greene, 1952). More extensive studies in mice (Furth and Sobel, 1946; Li and Gardner, 1947, 1949; Klein, 1952, 1953; Gardner, 1955; Guthrie, 1957) demonstrated a similar form of tumorigenesis.

Biskind had shown in 1941 that if pellets containing estradiol, either free or as the benzoate, were implanted into the spleens of ovariectomized rats, there was complete vaginal anestrus after a short period of estrus. Pellets implanted subcutaneously produced constant estrus, as did the estrogen-containing spleens when they were surgically removed and implanted subcutaneously into the same rats. The role of the liver in estrogen deactivation was thus apparent. It was postulated that estrogens, secreted by the intrasplenic ovary, were carried in the hepatic portal circulation directly to the liver where they were inactivated. The importance of the graft having only hepatic portal circulation was emphasized by the fact that intrapancreatic ovaries also became tumorous (Li and Gardner, 1952) but no tumors occurred if there were adhesions between the ovarian graft and the general circulation. In the absence of circulating estrogen, the secretion of the host's anterior pituitary is uninhibited, resulting in a high level of gonadotropins which stimulate growth and secretion in the intrasplenic ovary (Fig. 4). There is thus a chronic hormonal imbalance (Biskind and Biskind, 1944; Furth, 1967) which results in neoplasia in the target organ. This interpretation is supported by the fact that chronic treatment with estrogen or androgen will completely inhibit tumorigenesis in intrasplenic ovaries in mice, even although this treatment is not started until 92 to 138 days after implantation (Li and Gardner, 1949). However, Biskind *et al.* (1953) found some increase in tumor incidence in intrasplenic implants in castrated rats which were given luteotropin (Squibb) or PMSG three times weekly. This augmentation occurred despite the fact that heterologous gonadotropins have an antigenic potential which tends to reduce their effectiveness in an experimental system such as this.

Spontaneous ovarian tumors are not often observed in old mice in which it might be expected that a rise in gonadotropin would occur subsequent to ovarian atrophy. It was therefore suggested that the ovaries of old mice are less susceptible to tumor formation, but this does not appear to be so (Li and Gardner, 1952; Klein, 1952). The other possibility, that the gonadotropins in old mice are different in quality or quantity, was also disproved

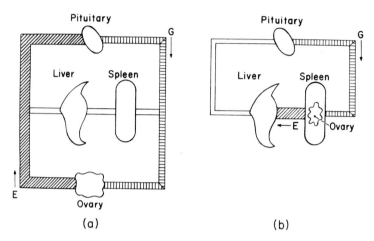

Fig. 4. Mechanism involved in tumorigenesis in the intrasplenically implanted ovary. (a) Normal situation: gonadotropin (G) stimulates estrogen escretion (E) by ovary which feeds back to inhibit pituitary secretion of G. (b) Intrasplenic ovary is stimulated by gonadotropin (G) to secrete estrogen (E) which is directly transported to the liver where it is metabolized to products which cannot inactivate secretion of G.

(Klein, 1953) when it was shown that transplants of ovaries from young mice to the spleen of castrated recipients 498 to 575 days old developed the same incidence of tumors as did intrasplenic transplants into mice aged 40–63 days (see also Volume II, Chapter 5).

In view of the high incidence of spontaneous ovarian tumors in C3H mice over 2½ years of age, it seems likely that the mouse ovary retains significant endocrine function for longer than realized. Another factor which influences the level of gonadotropins in old females of all species is the capacity of the adrenal to secrete significant amounts of sex hormones and so depress the amounts of endogenous gonadotropins. The relationship of adrenal secretion to the incidence of cancer in the ovary and other hormone-dependent tissues is certainly important and awaits clarification.

It would be expected that the retention of a normal ovary would inhibit the secretion of the high levels of gonadotropins necessary for neoplasia in an intrasplenic ovary and this has been proved (Biskind and Biskind, 1948; Biskind *et al.*, 1950). In unilaterally castrated rats, an intrasplenic ovarian graft remained atrophic until the ovary in the systemic circulation was removed. A similar inhibition by normal ovarian tissue of tumorigenesis in irradiated or DMBA-treated ovaries underlines the dependence on elevated levels of gonadotropins of all forms of granulosa cell neoplasia. Alterations in thyroid function have also been shown to affect the development of tumors in intrasplenic ovarian grafts (Miller and Gardner, 1954). Feeding of

0.2% thyroid powder in the diet reduced the tumor incidence to 0/3 in castrated males and 1/17 in ovariectomized females compared to 3/4 and 11/17 in the respective control groups. Thiouracil feeding and limitation of the food intake affected the tumor incidence slightly but not significantly. Inhibition of the hyperplasia of the adrenal cortex which normally follows castration occurred in those groups of mice in which there was inhibition of ovarian tumorigenesis. This finding suggests that the adrenal hyperplasia in castrates is due to elevated levels of gonadotropins, not of ACTH. Mice which had intrasplenic ovarian tumors frequently showed evidence of high systemic levels of ovarian hormones, indicating that the capacity of the liver to inactivate these steroids was exceeded.

In a detailed study of transplantable luteomas (induced intrasplenically) in mice (Furth and Sobel, 1947), progestin activity of the tumor secretion was shown by the repair of castration atrophy in males, and by uterine hyperplasia, absence of corpora lutea in the ovaries, atrophy of the vagina, and enlargement of the clitoris in intact females. All tumor-bearing mice showed excessive weight gain and obesity. The morphological similarity between the luteomas and adrenal cells was emphasized by the authors, and the possibility was raised that there could be androgen as well as progestin secretion. Transformation of luteomas into granulosa cell tumors did not occur and luteinization of granulosa cell tumors was not observed.

The Influence of Parabiotic Union

Hormonal imbalance, in which there is secretion of pituitary gonadotropins uninhibited by the feedback of ovarian hormones, can be achieved by the parabiotic union of an intact female with a castrated male or female partner (Fig. 5a). It was found in practice (Mühlbock, 1953) that, with hybrid 020 × dba or C57BL × dba mice, a granulosa cell tumor of the ovary was obtained in only one pair after 9 months, but no other pairs had actually survived for this time. This was due to the onset of hydronephrosis caused by the high estrogen secretion from the ovaries of the intact partner in response to the elevated gonadotropins.

Transplantation of the ovary to the spleen and subsequent parabiosis to one castrated partner (Fig. 5b) initially reduced the levels of estrogen in the mouse with the intrasplenic ovary. This reduction was sufficient to obviate the toxic effects of the high levels of estrogen. The capacity of the liver to inactivate ovarian hormones was eventually surpassed, however, and estrogenic activity was detected in both of the parabionts. As a result of this small feedback there was some inhibition of gonadotropin secretion and the appearance of ovarian tumors was delayed for 10 to 12 months. When the mouse bearing the intrasplenic ovarian transplant was united with two cas-

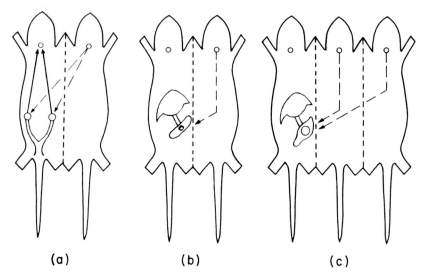

Fig. 5.   Examples of parabiotic union. (a) Normal to castrated rat; (b) ovariectomized rat with intrasplenic ovary coupled to castrate; (c) ovariectomized rat with intrasplenic ovary coupled to two castrates.

trates (Fig. 5c), inactivation of estrogens was complete in the first two mice, and gonadotropin secretion by the third mouse was unimpeded (Mühlbock, 1953). In these circumstances, the time for ovarian tumors to appear was reduced to 5 to 6 months. Thus, in the absence of an initial carcinogenic stimulus, such as chemicals or irradiation, persistent elevation of tropic hormones alone is sufficient to induce tumorigenesis.

In another series of experiments, female mice were irradiated with 200 rads before parabiosis to a castrated male. Ovarian tumors arose in the females of such pairs more frequently and sooner than after irradiation alone without parabiosis. Experiments in which testosterone, estrone, or progesterone were given to female mice bearing an intrasplenic ovary and united to two castrated partners (Fig. 5c) showed that there was no effect of these hormones on ovarian tumor development in these circumstances (Mühlbock *et al.*, 1958). The secretion of gonadotropin was only partly affected by feedback inhibition of the administered steroid hormone in the second mouse and not at all in the third, and any direct effect of the steroids on the development of neoplasia would have been apparent.

When 2-fluorenylacetamide was fed to female rats joined parabiotically to castrated male or female partners, ovarian tumors were induced in 50% of the pairs (Bielschowsky and Hall, 1951). No similar lesions occurred in single females fed this carcinogen. These experiments underline the fact that adequate hormonal stimulation is mandatory after the initial action of the carcinogenic agent.

F. Induction by Progestins

Lipschütz *et al.* (1963) reported the results of experiments in Balb/C mice that were designed to determine the effects of protracted periods of steroid-induced sterility on the subsequent capacity of the mice to breed. The steroid used as the antifertility agent was 19-norprogesterone implanted subcutaneously as a pellet. Unilateral ovarian tumors weighing 4, 34, 140, and 1240 mg were found in 4/18 mice at ages ranging from 532 to 718 days after periods of steroid treatment from 395 to 402 days. In later experiments (Lipschütz *et al.*, 1967b), similar tumors were found in 13/25 mice implanted with pellets of norethindrone (17α-ethinylnortestosterone) and 2/24 mice which had received pellets of norethynodrel (17α-ethinyl-$\Delta^{5,10}$-19-nortestosterone). When progesterone was used as the implanted steroid (Lipschütz *et al.*, 1967a), it was found that the incidence of these tumors was much less and depended on the amount of progesterone administered and the duration of treatment. Many of the tumors were small (< 0.5 mm^2 cross section), and the autonomy of these and many of the tumors of intermediate size (0.5–5 mm^2 cross section) is questionable. There were, however, larger tumors representing more advanced stages of neoplasia in all groups of mice. The tumors seemed to arise at the periphery of the ovary, either from follicles or from hyperplastic germinal epithelium; they were more like the latter in appearance. There was no obvious correlation between tumorigenic and antiluteinizing potency of the different compounds, but, in the case of progesterone, the daily amounts necessary for the suppression of corpus luteum formation and tumor induction were similar. Although the prolonged progestin treatment used by Lipschütz *et al.* (1963, 1967a,b) undoubtedly evoked hormonal imbalance, it does not seem possible that significant amounts of ovarian steroids or pituitary gonadotropins could have been present. The histology of the neoplasms illustrated shows no intermediate thecal cell hyperplasia or luteinization. It is reminiscent, however, of the evolution of the hyperplasia of capsular cells in the adrenal after castration.

G. Histogenesis of Induced Ovarian Tumors

The histological changes which precede and accompany the emergence of tumors in the ovary of the rat and the mouse during various experimental procedures have been described by many workers. The extent of their recorded observations varies considerably and there are minor differences regarding the sequence and nature of the events. It is not within the scope of this review to make a detailed analysis of individual contributions, but it is essential that the major trends of opinion which have evolved should be summarized. It should be borne in mind that few workers have used the same strain of mice or rats, and usually the age at which the ovary has been exposed to a carcinogenic trauma has varied considerably. The histological changes following DMBA administration or irradiation will first be considered. This will be succeeded by the major differences in the evolution of tumors in intrasplenic ovarian transplants. Finally, some comparisons will be made with the early changes occurring in mouse ovaries genetically deficient in oocytes.

In Sections II,G,3 and II,G,4 the term "interstitial cells" is used to describe ovarian cells not within organized follicles or other structures, but nevertheless distinct from an undifferentiated, connective tissue stroma. In those instances in which we have used the term, we are of the opinion that it describes cells derived from the theca interna of atretic follicles (see Volume I, Chapter 4). As this is not necessarily the interpretation of those whose work is reviewed, however, we have used their designation of "interstitial cells."

1. DMBA-Treated Ovaries

The cellular changes which follow exposure of the mouse ovary to DMBA have been described by a number of workers (Howell *et al.*, 1954; Mody, 1960; Biancifiori *et al.*, 1961; Krarup, 1970a). Apart from lysis of corpora lutea, which has been described on one occasion (Wong *et al.*, 1962), DMBA has no effect on the rat ovary at any age. Ovarian tumors followed administration of DMBA to *Mastomys natalensis*, but no effects of DMBA were seen in rabbits or sheep after periods of up to $3\frac{1}{2}$ years (W. J. Jull, unpublished).

The primary result of exposure of the mouse ovary to DMBA is the destruction of small follicles ($< 100~\mu$m in diameter). Whether this follicular destruction is initiated by the death of the oocyte, or whether oocyte death follows alteration in the function of the follicles is not apparent. After death of the oocyte, its remnants may be seen for several weeks within degenerating follicular structures in which the granulosa cells have disappeared, or the follicular remnants may enclose empty spaces. In the C3H strain of mouse the oocytes become calcified and have been found in this condition months after exposure to the carcinogen. The empty follicles tend to shrink and become merged with the stroma. Under normal conditions, the theca interna in the mouse is difficult to identify, and in these atretic follicles it cannot be distinguished with certainty from the stromal cells. Further changes in the ovary are apparently consequent upon the disappearance of any surviving follicles which, judging from continued breeding potential (Krarup, 1970b), function and ovulate normally. When both ovaries are devoid of any normal function, a phase of diffuse luteinization begins and the thecal remnants proliferate and luteinize. Old corpora lutea tend to shrink and become less distinct and the ovary presents an appearance of relatively homogeneous luteinization with occasional cystic or hemorrhagic follicles. At a later stage, tubular proliferations are common in some strains of mouse, especially the C3H; the tubules are possibly derived from downgrowths of the germinal epithelium but probably not from follicular components. These tubular elements sometimes form tumors which are similar to the tubular adenomata described by Gardner (1955). In our experience, the diffusely luteinized ovary proceeds to tumor formation by one of two pathways, depending on the genetic constitution of the mouse. In strains exemplified by CAF1 (Balb/C × A), there is progressive hyperplasia of luteal elements frequently culminating in luteomata, which may be very large, comprised of eosinophilic cells with abundant cytoplasm and large nuclei with diffuse chromatin (Fig. 6). Within these areas of luteal hyperplasia, there may be a progressive change throughout the lesion to smaller cells (Fig. 7) with more compact nuclei. Ultimately these smaller cells are indistinguishable from those of granulosa cell tumors, but within a group of these tumors there are gradations from the purely luteal to the distinctly granulosa cell. Alternatively, foci of "granulosa cells" proliferate within an area of luteal hyperplasia (Fig. 8) to form a granulosa-cell tumor (Figs. 9–11) in the manner described for intrasplenic ovaries by Myhre (1962). The latent period before the appearance of tumors of this type is usually fairly long (>10 months). Ovaries of some mouse strains, such as CNZ hybrids (C3H × NZY), do not seem to undergo an interim period of luteal proliferation. Instead, foci of thecal cell proliferation, often with cyst formation, arise within the atrophic ovary and these proceed to differentiate into thecal, granulosa, or mixed thecal/granulosa cell tumors. In the genesis of these tumors there is often slight or marked luteinization of the thecal cells. A large variety of tumor types, which are basically derived from thecal cells, can thus arise. On serial transplantation most of these tumors become progressively anaplastic and sarcomatous in appearance. Luteinization with increasing maturity of the tumor has never been seen. It is our view that increasing malignancy is invariably the rule and that a luteal structure is associated with more benign lesions.

2. Irradiated Ovaries

The histological events following irradiation of the ovary in mice or rats are basically the same as after DMBA treatment (Drips and Ford, 1932; Furth and Butterworth, 1936; Geist *et al.*, 1939; Guthrie, 1957; Peters, 1969). The small follicles disappear more rapidly and many of the larger

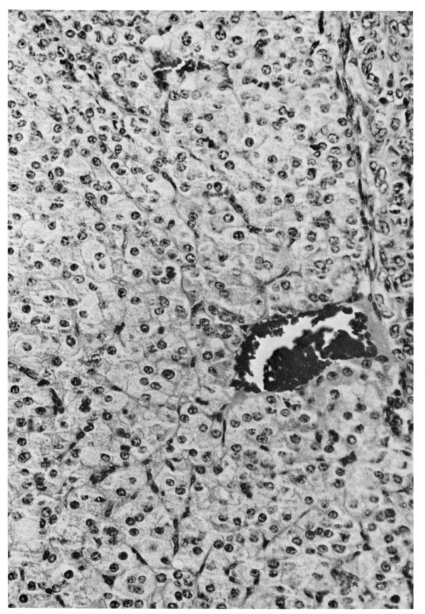

Fig. 6.   Luteoma (×110).

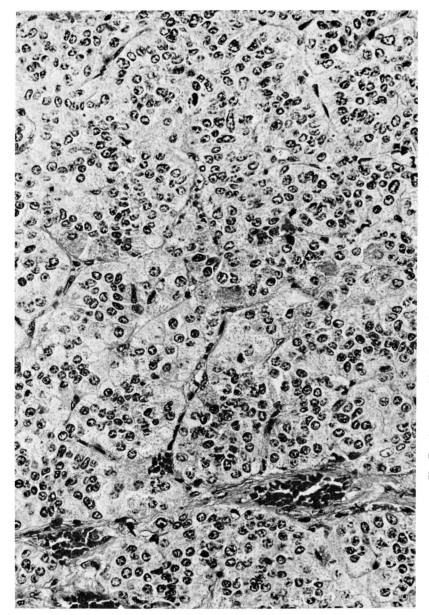

Fig. 7. Luteoma with early transition to granulosa-like cells (×110).

167

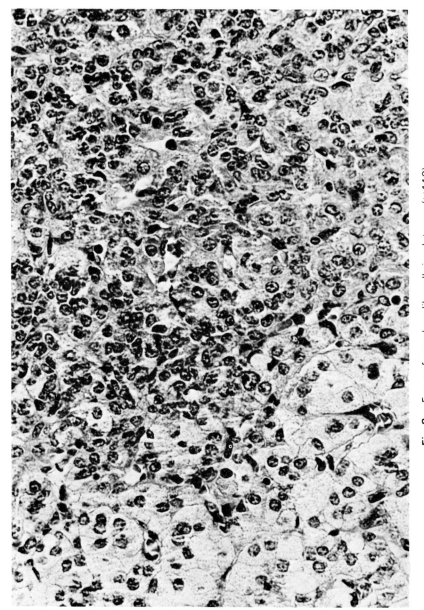

Fig. 8. Focus of granulosa-like cells in a luteoma (×110).

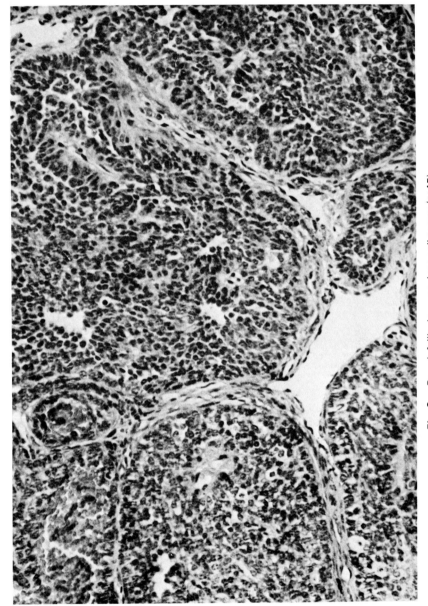

Fig. 9.   Pseudofollicular granulosa cell tumor ($\times$45).

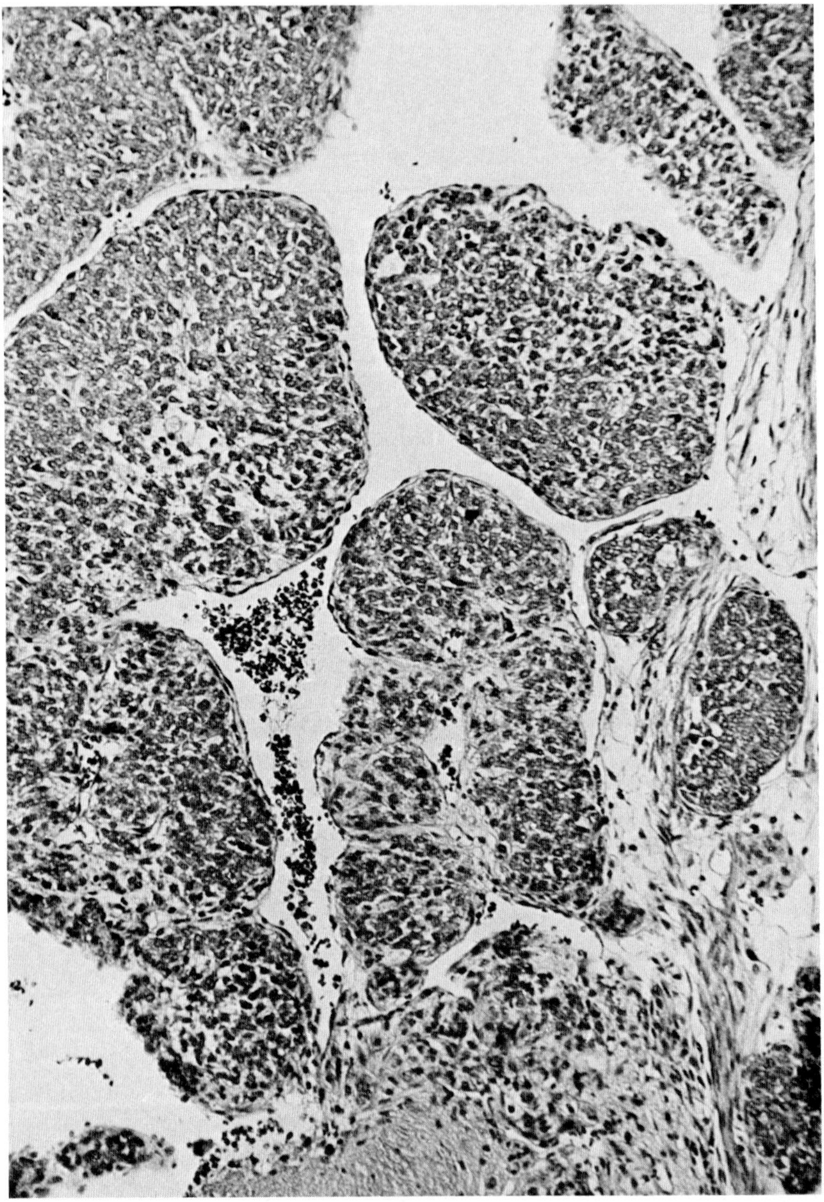

Fig. 10. Papillary granulosa cell tumor ($\times$45).

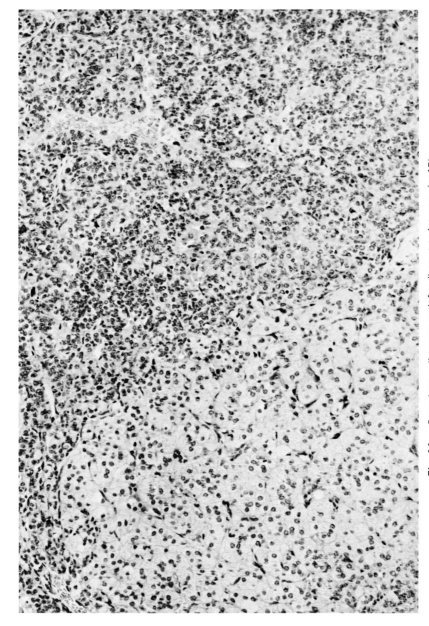

Fig. 11. Granulosa cell tumor on left adjacent to luteoma (×45).

follicles are also destroyed. The granulosa cells degenerate but corpora lutea are maintained. After all the surviving follicles have disappeared the remaining theca interna proliferates. There is then progressive diffuse thecal luteinization. Thecal, luteal, and granulosa cell tumors develop within this pre-neoplastic ovary. Irradiated ovaries differ from those treated with DMBA in the degree of response of the germinal epithelium. Several months after irradiation, the surface epithelium becomes columnar and thickened. Folds are formed, possibly due to proliferation of cells in the cortical zone, and downgrowth of germinal epithelium occurs to form numerous ramifying crypts. Extension of these downgrowths form irregular tubular masses occupying the whole of the ovary. There appears to be considerable similarity between this stage of tubular adenoma development and the lesions which occur more rarely in DMBA-treated ovaries, but the intermediate stages seem to differ. The fact that this type of neoplasia has not been seen commonly after DMBA treatment may reflect differences in the genetic constitution of the mice used rather than differences in the effects of the agent.

3. Intrasplenic Ovaries

A detailed study was made by Guthrie (1957) of the fate of mouse ovaries of the MA_f/Sp strain implanted into the spleens of ovariectomized mice at the age of 4 days. The transplants were vascularized within 3 days, and at the end of the first week of the experiment the number of follicles was maintained, regression being noted only in an occasional oocyte. Atrophy of the germinal epithelium occurred where it was in contact with the spleen. In the following weeks there was a reduction in the number of follicles, compared with those in normally situated ovaries of the same age, although there was an increase in size of some of the remaining follicles. By the end of the fourth week in the spleen, premature luteinization of many follicles occurred accompanied by the formation of hemorrhagic cysts. In other degenerating follicles, the granulosa became scanty and the follicle formed a cyst which progressively enlarged with a cellular rim of decreasing thickness. Concurrently there was proliferation of interstitial cells and an invasion of degenerating follicles by fibroblasts. The size of the intrasplenic ovary increased rapidly within the next few weeks, and where it broke through the splenic capsule there was a burst of cystic proliferation outside the confines of that tissue. Between 6 and 10 weeks there was growth and recession of many new follicles with frequent luteinization and cyst development. The lining cells grew out into the cysts to form arborizing partitions within them. Projections from the cysts extended into the surrounding ovarian tissue creating finer and more complex tubular systems, around which there was increased proliferation of interstitial tissue. Interstitial cells also invaded and proliferated within the cystic spaces. As the density of cells increased, the lumina of the cysts and tubules were reduced or obliterated with the formation of compact masses of tissue. Such aggregates of tubular and interstitial tissue were termed "tubular adenomas" and formed the foci from which various morphological types of "granulosa cell" tumors emerged. Hyperplastic and neoplastic changes did not take place evenly throughout the graft which was usually a composite of many types of cellular proliferation.

The tubular adenomas described by Guthrie (1957) do not seem to be the same lesion as that which occurs in irradiated ovaries. In the latter case, a derivation from ingrowths of the germinal epithelium is probable. There is no parallel in irradiated or DMBA-treated ovaries for the cystic degeneration which occurs in ovaries implanted intrasplenically. Differences due to species and strain are, however, qualifications of paramount importance in assessing the variations in the histogenesis of tumors induced by means of a variety of agents.

4. Ovaries Congenitally Deficient in Oocytes

Genes of the W-series in the mouse have well defined effects on viability, blood count, pigmentation, and fertility (Coulombre and Russell, 1954; Mintz and Russell, 1957). The reason for the impaired fertility was shown to be due to a congenital deficiency of germ cells in both males and females. Between the ninth and twelfth day of intrauterine life, the number of germ cells normally increases by a factor of between ten and thirty during their migration from the primitive streak to the germinal ridge. In sterile mice with the WW genotype, however, migration takes place without this increase in numbers.

The early development of bilateral ovarian tubular adenomas has been described in C57BL/6J-W^v/W^v females (Russell and Fekete, 1958) and recently in female mice of the hybrid (C57BL/6J $\times$ C3H/HeJ)F_1-W^x/W^v (Murphy, 1972). In the latter report, the histological changes occurring between 1 day and 22 weeks of age in the oocyte-deficient mouse ovaries were compared with the normal sequence of changes found in heterozygous mice of the same derivation. The predominant tissue at birth in the deficient ovaries was poorly defined cords of pregranulosa cells. There were very few follicles and only a small number of these underwent growth, ovulation, and corpus luteum formation. These ovaries were, therefore, virtually deficient in normal follicles from the time of birth. Thickening and tubular ingrowths of the germinal epithelium occurred early and were accompanied by marked hyperplasia of the interstitial cells. Extensions of the rete ovarii merged with the downgrowths of the germinal epithelium. The granulosa cell cords originally present became infiltrated with stromal cells, compressed, and were eventually indistinguishable from interstitial cells. Finally there was a mixed tubular and interstitial cell adenoma which occupied all of the ovary. These complex tubular adenomas can be grafted into $W^x W^v$ females and ovariectomized but not intact normal females carrying WW genes. After a long latent period in these situations granulosa cell tumors develop.

The intermediate stage of tubular adenoma described in these mice is comparable with that in intrasplenic ovarian transplants. There is the major difference, however, that in the present case the tubular proliferation is clearly seen to be from germinal epithelium, whereas, in the case of intrasplenic tumors, the tubular proliferations are from the walls of cystic follicles (Guthrie, 1957). The origin of the interstitial cells which become hyperplastic in the intrasplenic and genetically deficient mice is not known. They could come from the theca interna, corpora lutea, rete ovarii, from the stroma, or from downgrowths of the germinal epithelium (Murphy, 1971). In contrast, the corresponding cells in DMBA-treated and irradiated ovaries almost certainly arise in the theca interna of atretic follicles.

It was demonstrated (Russell and Russell, 1948) that replacement of the host ovaries in $W^v W^v$ females with fertile ovaries carrying the normal genes at this locus resulted in undisturbed development of the grafts followed by ovulation, pregnancy, and parturition. It is clear, therefore, that the pituitary function of $W^v W^v$ mice is compatible with normal ovarian function. The ovarian disturbances occurring in the deficient mice are, therefore, due to intrinsic ovarian defects, not to altered pituitary stimulation.

III. COMPARISON OF EXPERIMENTAL TUMORS IN ANIMALS WITH SPONTANEOUS TUMORS IN MAN

The ovarian tumors induced in rats and mice have been intensively investigated, with regard to function and histogenesis, in a manner which is not possible with human material. In man, in which every individual is genetically unique, the ovarian neoplasia is of unknown etiology, the age at presentation cannot be related to that of mice or rats, tumors occur unpredictably in a large population, they are usually observed only at a late stage of development, and induction or manipulation of the lesions is impossible.

In rodent ovaries there are at least two classes of tumors. The so-called granulosa cell tumors, theca cell tumors, and luteomas are all probably of thecal origin. As considered earlier, gradations in the degree of differentiation between these cell types, the coexistence of more than one cell type, and the development of degenerative, hemorrhagic and cystic areas leads to a profusion of histological appearances which might be interpreted as evidence of various cellular origins. Pituitary gonadotropin stimulation plays a dominant role in the induction and early growth of this class of neoplasms, although they attain autonomy and metastasize in advanced cases. These tumors usually secrete steroid hormones initially, but this capacity is lost with increasing anaplasia after serial transplantation. In man, functional ovarian tumors with a structure comparable with that in rodents comprise only a small number of the total. It is quite possible, however, that human ovarian tumors with no obvious hormonal activity have progressed beyond this stage by the time of presentation. Alternatively, hormonal activity might be detected in more cases with intensive investigation.

Many of the tubular adenomas described in rats and mice, and those tumors following prolonged progestin treatment of mice (Section II,F), could originate as downgrowths of the germinal epithelium. They comprise only a small proportion of the experimental ovarian tumors, however, and in individual instances the origin cannot be identified with certainty. There is at present no evidence that they have secretory functions, nor is it known whether they have any stage of dependence on gonadotropin stimulation. Experimental data regarding them are relatively scanty and, to be useful, biochemical investigations would have to be closely related to anatomical criteria. Such investigations have not yet been done. Unlike the situation in experimental animals, the majority of human ovarian tumors are nonfunctional and are considered to be derived from the germinal epithelium.

The correlations of ovarian tumor etiology in man and animals is a potentially rewarding field. The complexity of the neoplastic and degenerative lesions which occur in the ovary appears to be greater than in other organs. However, the difficulties of interpretation and investigation in this complex of tissues are not necessarily greater because they are obvious. It is probably a fallacious assumption that other systems, such as liver or skin, are less complex and therefore more easily analyzed.

REFERENCES

Allibone, E. C., and Collins, D. H. (1951). Symptomatic haemolytic anaemia associated with ovarian teratoma in a child. *J. Clin. Pathol.* **4,** 412–420.

Anderson, C., Cameron, H. M., Neville, A. M., and Simpson, H. W. (1967). Mixed mesodermal tumours of the ovary. *J. Pathol. Bacteriol.* **93,** 301–307.

Anderson, L. J., and Sandison, A. T. (1969). Tumours of the female genitalia in cattle, sheep and pigs found in a British abattoir survey. *J. Comp. Pathol.* **79,** 53–63.

Andrews, J. (1971). Streak gonads and the Y chromosome. *J. Obstet. Gynaecol. Br. Commonw.* **78**, 448–457.

Asadourian, L. A., and Taylor, H. B. (1969). Dysgerminoma—an analysis of 105 cases. *Obstet. Gynecol.* **33**, 370–379.

Atkin, N. B. (1971). Modal DNA value and chromosome number in ovarian neoplasia. *Cancer (Philadelphia)* **27**, 1064–1073.

Awadhiya, R. P., and Jain, S. K. (1967). Studies on the pathology of neoplasms in animals. 1. Ovarian tumours in fowls. *Indian Vet. J.* **44**, 917–919.

Azzopardi, J. G., and Hou, L. T. (1965). Intestinal metaplasia with argentaffin cells in cervical adenocarcinoma. *J. Pathol. Bacteriol.* **90**, 686–690.

Bali, T., and Furth, J. (1949). Morphological and biological characteristics of X-ray induced transplantable ovarian tumors. *Cancer Res.* **9**, 449–472.

Barzilai, G. (1943). "Atlas of Ovarian Tumors." Grune & Stratton, New York.

Benedict, W. F., Rosen, W. G., Brown, C. P., and Porter, I. H. (1969). Chromosomal aberrations in an ovarian cystadenoma. *Lancet* **2**, 640.

Berenblum, I., and Shubik, P. (1947). A new quantitative approach to the study of the stages of chemical carcinogenesis in the mouse's skin. *Br. J. Cancer* **1**, 383–390.

Biancifiori, C., Bonser, G. M., and Caschera, F. (1961). Ovarian and mammary tumors in intact C3Hb virgin mice following a limited dose of four carcinogenic chemicals. *Br. J. Cancer* **15**, 270–283.

Bielschowsky, F., and Hall, W. H. (1951). Carcinogenesis in parabiotic rats. Tumours of the ovary induced by acetylaminofluorene in intact females joined to gonadectomised litter-mates and the reaction of their pituitaries to endogenous oestrogens. *Br. J. Cancer* **5**, 331–344.

Biskind, G. R. (1941). The inactivation of estradiol and estradiol benzoate in castrate female rats. *Endocrinology* **28**, 894–896.

Biskind, G. R., and Biskind, M. S. (1948). Atrophy of ovaries transplanted to the spleen in unilaterally castrated rats; proliferative changes following subsequent removal of intact ovary. *Science* **108**, 137–138.

Biskind, G. R., Kordan, B., and Biskind, M. S. (1950). Ovary transplanted to spleen in rats: the effect of unilateral castration, pregnancy and subsequent castration. *Cancer Res.* **10**, 309–318.

Biskind, G. R., Bernstein, D. E., and Gospe, S. M. (1953). The effect of exogenous gonadotrophins on the development of experimental ovarian tumors in rats. *Cancer Res.* **13**, 216–220.

Biskind, M. S., and Biskind, G. R. (1944). Development of tumors in the rat ovary after transplantation into the spleen. *Proc. Soc. Exp. Biol. Med.* **55**, 176–179.

Bloom, F. (1954). Quoted by Cotchin, E. (1961). "Pathology of the Dog and Cat," p. 400. Am. Vet. Publ., Evanston, Illinois.

Bowey, C., and Spriggs, I. A. (1967). Chromosomes of human endometrium. *J. Med. Genet.* **4**, 91–95.

Brandly, P. J., and Migaki, G. (1963). Types of tumors found by federal meat inspectors in an eight-year survey. *Ann. N.Y. Acad. Sci.* **108**, 872–879.

Bruk, J., Dancaster, C. P., and Jackson, W. P. U. (1960). Granulosa-cell tumours causing precocious puberty. Oestrogen fractionations in two patients. *Br. Med. J.* **2**, 26–28.

Buergelt, C. D. (1968). Dysgerminomas in two dogs. *J. Am. Vet. Med. Assoc.* **153**, 553–555.

Burrows, H. (1943). The nomenclature of hormone-producing tumours of the ovary. *J. Obstet. Gynaecol. Br. Emp.* **50**, 430–432.

Campbell, J. G. (1951). Some unusual gonadal tumours in the fowl. *Br. J. Cancer* **5,** 69–82.

Carter, R. L. (1968). Pathology of ovarian neoplasms in rats and mice. *Eur. J. Cancer* **3,** 537–543.

Chesterman, F. C., and Pomerance, A. (1965). Spontaneous neoplasms in ferrets and polecats. *J. Pathol. Bacteriol.* **89,** 529–533.

Cordes, D. O. (1969). Equine granulosa tumours. *Vet. Rec.* **85,** 186–188.

Cotchin, E. (1960). Tumours of farm animals: a survey of tumours examined at the Royal Veterinary College, London, during 1950–1960. *Vet. Rec.* **72,** 816–822.

Cotchin, E. (1961). Canine ovarian neoplasms. *Res. Vet. Sci.* **2,** 133–142.

Coulombre, J. L., and Russell, E. S. (1954). Analysis of the pleiotropism at the W-locus in the mouse: effects of *W* and *W* substitution upon postnatal development of germ cells. *J. Exp. Zool.* **126,** 277–296.

Czernobilsky, B., Silverman, B. B., and Enterline, H. T. (1970). Clear-cell carcinoma of the ovary. A clinicopathologic analysis of pure and mixed forms and comparison with endometrial carcinoma. *Cancer (Philadelphia)* **25,** 762–772.

Davis, C. L., Leeper, R. B., and Shelton, J. E. (1933). Neoplasms encountered in federally inspected establishments in Denver, Colo. *J. Am. Vet. Med. Assoc.* **83,** 229.

Dehner, L. P., Norris, H. J., Garner, F. M., and Taylor, H. B. (1970). Comparative pathology of ovarian neoplasms. III. Germ cell tumours of canine, bovine, feline, rodent and human species. *J. Comp. Pathol.* **80,** 229–306.

Deringer, M. K., Lorenz, E., and Uphoff, D. E. (1954/55). Fertility and tumor development in (C57L×A) F_1 hybrid mice receiving X-radiation to ovaries only, to whole body, and to whole body with ovaries shielded. *J. Natl. Cancer Inst.* **15,** 931–941.

Dow, C. (1960). Ovarian abnormalities in the bitch. *J. Comp. Pathol.* **70,** 59–69.

Drips, D. G., and Ford, F. A. (1932). The study of the effects of roentgen rays on the estrual cycle and the ovaries of the white rat. *Surg. Gynecol. Obstet.* **55,** 596–606.

Dunnihoo, D. R., Grieme, D. L., and Woolf, R. B. (1966). Hilar-cell tumors of the ovary. Report of 2 new cases and a review of the world literature. *Obstet. Gynecol.* **27,** 703–713.

Engelbreth-Holm, J., and Lefevre, H. (1941). Acceleration of the development of leukaemias and mammary carcinomas in mice by 9:10-dimethyl-1:2-benzanthracene. *Cancer Res.* **1,** 102–108.

Essenberg, J. M. (1949). Response of germ cells to gonadotrophic hormones in X-ray injured ovaries of young white mice. *West. J. Surg., Obstet. Gynecol.* **57,** 61–66.

Federer, O. (1958). "Über hormonal wirksame Ovarialblastome bei Hund und Katze." Inaugural-dissertation zur Erlangung der Doktor würde der Veterinär-medizinischen Fakultät der Universität Bern.

Finckh, E. S., and Bolliger, A. (1963). Serous cystadenomata of the ovary in the koala. *J. Pathol. Bacteriol.* **85,** 526–528.

Flaks, J. (1948). The inhibitory influence of 20-methylcholanthrene on the action of chorionic gonadotrophin on the ovaries of immature mice. *Br. J. Cancer* **2,** 395–401.

Fortner, J. G. (1957). Spontaneous tumors including gastrointestinal neoplasms in the Syrian hamster. *Cancer (Philadelphia)* **10,** 1153–1156.

Fox, H. (1968). Ovarian tumors—histogenesis and systemic effects. *Calif. Med.* **109,** 295–300.

Fox, H., Kazzaz, B., and Langley, F. A. (1964). Argyrophil and argentaffin cells in the female genital tract and in ovarian mucinous cysts. *J. Pathol. Bacteriol.* **88,** 479–488.

Fracarro, M., Mannini, A., Tiepolo, L., Gerli, M., and Zara, C. (1968). Karyotypic clonal evolution in a cystic adenoma of the ovary. *Lancet* **1,** 613–614.

Frost, R. C. (1963). Observations concerning ovarian and related conditions in bitches kept as domestic pets. *Vet. Rec.* **75,** 653–654.

Furth, J. (1949). Relationship of pregnancies to induction of ovarian tumors by X-rays. *Proc. Soc. Exp. Biol. Med.* **71,** 274–277.

Furth, J. (1967). Hormones and neoplasia. *Cancer Aging, Thule Int. Symp., 2nd, 1967.*

Furth, J., and Butterworth, J. S. (1936). Neoplastic diseases occurring among mice subjected to general irradiation with X-rays. *Am. J. Cancer* **28,** 66–95.

Furth, J., and Furth, O. B. (1936). Neoplastic diseases produced in mice by general irradiation with X-rays. *Am. J. Cancer* **28,** 54–65.

Furth, J., and Sobel, H. (1946). Neoplastic transformation of granulosa cells in grafts of normal ovaries into spleens of gonadectomised mice. *J. Natl. Cancer Inst.* **8,** 7–16.

Furth, J. and Sobel, H. (1947). Transplantable luteomata in mice and associated secondary changes. *Cancer Res.* **7,** 246–262.

Gardner, W. S. (1938). Normal and pathological developments from the cells lining the Graafian follicle. *Surg., Gynecol. Obstet.* **67,** 455–466.

Gardner, W. U. (1955). Development and growth of tumors in ovaries transplanted into the spleen. *Cancer Res.* **15,** 109–117.

Geist, S. H., Gaines, J. A., and Pollack, A. D. (1939). Experimental biologically active tumors in mice. Histogenesis and relationship to similar human ovarian tumors. *Am. J. Obstet. Gynecol.* **38,** 786–797.

Gerli, M., Zara, C., and Mannini, A. (1968). Analisi cromosomica in un caso di cistoadenoma papillifero dell'ovaio. *Minerva Ginecol.* **20,** 1555–1557.

Goldzieher, J. W. (1968). The interplay of adrenocortical and ovarian function. *In* "Proceedings of the Second Annual Symposium on the Physiology and Pathology of Human Reproduction" (H. C. Mack, ed.), pp. 106–129. Thomas, Springfield, Illinois.

Gon, F., Edelstein, J., and Griffiths, S. B. (1965). An immune reaction in man against dysgerminomas? *Lancet* **1,** 710–711.

Gray, L. A., and Barnes, M. L. (1966). Relation of endometriosis to carcinoma of the ovary. Report of seven cases and literature review. *Ann. Surg.* **163,** 713–724.

Guthrie, M. J. (1957). Tumorigenesis in intrasplenic ovaries in mice. *Cancer (Philadelphia)* **10,** 190–203.

Guthrie, M. J. (1958). Tumorigenesis in ovaries of mice after X-irradiation. *Cancer (Philadelphia)* **11,** 1226–1235.

Haddow, A., Scott, C. M., and Scott, J. D. (1937). The influence of certain carcinogens and other hydrocarbons on body growth in the rat. *Proc. R. Soc. London, Ser. B* **122,** 477–507.

Harvey, W. F., Dawson, E. K., and Innes, J. R. M. (1940). "Debatable Tumours in Human and Animal Pathology." Oliver & Boyd, Edinburgh.

Hipkin, J. L. (1966). The lack of effect of 7,12-dimethylbenz(a)anthracene on the endocrine function of the gonads. *Cancer Res.* **26,** 89–91.

Horn, H. A., and Stewart, H. L. (1952–1953). A review of some spontaneous tumors in noninbred mice. *J. Natl. Cancer Inst.* **13**, 591–603.

Howell, J. S., Marchant, J., and Orr, J. W. (1954). The induction of ovarian tumours in mice with 9:10-dimethyl-1:2-benzanthracene. *Br. J. Cancer* **8**, 635–646.

Hughesdon, P. E. (1958). The structure and origin of theco-granulosal tumours. *J. Obstet. Gynaecol. Br. Emp.* **65**, 540–552.

Hughesdon, P. E. (1959). Structure, origin and histological relations of dysgerminoma. *J. Obstet. Gynaecol. Br. Emp.* **66**, 566–576.

Hughesdon, P. E., and Kumarasamy, T. (1970). Mixed germ cell tumours (gonadoblastomas) in normal and dysgenetic gonads. *Virchows Arch. A* **349**, 258–280.

Jabara, A. G. (1959). Canine ovarian tumours following stilboestrol administration. *Aust. J. Exp. Biol. Med. Sci.* **37**, 549–565.

Jabara, A. G. (1962). Induction of canine ovarian tumours by diethylstilboestrol and progesterone. *Aust. J. Exp. Biol. Med. Sci.* **40**, 139–152.

Jain, S. K., Singh, D. K., and Rao, U. R. K. (1970). Granulosa cell tumour in a guinea pig. *Indian Vet. J.* **47**, 594.

Jones, T. C., and Gilmore, C. E. (1963). The ovaries in animals. *In* "The Ovary" (H. G. Grady and D. E. Smith, eds.), *Int. Acad. Pathol. Monogr.*, pp. 255–265. Williams & Wilkins, Baltimore, Maryland.

Jull, J. W. (1969a). Mechanism of induction of ovarian tumors in the mouse by 7,12-dimethylbenz(a)anthracene. V. Effect of metabolic inhibitors. *J. Natl. Cancer Inst.* **42**, 961–966.

Jull, J. W. (1969b). Mechanism of induction of ovarian tumors in the mouse by 7,12-dimethylbenz(a)anthracene. VI. Effect of normal ovarian tissue on tumor development. *J. Natl. Cancer Inst.* **42**, 967–972.

Jull, J. W., and Jellinck, P. H. (1968). Mechanism of induction of ovarian tumors in the mouse by 7,12-dimethylbenz(a)anthracene. *J. Natl. Cancer Inst.* **40**, 707–712.

Jull, J. W., and Phillips, A. J. (1969). The effects of 7,12-dimethylbenz(a)anthracene on the ovarian response of mice and rats to gonadotrophins. *Cancer Res.* **29**, 1977–1987.

Jull, J. W., and Russell, A. (1970). Mechanism of induction of ovarian tumors in the mouse by 7,12-dimethylbenz(a)anthracene. VII. Relative activities of parent hydrocarbon and some of its metabolites. *J. Natl. Cancer Inst.* **44**, 841–844.

Jull, J. W., Streeter, D. J., and Sutherland, L. (1966). The mechanism of induction of ovarian tumors in the mouse by 7,12-dimethylbenz(a)anthracene. 1. Effect of steroid hormones and carcinogen concentration *in vivo*. *J. Natl. Cancer Inst.* **37**, 409–420.

Jull, J. W., Hawryluk, A., and Russell, A. (1968). Mechanism of induction of ovarian tumors in the mouse by 7,12-dimethylbenz(a)anthracene. III. Tumor induction in organ culture. *J. Natl. Cancer Inst.* **40**, 687–706.

Kanagawa, H., Kawata, K., Nakao, N., and Sung, W.-K. (1964). A case of granulosa cell tumor of the ovary in a newborn calf. *Jpn. J. Vet. Res.* **12**, 7–11.

Kaplan, H. S. (1950). Influence of ovarian function on incidence of radiation-induced ovarian tumors in mice. *J. Natl. Cancer Inst.* **11**, 125–132.

Klein, M. (1952). Ovarian tumorigenesis following intrasplenic transplantation of ovaries from young weanling, young adult and senile mice. *J. Natl. Cancer Inst.* **12**, 877–881.

Klein, M. (1953). Induction of ovarian neoplasia following intrasplenic transplantation of ovarian grafts to old castrate mice. *J. Natl. Cancer Inst.* **14**, 77–82.

Krarup, T. (1967). 9:10-Dimethyl-1:2-benzanthracene induced ovarian tumours in mice. *Acta Pathol. Microbiol. Scand.* **70**, 241–248.

Krarup, T. (1969). Oocyte destruction and ovarian tumorigenesis after direct application of a chemical carcinogen (9:10-dimethyl-1:2-benzanthracene) to the mouse ovary. *Int. J. Cancer* **4**, 61–75.

Krarup, T. (1970a). Effect of 9:10-dimethyl-1:2-benzanthracene on the mouse ovary. Ovarian tumorigenesis. *Br. J. Cancer* **24**, 168–186.

Krarup, T. (1970b). Effect of 9:10-dimethyl-1:2-benzanthracene on the mouse ovary. Impaired fertility and its relation to ovarian tumourigenesis. *Acta Endocrinol. (Copenhagen)* **64**, 489–507.

Krediet, G. (1931). Vermannelijking door een ovariale tumor bij een paard. *Tijdschr. Diergeneesk.* **58**, 1–14.

Kupperman, H. S. (1969). Steroidogenesis in ovarian malignancy. *Clin. Obstet. Gynecol.* **12**, 1018–1024.

Kuwara, I. (1967). Experimental induction of ovarian tumors in mice treated with single administrations of 7,12-dimethylbenz(a)anthracene and its histopathological observations. *Gann* **58**, 253–266.

Leopold, A. (1967). Tumore a cellule della granulosa dell'ovaio destro in una cavalla. *Nuova Vet.* **43**, 169–177.

Li, M. H., and Gardner, W. U. (1947). Experimental studies on the pathogenesis and histogenesis of ovarian tumors in mice. *Cancer Res.* **7**, 549–566.

Li, M. H., and Gardner, W. U. (1949). Further studies on the pathogenesis of ovarian tumors in mice. *Cancer Res.* **9**, 35–41.

Li, M. H., and Gardner, W. U. (1952). Influence of the age of the host and ovaries on tumorigenesis in intrasplenic and intrapancreatic ovarian grafts. *Cancer Res.* **10**, 162–165.

Lick, L., Kirschbaum, A., and Mixer, H. (1949). Mechanism of induction of ovarian tumors by X-rays. *Cancer Res.* **9**, 532–536.

Lipschütz, A., Iglesias, R., and Salinas, S. (1963). Further studies on the recovery of fertility in mice after protracted steroid-induced sterility. *Endocrinology* **6**, 99–113.

Lipschütz, A., Iglesias, R., Panasevich, V. I., and Salinas, S. (1967a). Granulosa-cell tumours induced in mice by progesterone. *Br. J. Cancer* **21**, 144–152.

Lipschütz, A., Iglesias, R., Panasevich, V. I., and Socorro, S. (1967b). Ovarian tumours and other ovarian changes induced in mice by two 19-norcontraceptives. *Br. J. Cancer* **21**, 153–159.

Lombard, C., and Havet, J. (1962). Premier cas de seminome ovarien chez la truie. *Bull. Acad. Vet. Fr.* **35**, 135–137.

Lucké, B. (1934). A neoplastic disease of the kidney of the frog, *Rana pipiens*. *Am. J. Cancer* **20**, 352–386.

McEntee, K., and Zepp, C. P. (1953). Quoted by Anderson and Sandison (1969). *Proc. World Congr. Fertil. Steril., 1st* Vol. 14, p. 649.

Mandl, A. M., and Zuckerman, S. (1956). Changes in the mouse after X-ray sterilisation. *J. Endocrinol.* **13**, 262–268.

Marchant, J. (1957). Chemical induction of ovarian tumours in mice. *Br. J. Cancer* **11**, 452–464.

Marchant, J. (1959a). Influence of the strain of ovarian grafts in the induction of breast and ovarian tumours in F1 C57BL×1F hybrid mice by 9:10-dimethyl-1:2-benzanthracene. *Br. J. Cancer* **13**, 306–312.

Marchant, J. (1959b). Breast and ovarian tumours in Fl C57BL×lF hybrid mice after reciprocal exchange of ovaries between normal females and females pretreated with 9:10-dimethyl-1:2-benzanthracene. *Acta Unio Int. Cancrum* **15,** 196–199.

Marchant, J. (1960a). The development of ovarian tumours in ovaries grafted from mice pretreated with dimethylbenzanthracene. Inhibition by the presence of normal ovarian tissue. *Br. J. Cancer* **14,** 514–518.

Marchant, J. (1960b). The development of ovarian tumours in ovaries grafted from mice pretreated with dimethylbenzanthracene. The effects of the grafting operation itself and the effects of grafting ovaries from mice at early stages in pretreatment with the carcinogen. *Br. J. Cancer* **14,** 519–523.

Marchant, J. (1961a). The effect of hypophysectomy on the development of ovarian tumours in mice treated with dimethylbenzanthracene. *Br. J. Cancer* **15,** 821–827.

Marchant, J. (1961b). Pathogenesis of ovarian tumours in mice treated with chemical carcinogens. *In* "The Morphological Precursors of Cancer" (L. Severi, ed.), pp. 709–714. Division of Cancer Research, University of Perugia.

Marchant, J., Orr, J. W., and Woodhouse, D. L. (1954). Induction of ovarian tumours with 9:10-dimethyl-1:2-benzanthracene. *Nature (London)* **173,** 307.

Marsh, J. M., Savard, K., Baggett, B., Van Wyk, J. J., and Talbert, L. M. (1962). Oestrogen synthesis in a feminising ovarian granulosa cell tumor. *J. Clin. Endocrinol. Metab.* **22,** 1196–1200.

Marshall, A. H. E., and Dayan, A. D. (1964). An immune reaction in man against seminomas, dysgerminomas, pinealomas and the mediastinal tumours of similar histological appearance. *Lancet* **2,** 1102–1104.

Martin, C. R., Jr., Misenhimer, H. R., and Ramsey, E. M. (1970). Ovarian tumors in rhesus monkeys (*Macaca mulatta*). *Lab. Anim. Care* **20,** 686–692.

Mawdesley-Thomas, L. E., and Cooke, L. (1967). Ovarian agenesis in a rat. *J. Pathol. Bacteriol.* **94,** 467–469.

Meyer, R. (1931). The pathology of some special ovarian tumors and their relation to sex characteristics. *Am. J. Obstet. Gynecol.* **22,** 697–713.

Miller, O. J., and Gardner, W. U. (1954). The role of thyroid function and food intake in experimental ovarian tumorigenesis in mice. *Cancer Res.* **14,** 220–226.

Mintz, B., and Russell, E. S. (1957). Gene-induced embryological modifications of primordial germ cells in the mouse. *J. Exp. Zool.* **134,** 207–237.

Misdorp, W. (1967). Tumours in large domestic animals in the Netherlands. *J. Comp. Pathol.* **77,** 211–216.

Mody, J. (1960). The action of four carcinogenic hydrocarbons on the ovaries of 1F mice and the histogenesis of induced tumours. *Br. J. Cancer* **14,** 256–266.

Monlux, A. W., Anderson, W. A., and Davis, C. L. (1956). A survey of tumors occurring in cattle, sheep and swine. *Am. J. Vet. Res.* **17,** 646–677.

Moon, R. C. (1964). Prolactin-releasing activity of 3-methylcholanthracene. *J. Natl. Cancer Inst.* **32,** 461–467.

Mühlbock, O. (1953). Ovarian tumours in mice in parabiotic union. *Acta Endocrinol. (Copenhagen)* **12,** 105–114.

Mühlbock, O., Van Nie, R., and Bosch, L. (1958). The production of oestrogenic hormones by granulosa cell tumours in mice. *Ciba Found. Colloq. Endocrinol.* [*Proc.*] **12,** 78–93.

Murphy, E. D. (1971). Hyperplastic and early neoplastic changes in the ovaries of mice following genic depletion of germ cells. *J. Natl. Cancer Inst.* (in press).

Myhre, E. (1962). The histogenesis of granulosa cell tumours an autoradiographic study of intrasplenic ovarian transplants in gonadectomised rats. *Acta Unio Int. Cancrum* **18**, 50–53.

Nelson, L. W., Todd, G. C., and Migaki, G. (1967). Ovarian neoplasms in swine. *J. Am. Vet. Med. Assoc.* **151**, 1331–1333.

Norris, H. J., Taylor, H. B., and Garner, F. M. (1968). Equine ovarian granulosa tumours. *Vet. Rec.* **82**, 419.

Norris, H. J., Garner, F. M., and Taylor, H. B. (1969a). Pathology of feline ovarian neoplasms. *J. Pathol. Bacteriol.* **97**, 138–143.

Norris, H. J., Taylor, H. B., and Garner, F. M. (1969b). Comparative pathology of ovarian neoplasms. II. Gonadal stromal tumors of bovine species. *Pathol. Vet.* **6**, 45–58.

Norris, H. J., Garner, F. M., and Taylor, H. B. (1970). Comparative pathology of ovarian neoplasms. IV. Gonadal stromal tumours of canine species. *J. Comp. Pathol.* **80**, 399–405.

O'Shea, J. D., and Jabara, A. G. (1967). The histogenesis of canine ovarian tumours induced by stilboestrol administration. *Pathol. Vet.* **4**, 137–148.

Owen, L. N., and Hall, L. W. (1962). Ascites in a dog due to a metastasis from an adenocarcinoma of the ovary. *Vet. Rec.* **74**, 220–223.

Peckham, B. M., and Greene, R. R. (1952). Experimentally produced granulosa-cell tumors in rats. *Cancer Res.* **12**, 25–29.

Pedersen, T., and Krarup, T. (1969). Cell population kinetics in the mouse ovary after treatment with a chemical carcinogen (DMBA). *Int. J. Cancer* **4**, 495–506.

Pedersen, T., Krarup, T., Peters, H., and Faber, M. (1969). Growth pattern of follicles in mice after X-ray or DMBA induction of ovarian tumours. *Radiat. Induced Cancer, Proc. Symp., 1969* pp. 117–133.

Pedowitz, P., Felmus, L. B., and Mackles, D. (1955). Precocious pseudopuberty due to ovarian tumors. *Obstet. & Gynecol. Surv.* **10**, 633–653.

Peters, H. (1969). The effect of radiation in early life on the morphology and reproductive function of the mouse ovary. *Adv. Reprod. Physiol.* **4**, 149–185.

Pfleiderer, A., Teufel, G., and Braun, R. (1968). Incidence and histochemical investigation of enzymatically active cells in stroma of ovarian tumors. *Am. J. Obstet. Gynecol.* **102**, 997–1003.

Plummer, P. J. (1956). A survey of 636 tumours from domesticated animals. *Can. J. Comp. Med.* **20**, 239–251.

Rashad, M. N., Fathalla, M. F., and Kerr, M. G. (1966). Sex chromatin and chromosome analysis in ovarian teratomas. *Am. J. Obstet. Gynecol.* **96**, 461–465.

Rewell, R. E. (1954). Uterine fibromyomas and bilateral ovarian granulosa-cell tumours in a senile Squirrel monkey, *Saimiri sciurea. J. Pathol. Bacteriol.* **68**, 291–293.

Rewell, R. E. (1958). The histogenesis of ovarian tumours. *J. Obstet. Gynaecol. Br. Emp.* **65**, 718–726.

Rewell, R. E. (1960). "Obstetrical and Gynaecological Pathology," pp. 162–215. Livingstone, Edinburgh.

Rewell, R. E., and Willis, R. A. (1962). Some tumours found in whales. *J. Pathol. Bacteriol.* **61**, 454–456.

Roberts, S. J. (1953). An ovarian tumor in a heifer. *Cornell Vet.* **43**, 531–536.

Ruffolo, E. H., and Withers, R. W. (1966). Argentaffin cells in mucinous cystadenoma of the ovary. *Am. J. Obstet. Gynecol.* **96**, 292–294.

Russell, E. S., and Fekete, E. (1958). Analysis of W-series pleiotropism in the mouse. Effect of the W^vW^v substitution on definitive germ cells and on ovarian tumorigenesis. *J. Natl. Cancer Inst.* **21**, 365–381.

Russell, W. L., and Russell, E. S. (1948). Investigation of the sterility-producing action of the W^v gene in the mouse by means of ovarian transplantation. *Genetics* **33**, 122–123.

Salm, R. (1967). Pure and mixed hilus-cell tumours of ovary. *Ann. R. Coll. Surg. Engl.* **41**, 344–363.

Sastry, S. A., and Tweihaus, M. J. (1964). A study of the animal neoplasms in Kansas State. II. Bovine. *Indian Vet. J.* **41**, 454–459.

Scott, J. S., and Bain, A. D. (1965). An immune reaction in man against dysgerminomas. *Lancet* **1**, 1068–1069.

Scott, J. S., Lumsden, C. E., and Levell, M. J. (1967). Ovarian endocrine activity in association with hormonally inactive neoplasia. *Am. J. Obstet. Gynecol.* **97**, 161–170.

Scully, R. E. (1953). A gonadal tumor related to the dysgerminoma (seminoma) and capable of sex-hormone production. *Cancer* (*Philadelphia*) **6**, 455–463.

Shisa, H., and Nishizuka, Y. (1968). Unilateral development of ovarian tumour in thymectomized Swiss mice following a single injection of 7:12-dimethylbenz(a)anthracene at neonatal stage. *Br. J. Cancer* **23**, 70–76.

Short, R. V., Shorter, D. R., and Linzell, J. L. (1963). Granulosa cell tumour of the ovary in a virgin heifer. *J. Endocrinol.* **27**, 327–332.

Stern, E., Mickey, M. R., and Osvaldo, L. (1965). Lactogenesis in the androgen-sterile rat: augmentation following 7,12-dimethylbenz(a)anthracene. *Nature* (*London*) **206**, 369–371.

Studer, E. (1967). Mucinous adenocarcinoma of the bovine ovary. *J. Am. Vet. Med. Assoc.* **151**, 438–441.

Stünzi, B., and Stünzi, H. (1950). Hermaphroditismus verus unilateralis beim hund mit blastomatöser entartung des rudimentären ovars. *Schweiz. Arch. Tierheilkd.* **92**, 67–82.

Taylor, D. O. N., and Dorn, C. R. (1967). Dysgerminoma in a 20-year-old female German shepherd dog. *Am. J. Vet. Res.* **28**, 587–591.

Taylor, H., Barter, R. H., and Jacobson, C. B. (1966). Neoplasms of dysgenetic gonads. *Am. J. Obstet. Gynecol.* **96**, 816–821.

Teilum, G. (1958). Classification of testicular and ovarian androblastoma and Sertoli cell tumors. *Cancer* (*Philadelphia*) **11**, 769–782.

Teoh, T. B. (1953a). Structure and development of Walthard's nests. *J. Pathol. Bacteriol.* **66**, 433–439.

Teoh, T. B. (1953b). Histogenesis of Brenner tumors of the ovary. *J. Pathol. Bacteriol.* **67**, 441–456.

Teoh, T. B. (1959). Further observations on the histogenesis of minute and small Brenner tumours. *J. Pathol. Bacteriol.* **78**, 145–150.

Teter, J. (1969). Rare gonadal tumors occurring in intersexes and their classification. *Int. J. Gynaecol. & Obstet.* **7**, 183–198.

Theiss, E. A., Ashley, D. J. B., and Mostofi, F. K. (1960). Nuclear sex of testicular tumors and some related ovarian and extragonadal neoplasms. *Cancer* (*Philadelphia*) **13**, 323–327.

Thomson, J. G. (1945). Fatal bronchial asthma showing the asthmatic reaction in an ovarian teratoma. *J. Pathol. Bacteriol.* **57**, 213–219.

Tortora, M. (1967). Chromosome analysis of four ovarian tumors. *Acta Cytol.* **11**, 225–228.

Trotter, A. M. (1911). Malignant diseases in bovines. *J. Comp. Pathol. Ther.* **24,** 1–20.

Uematsu, K., and Huggins, C. (1968). Induction of leukaemia and ovarian tumors in mice by pulse doses of polycyclic aromatic hydrocarbons. *Mol. Pharmacol.* **4,** 427–434.

VanEck-Vermande, G. J., and Freud, J. (1949). Action of gonadotrophins and of estrogens on X-rayed mouse ovaries. *Arch. Int. Pharmacodyn. Ther.* **78,** 67–68.

Weir, B. J. (1971). Some observations on reproduction in the female agouti, *Dasyprocta aguti. J. Reprod. Fertil.* **24,** 203–211.

Weiss, R. P., Richart, R. M., and Okagaki, T. (1969). DNA content of mucinous tumors of the ovary. *Am. J. Obstet. Gynecol.* **103,** 409–424.

Willis, R. A. (1962). Ovarian teratomas in guinea-pigs. *J. Pathol. Bacteriol.* **54,** 237–239.

Willis, R. A. (1967). "Pathology of Tumors," 4th ed., pp. 486–524. Butterworth, London.

Wong, T. W., Warner, N. E., and Yang, N. C. (1962). Acute necrosis of adrenal cortex and corpora lutea induced by 7,12-dimethylbenz(a)-anthracene and its implications in carcinogenesis. *Cancer Res.* **22,** 1053–1057.

Yamauchi, S. (1963). A histological study on ovaries of aged cows. *Jpn. J. Vet. Sci.* **25,** 315–322.

Zaldivar, R. (1967). Incidence of spontaneous neoplasms in beagles. *J. Am. Vet. Med. Assoc.* **151,** 1319–1321.

5

Clinical Manifestations of Disorders of the Human Ovary

P. M. F. Bishop

I. INTRODUCTION

Normal ovarian function depends on complete differentiation of the gonad during embryonic life, the awakening of gonadotropic activity at puberty, the integrative action of the endocrine system, and the absence of endogenous abnormalities affecting the production of its hormones.

Derangements of ovarian function must therefore be considered under the headings of those that arise during embryonic life, those that result from abnormal activity of other parts of the endocrine system, and abnormal behavior of the ovary itself. This may be due to the development of an ovarian tumor which is producing excessive quantities of one or other of the three principal ovarian hormones, namely, estrogen, progestagen, or androgen, or else producing them in an abnormal ratio, one to another.

This chapter deals exclusively with derangements of the human ovary and is mainly concerned with the clinical manifestations of disorders affecting the normal production of the ovarian hormones.

In describing these clinical pictures it is necessary to discuss their differential diagnosis. This could lead to a detailed enumeration of the features of disorders not directly concerned with abnormalities of ovarian function. To avoid this, the various hypothalamic syndromes and some pituitary tumors have been mentioned only briefly.

II. DISORDERS OF EMBRYONIC ORIGIN

The sex of the embryo is determined by the sex chromosome pattern of the zygote; it should be XX for the female and XY for the male. In the earliest stages of fetal development the gonad is "indifferent" but consists of a cortical and medullary portion. If the chromosome pattern is XY, the medulla will develop at the expense of the cortex and will eventually become a testis. In the absence of a Y chromosome, the cortex will obliterate the medulla to become an ovary. During the fetal life of the female, neither the ovary and its secretions nor maternal estrogens play a part in the differentiation of the Müllerian duct system or of the external genitalia. In the male, a chemical substance, which is not testosterone or any other steroid but may be a protein produced by the fetal Sertoli cells (Josso, 1972), suppresses the Müllerian duct system; the Leydig cells of the fetal testis are influenced by maternal (placental) gonadotropins and produce testosterone which furthers the differentiation of the Wolffian ducts and external genitalia in a male direction (Jost, 1953; and Volume 1, Chapters 3 and 8 and Volume II, Chapter 2).

Cases of true agonadism seem to be confined to XY individuals but there is usually some evidence of Müllerian regression. Medullary differentiation into a testis usually commences after the fifth week and, should gonadal resorption occur before that time, then the internal and external genitalia will passively develop as phenotypically female. The Müllerian suppressor exerts its influence during the tenth to twelfth week and, if resorption occurs during that time, Müllerian tract structures will have been suppressed and the internal genitalia will be male, though there may be female or ambiguous external genitalia. The influence of fetal testosterone overlaps the period of Müllerian regression and if testicular resorption commences after the sixteenth week the internal and external genitalia will be phenotypically male.

"Streak" gonads may be found bilaterally in the broad ligaments of a phenotypic female with no secondary sex characters, and the condition is often referred to as gonadal agenesis. The term "gonadal dysgenesis" is preferred because serial sections of the streaks would need to be cut to determine whether they contain gonadal stroma and even primordial follicles, or rete testis remnants. The 45,X chromosome complement (XO chromosome) of this condition may be due to loss of the second X chromosome of the female or the Y chromosome of the male.

This background permits an appreciation of the clinical manifestations arising from various degrees of disorder of the ovary.

A. Ovarian Dysgenesis

The most classical and most striking example of this condition is shown by "Turner's Syndrome" (Turner, 1938) which is typified by short stature (< 5 ft), infantilism (retardation of growth and sexual development), and congenital anomalies such as webbing of the skin of the neck. Some patients may have few, or no, congenital anomalies and be short in stature though usually they are taller than cases of Turner's syndrome; this clinical syndrome is often referred to as "ovarian dysgenesis without neck webbing": ("ovarian dysgenesis" is an anatomical diagnosis). Women who are of normal or tall stature and have no congenital anomalies are considered by Bishop et al. (1960) to be cases of "pure gonadal dysgenesis"; the condition probably occurs in 1 in 60,000 women.

1. The Ovaries and Secondary Sex Characters

Up to at least the third month of fetal life the development of the ovaries in the XO embryo and fetus is normal and the appearance and number of the primordial germ cells are similar to those of any other female (Singh and Carr, 1966). However, following this, the ovaries of individuals not possessing the full XX complement soon become atretic and degeneration of primordial germ cells is rapid. Germ cells have been found in some newborn XO infants but their number is less than 10% of the normal value (Polani, 1969) and by puberty there will be very few if any primordial follicles (Greenblatt et al., 1967). Very rarely have ova been found to persist well past puberty. Hoffenberg et al. (1957) and Stewart (1960) have described chromatin-negative XO individuals with Turner's syndrome who have menstruated over a period of several years. Bahner et al. (1960) described a patient with ovarian dysgenesis without neck webbing, with normal breast and pubic and axillary hair development and a 45,X chromosome complement without evidence of mosaicism, who gave birth to a normal male child at the age of 31 after menstruating regularly since the age of 16; other cases have been subsequently reported.

The majority of cases, however, have only bilateral fibrous streaks lying in the broad ligament at the time when the ovaries are looked for. Secondary sex characters are absent in most cases of Turner's syndrome, ovarian dysgenesis, and pure gonadal dysgenesis. The patients present with primary amenorrhea (see Section III), scanty pubic and axillary hair, immature appearance of the labia and vagina with "infantile" vaginal smears, and complete lack or scanty development of breast tissue. In Turner's syndrome the small nipples and areolae are widely spaced because of the shield-shaped chest. Treatment with estrogen may not only give rise to psychologically desirable withdrawal bleeds but also to variable degrees of breast development. Epiphysial ossification in Turner's syndrome is normal until puberty (van der Werff ten Bosch, 1959) but then becomes retarded so that at the age of 22 the bone age is only 13 or 14 (Acheson and Zampa, 1961). This is unlike the situation in pituitary infantilism in which epiphysial union is retarded from birth and is seldom, if ever, completed. In spite of claims that growth hormone may be present in excessive amounts or conversely deficient, there is no convincing evidence for either and growth hormone administration seems without effect on ultimate height.

Silver and Dodd (1957) and Grossman (1960) found greatly elevated levels of pituitary gonadotropins in the urine of 2½ to 9-year-old children with Turner's syndrome, and menopausal or high normal values of LH were found in urine by Papanicolaou *et al.* (1969).

The condition of 45,X and related ovarian dysgenetic streaks (but not 45,X/46,XY mixed gonadal dysgenesis) *may* carry a slightly increased risk of malignancy. This does not, however, warrant laparotomy and, if the condition is confirmed, gonadectomy. Nevertheless any coincidental laparotomy for other reasons warrants removal of the "gonads" in these cases. The presence of a Y or ?Y in any "woman" constitutes a more serious risk of malignancy (say 1 in 8,000 or 10,000) and demands castration unless otherwise contraindicated (for instance, if the subject was raised as a male and has reasonable testicular endocrine activity). Some of the conditions for which castration is indicated are testicular feminization, XY pure gonadal dysgenesis, and mixed gonadal dysgenesis raised as a female.

2. Height

In the extreme cases of Turner's syndrome with 45,X chromosome patterns, the birth weight and height are usually low and the adult height averages only 140 cm (4 ft 7 in). Patients with less extreme ovarian dysgenesis associated with chromosomal anomalies have an average height of only 145 cm (4 ft 9 in), but all cases of pure gonadal dysgenesis, as at present defined, are of normal height or taller than normal with disproportionately long legs since they have the normal 46,XX karyotype (Polani, 1969).

3. Congenital Anomalies

These may be numerous and result from abnormalities of the sex chromosome patterns although, at present, there is insufficient information to allocate individual anomalies to particular chromosomal deviations. The extent of the somatic anomalies varies considerably. Among the less malformed subjects a 45,X complement occurs in only about 50% whereas mosaicism or structural sex chromosome anomalies are common. These anomalies are rare in Turner's syndrome (3–5%) and are found in only a few pure gonadal dysgenesis cases (as defined above).

Neck webbing was first described by Turner (1938) and the syndrome bearing his name applies only to cases in which this anomaly is present. It can be seen at birth in the form of loose skin at the nape due to hypoplasia of the lymphatics. It is part of the peripheral lymphedema which is conspicuous in infancy in these cases and affects the hands and feet. It is related to malformations of the lymphatic blood vessels. One of the striking anomalies of many aborted 45,X conceptuses is the presence of cystic hygrometa, most often seen as huge lymph-filled sacs at the sides of the neck. This lymphedema of the nape gradually disappears and the neck webbing then consists of triangular folds of skin which stretch from the mastoid to the acromion process. The trapezius is not involved. In some cases there is webbing of the digits and anterior axilla. The hairline at the back of the neck may be M-shaped. It descends to a peak in the midline and extends down the lateral margin of the webbing on each side.

The epicanthic folds are prominent and this exaggerates the flattening of the bridge of the nose and contributes to the characteristic facial features which are instantly recognizable. Underdevelopment of the chin and mandible results in crowding of the teeth and the neck slopes obliquely backward from the chin to the sternal notch. There is an anti-mongoloid slant of the eyes and the ears are low set and protruding with the lobes tending to fuse with the neck.

Pigmented moles are the most invariable of the congenital anomalies; they may be the only ones. Pigmented naevi, café-au-lait patches of the skin, vitiligo, and even von Recklinghausen's neurofibromatosis may be present. Scars on the skin are liable to become keloid and this should be borne in mind by the plastic surgeon repairing a webbed neck.

Bony anomalies are numerous but may include a shield-shaped chest, pectus excavatus, pigeon chest, cubitus valgus, Madelung's deformity of the wrist, short and small fingers and toes, syndactyly, and hallux valgus accompanied by shortening of metacarpals, particularly the fourth and fifth.

Coarctation of the aorta occurs in one-quarter of the patients with Turner's syndrome. This condition is normally more common in males than in females. This is true also for cases of red–green color blindness (see Bishop, 1962), but this abnormality is known to be genetically determined by a mutant allele at one or two color vision loci on the X chromosome. The high frequency of color blindness [like that of the Xg(a-) X-linked blood group antigen] in Turner's syndrome females is related to their generally possessing a single X chromosome, as do normal males. The frequency of recessively determined X-linked genetic characters in 45,Xs is thus similar to the population incidence in normal males. Renal anomalies such as horse-shoe kidney and double ureter are common in Turner's syndrome. Intestinal telangiectases, occasional fusion of cervical vertebrae, lumbosacral spina bifida, and severe scoliosis may occur. In about one-half the cases, there is rarefaction of bone due to thinness of the bony matrix. Osteoporosis may be superimposed later and may cause vertebral collapse. There is an increased incidence of diabetes mellitus, a tendency to Hashimoto's disease, and a familial likelihood of autoantibodies to thyroglobulin. This may be specific for cases in which isochromosomes of the X chromosome are present (see below).

4. Chromosomal Anomalies

In Polani's review (1970) of 450 cases from a number of series of women with one of the major syndromes of dysgenetic ovaries, 62% were Barr-body negative and 51% were shown to have a 45,X chromosome pattern. This occurred in 86% of cases with Turner's syndrome, 45% of cases with ovarian dysgenesis without neck webbing, and in only 4 of 59 (7%) cases of pure gonadal dysgenesis. Though it is the most common chromosome anomaly occurring with an estimated frequency of 0.75% in all conceptions, it has a prenatal mortality of about 97% and among survivors at birth it, therefore, has a frequency of only 1 in about 5000. Structural abnormalities of the X chromosome consist of partial deletion of either the short or long arm of this chromosome, ring chromosome formation (following deletion of both chromosome ends), or formation of isochromosomes of the long or short arm. These occur when division of the chromosome through the centromere takes place horizontally rather than vertically and so each daughter cell contains either two long arms or two short arms. Isochromosomes of the long arm generally occur with mosaicism of an XO cell line together with the line with 46 chromosomes or the normal X and the isochromosome. Isochromosomes for the short arm of the X occur only rarely.

Structural abnormalities occur in rather less than one-quarter of women with dysgenetic ovaries. The majority of these structural abnormalities of one of the X chromosomes (chiefly isochromosomes for the long arm of the X) occur in the ovarian dysgenesis group, and there is an increased disposition to Hashimoto's disease in these cases. If the short arm is missing, the case is likely to develop the classical Turner phenotype; if the long arm is missing the case is likely to belong to the group of pure gonadal dysgenesis. Thus dwarfism and congenital anomalies are due to absence of circumstances controlling these somatic factors normally found on the short arm. Most cells with structurally abnormal X chromosomes are chromatin positive, though the X chromatin mass may be of unusual size.

Most chromatin-positive cases have structurally normal X chromosomes and they are mosaics of the XX/XO type. Such mosaics do not give rise to the more extreme cases of Turner's syndrome. Some XO/XX and XO/XXX and even XO/XX/XXX cases have germ cells in the gonads and may achieve a normal menarche and be fertile. About one in twenty of mosaics show an XO/XY pattern. These have bilateral streaks but tend to overlap with the cases of mixed gonadal dysgenesis, with a streak and testis. These cases are related to true hermaphrodites among whom about 10% have XO/XY sex chromosome mosaicism. There are also mosaic types which have structurally abnormal sex chromosomes and the most common of these is XO/X-iso-X. Another pattern is XO/X deleted X, the deletion occurring on the long or the short arm of the X chromosome. The long arm and the short arm are necessary for normal ovarian maintenance, probably because two X chromosomes must be functional in oocytes if these are to grow and develop normally. However, it seems that the long arm is not important in preventing normal statural growth.

Only 4.1% of Polani's (1969) 461 surveyed cases showed a 46/XX karyotype and 2.2% were 46/XY. All twenty-nine of these cases belonged to the group with pure gonadal dysgenesis and streak gonads.

The triple X anomaly is, of course, abnormal but not strictly a condition of ovarian dysgenesis. It is found in 1 in 1250 newborn girls. Only one-half to three-quarters of them menstruate regularly and the ovaries tend to be small and hypoplastic, with few follicles (Johnston et al., 1961). Though some of these women are fertile, many of them present with secondary amenorrhea and have delayed menarche. The children produced by XXX women generally have normal sex chromosomes. They have no distinctive physical features and unless their sex chromosomes are analyzed and found to be XXX (triple X) they may escape notice as one of the two sex chromosome trisomies, the other being the male trisomy XXY (Klinefelter's syndrome). In both these conditions however, there may be some degree of mental impairment and this is especially true of other women with more than three X chromosomes.

B. Mixed Gonadal Dysgenesis

In 1958, Greenblatt described a phenotypic female with bilateral vestigial streaks containing ova and occasional seminiferous tubules. A rudimentary uterus and tubes were present and the clitoris was not enlarged.
This is the most common type of what Sohval (1963, 1964) subsequently
called "pure" but not "mixed" gonadal dysgenesis. However, Federman
(1967), reviewing forty-two cases, has mentioned other combinations, such
as, an immature inguinal testis and a streak, unilateral gonadal agenesis,
bilateral hypoplastic abdominal gonads with one containing rudimentary
testicular elements, streak and gonadal tumor, or tumor only; these may be
called "mixed gonadal dysgenesis." In all cases but one in which "testes"
were present, they were dysgenetic to some extent. The "ghost" tubules of
Klinefelter's syndrome are not seen but the "Sertoli cell only" picture is
found with clumps of Leydig cells and hyalinization of the tubular basement
membrane. One patient showed spermatogenesis and there are cases in the
literature which were said to be fertile and may have been examples of
mixed gonadal dysgenesis. The streaks usually contained remnants of the
medullary component of the indifferent gonad. Nearly one-half of the
reported postpubertal cases which had been exposed to endogenous
gonadotropin developed gonadal tumors in the streak, the gonad, or
bilaterally. Two patients, one of whom already had a seminoma, developed
endometrial carcinoma. The internal genitalia were chiefly female, indicating lack of the male inducer though some male internal genital structures
were found, particularly on the side on which the testis rather than the
streak was found. Every case had a uterus, sometimes unicornuate, and at
least one Fallopian tube; two-thirds of the cases had bilateral tubes. In most
cases there was also an upper portion of the vagina. A vas deferens was
found on the side of the "testis" in about one-third of the cases and was
very occasionally bilateral. The external genitalia varied from the normal
male, through hypospadias, complete ambiguity with incomplete fusion of
the labioscrotal folds, and from enlargement of the clitoris to that of a
normal female. The external genitalia were ambiguous at birth in more than
one-half the cases and about two-thirds of the cases were reared as females.
The height was short in about one-half and stigmata of Turner's syndrome
were present in about one-third. Of the patients reared as female who had
subsequently reached puberty, about one-half showed some signs of virilism
such as an enlarged clitoris, a deep voice, and hirsutism. A limited number
of cases showed breast development and all of them developed gonadal
tumors. No patient raised as a male developed gynecomastia. Mental retardation was noted in about one-third of the cases in which the mental state
was investigated. Most, if not all of the cases reported have been mosaics

and have had an XO stem line, while one case had a Y chromosome; according to Polani (1970) the most usual sex chromosome pattern is XO/XY and all the cases are chromatin-negative. From the phenotypic characteristics it is suggested that the individual is what Jones and Scott (1958) refer to as "a transition to a male pseudohermaphrodite," in whom the masculinizing substance is absent or deficient so that it cannot suppress the Müllerian tract and the interstitial testicular tissue cannot produce sufficient fetal testosterone to complete male differentiation. At any rate, it is clear that the streak gonad in mixed gonadal dysgenesis had probably differentiated as an ovary the same way the XO gonad does. Its subsequent germ cell atresia is in keeping with the fate of the XO ovary. Thus mixed gonadal dysgenesis is really a type of true hermaphroditism. As Polani (1970) points out, XO/XY is a relatively common pattern in true hermaphroditism (see Section II,C). Wilkins (1965), who refers to the condition as "unilateral male pseudohermaphroditism," mentions other chromosomal patterns such as XO/XY/XX, XO/XY/XXX and XO/X deleted XY, as well as XY and XO, though XO/XY seems to be the most common.

C. True Hermaphroditism

The classification originally suggested by Klebs (1876) of true hermaphroditism and male and female pseudohermaphroditism seems to be less logical today than it was then, for true hermaphroditism is the only type that is due to chromosomal or genetic abnormalities whereas female pseudohermaphroditism is almost always due to congenital adrenal hyperplasia and male pseudohermaphroditism to an inability, of variable degree, to respond to endogenous testosterone. Some authors, therefore, such as Dewhurst and Gordon (1969), now prefer to speak of "hermaphroditism" and "male" and "female intersex."

The condition is said to be "lateral" when there is an ovary on one side and a testis on the other, which occurs in about one-third of the cases, "unilateral" when there is an ovotestis on one side and an ovary or testis on the other, and bilateral when there are ovotestes on both sides or, as described in one case by Clayton et al. (1957), an ovary and a testis on both sides. In some cases the ovarian and testicular elements of an ovotestis are clearly separated by fibrous tissue. In about one-half the cases the left gonad is an ovary as compared with about one-fifth in which the right gonad is an ovary (Polani, 1970). It is interesting that in most birds it is the left ovary that is functional while the right gonad is a rudimentary structure.

In man, one ovary always remains abdominal and the testis usually descends into the scrotum, or labioscrotal folds. An ovotestis may be abdominal or inguinoscrotal. A uterus is always present but it may be unicornuate. There is always some degree of masculinization of the external genitalia and three-quarters of the patients are brought up as males. There is usually some degree of hypospadias, though Wilkins (1965) states that some patients have a normally formed penis and urethra. There may be partial or complete failure of fusion of the labioscrotal folds and in about one-half the cases there is some degree of cryptorchidism and an inguinal hernia. The existence of a vas deferens or a Fallopian tube largely depends on the sex of the gonad, though in bilateral cases there is usually a tube only on each side (Federman, 1967). In unilateral cases involving an ovary and ovotestis there may be a vas deferens as well as bilateral tubes, and in the rarest combination of all, a testis and an ovotestis; the testis is usually scrotal with an adjacent vas deferens and the ovotestis is abdominal with a Fallopian tube. Despite the fact that the majority of patients are reared as boys, 80% develop gynecomastia at puberty and 50% menstruate or have cyclical hematuria. Ovulation actually takes place in 25% of the cases, but spermatogenesis is very rare. No patient has ever been reported to be fertile.

Dewhurst and Gordon (1969) point out that congenital adrenal hyperplasia should first be excluded by estimations of 17-oxosteroids and pregnanetriol. Contrast radiography to determine the arrangement of the internal genitalia and a laparotomy should be undertaken as soon after birth as possible and a decision made at the time of this operation on the sex of rearing. All inappropriate gonadal tissue should be excised even if this means relying on replacement therapy after puberty. If it is decided to rear the child as a male, the uterus should be removed. Important factors in the decision on rearing are the likely eventual size of the penis and the functional potentialities of the external genitalia.

The chromosomal pattern of the normal female is 46,XX and of the normal male 46,XY. It is generally assumed that the presence of a Y chromosome is necessary for the development of testicular tissue, that the presence of more than one X chromosome impairs the structure and function of the testis in the presence of a normal Y, and that two complete and normal X chromosomes are essential for the development of a functional ovary. Thus the patient with Klinefelter's syndrome (seminiferous tubule dysgenesis) with a chromosome pattern of 47/XXY has infertile testes in which the amount of tubular tissue is diminished and spermatogenesis is absent. The patient with Turner's syndrome (ovarian dysgenesis), with a chromosome pattern of 45/XO, has no recognizable ovarian tissue, but merely a streak in the broad ligament. It is clear now that XX (or XXY) germ cells cannot survive in a testis, though XO germ cells seem to be able to do so to some extent. Equally, XY germ cells cannot survive in an ovary. The same is generally true for XO cells but XXY germ cells in an ovarian environment might well differentiate into effective oocytes. Mosaics arise as the result of abnormal distribution of the sex chromosome material between the daughter cells of, usually, the first mitotic division of the zygote. Thus one daughter cell may contain two X chromosomes and the other only one. There are many combinations which can be envisaged to form mosaics, and it might be reasonable to assume that most cases of true hermaphroditism would have an XX/XY sex chromosome pattern indicative of chimerism. (A chimera is defined as a chromosomally mosaic subject derived, however, from the fusion of two separate zygotes, or at least of two cells that differ genetically as well as in respect to one or two chromosomes.) However, Polani (1970) reported the following patterns:

XX	59
XY	21
XO/XY	11
XX/XY	6
XX/XXY	5
Mosaic not satisfactorily identified	4
Mosaics without Y	2
TOTAL	108

Blank *et al.* (1964) refer to a triple mosaic XX/XXY/XXYY, to XX/XXX, and an XX/XXYY mosaic. An XX/XY/XO, an XX/XX plus fragment, and an XO/XX plus fragment (possibly an X/XX^y) are mentioned by Federman (1967), and Ribas-Mundo and Prats (1965) report an XX/XY/XXY.

Mosaics can obviously produce very complicated combinations of sex chromosomes and, provided they contain more than one X chromosome in one cell line and a Y in another, ovarian and testicular tissue could be present. It is difficult, however, to explain how a true hermaphrodite could show the sex chromosome pattern of a normal female or a normal male. Yet this was true in 80 out of Polani's collected series of 108 cases. It is, of course, well recognized that it may be difficult to disclose a mosaic, that more than one tissue has to be studied to do so, and that the second cell line may be found only in the gonads themselves. Thus a variable number of these 80 cases may have been hidden or undetected mosaics. Ferguson-Smith (1966) has suggested that one of the X chromosomes of a 46/XX true hermaphrodite contains the testis-determining gene(s) of a Y chromosome. This chromosome could, therefore, be designated X^y and the patient referred to as XX^y. Some cases are probably due to double fertilization of an ovum (Gartler *et al.*, 1962) or fertilization of an ovum and its polar body (Zuelzer *et al.*, 1964) with an X- and a Y-bearing spermatozoon. Rosenberg *et al.* (1963) have reported a family in which bilateral ovotestes were present in sibs, all of whom had XX karyotypes without demonstrable mosaicism. This suggests that in some instances there is a familial tendency to this hermaphroditic pattern which could be due to an aberrant autosomal gene with male potential that has interfered with the gonad being an ovary. Such autosomal genes are known in other mammals. It is interesting that in some races (for example, in the Bantu) true hermaphrodites are relatively common and are generally of the 46,XX type.

III. MENSTRUAL DISORDERS

Amenorrhea is a manifestation of primary ovarian deficiency or secondary failure of pituitary gonadotropic or hypothalamic luteinizing hormone-releasing hormone (LH-RH). This failure or deficiency may, however, be a matter of degree. It is of considerable practical value to classify cases into primary or secondary amenorrhea. Patients complaining of *primary amenorrhea* have never experienced a spontaneous menstrual period, though they may have responded to courses of estrogen or progestagen or combinations of the two with artificial periods that are, in fact, withdrawal bleeding. It is, therefore, important to establish the fact that there has been no spontaneous menstrual period before labeling the case as primary amenorrhea. *Secondary amenorrhea* is a condition in which spontaneous menstrual periods have been experienced but not for a considerable time.

In some patients, a spontaneous menstrual period may occasionally be evoked and may be preceded by ovulation (*oligomenorrhea*). On the other hand, ovulation may fail to occur; but prolonged and heavy bleeding may take place as the result of shedding of an hyperplastic endometrium which has been under the sole and continuous influence of estrogen (*metropathia hemorrhagica*). A milder degree of ovarian failure may result in excessive and prolonged menstrual bleeding due to inadequate progesterone stimulation of the endometrium (*ovulatory menorrhagia*).

A. Primary Amenorrhea

Menstruation usually commences at about 13 years of age, and advice is usually sought if it has not begun by the age of 15 or 16 years. In about 90% of girls presenting with primary amenorrhea, the diagnosis is idiopathic primary amenorrhea, delayed menarche, or ovarian dysgenesis—in that order. However, the first or second diagnoses cannot be accepted until all the other rare causes have been excluded.

1. Congenital Malformations of the Genital Tract

These are rare but may consist of absence of the uterus or the vagina, or an imperforate hymen. Suitable pelvic examination will reveal these causes of amenorrhea or *cryptomenorrhea.*

2. Debilitating Conditions

Diseases such as tuberculosis may induce amenorrhea but are seldom the cause of primary amenorrhea. However, diabetes in childhood and adolescence results in amenorrhea in about 50% of the cases (Jeffcoate, 1957). Emaciation due to loss of weight from anorexia nervosa very seldom occurs before the menarche though it is a common cause of secondary amenorrhea.

3. Sex Chromosome Anomalies

These result in cases of ovarian dysgenesis of various degrees (Section II,A) and the chromosomal aberrations have been discussed in detail (Section II,A,4).

4. Testicular Feminization

Testicular feminization is the extreme example of male pseudohermaphroditism which is due to various degrees of failure of the target organs responsible for developing the secondary sex characters to respond to normal quantities of endogenous androgen. In such cases, the phenotype is female and these individuals often look attractive. Sometimes the breast development is rather poor and usually there is no pubic and axillary hair, but otherwise the secondary sex characters are normal. However, the gonads are testes. They may be felt as firm lumps in the inguinal canals, or there may be a history of bilateral inguinal hernias operated upon in early childhood, an unaccountable mass having been found in each hernial sac and excised or returned to the abdomen. The uterus is either absent or vestigial and the vagina is represented by a blind pouch of variable length which may require dilatation or vaginoplasty shortly before marriage. This condition is not due to a sex chromosome anomaly and the nuclear sex is, therefore, chromatin-negative and the sex chromosome pattern is 46,XY—that of a normal male.

5. Pituitary Deficiency

a. Pituitary Infantilism. This condition is exceedingly rare in women and should never be diagnosed without very good reason. It is due to deficiency of all the anterior pituitary hormones. The most striking clinical signs are the miniature but well-proportioned stature, the sexual infantilism with failure to achieve secondary sex characters, and the childish, immature features. There are, however, no congenital anomalies and so it is usually easy to distinguish these cases from those of Turner's syndrome. The simplest and most revealing test is that of bone age, which is grossly retarded. In most cases the epiphyses of the long bones never unite, because of the lack of estrogen, but in the absence of growth hormone the height is short. The diagnosis is not easy to establish beyond doubt during the adolescent years when the patient is likely to present complaining of primary amenorrhea. However, failure to detect gonadotropins in a 24-hour specimen of urine on more than one occasion is indicative, especially if the urinary 17-oxosteroid levels are low and the nuclear sex is chromatin positive. It is rare to find clinical evidence of hypothyroidism or adrenocortical deficiency, though there may be a history of hypoglycemic attacks. Radioimmunoassays of growth hormone may show significantly low values. Treatment with human growth hormone may help these patients to attain reasonable statural growth and administration of gonadotropins may enable them to conceive.

b. Selective Gonadotropic Failure. In patients with pituitary deficiency there is a greater tendency for congenital and acquired failure to affect the hypophysial–gonadal axis than the other two target organ complexes, thyrotropic and adrenocorticotropic. Thus the condition of selective gonadotropic failure may not be revealed until the second decade. In the absence of estrogen, puberty and fusion of the epiphyses of the long bones are delayed. Nevertheless growth hormone is being actively secreted and, therefore, the bones of the extremities continue to lengthen. This results in eunuchoid skeletal proportions with long thin fingers and hands and slender toes and narrow feet. These girls are tall and lean. This condition should not be confused with pure gonadal dysgenesis (see Section II,A,) though the distinction may be difficult unless, in the latter case, the nuclear sex is chromatin negative and the sex chromosome pattern is 45,XO or some other abnormal combination. Lack of LH-RH and low or absent amounts of gonadotropin determined by radioimmunoassay would favor the diagnosis of selective pituitary failure, whereas their presence in normal or even excessive quantities would suggest that the case was one of pure gonadal dysgenesis. Treatment involves stimulating development of secondary sex characters with estrogens and progestagens and inducing ovulation with gonadotropins therapeutically administered as human pituitary gonadotropin or menopausal gonadotropin (HMG) and human chorionic gonadotropin (HCG) or by stimulating the production of the patient's own pituitary with LH-RH, when conception is required.

c. Craniopharyngioma. A history of brain injury, fracture of the skull or encephalitis suggests a possible cause of primary amenorrhea, but undoubtedly the most common acquired cause is craniopharyngioma. This is a rare tumor that develops before the age of 15 in 50% of cases. It is a neoplasm arising in a remnant of Rathke's (craniopharyngeal) pouch. It may take the form of a suprasellar cyst interfering with the hypothalamic–pituitary pathway, and may, therefore, be associated with diabetes insipidus, obesity, somnolence, and other manifestations of the hypothalamic syndrome, as well as amenorrhea because the gonadotropin-releasing factor cannot reach the pituitary. It may cause pressure on the optic chiasma or invade the 3rd ventricle or involve the 3rd, 4th, and 6th cranial nerves and lead to a number of visual field defects or diplopia. A suprasellar cyst may not deform the pituitary fossa but suprasellar flecks of calcification can be identified, with difficulty, in about two-thirds of the cases. When the tumor is intrasellar the degree of pituitary destruction may be so gross that the clinical picture resembles that of pituitary infantilism with short stature due to lack of growth hormone. These intrasellar tumors may become cystic and calcified and expand the pituitary fossa. Many cases of craniopharyngioma present, therefore, with obvious neurological or hypothalamic signs of manifestations of pituitary deficiency. Some cases present as primary (or occasionally secondary) amenorrhea. If it is decided to X ray the skull to confirm or exclude this diagnosis, the radiograph should be interpreted, if possible, by an expert neuroradiologist, since such positive signs as suprasellar flecks of calcification may be difficult to detect. The tumor should be removed surgically. If this is not possible, a combination of surgery and supervoltage irradiation may be the method of choice, though this highly differentiated tumor is rather radioresistant. In some cases aspiration of the cyst is all that can be achieved surgically and this may have to be repeated.

6. Adrenal Overactivity

It is rare for cases of this type to present with primary amenorrhea; the presenting clue is usually hirsutism or virilism.

In cases of "congenital adrenal hyperplasia" that reach adolescence without being detected, there is obvious stunting of growth because the epiphyses of the long bones unite prematurely, and the child exhibits almost achondroplastic proportions. The voice is gruff, hirsutism is marked, the physique is muscular and masculine, and there is no female breast development. The "adrenogenital syndrome" may develop during childhood as the result of an adrenocortical tumor producing an excess of androgens. Primary amenorrhea may be a coincidental result, but virilism will be the presenting feature. "Cushing's syndrome" in childhood is nearly always due to a tumor of the adrenal cortex, usually histologically malignant. These cases tend to display the florid signs of the condition and primary amenorrhea is coincidental.

7. Thyroid

The untreated cretin may not attain sexual maturity, but primary amenorrhea is most unlikely to be the presenting complaint. Juvenile myxedema may be a cause of primary amenorrhea and thyroid tests should be performed in case the classical signs of hypothyroidism are not clinically obvious.

8. Delayed Menarche

This is certainly a common cause of primary amenorrhea, and after exclusion of all the foregoing causes, which are either obvious or extremely rare, this diagnosis is the most probable. Late menarche is a familial characteristic, and the family history should therefore be investigated. The age of menarche and of social and sexual sophistication is getting earlier and earlier and against this background advice is usually sought if menstruation has not started by about the fifteenth year. Many girls, however, do not start menstruating until the late 'teens and they then have regular ovulatory cycles. Thus, up to the age of 18 or 20 years elaborate endocrine investigations are neither necessary nor justifiable.

9. Idiopathic

In a large number of women no cause for the amenorrhea can be found and this is called "idiopathic primary amenorrhea," although it may in fact be hypothalamic in origin.

B. Secondary Amenorrhea

As for primary amenorrhea, there are numerous possible causes of cessation of menstruation.

1. Physiological

a. Pregnancy. In all cases of secondary amenorrhea of relatively short duration, the possibility of pregnancy should be borne in mind.

b. Postpartum. It is not uncommon for there to be a delay in the return of menstruation after parturition.

The Chiari–Frommel syndrome (Chiari, 1855; Frommel, 1882) is a rare condition of persistent lactation and amenorrhea. It is associated with continued prolactin secretion which prevents the release of FSH, and clomiphene treatment has led to ovulation in many cases (Kaiser, 1963).

c. Menopause. From the age of 40 onward it is reasonable to suspect that secondary amenorrhea is due to the menopause (Section IV). However, although the menarche is getting earlier the menopause is getting later and very often does not occur until the age of 50 or after. Hot flushes are the most valuable clue, because most women suffer from these at the menopause, though flushes are merely a manifestation of ovarian deficiency and an indication that the ovary is producing diminishing amounts of estrogen. Consequently, the pituitary is releasing increasing quantities of gonadotropins (mainly FSH). The presence of hot flushes in a young woman suffering from amenorrhea does not necessarily indicate that she has reached a premature and permanent menopause, for her periods may return later spontaneously. Nevertheless, quite a number of cases of idiopathic secondary amenorrhea are probably due to a premature menopause, the ovaries having an inadequate number of primordial follicles to carry them through the normal span of menstrual life. High urinary gonadotropin titers, excessive LH-RH, vaginal smears indicating atrophy, and hot flushes are typical accompaniments of menopausal amenorrhea, whereas in many cases of idiopathic secondary amenorrhea the urinary gonadotropins and LH-RH are at levels within the normal range and vaginal smears indicate some degree of estrogenic activity and desquamation.

2. Systemic Disease

It is remarkable how seldom so-called debilitating diseases are accompanied by amenorrhea although they may directly affect the ovaries (e.g., bilateral ovarian abcesses) or the pelvic organs. Jeffcoate (1957) has pointed out that anemia is never a cause of amenorrhea and that almost the only debilitating disease which in its early stages leads to amenorrhea is tuberculosis.

a. Tuberculosis. Amenorrhea, according to Jeffcoate (1957), is seen in 50% of young women suffering from pulmonary tuberculosis and may indeed be the presenting symptom. A routine chest X ray should, therefore, be requested in all cases in which the cause of the amenorrhea is not immediately evident.

b. Diabetes. Jeffcoate (1957) also points out that juvenile diabetes is often accompanied by amenorrhea and that the urine should be routinely tested for sugar in young amenorrheic women.

c. Weight Changes. Any significant deviation from the normal or steady weight for the particular individual may give rise to infrequent periods or amenorrhea. This is less often true in cases of gain in weight than it is when there has been loss of weight, which is one of the most common and most overlooked causes of amenorrhea.

3. Psychogenic

a. Environmental Amenorrhea. This is quite common, though fortunately usually transient. Sometimes girls at boarding school have no periods while at school although they occur regularly during the holidays or vice versa. Women traveling to one country from another frequently develop amenorrhea which may persist until they return home. In the majority of these obvious cases of environmental amenorrhea, it is necessary only to recognize the cause and explain it to the patient and thus avoid any elaborate investigations.

b. Emotional Amenorrhea. The shock of a parent's death, the grief of a broken love affair, an unhappy home life, or the stress of working for examinations can induce amenorrhea, probably as the result of some inhibitory influence from the higher centers on the gonadotropin-releasing factors of the hypothalamus.

c. Anorexia Nervosa. This is a special example of hypothalamic amenorrhea due to psychogenic causes. In the classical case a teenage girl suddenly decides that she is overweight and must become slim. Sometimes the loss of weight is so rapid and so severe as to become positively alarming and the weight may even drop to less than 70 pounds. The victim may acquire the "will to die" and a few do so. Amenorrhea is instant and the periods cease from the time the decision to starve is made. It is, therefore, of psychopathic origin and not due to malnutrition or endocrine deficiency. Eventually, the whole of the endocrine system is affected by the malnutrition and endocrine investigations will suggest a marked degree of panhypopituitarism, with thyroid function tests giving hypothyroid levels, and low 17-oxosteroids and 17-oxogenic steroids; urinary gonadotropins are absent. These cases are often mistaken for Simmonds' disease, though, in fact, the typical patient suffering from Simmonds' disease does not lose weight but presents as a case of myxedema. Most cases of Simmonds' disease (and myxedema) lose their body hair—pubic, axillary, and outer third of eyebrows. Patients with anorexia nervosa tend to grow hair, especially on their arms and legs. The majority of cases, however, do not present this classical picture, and without specific enquiry as to whether there has been any loss of weight it is easy to miss the essential cause of the amenorrhea. The anorexia may even be quite unintentional. A girl leaves home and lives by herself in a bed-sitting room. She gets up too late to have breakfast;

she uses her lunch break to do her shopping and goes home in the evening too tired to bother about a decent meal and opens a tin of something. Imperceptibly she loses weight and develops amenorrhea. This sort of story is much more common than one may imagine and constitutes the vast majority of cases of "weight-loss amenorrhea."

4. Pituitary

a. Simmonds' Disease. This extremely rare condition is almost always a consequence of postpartum hemorrhage, as Sheehan (1937) has pointed out (the condition is often referred to as Sheehan's syndrome). As the result of a severe postpartum hemorrhage, thrombosis of the pituitary vessels may take place, leading to ischemic necrosis which may destroy 95% or more of the anterior pituitary gland and evidence of panhypopituitarism will accumulate. The clinical picture is that of classical myxedema, with a slight tendency to weight gain because of the subcutaneous myxedematous deposits and fluid retention. The woman is unable to breast-feed the infant because of failing lactation. The breasts atrophy and all body hair disappears. The skin, dry and scaly because of the hypothyroidism, because pale (depigmented) and cold. Slothful, lethargic, pale-skinned, and intolerant to cold, the woman may sit silent and immobile in a chair sunning herself like some amiable alligator, or in front of the fire burning her shins and acquiring thereby the classical cutis marmorata. The uterus is atrophied, the vaginal epithelium atrophic, the thyroid function tests indicate a gross degree of hypothyroidism, and yet these patients respond poorly to thyroid therapy. Urinary 17-oxosteroids may be less than 2 mg/24 hours and 17-hydroxycorticosteroid values are correspondingly low, so that cortisone therapy, in small physiological doses of 12.5 to 50 mg daily, is always indicated. Weakness of the muscles and loss of libido may suggest the advisability of anabolic hormone therapy. Some cases do spontaneously recover some degree of pituitary function, and there are rare reports of subsequent pregnancies with dramatic clinical improvement, though relapse may develop during the puerperium. This is the classical picture of Simmonds' disease and although cases are not always classical, it is probably true that Simmonds' disease is too frequently, and incorrectly, diagnosed.

b. Chromophobe Adenomas. Unlike the craniopharyngiomas in children, which typically lead to primary amenorrhea (see Section III,A), chromophobe adenomas most commonly develop in the third and fourth decades and comprise 70% of all pituitary tumors. They are one of the most common causes of hypopituitarism, but only 20% of cases present with abnormalities of sexual function; about 50% present with visual disturbances. An X ray will show enlargement of the pituitary fossa, in most cases, characteristically associated with atrophy of the clinoid processes. Supervoltage deep X-ray therapy is the treatment of choice and only about 20% of patients will require surgery, which is imperative if vision is materially affected.

About one-half the patients so far reported with the "Argonz and del Castillo syndrome" (1953), of persistent lactation and amenorrhea unassociated with the puerperium, have shown radiological evidence of enlargement of the pituitary fossa. When it has been possible to obtain histological evidence, the related tumor was a chromophobe adenoma as described by Forbes *et al.* (1954). The chromophobe cells can appear pre-eosinophilic. There are no demonstrable gonadotropins in the urine, though prolactin is usually secreted often in high amounts, and in some cases the galactorrhea and amenorrhea respond to clomiphene.

Persistent lactation associated with acromegaly is usually due to an eosinophil adenoma since the eosinophils secrete prolactin and growth hormone. In the early stages, the condition may not be easy to recognize clinically but enquiry may indicate a recent increase in shoe and glove size.

5. Adrenal Overactivity

a. Cushing's Syndrome. This is a rare cause of amenorrhea compared with simple obesity and secondary amenorrhea due to other causes, but it should be considered for obese patients presenting with amenorrhea.

b. Adrenogenital Syndrome. This is caused by an excess of adrenal androgens and, like Cushing's syndrome, may be due to an adrenocortical tumor or to bilateral hyperplasia. The excessive facial and body hair is coarse and dark. The clitoris may be moderately enlarged, the breasts shrunken, the body muscular, and the voice deep. Amenorrhea develops in due course but is seldom the presenting sign. The adrenogenital syndrome is usually associated with high levels of urinary 17-oxosteroids unlike the hirsutism or virilism of ovarian origin, in which conditions the 17-oxosteroid levels are usually within the normal range. Cases with 17-oxosteroid levels just above the upper limits of the normal range may show spontaneous ovulatory cycles after treatment with corticosteroids.

6. The Ovary

a. Masculinizing Tumors. These rarely occurring tumors may be classified pathologically, e.g., as arrhenoblastomas, masculinovoblastomas, dysgerminomas, Leydig cell tumors, and adrenal rest tumors (see Volume II, Chapter 4), but to clinicians they are simply virilizing or masculinizing tumors. They usually cause intense hirsutism and virilization without much, if any, rise in the urinary 17-oxosteroid levels. Gonadal androgens tend to give rise to dramatic biological signs of virilism, whereas adrenal androgens result in the excretion of high concentrations of androgenic metabolites (17-oxosteroids). These cases, therefore, primarily present clinically as virilism and the amenorrhea is an expected accompanying factor.

b. Polycystic Ovary. This is a relatively common cause of secondary amenorrhea. The classical features of the Stein–Leventhal syndrome (Stein and Leventhal, 1935) are amenorrhea, bilateral enlargement of the ovaries, hirsutism, obesity, and vague abdominal pain. The ovaries have a thickened capsule of pearly blue color and are much flattened and enlarged. Small cystic follicles protrude through the capsule but do not rupture to produce a corpus luteum. However, only one-half the patients are obese, only one-half are hirsute and many, particularly those that are obese, do not have palpably enlarged ovaries. Some may show normal ovulatory cycles with corpus luteum formation and the presenting feature may be metropathia hemorrhagica or oligomenorrhea. Therefore, in most cases of secondary amenorrhea for which the cause is not obvious and in which the patient complains of infertility, the ovaries should be visualized.

Supporting evidence of the diagnosis can be obtained from a study of the vaginal cytology and the urinary steroid metabolites. In many cases the urinary 17-oxosteroids are within the normal range though in some they are slightly though significantly raised, suggesting a mild degree of adrenocortical overactivity as well as abnormal function of the ovaries. Fractionation studies of the total 17-oxosteroids show an elevation of androsterone and etiocholanolone, presumably of ovarian origin, and of dehydroepiandrosterone, which is derived mainly from the adrenals. In addition, abnormally high quantities of Δ^5-pregnanetriol are found. There is evidence that the condition is associated with an enzyme defect in the biosynthesis of the ovarian hormones, at any rate in the affected cysts, for Δ^4-androstenedione has been isolated from them (Short and London, 1961) and this compound is the androgen that is converted via 19-hydroxyandrostenedione to estrone. It is to be assumed that this is only a partial defect and not a total enzyme block, for some estrogen activity is found in the normal range. Nevertheless the androgenic activity is sufficient to account for the hirsutism, sometimes marked, that occurs in many if not all the typical cases. Bilateral wedge resection of the ovaries frequently cures the condition, at least temporarily, as far as the amenorrhea is concerned and many patients thereafter ovulate and conceive. This was discovered by Stein and Leventhal (1935) in some of their original cases in which they merely performed bilateral ovarian biopsies in order to study the microscopic appearance of the ovaries. The biopsy is characterized by cortical follicular cysts and hyperthecosis of most of the follicles. However, resection seldom has any beneficial effect on the hirsutism. Clomiphene should be the treatment of first choice for it often induces ovulation and may, therefore, enable conception to take place.

c. Inadequate Luteal Activity. A short postovulatory phase may not always be associated with menorrhagia (see Section III,E). The condition may be indicated by a raised postovulatory temperature for only 10 or less days, a poor progestational vaginal smear response, or a low pregnanediol output. Recent studies (I. D. Cooke, personal communication) indicate diminished FSH activity followed by a low preovulatory estrogen peak and suggest that, if the subsequent inadequate luteal phase is associated with sub-fertility, the administration of human menopausal gonadotropin (HMG) may be more effective than trying to supplement the defective endogenous progesterone by oral progestagens or to stimulate greater luteal activity by injecting human chorionic gonadotropin (HCG). Cases of infertility associated with amenorrhea and lack of ovulation made to ovulate with clo-miphene sometimes have a poor luteal phase. These women may be prevented from aborting by being treated with HCG after ovulation (Bishop *et al.,* 1966; Kistner, 1966; Murray, 1974).

7. Thyroid

Thyrotoxicosis is seldom associated with amenorrhea, nor is myxedema, which most often presents in gynecological clinics as menorrhagia (see Section III,E).

8. Idiopathic

There is a considerable number of cases which cannot be diagnosed on any of the above criteria; they must, therefore, be labeled as idiopathic.

C. Oligomenorrhea

This term indicates that menstruation takes place less frequently than is considered normal. Nevertheless the menstruation, when it does arrive, is usually normal in duration and amount of loss, or perhaps rather scanty. If it is heavy and prolonged, a metropathic pattern may be suspected. It is usually found that the cycle, long though it may have been, was in fact ovulatory as indicated by the rise in temperature 14 days before the bleeding commenced. There are occasional reports of patients complaining of long-standing amenorrhea getting married and conceiving without having a menstrual period following their marriage. These must obviously be cases of ovulatory oligomenorrhea with very long intervals between periods. Hormone therapy is unnecessary and officious unless the problem is one of urgent infertility in which case clomiphene, possibly followed by HCG, given at monthly or 2-monthly intervals may increase the number of ovulations and thereby the chance of conception.

D. Metropathia Hemorrhagica

This term refers to the condition when excessive menstrual bleeding occurs but ovulation does not take place. Thus, instead of the alternating influence of estrogen before and estrogen and progesterone after ovulation, the endometrium is under the constant influence of estrogen alone. This produces an exaggerated proliferation of the endometrium giving rise to the characteristic histological appearance of cystic glandular hyperplasia. The endometrium is thickened, the glands vary considerably in diameter, some having a very dilated or cystic lumen. The epithelial cells are often duplicated or triplicated instead of consisting of a single layer lining the glands and the stroma may also show evidence of excessive mitotic activity. The blood vessels are engorged and dilated. Eventually, this juicy, almost polypoid, endometrium is shed piecemeal and accompanied by bleeding which sometimes amounts to flooding and sometimes is a mere trickle and which may continue for many days or even weeks. Obviously the length of time that it takes to build up the endometrium to the stage of repeating this tiresome and irregular shedding process varies, but usually the cystic glandular hyperplasia does not develop again for a considerable period, usually more than 4 weeks after the end of the last flow. The next bleeding episode can, therefore, be forestalled by artificially denuding the endometrium. Administration of an oral progestagen for 4 days will oppose the estrogenic influence on the endometrium and it will shed cleanly and appear to be a normal 4- or 5-day period. This procedure can be repeated month after month for years, and will produce an artificial menstrual pattern which will simulate normal, regular, menstrual cycles. Cases of metropathia also respond reasonably well to treatment with clomiphene, and it should be considered as an alternative to cyclical progestagen therapy, especially in married women who are anxious to conceive.

E. Menorrhagia

In menorrhagic women heavy bleeding lasts longer than in a normal menstruation and continues for 7 to 10 days. The intervals of bleeding are regular though they may be shorter than normal 28-day cycles. This is due to a short luteal phase resulting in an inadequate progesterone influence on the endometrium so that when it sheds it does not do so with the cleanness typical of normal progesterone-withdrawal bleeding. An endometrial biopsy on the first day of bleeding may reveal this inadequate luteal effect and basal temperatures may pinpoint the time of ovulation and indicate the existence of a short luteal phase. Serial vaginal smears may show poor progesterone influence and daily urinary pregnanediol estimations in the second half of the cycle may indicate low levels. The obvious treatment is supplementary progestagen, such as dydrogesterone or a 17α-acetoxy derivative, during the second half of the cycle.

F. Induction of Ovulation

Early attempts to stimulate ovulation in women infertile due to failure to ovulate consisted of efforts to mimic the natural cycle by a sequential administration of estrogen and estrogen and progestagen and were seldom successful. Attempts to stimulate the pituitary and/or ovaries with small doses of deep X rays (Kaplan, 1958) were largely abandoned or else were not indulged in because of the possibility of causing subsequent genetic damage, though so far as is known this has not occurred.

More logical assaults on the problem consisted of administering gonadotropins in order to start the normal ovulatory cycle of estrogen and progesterone secretion from the patient's ovary. For many years clinicians had to be content with human chorionic gonadotropin (HCG) derived from the urine of pregnant women. HCG is principally luteinizing in action and capable of rupturing a fully mature Graafian follicle which has developed under the influence of follicle-stimulating hormone (FSH). The timing and dosage, however, was difficult to gauge and HCG could only be used for those women capable of developing reasonably mature follicles because of adequate amounts of FSH. However, Gemzell *et al.* (1959), Gemzell (1965), and Buxton and Herrmann (1960) showed that ovulation followed by conception could be induced by treatment with extracts of human pituitary gonadotropin (HPG) followed by HCG. This therapeutic regimen is still followed but human menopausal gonadotropin (HMG) is a more readily obtainable source of FSH. Lunenfeld *et al.* (1963) induced ovulation with pregnant mares' serum gonadotropin (PMSG) but this produces antibodies after one or two courses and is seldom successful in leading to conception. The triphenylethylene compound, clomiphene, is strongly antiestrogenic in man and acts by blocking the central and peripheral receptors of natural estrogens. It was once thought that clomiphene might act as an oral contraceptive, but because the estrogen concentrations at the hypothalamus of nonovulating women are lowered, FSH and LH are released in ratios which stimulate ovulation. Greenblatt *et al.* (1961) and Tyler *et al.* (1962) reported ovulation and pregnancy with clomiphene administered orally. Multiple pregnancies and overstimulation of the ovaries are far less likely to occur than with gonadotropins, and clomiphene is preferred in the first instance (Bishop, 1972). Another triphenylethylene compound [1(*p*-dimethylaminoethoxyphenyl)-1,2-diphenylbut-1-ene: Tamoxifen] is also an antiestrogen and competes with estradiol-17β for binding sites on the receptor protein in the endometrial cytosol and is selectively taken up by the corpus luteum. Treatment with Tamoxifen has induced ovulation in cases of anovulatory menstrual disorders and conception has also occasionally occurred (Walpole, 1968; Klopper and Hall, 1971; Williamson and Ellis, 1973). Cyclofenil, a diphenylethylene, probably induces ovulation in a manner similar to clomiphene by acting as an antiestrogen or "impeded" estrogen and thus releasing FSH and LH by stimulating LH-RH (Miguel *et al.*, 1963; Persson, 1965).

G. Luteinizing Hormone–Releasing Hormone (LH-RH)

Hypothalamic factors capable of releasing luteinizing hormone and follicle-stimulating hormone from the pituitary were first demonstrated by McCann *et al.* (1960). In 1971, porcine luteinizing hormone-releasing factor (LH-RH) was structurally identified as a decapeptide by Schally *et al.* (1971) and Matsuo *et al.* (1971) and ovine LH-RF was found to have the same structure (Amoss *et al.*, 1971). A synthetic decapeptide having the same properties as the natural compound (Matsuo *et al.*, 1971) was produced in many laboratories and is now available for use clinically and is referred to as LH-RH, synthetic LH-RH, FSH/LH-RH or gonadotropin-releasing hormone (Gn-RH).

The prinicipal effect of LH-RH is to raise the blood level of LH in males and females. A peak is reached within 30 minutes of administration and falls off rapidly during the next 2 or 3 hours because of the short half-life of only 4 minutes in man. There is also a rise in FSH, though this response is smaller and more variable. The LH-RH may be administered intravenously, intramuscularly, subcutaneously, or intranasally, though with the last method five to ten times as much material is required. Though exogenously administered estrogen and progestagen depress the pituitary response to LH-RH, the sensitivity of the pituitary to LH-RH is at its maximum in the late follicular and early luteal phase of the normal menstrual cycle, when there are peaks of LH.

At present LH-RH is more useful in the diagnosis of hypothalamic–pituitary–gonadal disorders than in their treatment. In cases in which the primary lesion is due to pituitary disorder there is no, or only small amounts of, LH to release and therefore there is a diminished or absent response to LH-RH stimulation. Thus in the amenorrhea–galactorrhea syndrome associated with a pituitary tumor there is little or no response to LH-RH, though there may be high prolactin levels (Zarate *et al.*, 1973). Similar results are obtained in Sheehan's syndrome and panhypopituitarism (Aono *et al.*, 1973; Coscia *et al.* 1974), and in anorexia nervosa and amenorrhea associated with severe dietary deficiency lowered LH-RH responses are obtained (Warren *et al.*, 1974; Patton *et al.*, 1974). In cases of amenorrhea in which the pituitary is capable of normal function and the urinary estrogen excretion was between 5 and 21 $\mu g/24$ hours, there is an enhanced response to LH-RH (Nillius and Wide, 1972). With lower estrogen excretion levels the response to LH-RH is less. In cases of Stein–Leventhal syndrome (amenorrhea and enlarged polycystic ovaries) there is a high LH but a low FSH response (Taymor, 1974).

At present infertility treatment with LH-RH is disappointing because the ovulation rate is less than that achieved with clomiphene, HCG, or HMG. Where few or no follicles are present in the ovary, multiple injections have to be given to prepare some follicles for ovulation and, because of the short half-life of the LH-RH injections, may need to be given more than once daily. In some cases of successful ovulation and pregnancy, clomiphene or HMG have been combined with the LH-RH treatment.

IV. THE MENOPAUSE

Complete cessation of menstruation is referred to as the menopause but this is only one manifestation of the climacteric, and is not now considered to be the direct result of loss of ovarian function. The ovaries may be active for some time after the menopause and ovulation resulting in conception has been recorded (Sharman, 1962). Other physical and clinical signs of this condition, which is unique to man, are described by Sharman and the reader is referred to the earlier edition of this treatise for details.

The management of the condition is still, however, a matter of considerable contention among general practitioners, physicians, gynecologists, and patients who have to pass through it. Three chief attitudes can be considered. The first is that of leaving well alone, and, since the majority of climacteric women do not suffer from tiresome menopausal symptoms for anything other than a short time, this is not unreasonable. However, to deny treatment to women on the grounds that addiction to the hormonal therapy will occur seems unfair to the small proportion of patients (say 15–20%) that are bewildered by the usual hot flushes and night sweats, and the dizziness, headaches, lack of self-confidence, loss of memory, and general feeling of inability to cope. Treatment with small doses of estrogens can be of enormous benefit to these women for relief of climacteric symptoms. It has been indicated that the optimal dose of estrogen is a low one. It can be given in its synthetic form of diethylstilbestrol or its equivalent of natural estrogen such as estrone sulfate from pregnant mare's urine, and it is usual to start with a low dose such as 0.1 mg of diethylstilbestrol or 0.625 mg of Premarin. This can be raised until the number of hot flushes is reduced to an occasional one daily or even weekly; no more than 0.25 mg or 0.5 mg or 1.25 mg Premarin is usually required. The optimum dose should be given for 3 weeks, starting on the first day of the month. The next course starts on the first day of the following month and after the third course the patient waits until the symptoms return and become troublesome before undertaking another 3-week course. Some bleeding may take place after the estrogen withdrawal and, though orthodox gynecological teaching necessitates diagnostic curettage, this should not invariably be repeated if further estrogen withdrawal bleeds occur. The use of ethinyl estradiol or mestranol as the estrogen of the contraceptive pill has made the former a popular alternative to diethylstilbestrol and estrone sulfate in the form of Premarin (*pregn*ant) *mare* ur*ine* extract) is frequently used for those who prefer a natural estrogen. The use of estrogen–androgen tablets for therapy is not advised as most marketed combinations contain enough androgen to produce virilizing side effects.

It is questionable, however, whether it is necessary to continue estrogen therapy for the rest of the patient's life because it will keep her young, prevent the skin wrinkling, and boost morale. Nevertheless such is becoming the modern trend and hormone replacement therapy (HRT) clinics are becoming established throughout the world.

Estrogens and Cancer

There is still considerable argument as to whether estrogens can cause cancer and should therefore not be administered at all or whether they are perfectly safe.

Some gynecologists advocate induction of an artificial menopause by prophylactic removal of the ovaries at hysterectomy over the age of 40 lest they subsequently become malignant. Carcinoma of the ovary is a rare though unpleasant condition, but most people would willingly take the risk of developing it as an alternative to the certainty of the horrors of an indefinite artificial menopause. The argument that estrogen therapy prolongs the menopause is ill founded in my experience, but the fear of development of a gynecological cancer must be carefully considered. This fear has been fostered in recent years because of the increased awareness of cancer as a killing condition.

That estrogen treatment of the menopause might cause cancer of the breast or uterus in some patients was standard teaching for many years though there was in fact singularly little actual evidence for it and indeed it seems likely that women who have been treated with estrogen are *less* likely to develop breast cancer (Vessey *et al.* 1972, 1975). A report appeared that carcinoma of the vagina had developed in the daughters of women who were taking high doses of synthetic estrogens for threatened or habitual abortion during the pregnancy in which they, the daughters, had been conceived (Herbst *et al.*, 1971). This certainly seemed significant although in a study of the 818 daughters of 1719 women treated with estrogen during pregnancy at the Mayo Clinic no case of vaginal cancer was found (Lanier *et al.*, 1973). Estrogens are seldom given in pregnancy nowadays.

More pertinent to the problem, however, was the finding of Ziel and Finkle (1975) that 57% of 94 patients with endometrial cancer had been treated with estrogens as compared with 15% who had not. Weiss (1975) has calculated that 1 in every 1000 postmenopausal women with an intact uterus develops carcinoma of the endometrium. Estrogen therapy would according to their results raise this figure to between 4 and 8 per 1000. Kinlen and Doll (1973), however, found no such increase in the United Kingdom.

Thus, though the carcinogenic propensities of estrogens are still by no means universally denied, each of the principal claims is refuted by highly reliable and distinguished workers. The claim of the tendency of estrogen to give rise to breast cancer is reversed by Vessey and Doll (Vessey *et al.*, 1976) from the Department of the Regius Professor of Medicine at Oxford; Sir Richard Doll and his colleagues could also find no increase in postmenopausal endometrial cancer in women who had been using oral contraceptives or taking estrogen for other reasons; and at the Mayo Graduate School of Medicine no case of vaginal cancer has been found in the daughters of women who had been treated with estrogen during the pregnancy which had accounted for the birth of these girls. On the other hand, hormone replacement therapy after the clinical effects of the menopause have disappeared is a relatively new concept which I still consider open to argument.

ACKNOWLEDGMENTS

During the period when the text of this chapter was being prepared in manuscript, Dr. Bishop was prevented by indisposition from giving his full and up to date attention to it. He therefore wishes to acknowledge the help he received from Professor P. E. Polani, FRS, Prince Philip Professor of Pediatric Research at Guy's Hospital Medical School, in bringing up to date the section on Disorders of Embryonic Origin. All the comments he made on the author's manuscript have been entered verbatim in the text and, consequently, Professor Polani's contributions to this section have been considerable. Professor I. D. Cooke, FRCOG, Professor of Obstetrics and Gynaecology of Sheffield University, kindly read the script of the whole chapter and made a number of helpful suggestions for which the author is most grateful.

REFERENCES

Acheson, R. M., and Zampa, G. A. (1961). Skeletal maturation in ovarian dysgenesis and Turner's syndrome. *Lancet* **1**, 917–920.

Amoss, M., Burgus, R., Blackwell, R., Vale, W., Fellows, R., and Guillemin, R. (1971). Purification, amino acid composition and N-terminus of the hypothalamic luteinizing hormone release factor (LRF) of ovine origin. *Biochem, Biophys. Res. Commun.* **44**, 205–210.

Aono, T., Minagawa, J., Kinugasa, T., Tanizawa, O., and Kurachi, K. (1973). Response of pituitary LH and FSH to synthetic LH-releasing hormone in normal subjects and patients with Sheehan's syndrome. *Am. J. Obstet. Gynecol.* **117**, 1046–1052.

Argonz, J., and del Castillo, E. B. (1953). A syndrome characterized by estrogenic insufficiency, galactorrhea and decreased urinary gonadotropin. *J. Clin. Endocrinol. Metab.* **13**, 79–87.

Bahner, F., Schwarz, G., Harnden, D. G., Jacobs, P. A., Heinz, H. A., and Walter, K. (1960). A fertile female with XO sex chromosome constitution. *Lancet* **2**, 100–101.

Bishop, P. M. F. (1962). Derangements of ovarian function. *In* "The Ovary" (S. Zuckerman, ed.), 1st ed., pp. 553–575. Academic Press, New York.

Bishop, P. M. F. (1972). Clomiphene. *Br. Med. Bull.* **26**, 22–25.

Bishop, P. M. F., Lessof, M. H., and Polani, P. E. (1960). Turner's syndrome and allied conditions. *Mem. Soc. Endocrinol.* **7**, 162–172.

Bishop, P. M. F., Murray, M., and Osmond-Clarke, F. (1966). Short luteal phase following ovulation induced by Clomid. *Excerpta Med. Found., Int. Congr. Ser.* **111**, 207 (Abstr. 357).

Blank, C. E., Zachary, R. B., Bishop, A. M., Emery, J. L., Dewhurst, C. J., and Bond, J. H. (1964). Chromosome mosaicism in a hermaphrodite. *Br. Med. J.* **2**, 90–93.

Buxton, C. L., and Herrmann, W. (1960). Induction of ovulation in the human with human gonadotropins. *Yale J. Obstet. Gynecol.* **76**, 447–453.

Chiari, J. (1855). Bericht über die in den Jahren 1848 bis inclusive 1851 an der gynäkologischen Abtheilung in Wien beobachteten Frauenkrankheiten im engern Sinne des Wortes. *In* "Klinik der Geburtshilfe und Gynaekologie" (J. Chiari, C. Braun, and J. Spaeth, eds.), pp. 362–415. Enke, Stuttgart.

Clayton, G. W., O'Heeron, M. K., Smith, J. D., and Grabstald, H. (1957). A case of true hermaphroditism: possible relationship to Klinefelter's syndrome. *J. Clin. Endocrinol. Metab.* **17**, 1002–1005.

Coscia, A. M., Fleischer, N., Besch, P. K., Brown, L. P., and Desiderio, D. (1974). The effect of synthetic luteinizing hormone-releasing factor on plasma LH levels in pituitary disease. *J. Clin. Endocrinol. Metab.* **38**, 83–88.

Dewhurst, C. J., and Gordon, R. R. (1969). "The Intersexual Disorders." Baillière, London.

Federman, D. D. (1967). "Abnormal Sexual Development, a Genetic and Endocrine Approach to Differential Diagnosis." Saunders, Philadelphia, Pennsylvania.

Ferguson-Smith, M. A. (1966). X-Y chromosomal interchange in the aetiology of true hermaphroditism and of XX Klinefelter's syndrome. *Lancet* **2**, 475–476.

Forbes, A. P., Henneman, P. H., Griswold, G. C., and Albright, J. (1954). Syndrome characterized by galactorrhea and low urinary FSH: comparison with acromegaly and normal lactation. *J. Clin. Endocrinol. Metab.* **14**, 265–271.

Frommel, R. (1882). Ueber puerperale Atrophie der Uterus. *Z. Geburtshilfe Gynaekol.* **7**, 305–313.

Gartler, S. M., Waxman, S. H., and Giblett, E. (1962). An XX/XY human hermaphrodite resulting from double fertilization. *Proc. Natl. Acad. Sci. U.S.A.* **48**, 332–335.

Gemzell, C. (1965). Induction of ovulation with human gonadotropins. *Recent Prog. Horm. Res.* **21**, 179–204.

Gemzell, C., Diczfalusy, E., and Tillinger, K. G. (1959). Further studies of the clinical effects of human pituitary follicle-stimulating hormone (FSH). Lack of response in ovarian insufficiency. *Acta Obstet. Gynecol. Scand.* **38**, 465–476.

Greenblatt, R. B., Barfield, W. E., Jungck, E. C., and Ray, A. W. (1961). Induction of ovulation with MRL/41. *J. Am. Med. Assoc.* **178**, 101–104.

Greenblatt, R. B., Byrd, J. R., McDonough, P. G., and Mahesh, V. B. (1967). The spectrum of gonadal dysgensis. *Am. J. Obstet. Gynecol.* **98**, 151–172.

Grossman, E. R. (1960). Pituitary gonadotropins in gonadal dysgenesis. *Pediatrics* **25**, 198–303.

Herbst, A. L., Ulfelder, H., and Poskanzer, D. C. (1971). Adenocarcinoma of the vagina: association of maternal stilboestrol therapy with tumor appearance in young women. *N. Eng. J. Med.* **284**, 878–881.

Hoffenberg, R., Jackson, W. P. U., and Muller, W. H. (1957). Gonadal dysgenesis with menstruation: a report of two cases. *J. Clin. Endocrinol. Metab.* **17**, 902–907.

Jeffcoate, T. N. A. (1957). "Principles of Gynaecology." Butterworth, London.

Johnston, A. W., Ferguson-Smith, M. A., Handmaker, S. D., Jones, H. W., and Jones, G. S. (1961). The triple X syndrome. Clinical, pathological, and chromosomal studies of three mentally retarded cases. *Br. Med. J.* **2**, 1046–1052.

Jones, H. W., and Scott, W. W. (1958). "Hermaphroditism, Genital Abnormalities and Related Endocrine Disorders." Williams & Wilkins, Baltimore, Maryland.

Josso, N. (1972). Permeability of membranes to the Müllerian-inhibiting substance synthesized by the human fetal testis *in vitro*: a clue to its biochemical nature. *J. Clin. Endocrinol. Metab.* **34**, 265–270.

Jost, A. (1953). Problems of fetal endocrinology: the gonadal and hypophyseal hormones. *Recent Prog. Horm. Res.* **8**, 379–413.

Kaiser, I. H. (1963). Pregnancy following clomiphene induced ovulation in Chiari-Frommel syndrome. *Am. J. Obstet. Gynecol.* **87**, 149–151.

Kaplan, I. I. (1958). The treatment of female sterility with X-ray therapy directed to the pituitary and ovaries. *Am. J. Obstet. Gynecol.* **76**, 447–453.

Kinlen, L. J., and Doll, R. (1973). Trends in mortality from cancer of the uterus in Canada and in England and Wales. *Br. J. Prevent Soc. Med.* **27**, 146–149.

Kistner, R. W. (1966). Use of clomiphene citrate, human chorionic gonadotropin, and human menopausal gonadotropin for induction of ovulation in the human female. *Fertil. Steril.* **17**, 569–604.

Klebs, E. (1876). "Handbuch der pathologischen Anatomie," Vol. 1, Sect. 2. Geschlechtsapparat. Hirschwald, Berlin.

Klopper, A., and Hall, M. (1971). New synthetic agent for the induction of ovulation: preliminary trials in women. *Br. Med. J.* **1**, 152–154.

Lanier, A. P., Noller, K. L., Decker, D. G., Elveback, L. R., and Kurland, L. T. (1973). Cancer and Stilboestrol. *Mayo Clin. Proc.* **48**, 793–799.

Lunenfeld, B., Sulimovici, S., and Rabau, E. (1963). Mechanism of action of anti-ovulatory compounds. *J. Clin. Endocrinol. Metab.* **23**, 391–395.

McCann, S. M., Taleisnik, S., and Friedman, H. M. (1960). LH-releasing activity in hypothalamic extracts. *Proc. Soc. Exp. Biol. Med.* **104**, 432–434.

Matsuo, H., Baba, Y., Nair, R. M. G., Arimura, A., and Schally, A. V. (1971). Structure of the porcine LH- and FSH-releasing hormone. I. The proposed amino acid sequence. *Biochem. Biophys. Res. Commun.* **43**, 1334–1339.

Miguel, J. F., Wahlstam, H., Olsson, K., and Sundebeck, B. (1963). Synthesis of unsymmetrical diphenylalkenes. *J. Med. Chem.* **6**, 774–780.

Murray, M. (1974). Inadequate luteal phase as a cause of infertility. *Proc. R. Soc. Med.* **67**, 937–938.

Nillius, S. J., and Wide, L. (1972). The LH-releasing hormone test in 31 women with secondary amenorrhoea. *J. Obstet. Gynaecol. Br. Commonw.* **79**, 874–882.

Papanicolaou, A. D., Loraine, J. A., and Charles, D. (1969). Pituitary-gonadotrophic function in Turner's syndrome. *Lancet* **1**, 184–186.

Patton, W. C., Thompson, I. E., Berger, M. J., Chong, A. P., and Taymor, M. L. (1974). Synthetic luteinizing hormone releasing hormone. *Obstet. Gynecol.* **44**, 823–829.

Persson, B. H. (1965). The effect of bis(*p*-acetoxyphenyl) cyclohexylidene methane (compound F6066) on hormone excretion in postmenopausal women. *Acta Soc. Med. Ups.* **70**, 1–16.

Polani, P. E. (1969). Chromosome phenotypes—sex chromosomes. *In* "Congenital Malformations" (F. C. Fraser and V. A. McKusick, eds.), Int. Congr. Ser. No. 204, pp. 233–250. Excerpta Med. Found., Amsterdam.

Polani, P. E. (1970). Hormonal and clinical aspects of hermaphroditism and the testicular feminizing syndrome in man. *Philos. Trans. R. Soc. London, Ser. B* **259**, 187–204.

Ribas-Mundo, M., and Prats, J. (1965). Hermaphrodite with mosaic XX/XY/XXY. *Lancet* **2**, 494.

Rosenberg, H. S., Clayton, G. W., and Hsu, T. C. (1963). Familial true hermaphroditism. *J. Clin. Endocrinol. Metab.* **23**, 203–206.

Schally, A. V., Baba, Y., Arimura, A., and Redding, T. W. (1971). Evidence for peptide nature of LH and FSH-releasing hormones. *Biochem. Biophys. Res. Commun.* **42**, 50–56.

Sharman, A. (1962). The menopause. *In* "The Ovary" (S. Zuckerman, ed.), 1st ed., pp. 539–551. Academic Press, New York.

Sheehan, H. L. (1937). Post-partum necrosis of the anterior pituitary. *J. Pathol. Bacteriol.* **47**, 189–214.

Short, R. V., and London, D. R. (1961). Defective biosynthesis of ovarian steroid hormones in the Stein-Leventhal syndrome. *Br. Med. J.* **1**, 1724–1727.

Silver, H. K., and Dodd, S. G. (1957). Gonadal dysgenesis. Bonnevie-Ulrich-Turner syndrome with elevated gonadotropins in a nine-year-old child. *Am. J. Dis. Chil.* **94**, 702–707.

Singh, R. P., and Carr, D. H. (1966). The anatomy and histology of XO human embryos and fetuses. *Anat. Rec.* **155**, 369–384.

Sohval, A. R. (1963). "Mixed" gonadal dysgenesis: a variety of hermaphroditism. *Am. J. Hum. Genet.* **15**, 155–158.

Sohval, A. R. (1964). Hermaphroditism with "atypical" or "mixed" gonadal dysgenesis. *Am. J. Med.* **36**, 281–292.

Stein, J. F., and Leventhal, M. L. (1935). Amenorrhea associated with bilateral polycystic ovaries. *Am. J. Obstet. Gynecol.* **19**, 181–191.

Stewart, J. S. S. (1960). Gonadal dysgenesis: the genetic significance of unusual variants. *Acta Endocrinol. (Copenhagen)* **33**, 89–102.

Taymor, M. L. (1974). The use of luteinizing hormone-releasing hormone in gynecologic endocrinology. *Fertil. Steril.* **25**, 992–1005.

Turner, H. H. (1938). A syndrome of infantilism, congenital webbed neck, and cubitus valgus. *Endocrinology* **23**, 566–574.

Tyler, E. T., Winer, J., Gotlib, M., Olson, H. J., and Nakabayashi, N. (1962). Effects of MRL-41 in human male and female fertility studies. *Clin. Res.* **10**, 119.

van der Werff ten Bosch, J. J. (1959). "Normale en abnormale geschlactsrijping." Leiden.

Vessey, M. P., Doll, R., and Sutton, P. M. (1972). Oral contraceptives and breast neoplasia: a retrospective study. *Br. Med. J.* **3**, 719–724.

Vessey, M. P., Doll, R., and Jones, K. (1975). Oral contraceptives and breast neoplasia: a retrospective study. *Lancet* **1**, 941.

Vessey, M. P., Doll, R., Peto, R., Johnson, B., and Wiggins, P. (1976). A long-term follow-up study of women using different methods of contraception—an interim report. *J. Biosoc. Sci.* **8**, 373–427.

Walpole, A. L. (1968). Non-steroidal drugs in relation to ovulation and implantation. *J. Reprod. Fertil., Suppl.* **4**, 3–14.

Warren, M. P., Jewelewicz, R., and Dyenfurth, I. (1974). The effect of weight loss on pituitary response to LH-RH in women with secondary amenorrhea. *Program Soc. Gynecol. Invest.*, Abstract 34.

Weiss, N. S. (1975). Risks and benefits of Estrogen use. *N. Z. J. Med.* **293**, 1200–1202.

Wilkins, L. (1965). "The Diagnosis and Treatment of Endocrine Disorders in Childhood and Adolescence," 3rd ed. Thomas, Springfield, Illinois.

Williamson, J. G., and Ellis, J. D. (1973). The induction of ovulation by Tamoxifen. *J. Obstet. Gynaecol. Br. Commonw.* **80**, 844–847.

Zarate, A., Jacobs, L. S., Canales, E. S., Schally, A. V., de la Cruz, A., Soria, J., and Daughaday, W. H. (1973). Functional evaluation of pituitary reserve in patients with the amenorrhea-galactorrhea syndrome utilizing luteinizing hormone-releasing hormone (LH-RH), L-DOPA and chlorpromazine. *J. Clin. Endocrinol. Metab.* **37**, 855–859.

Ziel, H. K., and Finkle, W. D. (1975). Increased risk of endometrial carcinoma among users of conjugated estrogens. *N. Engl. J. Med.* **293**, 1167–1170.

Zuelzer, W. W., Beattie, K. M., and Reisman, L. E. (1964). Generalized unbalanced mosaicism attributable to dispermy and probable fertilization of a polar body. *Am. J. Hum. Genet.* **16**, 38–51.

6

The Ovarian Cycle in Vertebrates

I. W. Rowlands and Barbara J. Weir

I. INTRODUCTION

The decade which followed the appearance of the first edition of this book has been a fruitful period in the accumulation of data about, and in the growth of our understanding of, the breeding season and its correlations with the ovarian cycle. The increasing volume of literature has not made the revision of this chapter easy. Published papers vary from relatively simple descriptions of breeding habits to sophisticated investigations of such mechanisms as the photoperiodic control of the ovarian cycle of birds. Between these two extremes lie reports of such differing quality in terms of content and presentation that accurate comparisons of data have rarely been possible.

An attempt has been made to describe the more significant advances, paying particular attention to studies of ovarian activity in the reproductive cycle of species that have been only recently investigated. Many of these species are comparatively rare and cannot be killed in numbers sufficiently great to provide anatomical correlations of reproductive activity. It has sometimes been necessary to resort to deductions from behavioral studies. For example, the periodicity of the estrous cycle has frequently been regarded as a measure of the length of the ovarian cycle; the relationship between these two cycles was described by Perry and Rowlands (1962). Repetition of material discussed by Perry and Rowlands (1962) has been kept to a minimum and is included only where relevant. The reader is referred to Volume I, Chapters 4 and 5 for descriptions of ovarian tissues and to Chapter 8 in this volume, for changes in the mammalian ovary during pregnancy. Another useful reference to ovarian cycles of mammals is that of Perry (1971). The distinction between ovarian cycles of mammals and those of other vertebrates is common. Barr (1968) has pointed out that the ovarian cycle of fishes and other lower vertebrates consists only of the follicular phase of the cycle that is found in mammals for the reason that the majority of these animals are either oviparous or ovoviviparous and, therefore, have no need of a corpus luteum for the purpose which it serves in mammals. The occurrence of corpora lutea, however, is not unknown in quite a wide selection of lower vertebrates and is not a correlate of the viviparous habit.

II. PISCES

The reproductive processes of fishes are extremely diverse, partly on account of their great phylogenetic age and partly because of "the freedom of the reproductive structures and processes from the demands of other organs and systems" (Hoar, 1965). An account of ovarian activity in cyclostomes and other fishes is provided by Barr (1968).

A. Cyclostomata

The cyclostomes are the only living representatives of the first vertebrate class to appear in the fossil record (see Dodd *et al.*, 1960). Busson-Mabillot (1967a) has shown that the gonad of *Lampetra planeri* remains undifferentiated for the first 2 years of larval life, and, following differentiation of the germ cells, another 2 years or more elapse before the onset of oocyte and follicular development. Metamorphosis in *L. planeri* and *L. fluviatilis* occurs during the autumn of the 4th year of larval life when gonadal development begins (Dodd *et al.*, 1960). The gonads mature in the following spring and produce large numbers of follicles all at the same stage of development. The rate of follicular development is controlled to a large extent by the water temperature and is variable in *L. planeri* and *L. fluvia-tilis* (Dodd *et al.*, 1960) or may be held in complete abeyance in the sea lamprey, *Petromyzon marinus*, in the Great Lakes (Lewis and McMillan, 1965). The eggs of the sea lamprey are extruded at spawning in spring and summer through the body wall via the gonopore because cyclostomes have no genital ducts. They are expelled in clutches every 5–10 minutes for periods of 16 hours to 3½ days, in which time a total of 25,000 to 100,000 eggs are laid. Hatching takes place 2 weeks later and, as would be expected, thermal control is maintained over the rate of larval development.

A common feature of cyclostomes is that during vitellogenesis and gonadal ripening the gut begins to shrink and, after spawning, it becomes atrophic and death follows soon afterward. During its lifetime each individual animal has only one ovarian cycle, and the full complement of follicles is produced simultaneously to reproduce an animal which spends about 80% of its life in a larval form (the ammocoete). Not all the follicles mature and ovulate; some degenerate before, and others after, the onset of vitellogenesis. The latter give rise to corpora atretica (Busson-Mabillot, 1967b).

The dependence of the cycle on pituitary activity has been demonstrated experimentally by Evennett and Dodd (1963).

B. Elasmobranchii

The elasmobranchs form only a relatively small group of fishes and very little additional to that reported by Perry and Rowlands (1962) has been added in recent years to our knowledge of the periodicity of ovarian function. Most of the species studied exhibit viviparity, but, irrespective of breeding habit, an ovarian feature common to all is the presence of large numbers of corpora lutea. Not all of these are true corpora lutea formed from spent follicles; the majority are the so-called corpora atretica which arise from unovulated follicles by atresia and persist for long periods. The fate of these structures, which have been associated with the occurrence of internal fertilization, has been reviewed by Chieffi (1967) and Barr (1968). The presence of estrogenic hormones and progesterone has been noted in both types of corpus luteum (Chieffi, 1967; Simpson *et al.*, 1968), but opinion is divided about the function of these hormones and their relationship to the animals' mode of reproduction (Lance and Callard, 1969). Much has yet to be learned before an accurate account can be given of the control of the ovarian cycle of these fishes, but at least it would appear that the presence of active corpora lutea does not prevent growth and development of follicles.

C. Crossopterygii

This class, the bony fishes, contains more than 20,000 species, which exceeds the total number of all other living vertebrate species. The majority of teleosts studied have a well-defined breeding season rigidly controlled by various external factors. A range of breeding habits is exhibited, from production of a single batch of eggs in a lifetime, as in the salmon, *Oncorhynchus* (Hoar, 1955), to an almost continuous flow of free swimming young throughout adult life as in *Brachyraphis* (Turner, 1938). Between these two extremes, every conceivable modification relating to the site of fertilization and embryonic development, mating patterns, and parental behavior is displayed, and, as Perry and Rowlands (1962) have pointed out, there is no close homology between these and the corresponding phenomena found in higher vertebrates.

1. Oviparous Forms

Most bony fishes are oviparous and shed a large number of small ova that are fertilized externally. In these animals the main purpose of the ovarian cycle is the transfer of materials to the oocytes during vitellogenesis. As in all oviparous lower vertebrates, the cycle is equivalent only to that part of the mammalian cycle which is commonly referred to as the follicular phase. Barr (1968) has pointed out that the cycle has four principal features: (i) the multiplication of oogonia by mitosis, (ii) their conversion to primary oocytes and the formation of the protective cellular covering to produce a follicle, (iii) the formation and deposition of yolk (vitellogenesis), and (iv) maturation culminating in ovulation and spawning. Vitellogenesis is accompanied by a rapid expansion of the follicles presumably under the stimulus of a pituitary gonadotropin. There is no postovulatory development of the spent follicle.

a. The Nature of the Cycle. The major overt event associated with the ovarian cycle is that of spawning, and there is a wide range in frequency with which it takes place in the life of the individual. For this reason it is not possible to estimate the length of the cycle on the basis of the interval between two succeeding spawning episodes in one season. Marza (1938), who is quoted extensively by Yamamoto and Yamazaki (1961) and Lehri (1968), has classified the differences as follows: (1) a synchronous development of all oocytes so that they develop, mature, ovulate, and spawn together and then the animals die, as do cyclostomes and the salmon; (2) group synchrony in which the ovaries at the time of ovulation contain a second group of developing, yolk-free oocytes which will mature and ovulate later in the same breeding season or early in the next season. This is the most common form of cyclical ovarian activity among these fishes and has been described in detail for the catfish, *Clarias batrachus* (Lehri, 1968). Oocyte development in the third group (3) is described as asynchronous because the ovary at any one time contains oocytes at all stages of development, and ovulation and spawning occur frequently throughout a long breeding season. Included in group (3) are the mackerel (Bara, 1960), carp (Sathyanesan, 1961), sardine (Farris, 1963), and goldfish (Yamamoto and Yamazaki, 1961). An extreme case of asynchrony is the medaka, *Oryzias latipes*, in which a clutch of eggs is spawned daily over long periods during spring and summer (Robinson and Rugh, 1943; Tsusaka and Nakano, 1965).

At the onset of the breeding season, which usually occurs in spring of each year, the oocytes of the bony fishes are about 100 μm in diameter. Yolk is secreted by the membrana granulosa and deposited in oocytes of 200 to 300 μm in diameter, which causes a considerable and rapid decrease in ovarian size and weight. Ovulation usually occurs when the oocytes are 700–1000 μm in diameter and results in a much more dramatic collapse of the ovary than it does in mammals. Spawning rapidly follows ovulation. In several species, the age at first spawning is 2–3 years, and good evidence has been provided to suggest that the cycle of oocyte development is spread over most of this time in several of the flat fishes, for example, the flounder, (Yamamoto, 1956), winter flounder, plaice, haddock, and hake (Dunn and Tyler, 1969). Bara (1960) showed, however, that in the mackerel, which also spawns first at the end of the second year of life, the process of oocyte development is not continuous, as maturation of the second set of oocytes stops during spawning of the previously ripened oocytes. According to Rousson (1957), the sturgeon, *Acipenser* (which is not strictly a teleost), first spawns at an average age of 18 years (range 14–23 years) and the interval between consecutive spawning periods can vary from 4 to 7 years. Magnin (1966) has reported that, if conditions are inclement, spawning in *A. fulvescens* in James' Bay, Canada can be delayed for up to 3 years, and he claims that females live for up to 80 years (males 55 years).

b. Factors Affecting the Cycle. The time of onset of spawning in fishes is controlled to a large extent by external factors of which water temperature is of considerable importance. In several species, spawning is known to occur only within a certain range of temperature, although the range varies considerably between species. Thus, the onset of the breeding season in the killifish, *Fundulus heteroclitus*, is earlier in the warmer waters of North Carolina (Brummett, 1966) than in the colder waters at Woods Hole, near Cape Cod (Matthews, 1938). In the brook stickleback, *Eucalia inconstans*, the growth of oocytes is inhibited in winter when the water temperature approaches 0°C; spawning in this species begins when water temperature reaches 10°C and stops when it exceeds 22°C (Braekevelt and McMillan, 1967). A similar control of spawning has been observed in the shield darter, *Percina peltata*, near New York (New, 1966). Spawning in the catfish, *Mystus seenghala*, in the Ganges, is dependent on a temperature of 25° to 30°C (Sathyanesan, 1962). The periodicity of spawning in the zebra fish, *Brachydanio rerio*, is closely controlled by temperature. At 26°C the cycle recurs every fifth day and at 29°C every other day, but at 22.5°C, spawning is inhibited. In this fish, the eggs are released only in response to stimulation by the male (Hisaoka and Firlit, 1962). Fluctuation in water temperature has been shown to have a significant effect on the time of spawning in the mud eel, *Amphipnous cuchia* (Sinha and Rastogi, 1967), and the bigmouth buffalo fish, *Ictiobus cyprinellus* (Johnson, 1963).

Rainfall is another factor that, to some extent, controls the onset of spawning, especially in some tropical fish. Thus, Sathyanesan (1961) has shown that several species of carp, e.g., *Barbus stigma* in the Ganges, spawn only during the early monsoon months. John (1963) found that spawning was similarly influenced in the speckled dace, *Rhinichthys osculus*, in Arizona, although basically the annual period of reproduction in this species is considered to be determined by the natural photoperiod. The photoperiod has been shown also to affect gonadal development in the catfish, *Heteropneustes fossilis* (Sundararaj and Sehgal, 1970), in the eastern brook trout, *Salvelinus frontinalis* (Henderson, 1963), and in *Tilapia* spp. (Hyder, 1970).

c. **The Fate of Unruptured Follicles.** Hypertrophy of the follicular membrane of unruptured follicles, followed by degeneration of the oocyte and resorption of yolk leading to the formation of luteal-like bodies, is common in crossopterygians (see Perry and Rowlands, 1962). These structures, variously called the preovulatory corpora lutea, corpora lutea atretica, or just corpora atretica, have been studied in detail in several species, particularly the catfishes, *Mystus seenghala* (Sathyanesan, 1962), *Heteropneustes fossilis* (Nair, 1963; Srivastava and Rathi, 1970), and *Clarias batrachus* (Lehri, 1968). It is generally considered that the function of these corpora is associated with the mechanism of the removal of yolk rather than with the secretion of hormones which occurs in mammals (Hoar, 1955).

When studying the production of preovulatory corpora lutea in the bitterling, *Rhodeus amarus*, Polder (1964) observed that, in ideal conditions for spawning, all large follicles ovulated, and none of the smaller follicles that remained unovulated was converted to corpora lutea atretica. He refuted the claim made by Bretschneider and Duyvené de Wit (1947) that the growth of the ovipositor in the bitterling was controlled by a substance "oviductin" secreted by the corpora atretica.

2. Viviparous Forms

Only a comparatively small number of bony fishes give birth to living young, and, as the ovary is closely involved in the mechanism of gestation, the few investigations that have been made provide much information about the ovarian cycle. Fertilization always takes place within the follicle, although Moser (1967) remarked that "ovulation" in the ovoviviparous *Sebastodes paucispinis* preceded fertilization as spermatozoa were not observed in follicles before or after that event. Since embryonic development in these viviparous forms is to some extent also intrafollicular, the follicles do not become converted into corpora lutea and gestation is not associated with a luteal phase of the ovarian cycle. Structures equivalent to the preovulatory corpora lutea of oviparous species are formed by "luteinization" of follicles other than those involved in the development of embryos.

The black surfperch, *Embiotoca jacksoni*, and the sea perch, *Cymatogaster aggregata*, produce one brood a year during July and August, after a gestation of 6 months. A delay of 3 months or more between mating and fertilization is a feature of these two species and of other embiotocids (Lagios, 1965; Wiebe, 1968). The ovarian cycle is a single annual event. Follicles in which the theca and granulosa membranes are not differentiated are found in the adult soon after the birth of the brood in August. Thecal differentiation and vitellogenesis follow quickly and the highest proportion of fully developed follicles occurs in November and December, although spermatozoa are present in the oviduct from August to December. Follicular degeneration increases during gestation and many corpora atretica of variable size are found, indicating that their formation is a continuous process during pregnancy. There is a rapid regression of ovarian activity after parturition.

Mendoza (1962) studied three species of the family Goodeidae which were all taken from small fresh-water lakes. *Alloophorus robustus* and *Goodea luitpoldii* have a single cycle which results in one brood of about twenty young from June to August, but in *Neoophorus diazi*, the cycle is repeated and young are born in three broods of about forty young each between April and February. *Neoophorus* matures after 1 year and *Alloophorus* and *Goodea* take 2 years. In all three species, the number of ovaries in a resting state was high during early spring, but the ensuing burst of oocyte growth ceased in June in *Alloophorus* and *Goodea* and in November in *Neoophorus*. There was no evidence of delayed fertilization as noted above for embiotocids, or of superfetation, as in some poecilid species (Turner, 1937). A detailed histological and histochemical study of the ovary of the guppy, *Poecilia reticulata*, has been reported by Lambert and van Oordt (1965).

The recent studies made on these viviparous forms are insufficient to confirm or refute the two trends of evolution, varying placental complexity and superfetation, that were discussed by Perry and Rowlands (1962).

III. AMPHIBIA

1. Oviparous Forms

The great majority of amphibians are oviparous and fertilization is external. Temperate zone species spend the winter months in a torpid state until emergence the following spring. The changes resulting in ovarian growth, follicular development, and ovulation occupy several months up to the time of spawning and are in sharp contrast to the very rapid loss in ovarian weight that follows ovulation (Jastrzebski, 1968). Usually only one large clutch of eggs is produced every year. Ovulation is followed by luteinization of the follicular epithelium, but, as Perry and Rowlands (1962) pointed out, it is probable that the postovulatory corpora lutea (as in certain teleosts) do not have as important a part to play as do the preovulatory corpora lutea produced by the degeneration of follicles during vitellogenesis. In amphibians, the latter are probably involved in the secretion of mucus which is deposited around the eggs during their passage through the oviduct to the exterior.

Oviparous amphibians return to water to breed, so that rainfall would be expected to be a very important factor in the control of their ovarian cycles. Much of the earlier work on reproduction in this vertebrate class was on species living in the temperate zone; the onset of spawning was convincingly shown to be rigidly controlled by rainfall, and the induction of embryogenesis to be determined by ambient water temperature (Balinsky, 1969). Within the temperate zone, inclement weather conditions, such as those prevailing at high altitude, may modify significantly the ovarian cycles of the same species living in less rigorous conditions nearer to sea level. Thus, in *Triton alpestris*, the periodicity of the ovarian cycle may be changed from an annual to a biennial event (Vilter and Vilter, 1963), and similar changes have been described in other species.

Studies made during the past decade on tropical amphibians, particularly *Bufo melanisticus* in Java (Church, 1960), *Rana erythraea* in Sarawak (Inger and Greenberg, 1963), and seven species near Singapore (Berry, 1964), have all indicated that rainfall probably accounts for the small variation in annual breeding periodicity. Other factors, such as the photoperiod and ambient temperature, remain constant throughout the year.

2. Viviparous Forms

Amphibians which are fertilized internally and are ovoviviparous have been reviewed by Amoroso (1952) and more recently some exhibiting viviparity have been described. A detailed description has been given of a viviparous anuran, *Nectophrynoides orientalis*, in West Africa, which is fully metamorphosed at birth (Lamotte *et al.*, 1964). The young are born in June, during the wet season, some mature rapidly and mate in September, and the rest remain immature. The proportion of each group depends on the time of onset of the dry season. Thus, if the rains stop in September, about 20% become pregnant before hibernation, but if they continue until November, the number pregnant may reach 80%. When the rains stop, all the frogs hibernate below ground, and, for the entire 6 months, the growth of those that mated and of the immature members of the population is at a standstill. When the pregnant females emerge and start eating again in September, embryonic growth recommences, thus completing a breeding cycle that recurs for 3 more years. About one-half of the total females born in any one year will breed in the same year (i.e., at 3 months of age) and the other half a year later (after hibernation) when 15 months old.

The black salamander, *Salamandra atra*, in the European alps is completely viviparous and, according to Vilter and Vilter (1964), usually gives birth to two young after a gestation period of 3 to 4 years, the young being completely adapted to terrestrial life. In this species, the supply of yolk is expended in 5–6 weeks and, thereafter, the embryos survive on the residual yolk in unfertilized eggs and follicles that undergo atresia, and by maternal mucus secreted by the gut. All follicles that ovulate (30–40) are converted into corpora lutea which decrease rapidly in size and number during the third year of gestation. Vilter and Vilter (1964) consider that the corpora lutea serve to maintain the activity of the pregnant uterus, but their contention is not substantiated by any endocrinological data. The corpora lutea inhibit oogenesis and follicular growth for a period of 2 years or more so that the ovarian cycle would probably be a triennial event. These authors were unable to account for the great disparity between the numbers of corpora lutea and embryos brought to term.

The presence of lipids, 3β-hydroxysteroid dehydrogenase and a steroid, probably progesterone, in the postovulatory follicle and in the fully developed corpora lutea of pregnancy in *Necturus maculosus* was reported by Kessel and Panje (1968) and in *Nectophrynoides occidentalis* by Xavier

(1969). These findings, together with the results of further experimental investigations by Xavier (1970) on the abortifacient effect of oophorectomy in the latter species, leave little room for doubt that the luteal phase of the cycle is intimately connected, endocrinologically, with the maintenance of gestation in these two amphibians.

IV. REPTILIA

1. Breeding Habits

Reptiles are distributed throughout the tropics and in temperate zones, and oviparity, ovoviviparity, and viviparity of varying complexity occur widely. A large number of species, particularly lizards, have been added in recent years to the list of those whose reproductive biology has been partly studied, but the lack of uniformity of the data that are presented renders any detailed comparisons of breeding habits and characteristics very difficult.

Factors affecting the mode of reproduction and the development of viviparity in snakes and lizards have been examined in several species. Temperature differences caused by altitude or latitude are clearly of great importance in the regulation of the breeding habits of some species. Weekes (1933) observed that the proportion of viviparous to oviparous species of snakes and lizards living at 1300 m or above in Europe and Australia was greater than that of others living at lower altitudes, and suggested that the colder air of the mountainous habitat was detrimental to the development of eggs in the nest. This observation has been confirmed by several authors. Thus, Neill (1964) reviewed the subject in considerable detail and noted, in particular, the change that occurred in the breeding habit of the North American colubrid snake, *Opheodrys vernalis*, throughout its natural ecological range. In specimens living in the extreme northern end of the distribution, the interval between lay and hatching was about 4 days, whereas at the southern end of its range, it was 1 month. Other examples have been provided by Greer (1966) for some Mexican and South American snakes. Greer (1968a,b) also analyzed the proportion of oviparous and live-bearing snakes and lizards in three ecological zones of different altitude in East Africa. He found that, on coastal plain lower than 300 m, 86.5% of 96 species were oviparous and 10.4% viviparous. A similar relationship held for another 123 species living on upland savanna (300–1800 m) but, in grassy uplands and alpine meadows varying between 1500 and 3500 m, eleven of the fourteen species (78.6%) were viviparous and the other three species were oviparous. Similar observations have been made for several iguanid lizards in Guatemala (McCoy, 1968). These differences are clearly adaptations to altitude.

Breeding in reptiles is a seasonal event and occurs annually. Tinkle (1962), however, has produced evidence which indicates that in at least two species of rattlesnake, *Crotalus atrox* and *C. viridis*, reproduction occurs only in alternate years, as was found by Rahn (1942) for *C. viridis* (see Perry and Rowlands, 1962). A large proportion of snakes and lizards live in tropical habitats where the annual rhythms in ambient temperature, photoperiod, and humidity may be reduced to a minimum as in some tropical forests, or in deserts where diurnal variations of temperature, at least, may be considerable. Of the twenty-three tropical species for which information about the female reproductive cycle is known, no fewer than nineteen have been recorded as breeding, more or less continuously, throughout the year (see Tinkle, 1969). Three species of Javanese house geckos were found to breed continuously in constant climatic and nutritional conditions (Church, 1962). Copulation may occur when ovarian eggs are quite small, indicating that spermatozoa are retained within the oviduct for long periods. This is a feature of reproduction that has long been known to occur in several other reptiles (see Kopstein, 1938). Neill (1962) has reported the occurrence of a well-defined breeding season in tropical snakes in British Honduras (15°N) and suggested that its limits were controlled by ambient temperature even though the mean annual minima and maxima over a 5-year period varied only from 75.3° to 85.5°F. A similar annual cycle occurs in the Indian house lizard, *Hemidactylus flaviviridis* (Sanyal and Prasad, 1967).

Rainfall in association with availability of food is another factor affecting the variability and productivity of the breeding season of some reptiles, particularly some desert lizards such as *Sceloporus orcutti* (Mayhew, 1963), *Uma inornata* (Mayhew, 1965), and *U. scoparia* (Mayhew, 1966). In these species in California, breeding is limited by the extent of the rainfall that occurred during the preceding winter. The lizards depend on various insects as a source of food and a rich supply in spring ensures full development of testicular and ovarian activity. The insects themselves depend on a good supply of plant growth which in turn is produced only by adequate rainfall over winter. Following a dry winter, therefore, gonadal activity is inadequate and the number of eggs produced is greatly reduced so that prolificacy is low compared with that in a breeding season after a wet winter. The East African lizard, *Agama agama lionotus*, at a latitude of 0°01′N where the annual variation of the photoperiod is only 2 minutes, has a breeding season which is controlled similarly by an increase in the insect population following the cessation of the "long rains" in spring (Marshall and Hook, 1960).

Reptiles in temperate zones commonly hibernate in winter. Reproductive activity begins soon after arousal in spring and continues throughout the summer months. Some recent studies have shown that the length of the annual breeding period may vary when the geographical distribution of the species covers a wide range of latitude. Thus, according to Fitch and Greene (1965), the ground skink, *Lygosoma laterale*, along the Gulf of Mexico has a breeding season which is twice as long as that of others in the latitude of Kansas City, and McCoy and Hoddenbach (1966) have also shown that the teiid lizard, *Cnemidophorus tigris*, has a longer season of reproductive activity in Texas than in Colorado. Unfortunately, the Japanese genus of iguanid lizard, *Takydromus*, has not been studied throughout its complete range of 54° of latitude.

2. The Follicle

The annual pattern of the ovarian cycle in adult reptiles is extremely variable, but in many oviparous and ovoviviparous species that have been investigated in recent years, especially those living in southern parts of the United States, it consists of an initial period of follicular activity in spring followed by ovulation, mating, and oviposition in early summer and ending in the autumn with the appearance of hatchlings after a variable period of incubation. Some follicular growth in hatchlings takes place before hibernation sets in, but is usually inhibited until after arousal in the following spring. The process of follicular growth is initiated every year and is usually completed in the same breeding season; occasionally, as in *Xantusia vigilis* (Miller, 1948), it extends over 3 years, but a new generation is produced annually.

The pattern in some viviparous species, e.g., *Sceloporus jarrovi* (Goldberg, 1970), differs considerably in that the follicular phase does not start until late summer and yolk deposition and ovulation take place in autumn, followed quickly by mating, fertilization, and early development to the blastoderm stage. Embryogenesis is suspended at this stage until the following April, and the young are born about mid-June. A similarly phased ovarian cycle involving pregnancy has been reported by McCoy (1968) for the Guatemalan iguanid lizard, *Corythophanes percarinatus*. The cycle begins in July or August and ends with parturition in the following May or June. McCoy does not mention the occurrence of a prolonged period of delay in embryogenesis in the latter species, but in both species the complete cycle involving gestation occupies about 10 months of each year. The North American turtle, *Chrysemys picta* (Gibbons, 1968), is an example of a species in which two ovarian cycles occur in a breeding season. Follicular activity, which commences in late summer, continues throughout the winter and by spring a group of six or seven follicles measuring 16 mm in diameter is present. Mating and ovulation follow and a second clutch of ovarian eggs develop when the first is in the oviducts. A third clutch does not develop when the second is in the oviducts. The ovary at this stage (July) regresses and remains inactive until September, when a new breeding season begins.

One of the most comprehensive studies of the ovarian cycle of a reptile has been provided by Telford (1969) on the Japanese lizard *Takydromus tachydromoides* from the island of Honshu (see Section IV,1). The study spanned three complete annual reproductive cycles and involved more than 2000 specimens, of which 75% were autopsied and examined biometrically. The ovaries of hatchlings examined in late summer contained a few follicles 0.1–1.3 mm in diameter and their growth was interrupted by hibernation. In the following March, follicular growth was resumed and, by May, vitellogenesis had commenced in six or eight follicles averaging about 1.0 mm in diameter. This resulted in rapid follicular enlargement and, by the time of ovulation in June, the follicles measured 7.6 mm in diameter. The majority of the first-year females ovulated from July to mid-August. This sequence of events was advanced by a few weeks in second-year and older lizards. One to four clutches, averaging 2.5 eggs each, are produced in first-year lizards. In older lizards, the number of clutches and the number of eggs per clutch are slightly greater.

Information about the frequency and size of the egg clutch is valuable as a means of determining the cyclical nature of ovarian activity and provides an estimate of the ovulation rate. Although it is believed that the annual reproductive activity of the majority of reptiles involves only one ovarian cycle, several species have been reported recently in which, on the above evidence, the cycle may be repeated and continue to do so for as long as environmental conditions permit. Thus, in the teiid lizard, *Cnemidophorus tigris* (McCoy and Hoddenbach, 1966), and in the skink, *Lygosoma laterale* (Fitch and Greene, 1965), the number of ovarian cycles is doubled in proportion to the duration of the breeding season within the range of latitude of their natural distribution.

The rapid growth of the follicle in its final stages is associated with the formation and deposition of yolk in the large macrolecithal egg which is characteristic of all reptiles. The secretion of yolk is a function of the granulosa membrane surrounding the egg, and the process of vitellogenesis is confined only to the eggs that are to be ovulated next. According to Tinkle (1961), follicular development and vitellogenesis rarely occur in the iguanid lizard, *Uta stansburiana stejnegeri*, while the previously ovulated clutch of fertilized eggs are in the oviduct. It is possible that the large yolk-filled eggs produce an inhibitory effect on the development of the follicles, and that this helps to regulate the follicular part of the ovarian cycle.

3. The Corpus Luteum

The macrolecithal egg of reptiles may attain a diameter of up to 20 mm and its ovulation involves far greater follicular trauma than is usual in mammals. Perry and Rowlands (1962) point out that the epithelium lining the follicle may become detached at ovulation, as was observed by Rahn (1939) and Bragdon (1952) in several genera of snakes. These workers, however, noted that the epithelium regains contact very quickly in the rapid conversion of the spent follicle into the corpus luteum. Several workers, among them Varma (1970) who studied the oviparous garden lizard, *Calotes versicolor*, have noted that the luteal cells are derived from granulosa cells which fill the cavity of the spent follicle, and that the vascular supply and the supporting fibrous connective tissues are derived from elements in the theca interna. The corpora lutea are much smaller than the follicles from which they developed and in this respect they differ from those of most mammals. Perry and Rowlands (1962) noted that the corpora lutea are large in relation to the body weight as compared with those of mammals.

As information becomes available on new species there seems to be increasing evidence of an association between the period of corpus luteum maintenance and that of retention of the fertilized eggs in the oviducts. Such a relationship is seen in several lizards such as *Uma scoparia* (Mayhew, 1966), *Dipsosaurus dorsatus* (Mayhew, 1971), *Uta stansburiana stejnegeri* (Tinkle, 1961), *Calotes versicolor* (Varma, 1970), *Leiolopisma rhomboidalis* (Wilhoft, 1963), and in *Takydromus tachydromoides* (Telford, 1969). In the last mentioned species, Telford found a close association between the color and size of the corpora lutea as they aged. In recently ovulated *Takydromus*, the corpus luteum varied in color from red to cream or white as it increased in diameter from 1.9 to 3.0 mm. During regression it became increasingly yellow and decreased in size to about 1 mm in diameter. Telford estimated from specimens killed at various times after oviposition that the corpora lutea were established in 5 days. At 60 days of age the corpora lutea had only just begun to regress, but their color had changed to deep yellow. The corpora lutea were deep orange at 108 days and were then regarded as corpora albicantia. These persist without further change throughout the life of the animal.

Evidence has been obtained for the persistence of the corpora lutea during gestation in several other viviparous species. The corpora lutea of *Sceloporus cyanogenys* are maintained for most of the 4-month gestation (Crisp, 1964), and in *S. jarrovi*, in which gestation is interrupted by a prolonged period of embryonic diapause (see Section IV,2), the corpora lutea persist until after parturition (about 7 months after mating) and then degenerate very rapidly. Badir (1968) reported that in another viviparous lizard, *Chalicides ocellatus*, the corpora lutea appear cytologically active for 6 to 8 weeks, a period equivalent to about half of gestation. However, the removal of the corpora lutea at any stage in pregnancy in this and other species quoted by Badir, does not interfere with the pregnancy. These lizards thus differ from the snakes in which the corpora lutea are necessary for the survival of the embryo, at least in the early stage of gestation (see Perry and Rowlands, 1962; and Volume II, Chapter 8).

V. AVES

1. Breeding Cycle

Breeding in birds is a seasonal event which is rigidly controlled by one or more external factors. In birds living in temperate zones the stimulation of gonadal activity is initiated, at least, by increase in day length, as shown originally by Rowan (1929) and in many it may be closely associated with the migratory pattern (Wolfson, 1966). In tropical conditions, where diurnal variation in the photoperiod is very slight, the onset of reproductive activity depends upon the advent of a regular period of high rainfall.

The advances made in the past decade or so in the knowledge of the mechanisms controlling the activity of the avian gonad have been among the most significant in the whole field of reproductive physiology. The stimulation of the female gonad is extremely complex and full gonadal development requires external stimuli other than those provided by photoperiodicity. In the white-crowned sparrow, *Zonotrichia leucophrys gambelii*, as in most birds, long daily photoperiods provide the initial stimulus to ovarian growth, and, according to Farner *et al.* (1966), the gonad of captive specimens of this particular subspecies does not develop beyond a weight of 50 mg unless the bird is stimulated by the presence of a fully sexually active male bird. The attainment of maximal ovarian weight resulting from full follicular development cannot be achieved without the provision of nesting sites and materials for nest building. The final burst of follicular growth in pigeons depends on stimulation evoked during behavioral displays in actual courtship. This bird is thus an induced ovulator. The role of the male in the maintenance of the oviduct and the frequency of ovulation in the ringdove, *Columba palumbus*, was demonstrated experimentally by Erickson and Lehrman (1964). Ovulation is also reported to be stimulated by the male and by nesting material in the Bengalese finch, *Lonchura striata* (Slater, 1969).

Food supply has been shown to be an important factor associated with periodicity in determining the duration of ovarian activity in the wood pigeon, *Columba palumbus* (Murton *et al.*, 1963), the rock dove, *C. livia*, and the common pigeon, *C. livia* (Lofts *et al.*, 1966). Feral populations of *C. livia* in Liverpool's dockland and in the city center of Leeds bred continuously throughout the year in the presence of abundant food. Dunmore and Davis (1963) reported a similar change of breeding pattern in feral pigeons having access to a plentiful food supply even in subzero temperatures in Pennsylvania. Lofts *et al.* (1966) postulated that gonadal regression in wild populations is necessary to ensure an adequate deposition of winter fat.

In tropical birds, some of which exhibit a well-defined annual breeding season, the absence of any significant variation in the natural photoperiod indicates that the periodicity of reproduction must be controlled in some other way. Snow and Snow (1964) studied the annual cycles in a variety of land birds in Trinidad over a 5-year period. The availability of food in some instances and rainfall level in others were clearly implicated. As a general rule, Snow and Snow (1964) believe that the breeding cycle is closely associated with the cycle of molt, and that the time of breeding depends upon the factors that initiate the molt.

Other studies, such as those carried out by Harris (1969) on the sea birds of the Galapagos Islands, and by Morel and Morel (1962) on 136 species in the Vale of Senegal, indicate that the majority of tropical birds have rigidly controlled breeding seasons. The nesting period may occur immediately after the end of the annual rains, during the dry season, or at any time during the year. Specific differences in this respect are very great.

2. Follicular Growth

The general picture that has emerged from a great variety of studies of many species living in different habitats is that during the winter follicles grow slowly over a long initial period and then pass through a short but much accelerated phase of active growth, due to the deposition of yolk. Ovulation and lay then occur.

Aspects of rapid follicular growth in the domestic hen and turkey have been reviewed by Bacon and Cherms (1968) who noted that its duration in the chicken had been reported by several previous workers to vary from 6 to 14 days and in the turkey from 9 to 18 days. The diameter of the follicle at ovulation, from 5 to 25 mm, depends on the size of the bird and the number of eggs forming the clutch. The number of clutches produced during a breeding season varies from one to four, or sometimes more, and depends on the length of the season. Thus, in several birds that breed in very high latitudes, such as the red-backed sand piper, *Calidris alpina*, and other species of the same genus (Holmes, 1966) in northwestern United States, and the white-crowned sparrow (King *et al.*, 1966) near Fairbanks, Alaska, there is only one follicular cycle in the short northern summer. Follicular growth in the latter species is closely correlated with seasonal changes in ovarian weight. According to Kern (1972) there is a slight enlargement of follicles during autumn and winter followed by accelerated growth in spring, and, as summer approaches, there is a third and final spurt beginning when the follicles measure 1–2 mm in diameter. The incubation period of *Zonotrichia* is a mere 20–22 days, and, in the event of premature loss of the clutch, a second clutch may be laid. The chukar, *Alectoris graeca*, which breeds in the southeastern region of Washington State, reaches full breeding condition and first lay is accomplished before the end of March (Mackie and Buechner, 1963). The follicles increase slowly in diameter to about 5 mm during a 4- to 7-week period. Within 1 week they attain a diameter of 22 to 27 mm and at this size ovulation occurs; this final spurt of growth represents an increase in diameter of about 4 mm per day. Subsequently, a new clutch of eggs is ovulated every other day during the early part of the season; the rate increases to a maximum of one a day during the remainder of the season. This frequency of lay is comparable to that in the domestic hen, which is a special case arising from intense domestication and intensive breeding.

In the event of a follicle failing to ovulate, it may degenerate or become atretic. Atresia (Volume I, Chapter 6) is an integral part of the ovarian cycle of birds and occurs chiefly in the period of rapid follicular growth. Several types of atresia have been described (see Volume I, Chapter 6) but the function(s) of the resulting bodies have not been determined. Yolky atretic follicles are found mainly during egg laying, incubation, and nestling periods. Lipoglandular atretic follicles occur mostly during transitional times such as that between winter and the pursuit periods to nest building, and from the latter to egg laying, and from then to incubation.

3. Clutch Size

The size of the clutch in birds provides a reasonable measure of the ovulation rate, as does litter size and number of corpora lutea in mammals. In normal circumstances, the size of the clutch and the number of clutches produced by any given species during the breeding season show little variation. It is well known, however, that the reduction in the number of the clutch or its complete removal from the nest, either by accident or by deliberate action, leads in many birds to the rapid restoration of the original number by new ovulations. Such birds are called determinate, as contrasted with indeterminate, layers. Dunham and Clapp (1962) quote the flicker,

Colaptes auratus, as an extreme case of a determinate layer. Seventy-two eggs were made to ovulate over 73 days (Phillips, 1887). Erpino (1969) comments on the determinate nature of the laying pattern of the black-billed magpie, *Pica pica*.

4. The Postovulatory Follicle

Many detailed accounts have been given of the mature avian follicle and the changes that occur at ovulation (blackbirds, *Agelaius tricolor* and *A. phoenicus*: Payne, 1966; black-billed magpie: Erpino, 1969; domestic pigeon: Dominic, 1960; domestic hen: Gilbert, 1968). The latter author provides information on the ultrastructure of the follicle.

In none of these species does the large spent follicle become converted into a structure comparable to the corpus luteum of mammals. A common feature of the follicle is the considerable shrinkage that occurs after ovulation. This is not altogether unexpected in view of the large size of the macrolecithal egg of birds and the extent of trauma caused during ovulation. The contraction of the follicle has been described by Payne (1966) in the blackbird as being probably aided by the fibrous elements of the theca externa and the sloughing of the granulosa layer whose cells quickly become pycnotic. Payne also describes the invasion of certain cell types that are typical of an inflammatory response and suggests, therefore, that the cellular components of the follicle wall are degenerating rather than developing a secretory function. Nevertheless, Payne demonstrated the presence of lipids and cholesterol in all three layers of the blackbirds' follicle for at least 10 days after ovulation. Dominic (1960) observed an initial hypertrophy of granulosa cells followed by breakdown of the membrana propria in the follicles of the pigeon, and the appearance of yolk granules indicated secretory activity occurring simultaneously with shrinkage.

In the hen, the egg remains in the oviduct for only 1 day before it is laid and although the life span of the spent follicle is much shorter than that of most, if not all, mammals, it is longer than the interval between ovulation and oviposition or lay, and is visible in the ovary for 7 to 10 days. The removal of the postovulatory follicle during the 24-hour period after ovulation delays the time of oviposition of the egg which arose from that follicle, and subsequent nesting behavior is unusual. Gilbert (1971) implies that the ruptured follicle has an important role to play in the time of lay. The hen, which has been domesticated by man for more than 5000 years and in recent times has been converted into a machine to lay an egg a day throughout its adult life or until it ceases to do so, obviously presents a special case and comparison with other avian species is probably unjustifiable. A three-volume compendium edited by Bell and Freeman (1971) on the physiology and biochemisty of *Gallus domesticus* should be consulted as the most comprehensive reference to its reproductive physiology.

VI. MAMMALIA

A. Prototheria

New information about the monotremes seems to be limited to some incidental observations of the single corpus luteum of two echidna, *Tachyglossus aculeatus multiaculeatus*, by Griffiths *et al.* (1969) during a study of the mammary glands. The ovaries were studied histologically 11 and 19 days after oviposition. The diameters of the older corpus luteum (4.6 × 3.0 mm) were twice as great as those of the younger one (2.2 × 1.5 mm), but the more advanced state of degeneration of the luteal cells of the former was more in keeping with its greater age. Griffiths *et al.* (1969) confirm that regression of the corpus luteum commences before egg laying (see Perry and Rowlands, 1962).

B. Metatheria

In contrast to the preceding group, numerous studies have been reported of the reproductive processes of the marsupial mammals. Unfortunately, only a few deal directly, or even indirectly, with the ovarian cycle. Many relate to the diprotodont group (kangaroos, phalangers, and wombats) which constitute the superfamily Phalangeroidea of Simpson (1945). The earlier work on this group was reviewed by Sharman *et al.* (1966) and a timely summary of various aspects of the evolution of reproduction in marsupials has been provided by Sharman (1970).

The majority of the species studied tend to breed in the first half of each year in Australia, but the breeding season may extend into September. Its duration is largely determined by environmental conditions, of which rainfall and nutrition are probably the most decisive. Breeding tends to be continuous throughout the year in the red kangaroo, *Megaleia rufa* (Sharman and Calaby, 1964), and in the marsupial mouse, *Antechinus stuartii* (Wolley, 1966), in controlled conditions. The latter species was found not to be strictly monestrous in the field, and in captivity there was a tendency toward greater prolificacy and an extended breeding season. In isolated females in captivity, the normal single cycle of development and regression of the marsupium was not repeated in the same breeding season so that it is unlikely that the basic ovarian cycle of the unmated animals is affected by domestic condition.

A large number of marsupials are polyestrous and monovular. One of the first to be described in any detail was the quokka, *Setonix brachyurus* (Sharman, 1955a). In the first cycle, at the onset of the breeding season, the single follicle matured rapidly to a diameter of about 3 mm and was accompanied by behavioral estrus lasting about 12 hours. Ovulation occurred 12–24 hours later and the next follicle began to enlarge about 18 days later when the corpus luteum of the first cycle was in decline. By day 26 a fully grown (3 mm in diameter) follicle was present and ovulation recurred. The degenerating corpus luteum was still recognizable and, in the absence of pregnancy, one or more corpora lutea were added in each succeeding cycle. The ovary of the brush possum, *Trichosurus vulpecula*, at estrus also contained only one large ovulable follicle (3 mm in diameter); neither ovary contained other follicles larger than 1 mm in diameter. In many of the latter, thecal elements were penetrating the granulosa to form corpora atretica. Shortly after estrus another group of two to seven follicles, 1–2 mm in diameter, had developed but of those only one was destined to ovulate during the next cycle (Pilton and Sharman, 1962). A similar situation has been reported for the marsupial glider, *Schoinobates volans*, in which the single follicle measured about 1.2 mm^3. The ovaries of these animals contained more than one corpus luteum although each was of a different age, thus confirming the polyestrous habit of this species (Smith, 1969).

Examples of polyestrous species are to be found in the families Phalangeridae and Dasyuridae. In the American opossum, *Didelphis virginiana*, which was one of the earliest marsupials to be studied in detail (Hartman, 1921), and in the marsupial mouse, *Sminthopsis larapinta* (Godfrey, 1969), the number of ova ovulated is larger than the litter size. Thus, in the latter species, the ovulation rate was thirty but the average litter size was only eight; the number of teats in the pouch is closer to the latter number. Polyovulation to this extent is unusual, but some extreme examples of it are recorded in some Eutheria (see Volume I, Chapters 4 and 6). Ovulation in all marsupials is spontaneous and occurs independently of any stimuli provided by mating.

The mode of reproduction adopted by all marsupials has dispensed with the need for their eggs to amass a large store of yolk as nutrient for the embryo, and in all species examined to date, the eggs are only slightly larger than those of eutherian mammals. The mature follicle of the marsupials, however, has not become correspondingly reduced in size as a fluid filled cavity or antrum develops during its growth and maturation. The events which have given rise to the antrum to form a Graafian follicle constitute a major change in the reproductive physiology of viviparous mammals from that of oviparous monotremes. Graafian follicles quickly become converted into fully formed corpora lutea. In didelphids and some dasyurids, in which gestation is extremely short, such as the American opossum (Hartman, 1923) and the marsupial mouse (Godfrey, 1969), the fully formed corpora lutea are produced in about 3 days. In other species, in which gestation is longer, the corpus luteum achieves its maximum size 7–9 days after ovulation. This early development of the corpus luteum is clearly correlated with the brief time that is taken by the egg to reach the uterus, e.g., only 24 hours in the quokka (Sharman, 1955b).

The corpus luteum persists throughout the very short gestation period, which rarely exceeds the length of one estrous cycle. However, it was shown by Tyndale-Biscoe (1963a) that, in the quokka, the corpus luteum is essential only during the first 7 days and later ovariectomy did not interrupt pregnancy (see Volume II, Chapter 8) although parturition failed to occur in the bilaterally ovariectomized animals. In the swamp wallaby, *Protemnodon bicolor,* in which gestation exceeds the length of one estrous cycle, the corpus luteum of pregnancy does not prevent the onset of estrus and ovulation a few days before parturition (Sharman *et al.,* 1966).

Mating at the postpartum estrus followed by conception in macropodid marsupials has the characteristic effect of prolonging the interval to the next parturition, if at the same time the young of the previous pregnancy are being suckled in the pouch. Sharman (1955a,b) ascribed this effect of lactation to an arrest of development of the blastocyst in the quokka, and halted luteal growth in the swamp wallaby. Removal of the young, either at the normal time or experimentally, led to the completion of luteal growth and the resumption of pregnancy, unlike the condition found in eutherians. The condition of arrested pregnancy in marsupials is due to the inhibitory effect of lactation and/or suckling on luteal growth and the term "embryonic diapause" was used by Tyndale-Biscoe (1963b) to describe the condition and to distinguish it from that of delayed implantation in eutherians; both conditions seem to be induced by an endocrine imbalance of ovarian origin. Tyndale-Biscoe found that the removal of the suckling stimulus caused the corpus luteum of lactation to increase in size, by hypertrophy and hyperplasia of the luteal cells, for a period of 17 days. By day 23, luteal regression was apparent and parturition occurred 1–2 days later, which is slightly shorter than the gestation period when pregnancy proceeds without interruption. The corpus luteum would therefore appear to be essential for embryonic development in marsupials.

One of the most prolonged periods of embryonic diapause occurs in the tammar wallaby, *Protemnodon eugenii.* Berger and Sharman (1969) showed that the first pregnancy of the breeding season was completed in 28 days and the length of the luteal phase was comparable to that in several other macropodids. The next pregnancy of the season, resulting from a post-partum ovulation, was interrupted by a period of diapause lasting on average 250 days, and the animal did not come to term until after the onset of the next breeding season. The corpus luteum remained dormant throughout this period and implantation followed the renewal of its activity. As in other species, the removal of the young from the pouch (see Volume II, Chapter 8, Table III, p. 361) or the injection of exogenous progesterone terminated the diapause and provided additional proof of the role of luteal activity in the regulation of the reproductive cycle in lactating macropodids.

C. Eutheria

The basic ovarian cycle of eutherian or placental mammals, uncompli-cated by pregnancy and/or lactation, is rarely encountered in wild popula-tions and is demonstrable only as an artifact in females isolated from males in captivity. However, the existence, in some rodents and other species, of two or more sets of corpora lutea of varied histological appearance and age between the time of onset of reproductive activity and of attainment of sexual maturity, or at the onset of a breeding season, indicates the rapid recurrence of ovulation with or without synchronous estrus (see Perry and Rowlands, 1962).

Three types of cycle (see Fig. 1) are evident among the commonly availa-ble laboratory animals. One type is that seen in the species which do not ovulate spontaneously (e.g., rabbit, Fig. 1a); the follicular phase is terminated by degeneration beginning with pycnosis and dehiscence of the membrana granulosa. The follicles are replaced by another set at intervals of 4 to 6 days without the intervention of a luteal phase. The other two types of cycle (Fig. 1b and c) are found in species in which ovulation occurs spontaneously and the difference between them relates to the degree of secretory activity of the corpora lutea. Studies on the laboratory rat, mouse, and hamster (see Perry and Rowlands, 1962: Fig. 1b) have shown that in sexually isolated females, ovulation is followed by the development of small, inactive corpora lutea having little physiological significance. Greenwald (1968), who has confirmed these findings in the golden hamster, referred to these corpora lutea as "merely the histological consequence of ovulation" having little, if any, ability to secrete progesterone. Luteal regression occurred on day 3 and atresia accounted for the reduction in the number of large follicles that would otherwise have ruptured on the follow-ing day. In the third type of cycle (Fig. 1c), spontaneous ovulation is followed by the development of fully functional corpora lutea which are

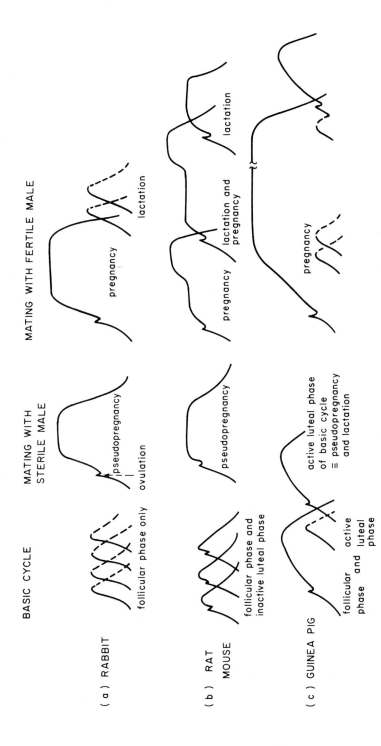

Fig. 1. Diagram of various types of ovarian cycle depicting the growth of the follicles and corpora lutea as modified in various reproductive states in (a) reflexly and (b, c) spontaneously ovulating mammals.

capable of inducing progestational changes in the reproductive tract. Among the small laboratory animals this type of cycle is to be found in the guinea pig and also occurs commonly in carnivores, ungulates, and primates. The corpus luteum is large, and persists for much longer periods than is usual in species like the hamster. Removal of the corpus luteum is usually followed by a quick return to estrus and ovulation, indicating the well-known suppressive influence of progesterone on these events.

When species that exhibit the first and second type of cycle mate but do not conceive, the ovulated follicle develops into a functional corpus luteum equivalent to those which characterize the cycle of type 3, and is associated with the condition known as pseudopregnancy. Conception in all eutherians, irrespective of the type of their ovarian cycle in the unmated state, leads to a further extension of the luteal phase in accordance with specific requirements for gestation, and in this respect eutherian differ fundamentally from marsupial mammals.

1. Rodentia

About one-third of all mammals are rodents and of this taxonomic order about 70% are myomorphs (rats, mice, hamsters, etc.), 20% sciuromorphs (squirrels), and 10% hystricomorphs (porcupine, guinea pigs, etc.). The cycle of the three best known myomorph rodents is essentially of the same pattern, but how far this may be regarded as representative of all the 1000 or more other species, and, similarly, to what extent the cycle of the guinea pig is characteristic of other hystricomorphs, is largely conjectural.

In myomorph rodents, the estrous cycle is usually 4–8 days and gestation lasts 20–25 days (see Asdell, 1964; Worden and Lane-Petter, 1957). One exception is the Australian native rat, *Mesembriomys gouldii,* in which the cycle is 27 days and pregnancy 44 days (Crichton, 1969). The corpus luteum appears to be active for 17 days of the cycle. Thus, the ovarian cycle of this myomorph is of the type shown in Fig. 1c and is unique for this suborder. A postpartum estrus and ovulation are reported to occur in *Mesembriomys* but gestation is not prolonged by a delay of implantation during lactation (see Salzmann, 1963), as is common in some other species (see Bland, 1969; Egoscue, 1970). The number of myomorph species in which mating or some other equivalent cervical stimulus is required to induce ovulation, thus indicating the existence of the first type of ovarian cycle (Fig. 1a), is growing; so far all are of the genus *Microtus.* Among those for which there is unequivocal evidence of reflex ovulation are *M. guentheri* (Bodenheimer and Sulman, 1946), *M. arvalis* (Delost, 1955), *M. californicus* (Greenwald, 1956), *M. agrestis* (Austin, 1957; Breed, 1967), *M. ochrogaster* (Richmond and Conaway, 1969), and *M. pennsylvanicus* (Clulow and Mallory, 1970).

A recent study on a laboratory-bred stock of *M. agrestis* has shown that luteal growth in pseudopregnancy is completed within 48 hours and regression, beginning on day 9, is rapid and is completed by day 12 (Breed and Clarke, 1970). The biphasic nature of the growth curve of the corpus luteum of pregnancy is similar to that of the laboratory rat (Perry and Rowlands, 1961), and although the onset of the second phase coincides with that of regression in pseudopregnancy in both species, it begins much later (13th day) in the gestating rat.

At the time of publication of the first edition of this book (Perry and Rowlands, 1962), the only significant information available on ovarian cycles of hystricomorphs related to the guinea pig. A little was known for the Canadian porcupine, *Erethizon dorsatum* (Mossman and Judas, 1949), and the Peruvian mountain viscacha, *Lagidium peruanum* (Pearson, 1949). The guinea pig was the first species to be used as a laboratory animal and the now classical studies of Loeb (1911) on the ovarian cycle, of Stockard and Papanicolaou (1917) on the estrous cycle, and Stockard and Papanicolaou (1919) on the vaginal closure membrane provide the foundation of our knowledge of the reproductive physiology of hystricomorphs. The occurrence of a vaginal closure membrane, unusually long gestation periods, and well developed young at birth appears to be characteristic of the suborder (Weir, 1974).

Since 1964, when interest in other hystricomorphs spread to the coypu, *Myocastor coypus* (Newson, 1966; Rowlands and Heap, 1966), several other species have been examined. The estrous cycle is generally 30–40 days, and gestation lengths are mostly over 90 days (see Weir, 1974). The shortest cycles (15–25 days) and pregnancies (52–70 days) occur in the cavies (Rood and Weir, 1970) and the casiragua, *Proechimys guairae* (Weir, 1973a), but even these periods are longer than those of myomorph rodents. Like the guinea pig, other hystricomorphs have a vaginal closure membrane, the perforation of which is an external indicator of ovulation. Ovulation is usually spontaneous, but the coypu and the tuco-tuco (*Ctenomys talarum*) may ovulate reflexly. Few female cuis, *Galea musteloides*, experience spontaneous estrus; the male evokes estrus, and ovulation is induced by coitus (Weir, 1971b, 1973b).

One of the most spectacular of the hystricomorphs is the plains viscacha, *Lagostomus maximus,* in which 200–800 extremely small follicles (200 μm) ovulate shortly after antrum formation (Weir, 1971a). The development of follicles to ovulable size seems to be a continuous process and ovulation occurs frequently during pregnancy. Numerous corpora lutea about 300 μm in diameter are formed from ovulated follicles, and similar numbers of accessory corpora lutea also develop. Follicular cycles terminating in degeneration, initiated by pycnosis of the granulosa cells, occur continuously in the pregnant and nonpregnant guinea pig (Perry and Rowlands, 1962). Very rarely do the follicles of the guinea pig become converted into accessory corpora lutea, as happens in almost all other nonpregnant and pregnant hystricomorphs (Perry and Rowlands, 1962; see also Volume I, Chapters 4 and 6 and Chapter 8 of this volume). The corpora lutea and the accessory corpora lutea have been shown to secrete progesterone (Tam, 1970). Different patterns in the rates of secretion of progesterone during the course of the extended luteal phase of gestation have been described in the guinea pig (Rowlands and Short, 1959), coypu (Rowlands and Heap, 1966), acouchi (Rowlands *et al.,* 1970), chinchilla (Tam, 1971), and cuis (Tam, 1973).

It is clear that very great differences exist between cyclic ovarian activity in hystricomorph and myomorph rodents. In fact, the diversification of the former is as striking as the similarity or conformity of the myomorphs to a common plan.

Our knowledge of the sciuromorphs is even more scanty than that of the other two suborders. Only about six of the 360 species of these rodents have been studied in any detail (see Seth and Prasad, 1969), of which four are North American and one each are British and Indian. The thirteen-lined ground squirrel, *Citellus tridecemlineatus* (Foster, 1934), and the Indian palm squirrel, *Funambulus pennanti* (Seth and Prasad, 1969), are considered to be reflex ovulators and their ovarian cycles are of the type (Fig. 1a) exemplified by the rabbit. In *C. beecheyi* (Tomich, 1962), *C. beldingi,* and *C. lateralis* (McKeever, 1966), ovulation is spontaneous, but the pattern of the ovarian cycle is unknown. The luteal phase extends throughout gestation, which for the majority of the sciuromorphs varies between 25 and 35 days (see Asdell, 1964). In *F. pennanti,* there is an extensive production of follicles during late pregnancy which are eliminated by atresia at parturition without ovulation and estrus taking place.

2. Artiodactyla

a. Domesticated Species. The estrous cycle in domesticated ungulates may be strictly seasonal in response to a decreasing daily photoperiod (sheep and goat). Although out-of-season breeding is encouraged in domestic animals for economic reasons, breeding is a seasonal event in all wild artiodactyls. A condition of anestrus brought about by ovarian hypoplasia is common in cattle, especially heifers, during the winter months. As the condition usually disappears spontaneously with the onset of spring, or may be corrected by exogenous gonadotropins (see Folley and Malpress, 1944), it may be environmentally determined, and could be a reflection of an ancestral type of restricted breeding. It would also seem significant that, although all sexual activity may cease in the sheep during spring and summer, the ovarian cycle, in particular cyclic follicular growth and regression, continues beyond the period of overt estrous behavior (see Robinson, 1951; Ortavant *et al.,* 1964). Such activity represents "silent" heats at the onset of each breeding season (Miller and Wiggins, 1964). Reeves and Ellington (1967) also provide evidence of cyclic ovarian activity in the absence of overt estrus. Some follicular growth occurs in late fetal life and in the neonatal cow (Marden, 1953) and giraffe (Kellas *et al.,* 1958; Kayanja and Blankenship, 1973). There is a recession of ovarian activity for some months before puberty, and during this period only small follicles (up to 2 mm in diameter in the cow) are formed (Morrow, 1969). According to Erickson (1966), the number of primordial follicles in the bovine ovary remains stable until the 4th year and then declines to zero at 15 to 20 years. The number of growing follicles increases rapidly between the 50th and 80th day and then more slowly up to 120 days. Vesicular follicles are most abundant at about 6 months of age; the numbers decline slightly until puberty (about 8 months) and remain unchanged until 10–14 years.

The pattern of follicular development during the estrous cycle has been reinvestigated in young, unmated heifers, by Rajakowski (1960) and in maiden ewes by Smeaton and Robertson (1971). In both studies some doubt has been cast on the widely held view that the follicular phase comprises the continuous growth of one or a group of follicles up to the point of ovulation at the next estrus. Rajakowski (1960) observed that between two succeeding periods of estrus and ovulation in Swedish Red and White cattle, two follicles grew, of which only the second ovulated. The first follicle became obvious during day 3 to 4 following the previous ovulation and it continued to grow, together with some others which remained smaller, until day 11 to 12, when it reached a diameter of nearly 5 mm. At this time all the smaller follicles became atretic, the large follicle became cystic and degenerated. Another wave of follicles developed between days 12 to 14, the most advanced of which underwent preovulatory changes within a few hours and ruptured on the second day of the 2-day period of estrus. Growth of the corpora lutea was complete by days 7 to 8 after ovulation and an increase in connective tissue and a thickening of the blood vessel walls on days 13 to 16 provided the first signs of luteal degeneration. Lutein cell regression, however, did not take place until days 17 to 18 and was virtually completed by day 21.

In the sheep, Smeaton and Robertson (1971) noted the separate development of three different groups of follicles during a single estrous cycle. The largest follicle of the first group attained a maximum diameter of 5 mm between days 6 to 9, and of the second between days 13 to 15. One follicle of the third group reached a size of 5 mm 36 to 48 hours before the onset of the next estrus and increased to about 12 mm by the early part of the second day of estrus and then ovulated. The 5 mm follicles of groups 1 and 2 became atretic during maximal luteal activity in the previous ovarian cycle. The preovulatory changes and extensive increase in size of the follicle which was to rupture occurred during the short period of luteal regression. A similar pattern of follicular development occurs in the mouse (Peters and Levy, 1966). It was suggested in much of the older literature that estrus and ovulation occurred only when the corpus luteum was waning. It is now apparent that ovulation is inhibited by progesterone in many species, particularly in sheep and cattle.

A day by day series of color photographs taken throughout the 20-day cycle of the gilt was used by Akins and Morrissette (1968) to show that there were three phases of ovarian change. The first (days 1–8) is a phase of rapid growth of the new corpus luteum, the regression of the previous set of corpora albicantia, and the initial growth of the next group of follicles. The second (days 9–14) is the period of maximal luteal activity, vascularity, and color of the corpus luteum; little change occurred in follicular growth. During the third phase (days 15–20) the corpus luteum regresses, and follicular growth recommences as estrus approaches. Akins *et al.* (1969) examined acid and alkaline phosphatase activities of nulliparous gilts during the cycle. Acid phosphatase activity in the luteal tissue rose sharply at day 14, indicating the onset of luteal degeneration. Alkaline phosphatase declined to a minimum at 10 days as the corpus luteum matured, but increased again from 10 to 15 days.

Several studies have been made of progesterone secretion and release during the ovarian cycle of domestic ungulates (cow: Hafs and Armstrong, 1968; Dobrowolski *et al.*, 1968; Stabenfeldt *et al.*, 1969a; sheep: see Moore *et al.*, 1969; pig: Gomes *et al.*, 1965). All provide evidence of the activity of the cyclic corpus luteum and, as judged by progesterone levels in peripheral plasma, a common feature in these ungulates is a precipitous decline in progesterone level just before the onset of the next estrus and ovulation (Stabenfeldt *et al.*, 1969a,b). This pattern contrasts markedly with the much slower decline noted in primates (Section VI,C,11,b).

Several other common but less familiar domesticated bovids have been studied in various parts of the world. Chief among these are the domestic zebu, *Bos indicus,* which has recently been studied in the Gulf Coast of the United States by Plasse *et al.* (1970) and in Central Africa by Rakha *et al.* (1970). The latter workers also studied the Angoni and Boran breeds of zebu and Sanga cattle. Striking differences were found between the reproductive characteristics of zebu in the two locations but details of their ovarian cycle are not yet available.

The reproductive pattern of the domestic buffalo, *Bubalus bubalis,* in

Egypt was described by Zaki *et al.* (1963) and an histological and biochemical study of the corpus luteum was made by El-Sheikh *et al.* (1969). Total output of progesterone by the single corpus luteum reaches a maximum of 200 to 300 μg and is maintained throughout the second quarter of pregnancy (gestation 240 days). The Murrah buffalo in India has a cycle length of 19 to 23 days, and in other features its reproductive pattern closely resembles that of European domestic breeds of cattle. For example, progesterone concentration in the corpus luteum rises to a peak in midcycle (41.7 ng/mg on day 12) and declines significantly before the onset of the ensuing estrus (Roy and Mullick, 1964).

b. Wild Species. The number of wild ungulates whose reproductive physiology has been studied has increased rapidly in recent years. In a review of African species, Fairall (1970) mentions more than a dozen for which information on mating and calving seasons, the recurrence and duration of estrus, and gestation provides some basis for a discussion of breeding patterns and cyclic ovarian activity. Among those studied in some detail are the impala, *Aepyceros melampus* (Kayanja, 1969), Uganda kob, *Adenota kob* (Buechner *et al.,* 1966), defassa waterbuck, *Kobus defassa* (Spinage, 1969), and the Serengeti wildebeest, *Connochaetes taurinus* (Watson, 1969). In these bovids, as in all wild mammals, a high pregnancy rate obscures the basic ovarian cycle that is observed in the domesticated species. The occurrence of "silent" heats suggests that, in wildebeest and Uganda kob, cycle length is about 15 and 20–26 days, respectively. The corpora lutea in these aberrant cycles are smaller and have a much shorter life-span than those formed during pregnancy. The mature follicle has a diameter of 10 to 12 mm and, in the wildebeest at least, is present only in the short mating period during May and June. Other normal follicles up to 15 mm occur throughout the spring and summer months and sporadically at other times of the year. This widespread or even continuous production of follicles may be associated with the relatively constant photoperiod of tropical and subtropical environments. Morrison (1971) and Morrison and Buechner (1971) have described the corpora lutea of the Uganda kob.

The ovaries of the fetal defassa waterbuck were similar to those of the domestic cow in that a conspicuous wave of follicular activity occurred before birth, leading to the production of many antral follicles up to 2.5 mm in diameter (Spinage, 1969). Their regression in early postnatal life and the time of resumption of subsequent ovarian activity were similar to those of the domestic calf and cow. The largest follicles found in mature animals measured about 11 mm.

The reproductive cycle of the few deer (family Cervidae) that have been examined conforms in its general pattern to that of other artiodactyls, such as the sheep. These animals have an autumnal mating season and drop their young in the spring; pregnancy is maintained throughout by an active luteal phase. The roe deer, *Capreolus capreolus,* is, as far as is known, unique among Artiodactyla in exhibiting delayed implantation, a phenomenon which was first described in this species. Short and Hay (1966) have pointed out that the corpus luteum remains fully active throughout the 5-month period of delay, and that following implantation it undergoes no change in histology or progesterone content. Neither the above authors nor Chaplin *et al.* (1966) were able to observe the continuation of luteal function to the end of gestation (about 40 weeks). Guinness *et al.* (1971) could not detect any evidence of silent heats with ovulation preceding the first estrus of yearling red deer, *Cervus elaphus,* in October. They assumed, therefore, that in the red deer conception occurs during the first and only ovarian cycle of the breeding season. The length of the cycle in the absence of a male was about 18 days, and most hinds experienced a so-called "accessory ovulation" early in pregnancy without any associated estrous behavior. Gestation lasts approximately 230 days. Although the corpus luteum was considered to remain functional throughout pregnancy, only insignificant amounts of progesterone were detected in the peripheral plasma. The incidence of silent heat in Canadian cervids was discussed by Simkin (1965), who noted its occurrence in a large proportion of moose, *Alces alces.* He was unable to provide any data about the length of the basic ovarian cycle.

The camelids (see Novoa, 1970) are of particular interest in that of the six species, at least the alpaca, *Lama pacos* (San-Martin *et al.,* 1968), and the llama, *L. glama* (England *et al.,* 1969), have been shown to be induced ovulators. Shalash and Nawito (1964) have suggested that the dromedary, *Camelus dromedarius,* might also ovulate reflexly, a view which Nawito *et al.* (1967) have amended by suggesting that although cyclical estrous activity does not occur, follicular growth recurs every 21–24 days. The bactrian camel, *C. bactrianus,* is stated by Asdell (1964) to be polyestrous and may ovulate spontaneously. There is no information about the guanaco, *L. guanicoe,* and vicuna, *Vicuna vicuna.* Luteal growth and progesterone content following ovulation induced by mating in the llama (England *et al.,* 1969) is maximal after 8 days and considerably reduced by day 16. The corpus luteum remains identifiable until day 120, but is then completely inactive. Pseudopregnancy, such as was reported in the alpaca (San-Martin *et al.,* 1968), does not occur in the llama. Although gestation periods have been recorded for these animals, little is known about the survival period of the corpus luteum of pregnancy.

Henry (1968) showed that the length of the estrous cycle in the European wild hog, *Sus scrofa,* is 21–23 days and the gestation period 115 ± 2.3 days; these values are not significantly different from those of the domestic sow. The warthog, *Phacochoerus aethiopicus,* breeds throughout the year in

Uganda and estrus, lasting 48 hours, recurs every 6 weeks; gestation is 173 days (Clough, 1969). A feature of the luteal phase is the rapid growth of the corpus luteum to a diameter of about 10 mm before the egg leaves the Fallopian tube. Regression begins only after parturition and the corpus albicans persists for many months. Ovulation does not occur at parturition or throughout lactation. In captivity the collared peccary, *Tayassu tajacu,* is polyestrous and the cycle recurs every 17–30 days. Pregnancy lasts 142–149 days, and estrus and ovulation occur about 8 days after parturition (Sowls, 1966). No direct observations have been made of the ovarian cycle of this species.

A comprehensive study of the hippopotamus, *Hippopotamus amphibius,* in Uganda has provided accurate data on the ovarian cycle (Laws and Clough, 1966). Reproduction usually follows a 2-year cycle. The gestation period is 240 days and the corpora lutea persist throughout pregnancy. They reach maximal size (47.6 mm, 88 gm) as early as 22 days, 50 mm (98 gm) at 120 days, and 46 mm (86 gm) at term. The corpus luteum regresses rapidly after parturition, and persists as a corpus albicans. Many follicles grow during pregnancy and become converted into accessory corpora lutea. There is a linear increase in the number of these structures as pregnancy advances.

3. Insectivora

One of the earliest detailed descriptions of the ovarian cycle in an insectivore was provided by Brambell (1935) for the common shrew, *Sorex araneus,* in Britain. This shrew is polyestrous and the second pregnancy of the breeding season is initiated at the postpartum estrus. Brambell considered that ovulation was coitus-induced, as in several other insectivores, especially soricids.

A feature common to several species of insectivores is the small size of the Graafian follicle, which is attributed to the meager development of the antrum. Thus, in the Asian house shrew, *Suncus murinus,* another induced ovulator, antrum formation occurred only after copulation (Dryden, 1969). In this species too, all adult females contain large nonantral follicles and there is no apparent cycle of ovarian activity. Another common character is the eversion of the spent follicles to produce fungiform corpora lutea situated within the periovarian capsule. This occurs in various elephant shrews (see Tripp, 1971) and tenrecs (see Dryden, 1969). Tripp examined eleven species of elephant shrew and showed that eversion of the corpora lutea is not correlated with the phenomenon of polyovulation which was first reported by van der Horst and Gillman (1940). Very little is known about the basic cycle in spontaneously ovulating insectivores, and, generally, luteal function does not extend throughout gestation, indicating that the later stages of pregnancy are maintained independently of ovarian function.

The ovarian interstitium of the mole, *Talpa europaea,* undergoes an annual cycle of activity (see Perry and Rowlands, 1962, and Volume I, Chapter 4). According to Deanesly (1966) the so-called interstitial gland develops from the expansion of the ovarian medulla during the winter anestrus as a result of an invasion and hypertrophy of the eosinophilic elements. When fully developed, this gland occupies most of the ovary, the quiescent ovigenic tissue being reduced to a thin cortical rind. A similar separation of the latter tissue from a medullary (interstitial) zone has been described in the white-tailed mole, *Parascaptor leucurus* (Duke, 1966), and the desman, *Galemys pyrenaicus* (Peyre, 1969). As far as is known, the short-tailed shrew, *Blarina brevicauda,* is unique among mammals in having intrafollicular fertilization followed by ovulation (Pearson, 1944).

4. Carnivora

The most studied family of this order is the Mustelidae, among which are some of the smallest carnivores, such as the weasel. In the wild, mustelids breed in spring or summer. In winter, some are in anestrus; others experience a delay of implantation which is a phenomenon frequently found in carnivores. The ferret is an albino mutant of the polecat, *Mustela putorius,* and has retained its seasonal breeding habit in captivity. The onset of gonadal activity can be altered by increasing the photoperiod during winter. In normal laboratory conditions, follicular activity commences in late winter and as ovulation is coitus-induced, the luteal phase is controlled by the time of mating. A recent study by Carlson and Rust (1969) showed that on day 30 of pseudopregnancy and of pregnancy (both 42 days) plasma progesterone levels were 8.9 and 8.4 ng/ml, respectively, which clearly indicates the similarity of luteal activity in these two reproductive states. Mustelids in which pregnancy proceeds without interruption, such as the ferret and weasel, *M. nivalis,* have gestation periods of 5 to 6 weeks and usually are able to produce two litters in a season. By the end of the first lactation the ovaries contain large follicles and estrus recurs rapidly after suckling stops. The immediate recurrence of estrus in lactating ferrets after the premature loss of some or all members of a litter is a clear indication of the suppressive action of the suckling stimulus on follicular maturation and estrus (Rowlands, 1957). In these species the second ovarian cycle is initiated rapidly after the end of lactation, but after the second lactation it is rare for a third set of follicles to mature. If this does happen and mating occurs, a third pregnancy is usually prevented by a failure of implantation. In most cases, follicular activity diminishes with the onset of anestrus.

The ovarian cycle of species in which implantation is delayed is protracted because of the very long life-span of the corpus luteum. The period of delay in implantation is commonly associated with the failure of the corpus luteum to develop fully. A similar pattern of increased luteal growth, vascularity, and of progesterone secretion following nidation has been described in many mustelids, including the American badger, *Taxidea taxus* (Wright, 1966), the otter, *Lutra canadensis* (Hamilton and Eadie, 1964), the fisher, *Martes pennanti* (Eadie and Hamilton, 1958), the mink, *Mustela vison,* the European badger, *Meles meles,* and marten, *Martes americana* (see Canivenc et al., 1969), the spotted skunk, *Spilogale putorius* (see Mead and Eik-Nes, 1969), and stoat, *Mustela erminea* (Gulamhusein, 1973). Considerable luteal growth occurs before implantation in the sea otter, *Enhydra lutris* (Sinha and Conaway, 1968).

The stoat has one of the longest periods of delayed implantation among mustelids (Deanesly, 1943). In a recent study of the ovarian cycle in wild-caught and laboratory-bred stock, Gulamhusein (1973) has observed a reciprocal relationship between the activity of the luteal cells and that of the interstitial tissue. The cells forming the latter are most active during the preimplantation period of 7 to 9 months, while the luteal cells are not maximally active until after implantation. Another remarkable feature of the stoat ovary is the very early postnatal development and maturation of the follicles, with an associated estrus some 2–3 weeks before weaning. The luteal phase persists for 8 to 11 months; it is remarkable in so far as the corpora lutea are stimulated to maximal size and secretory activity only for a period of about 5 weeks after implantation. In the stoat and most other species the corpora lutea regress rapidly after parturition. In the spotted skunk, *Spilogale,* there is a rapid rise in secretory activity soon after implantation, followed by an immediate decrease until parturition. Two eastern forms or subspecies of this species (*S. p. interrupta* and *S. p. ambarvalis*) would appear to ovulate spontaneously and not to have a delay of implantation, whereas in the western subspecies, *S. p. latifrons,* ovulation is controlled by mating, and there is a long period of delay in implantation. The eastern subspecies produce two litters in a breeding season and the western only one (Mead and Eik-Nes, 1969).

The ovarian cycle of the mink is very unusual and probably unique. Follicular maturation and mating, followed by ovulation and fertilization, normally occur during the delay of implantation. The period of delay is variable and depends largely on photostimulation (Enders and Enders, 1963). Within 8 to 14 days following mating, a new follicular cycle begins, another batch of mature follicles is formed, and estrus recurs. A second mating is a normal procedure in commercial mink breeding and the products of the second conception contribute 80–90% of the embryos that come to term.

The gestation period varies between 42 and 79 days; true gestation lasts about 28 to 32 days. During the period of delay, mature follicles are always abundant and atresia is common. After implantation, the atretic follicles form blocks of interstitial tissue and the growth of mature follicles seems to be inhibited.

Delayed implantation is not known to occur in Felidae (cats), Canidae (dogs), or Procyonidae (racoons, pandas). It probably occurs in Ursidae (bears), in which gestation lasts for 8 months in captivity. The fetuses at term are extremely small in proportion to adult weight (Wimsatt, 1963). The small size of the neonate presupposes a considerable period of lactation, so that the duration of the complete reproductive cycle is probably well in excess of 1 year. Craighead *et al.* (1969) suggest that bears produce a litter only biennially, and that the annual ovarian cycle is thus suppressed on alternate years by the inhibitory effect of lactation. Wimsatt (1963) has provided a detailed account of the ovarian cycle of the black bear, *Ursus americanus.* His evidence suggests a period of implantation delay of about 5 months and a period of postimplantation embryonic growth of only 6–8 weeks. He observed minimal luteal growth during the delay period, but following implantation there was a threefold increase in the size of the corpus luteum and a thirteen-fold enlargement of its component cells. The general trend of events is similar to that described for mustelids. Wimsatt (1963) regards ovulation as being induced by coitus, a characteristic which is common among delayed implanters. The dog, wolf, and fox are monestrous, the fox and wolf having one and the dog two cycles in a year. Both the follicular and luteal phases of the cycle last about 2 months, so that in the dog there is a short period of anestrus (about 6 weeks) twice a year, and a much longer one, of about 6 months, in the fox (Rowlands and Parkes, 1935). Ovulation is spontaneous, and the corpora lutea are very active, even in the absence of mating, and significant levels of progesterone are present at least 2 months after ovulation (see Christie *et al.,* 1971). These levels of progesterone may induce a physiological condition equivalent to pseudopregnancy which occurs following mating. The corpora lutea of pregnancy in the wolf are recognizable up to 5 months following parturition (Bonnin-Laffargue and Canivenc, 1970), although the decrease in progesterone secretion is more rapid than the regression of the corpora lutea.

5. Chiroptera

There are nearly 1000 extant species of bat. They form the second largest order of mammals, and are widely distributed throughout temperate and tropical regions. Many unusual features of their reproductive processes appear to have evolved in response to the demands of habitat and feeding habits. More than 80% of bats are insectivorous (Microchiroptera) and comprise all the species that live in temperate zones in which breeding is closely linked to the photoperiod. The remainder (all Megachiroptera and some Microchiroptera) are tropical and live on fruit. As far as can be determined, all bats are monotocous and only one follicle matures in each monestrous cycle of the breeding season.

The most noteworthy reproductive feature of temperate zone microchiropterans is delayed fertilization. This results from the separation in time between estrus and copulation (late summer and early autumn), and ovulation and fertilization in the following spring, 6–7 months later. The phenomenon probably evolved in association with hibernation as an adaptation to the lack of insect food in winter months. Thus, the ovarian cycle is unique because of its protracted follicular phase (see Perry and Rowlands, 1962; Volume I, Chapter 4). Exogenous gonadotropins are capable of inducing follicular maturation and ovulation in hibernating bats, and Wimsatt (1960) suggests that the long natural delay is caused by an inadequate production of the endogenous hormones. Pituitary–ovarian relationships in eight species of hibernating bats (Vespertilionidae) have been reviewed by Racey (1972). Irrefutable evidence of delayed fertilization, of the survival of the outstanding follicle, and of sperm viability throughout hibernation in all these species, was obtained by segregation of the sexes in the autumn (see also Racey and Kleiman, 1970). Young were born in the early summer some 8 months later. A detailed study of the development of the outstanding follicle in the little brown bat, *Myotis lucifugus,* was reported by Wimsatt and Parks (1966). Herlant (1968) has correlated the cycles of ovarian and pituitary activity in another vespertilionid, the common European bat, *Myotis myotis.*

The slow growth of the follicle throughout hibernation, the hypertrophy of the membrana granulosa, and the reduction of tension within the follicle leading to the apparent lack of an antrum are all characteristics of the preovulatory condition. Ovulation frequently results in the eversion of the corpus luteum as reported for insectivores (see Section VI,C,3). Racey (1972) observed that the corpus luteum grows slowly at first and that maximum size is attained by midpregnancy, following a dramatic hypertrophy of the luteal cells shortly before that time. In noctules, *Nyctalus noctula,* which failed to conceive in captivity, the corpus luteum continued to develop, suggesting a cycle of pseudopregnancy (Racey and Kleiman, 1970; Racey, 1972).

The diversity of ovarian patterns in neotropical and tropical micro- and megachiropterans is in marked contrast to the rigidly controlled annual cycle of temperate zone bats. Monotocy seems to be preserved, but the number of cycles per year may be increased to two or even three as in

Myotis nigricans (Wilson and Findley, 1970). In those species having a single pregnancy per year, gestation is usually completed during spring and summer and is followed by anestrus. In the Indian vampire bat, *Megaderma lyra,* and possibly other megadermids (Ramaswamy, 1961), and in the bent-winged bat, *Miniopterus australis,* in Northern Borneo (Medway, 1971), the single annual cycle commences in late autumn and parturition takes place $4\frac{1}{2}$–5 months later. Medway (1971) considers that this type of cycle is not uncommon in an equatorial environment with a minimum of seasonal change. The occurrence of seasonality in ovarian activity in a tropical environment (India) has also been reported by Anand Kumar (1965) for the rat-tailed bat, *Rhinopoma kinneari,* in which there is a well-defined period of anestrus from late summer to the following spring. Pregnancy in the long-tongued bat, *Glossophaga soricina,* may occur at all times of the year, and when kept in laboratory conditions this tropical bat is polyestrous, the modal length of the cycle being 24 days (range 22–26 days). Ovulation, which is spontaneous and monovular, is followed by an active luteal phase whose end is marked by vaginal bleeding resulting from the degeneration of an hypertrophied glandular endometrium (see Asdell, 1964; Rasweiler, 1972).

Gopalakrishna (1964, 1969) describes an unusual pattern of ovarian activity in the Indian fruit bat, *Rousettus leschenaulti.* Gestation lasts for 4 months and the corpus luteum of pregnancy persists until halfway through the next pregnancy. When a second pregnancy is initiated at the postpartum estrus in March, the life-span of the corpus luteum of the first pregnancy is 6 months. After parturition in July, there is a 4-month anestrus and the corpus luteum of the second pregnancy lasts for 10 months. The luteal phases of the two cycles overlap until midpregnancy when the corpus luteum of the previous pregnancy rapidly regresses. Follicular growth then occurs in this ovary, thus causing an alternation of ovulation. The first ovulation is random, but, in some bats, the ovarian cycle occurs exclusively or predominantly in one or other ovary throughout life (see Racey, 1972). A similar list on a much wider taxonomic basis, including some bats, is given by Pearson (1949).

6. Pinnipedia

A review by Harrison (1969) of the reproductive patterns and morphology of the female genitalia in a variety of seals and sea lions summarizes, in tabular form, information about age at sexual maturity, length of breeding season, and gestation and lactation periods.

The ovaries of several seals undergo marked development during the second half of embryonic life. This was described in detail for the gray seal,

Halichoerus grypus, and common seal, *Phoca vitulina,* by Amoroso *et al.* (1965) as an hypertrophy of the interstitial-like cells which is maximal before birth and regresses rapidly afterward. These changes are histologically similar to those occurring in various perissodactyls (see Section VI,C,9).

The general pattern of ovarian development is for estrus, accompanied by ovulation, to occur in late summer or autumn in the fourth or fifth years of life. Ovulation occurs spontaneously, and in many species pregnancy is interrupted by a delay in implantation lasting many months. During this period, the development of the corpus luteum is slow, and increases rapidly after implantation. In the fur seal, *Callorhinus ursinus,* secretory activity of the corpus luteum ceases soon after implantation, and luteal regression coincides with the establishment of the placenta (Craig, 1964). The luteal phase in the hooded seal, *Cystophora cristata,* persists throughout gestation and there is little follicular activity in the functional ovary (Øritsland, 1964). Both species have an ovarian cycle, including lactation, which occupies 1 year and alternates between the ovaries with great regularity. Craig (1964) noted in the fur seal that about fifteen follicles developed, and only four enlarged in late pregnancy. One ultimately survived and ovulated; the rest became atretic. The mature follicles measured about 10 mm in diameter and ruptured 3–5 days after parturition in mid-July. By contrast, ovulation in parous hooded seals occurs shortly after a lactation lasting only 10–12 days (Øritsland, 1964).

7. Cetacea

The cetaceans comprise the toothed whales, dolphins, and porpoises (Odontoceti) and the baleen or whalebone whales (Mysticeti). One of the main features of their ovarian cycles is the length of the luteal phase (see Harrison, 1969). The corpus luteum exists for more than 1 year, and then becomes converted into a corpus albicans which persists indefinitely. The corpora lutea accumulate in the ovary and, therefore, provide some indication of the age of the animal. Detailed studies have been made of the humpback whale, *Megaptera nodosa,* in Australian waters (Chittleborough, 1954), fin whale, *Balaenoptera physalus* (Laws, 1961), Sei whale, *B. borealis* (Gambell, 1968), and sperm whale, *Physeter catodon* (Best, 1968).

The reproductive cycle in cetaceans lasts a year and recurs annually when conception occurs at the postpartum estrus. The cycle is closely linked to an annual migration from the main breeding grounds in the southern Indian and Pacific Oceans to Antarctica to coincide with the maximum abundance of the planktonic food, krill (*Euphausia superba*), during the summer. On the journey south, the adults are lactating and some of those that mated at the postpartum estrus are also pregnant. The young are weaned in the Antarctic and return independently to tropical and subtropical waters to breed the following winter. On the northern journey, the adults are in the later states of pregnancy or their reproductive organs are quiescent.

Laws' (1961) account of the reproduction of the fin whale indicates that a large proportion of multiparous females experience a postpartum estrus and ovulation but that less than 20% conceive. Pregnancy, therefore, is an annual event in only a minority of whales. Another small proportion experience a postlactational estrus, with ovulation, but pregnancy rarely results. Gambell (1973) has summarized the effects of gross reduction in the numbers of three species of whale which have been exploited by man. The age at puberty is significantly earlier. Pregnancy rates have almost doubled through an increased occurrence of postpartum and postlactational ovulations resulting from additional cycles of follicular activity during the late luteal phase. Conception failed in a large proportion of whales undergoing these aberrant cycles.

Except for the bottle-nose dolphins, *Tursiops truncatus,* commonly kept in dolphinaria, the ovarian cycle of other delphinids has been little studied. Harrison and Ridgway (1971) record that gestation lasts for 12 months and lactation for 18 months, and that the ovaries of this dolphin contain only one corpus luteum, which is always associated with pregnancy. The corpus luteum remains functional throughout gestation and then becomes converted into a corpus albicans during the next 3 years. The frequency of the ovarian cycle of pregnancy in the bottle-nose dolphin is difficult to assess and, in the probable absence of a postpartum estrus and conception, it is unlikely to be less than biennial. A postpartum estrus with ovulation occurs in the common dolphin, *Delphinus delphis* (see Harrison, 1969, for references), and, perhaps, in the South American "bouto," *Inia geofroyensis.* In these species, the ovarian cycle lasts for 1 year. Ohsumi (1964) observed that in delphinids the left ovary matures before the right ovary, and, in later life, ovulation occurs more frequently from the right ovary.

8. Proboscidea

Some extensive investigations have been made of the ovaries of the elephant, *Loxodonta africana* (Short and Buss, 1965; Laws, 1969; Hanks and Short, 1972). It has been confirmed that luteal tissue and plasma in this species contain little, if any, progesterone (see Smith *et al.,* 1969). Although the significance of corpora lutea in this species is uncertain, it is generally agreed that corpora lutea are abundant in pregnant and nonpregnant animals. The corpora lutea are less numerous in pregnancy and there is no evidence that they become enlarged in this condition. Buss and Smith (1966) and Laws (1969) have indicated that the corpus luteum persists throughout gestation (22 months) and have been unable to confirm Perry's (1953) claim that a second crop of corpora lutea develop in midpregnancy (see Volume I, Chapter 4).

Accurate information about the periodicity of the estrous cycle in wild elephants is lacking, but in captivity, in the presence of a male, symptoms of estrus have been detected at monthly intervals, and a cycle of 3 to 4 weeks has been reported in the Asiatic elephant, *Elephus maximus* (Jainudeen *et al.*, 1971). However, there is some doubt whether or not the African elephant is polyovular (Perry, 1953; Laws, 1969) or monovular (Short, 1966). If all the corpora lutea in the ovary are coeval, polyovuly is implicit, but if the cycle is monovular, regular cyclical activity for 12 to 18 months must be postulated to account for the number of corpora lutea present at conception.

9. Perissodactyla

Several recent studies of the mare, *Equus caballus,* have dealt with ovulation during early pregnancy, the fate of the primary corpus luteum, and the role of the secondary corpora lutea during and just after implantation (Bain, 1967; van Rensburg and van Niekerk, 1968; Allen, 1971). Contrary to earlier opinion, van Rensburg and van Niekerk (1968) consider that the primary corpus luteum remains hormonally active for at least 3 months. A peak in estradiol secretion, as a measure of follicular activity, occurs at about the 25th day after mating, and maximum progesterone concentration in the primary corpus luteum was found 2 days later, just before implantation on the 28th day when the amount of follicular development is greater than at any other time during pregnancy. The cyclical development of follicles (Bain, 1967) is consistent with the appearance of numerous secondary corpora lutea during the third month of pregnancy. All these corpora lutea in the mare and other equids disappear completely by midpregnancy (see Rowlands, 1964). The high progesterone content of the fetal circulation in late pregnancy was considered by van Rensburg and van Niekerk (1968) to be of placental origin. A 21-day ovarian cycle in the unmated mare and the occurrence of ovulation 6–10 days after parturition was confirmed by Matthews *et al.* (1967). Usually only one follicle matures during or about the fourth day of a 6-day period of estrus and is one of the largest (6 cm) to be found in a mammal. Progesterone concentration reaches a peak in the corpus luteum of the cycle 12–14 days after ovulation, and in the primary corpus luteum of pregnancy at about the time of implantation. Data about the ovarian cycle in other equids, such as the donkey and zebra, have been reviewed by Rowlands (1964).

10. Edentata

Only the nine-banded armadillo, *Dasypus novemcinctus,* has been studied in any detail, chiefly because it is characterized by polyembryony and delayed implantation. The follicular phase occurs during late spring and early summer, and ovulation from a single large follicle takes place in July (Talmage *et al.,* 1954). The corpus luteum develops very rapidly and reaches maximum size in about 10 days in nonpregnant and pregnant animals. When the blastocysts are free in the uterus and after implantation, there is little change in the appearance of the corpus luteum (Enders, 1966), but secretion of progesterone is greatly increased after nidation (Labhsetwar and Enders, 1968). Implantation has been induced by ovariectomy and the significance of the ovarian steroids to this process in the intact armadillo is not yet understood.

11. Primates

a. Prosimii. The ovarian cycle of tree shrews, lemurs, lorises, and bush-babies has not been investigated in any detail, although studies have been made of the estrous cycle of several species maintained in captivity. Conaway and Sorenson (1966) reported that the estrous cycle of tree shrews (Tupaiidae) lasts 10–12 days. When ovulation occurs, the luteal phase is short and the corpora lutea are probably nonfunctional. Other estrous cycles with a periodicity of 18 to 28 days (mode 21 days) are similar to those in females mated by vasectomized males, and are considered equivalent to a state of pseudopregnancy. Gestation varies between 45 and 50 days, and according to Duke and Luckett (1965), the corpora lutea of pregnancy are much larger than those of pseudopregnancy, and persist until parturition. The short (10–12 days) cycle and the pseudopregnancy (21 days) are regarded by Conaway and Sorenson (1966) as artifacts of captivity, like those occurring in rats, mice, and other myomorph rodents. Conaway and Sorenson (1966) discuss the affinities of the ovarian cycle with that of insectivores and other primates.

Many tupaiids (Conaway and Sorenson, 1966), lorisoids (Manley, 1966; Charles-Dominique, 1969), and lemuroids (Petter-Rousseaux, 1969) are polyestrous throughout the year. The slender loris, *Loris tardigradus* (Manley, 1966), and the ring-tail lemur, *Lemur catta* (Evans and Goy, 1968), may be affected by local environmental conditions and breed seasonally. Captive specimens of the latter species kept in the natural photoperiod of Portland, Oregon, mate between October and February and females experience not more than three periods of estrus of a mean duration of 4.7 days at a mean interval of 39 days. For the remainder of the year the ovaries are quiescent. In the absence of conception, the mean interval between successive estrous periods in the bushbaby, *Galago senegalensis,* is 32 days but is considerably shorter (22–27 days) in pubertal females. In the potto, *Perodicticus potto,* and the angwantibo, *Arctocebus calabarensis,* the interval is 39 days and in the slow loris, *Nycticebus coucang,* 42 days. Manley's (1966) suggestion that the stimulus of copulation does not affect cycle length in *Galago* indicates that pseudopregnancy does not occur in at least some lorisoids. Thus, there appears to be a familial difference between tupaiid, on the one hand, and lorisoid and lemuroid primates, on the other. The concurrence of pregnancy and lactation in the angwantibo and potto probably indicates the existence of a postpartum estrus and ovulation (Charles-Dominique, 1969).

Butler (1967) studied the ovaries of Senegal bushbabies caught in the Sudan. Ovulation is spontaneous and the ruptured follicle quickly develops a central blood clot surrounded by a luteinizing membrana granulosa. Two weeks later, the corpus luteum, measuring 500–700 μm in diameter, was solid and gave no indication of hemorrhage having occurred. The corpus luteum in two specimens examined 26 days after ovulation was pedunculate and measured 1350 and 1200 μm in diameter. Most luteal cells appeared histologically normal although some had vacuolated cytoplasm. The appearance of the endometrium suggested that luteal activity was at its peak, and in another two animals examined 40 days after ovulation corpora lutea were not recognizable. Follicular activity was much in evidence throughout the luteal phase. Shortly after ovulation, large follicles of 600 μm diameter and smaller solid follicles 300 μm in diameter were abundant. When luteal activity was maximal at 26 days, the largest follicles measured 700 μm and the smallest ones, showing signs of antrum formation, were 300–400 μm in diameter. Follicles up to 600 μm in diameter were present throughout the luteal phase and the ensuing anestrus. Atresia was common in follicles at all stages, particularly of the largest follicles.

b. **Anthropoidea.** New information has been published about the macaque, *Macaca mulatta,* the baboon, *Papio* spp., and the squirrel monkey, *Saimiri sciureus,* which all breed well in captivity. *Macaca mulatta,* in free-range situations on small islands off Puerto Rico (see Conaway and Koford, 1964; Vandenbergh and Vessey, 1968), cycle between August and April, with a peak in November and December. Mating is not confined to any one part of the ovarian cycle (Phoenix *et al.,* 1968). Koering *et al.* (1968) investigated progesterone secretion in the macaque during the menstrual cycle. A detailed histological study of the macaque ovary (Koering, 1969) indicates that follicular growth begins soon after ovulation and is arrested during the luteal phase of the previous cycle. The follicle destined to rupture is visible and has reached maximum diameter (about 6.3 mm) 9–11 days after the start of bleeding. All other follicles, 1.0–3.1 mm in diameter, undergo atresia. The concentration of plasma progesterone reaches a maximum of 8.7 ng/ml and parallels the variation in size of the corpus luteum. The cycle of tumescence in the perineal skin of the baboons *Papio cynocephalus* and *P. anubis* (Hendrickx and Kraemer, 1969) has been studied to determine optimal dates for mating. The results agreed well with those of Zuckerman and Parkes (1932) and Zuckerman (1937). The menstrual cycles of baboons maintained in isolation are significantly shorter than those of the same animals when they are kept in groups (Rowell, 1969). The effect is a shortening of the period of tumescence, that is, of the follicular phase.

The squirrel monkey is being bred extensively in many laboratories but the seasonality of breeding, and the length of the cycle as determined by vaginal smears, is still undecided (Clewe, 1969). Much of the variation in the vaginal cytology could be due to differences in the environment, nutrition, and management. Clewe (1969) could not detect a regular cycle and quotes other workers' observations of cycles 1–4 weeks in length. Similar variations in cycle lengths in the cinnamon ringtail, *Cebus albifrons,* the woolly monkey, *Lagothrix lagotriche,* and in a large colony of *S. sciureus* have been reported by Castellanos and McCombs (1968). Variability in cycle length was much less (26 to 38 days) in the crab-eating macaque, *M. fascicularis*; 70% of cycles varied only between 28 and 30 days (Nawar and Hafez, 1972). The accompanying changes in vaginal cytology provided no indication of the occurrence of ovulation.

c. Man. Much of the recent work on the menstrual cycle and the endocrinology of gestation in man has concerned the measurement in the blood of various hormones, particularly progesterone. The data which have been collected make it possible for clinicians to characterize the normal menstrual cycle (Haller, 1968; Johansson, 1969) and to pinpoint aberrations, particularly those associated with ovulation (Johansson and Wide, 1969). In the past, menstruation has been the primary sign of normal activity of the ovary, but the development of contraceptive methods and techniques to induce fertility has led to greater interest in the time of ovulation. The mean length of the cycle in women aged 15–45 years is 28 ± 4 days, the normal range being 21–35 days. Cycles tend to be irregular in women under 25 years of age, most regular from 30 to 40 years, and irregular again in later life (Chiazze *et al.,* 1968). Variation within an individual is usually less than 8 days, but some longer or anovulatory cycles and periods of amenorrhea do occur (Brayer *et al.,* 1969), and accurate prediction of the time of ovulation (16–12 days before the next menstruation) is difficult.

Plasma progesterone levels have been used to indicate the extent and duration of luteal activity during pregnancy. Johansson (1969) has shown that there is a progressive decrease in progesterone concentration up to the ninth or tenth week. Progesterone secretion is then increased. These data suggest that luteal regression occurs early in gestation and that the placenta then becomes the major source of this hormone. Although a new phase of follicular activity commences at about the tenth week of pregnancy in man (Govan, 1968), maturation of these follicles and ovulation do not occur as they do in equids. The relationship between this cycle of events, in which there is much follicular atresia, and the presence of large amounts of HCG in the blood, is discussed by Johansson (1969).

REFERENCES

Akins, E. L., and Morrissette, M. C. (1968). Gross ovarian changes during estrous cycle of swine. *Am. J. Vet. Res.* **29**, 1953–1957.

Akins, E. L., Morrissette, M. C., and Cardeilhac, P. T. (1969). Luteal and endometrial phosphatase activities during the porcine estrous cycle. *J. Anim. Sci.* **28**, 51–56.

Allen, W. E. (1971). The occurrence of ovulation during pregnancy in the mare. *Vet. Rec.* **88**, 508–509.

Amoroso, E. C. (1952). Placentation. *In* "Marshall's Physiology of Reproduction" (A. S. Parkes, ed.), 3rd ed., Vol. 2, Chapter 15, pp. 127–311. Longmans Green, London and New York.

Amoroso, E. C., Bourne, G. H., Harrison, R. J., Matthews, L. H., Rowlands, I. W., and Sloper, J. C. (1965). Reproductive and endocrine organs of foetal, newborn and adult seals. *J. Zool.* **147**, 430–486.

Anand Kumar, T. C. (1965). Reproduction in the rat-tailed bat *Rhinopoma kinneari. J. Zool.* **147**, 147–155.

Asdell, S. A. (1964). "Patterns of Mammalian Reproduction," 2nd ed., Cornell Univ. Press, Ithaca, New York.

Austin, C. R. (1957). Oestrus and ovulation in the field vole (*Microtus agrestis*). *J. Endocrinol.* **15**, iv.

Bacon, W. L., and Cherms, F. L. (1968). Ovarian follicular growth and maturation in the domestic turkey. *Poult Sci.* **47**, 1303–1314.

Badir, N. (1968). Structure and function of corpus luteum during gestation in the viviparous lizard, *Chalicides ocellatus. Anat. Anz.* **122**, 1–10.

Bain, A. M. (1967). The ovaries of the mare during early pregnancy. *Vet. Rec.* **80**, 229–231.

Balinsky, B. I. (1969). The reproductive ecology of amphibians of the Transvaal highveld. *Zool. Afr.* **4**, 37–93.

Bara, G. (1960). Histological and cytological changes in the ovaries of the mackerel (*Scomber scomber* L.) during the annual cycle. *Istanbul Univ. Fen Fak. Mecm., Seri* B [N.S.] **25**, 41–91.

Barr, W. A. (1968). Patterns of ovarian activity. *In* "Perspectives in Endocrinology" (E. J. W. Barrington and C. Barker Jørgensen, eds.), pp. 163–238. Academic Press, New York.

Bell, D. J., and Freeman, B. H., eds. (1971). "Physiology and Biochemistry of the Domestic Fowl," 3 vols. Academic Press, New York.

Berger, P. J., and Sharman, G. B. (1969). Progesterone-induced development of dormant blastocysts in the tammar wallaby, *Macropus eugenii* Desmarest: Marsupialia. *J. Reprod. Fertil.* **20**, 201–210.

Berry, P. Y. (1964). Breeding patterns of seven species of Singapore Anura. *J. Anim. Ecol.* **33**, 227–243.

Best, P. B. (1968). The sperm whale (*Physeter catodon*) off the west coast of South Africa. 2. Reproduction in the female. *Invest. Rep., Div. Sea Fish S. Afr.* **66**, 1–32.

Bland, K. P. (1969). Reproduction in the female Indian gerbil (*Tatera indica*). *J. Zool.* **157**, 47–61.

Bodenheimer, F. S., and Sulman, F. (1946). The estrous cycle of *Microtus guentheri* D. and A. and its ecological implications. *Ecology* **27**, 255–256.

Bonnin-Laffargue, M., and Canivenc, R. (1970). Biologie lutéale chez le Renard (*Vulpes vulpes* L.). Persistance du corps jaune après la mise bas. *C. R. Acad. Sci. Ser. D* **271**, 1402–1405.

Braekevelt, C. R., and McMillan, D. B. (1967). Cyclic changes in the ovary of the brook stickleback *Eucalia inconstans* (Kirtland). *J. Morphol.* **123**, 373–396.

Bragdon, D. E. (1952). Corpus luteum formation and follicular atresia in the common garter snake, *Thamnophis sirtalis*. *J. Morphol.* **91**, 413–446.

Brambell, F. W. R. (1935). Reproduction in the common shrew (*Sorex araneus* Linnaeus). 1. The oestrous cycle of the female. *Philos. Trans. R. Soc. London, Ser. B* **225**, 1–50.

Brayer, F. T., Chiazze, L., and Duffy, B. J. (1969). Calendar rhythm and menstrual cycle range. *Fertil. Steril.* **20**, 279–288.

Breed, W. G. (1967). Ovulation in the genus *Microtus*. *Nature (London)* **214**, 826.

Breed, W. G., and Clarke, J. R. (1970). Ovarian changes during pregnancy and pseudopregnancy in the vole, *Microtus agrestis*. *J. Reprod. Fertil.* **23**, 447–456.

Bretschneider, L. H., and Duyvené de Wit, J. J. (1947). "Sexual Endocrinology of Non-mammalian Vertebrates." Elsevier, Amsterdam.

Brummett, A. R. (1966). Observations on the eggs and breeding season of *Fundulus heteroclitus* at Beaufort, North Carolina. *Copeia* pp. 616–620.

Buechner, H. K., Morrison, J. A., and Leuthold, W. (1966). Reproduction in Uganda kob with special reference to behavior. *Symp. Zool. Soc. London* **15**, 69–88.

Buss, I. C., and Smith, N. S. (1966). Observations on reproduction and breeding behavior of the African elephant. *J. Wildl. Manage.* **30**, 375–388.

Busson-Mabillot, S. (1967a). Gonadogenèse, différenciation sexuelle et structure de l'ovaire chez la lamproie de planer, *Lampetra planeri* (Bloch). *Arch. Zool. Exp. Gen.* **108**, 293–318.

Busson-Mabillot, S. (1967b). Structure ovarienne chez la lamproie de planer adulte *Lampetra planeri* (Bloch). *Arch. Zool. Exp. Gen.* **108**, 413–446.

Butler, H. (1967). The oestrus cycle of the Senegal bushbaby (*Galago senegalensis senegalensis*) in the Sudan. *J. Zool.* **151**, 143–162.

Canivenc, R., Bonnin-Laffargue, M., and Relexans, M.-C. (1969). Cycles génitaux de quelques Mustellidés européens. *In* "Cycles génitaux saisonniers de mammifères sauvages" (R. Canivenc, ed.), pp. 85–110. Masson, Paris.

Carlson, I. H., and Rust, C. C. (1969). Plasma progesterone levels in pregnant, pseudopregnant and anestrous ferrets. *Endocrinology* **85**, 623.

Castellanos, H., and McCombs, H. L. (1968). Reproductive cycle of the new world monkey: gynecologic problems in a breeding colony. *Fertil. Steril.* **19,** 213–227.

Chaplin, R. E., Chapman, D. I., and Prior, R. (1966). Examination of the uterus and ovaries of some female roe deer (*Capreolus capreolus*) from Wiltshire and Dorset, England. *J. Zool.* **148,** 570–574.

Charles-Dominique, P. (1969). Reproduction des lorisides Africains. *In* "Cycles Génitaux Saisonnièrs de Mammifères Sauvages" (R. Canivenc, ed.), pp. 2–9. Masson, Paris.

Chiazze, L., Brayer, F. T., Macisco, J. J., Jr., Parker, M. P., and Duffy, B. J. (1968). The length and variability of the human menstrual cycle. *J. Am. Med. Assoc.* **203,** 377–380.

Chieffi, G. (1967). The reproductive system of elasmobranchs: developmental and endocrinological aspects. *In* "Sharks, Skates and Rays" (W. Gilbert, ed.), pp. 553–580. Johns Hopkins Press, Baltimore, Maryland.

Chittleborough, R. G. (1954). Studies on the ovaries of the humpback whale, *Megaptera nodosa* (Bonnaterre), on the west Australian coast. *Aust. J. Mar. Freshwater Res.* **5,** 35–63.

Christie, D. W., Bell, E. T., Horth, C. E., and Palmer, R. F. (1971). Peripheral plasma progesterone levels during the canine oestrous cycle. *Acta Endocrinol. (Copenhagen)* **68,** 543–550.

Church, G. (1960). Annual and lunar periodicity in the sexual cycles of the Javanese toad, *Bufo melanostictus* Schneider. *Zoologica* (*N.Y.*) **45,** 181–188.

Church, G. (1962). The reproductive cycles of the Javanese house geckos, *Cosymbotus platyurus, Hemidactylus frenatus* and *Peropus mutilatus. Copeia* pp. 262–269.

Clewe, T. H. (1969). Observations on reproduction of squirrel monkeys in captivity. *J. Reprod. Fertil., Suppl.* **6,** 151–156.

Clough, G. (1969). Some preliminary observations on reproduction in the warthog, *Phacochoerus aethiopicus* Pallas. *J. Reprod. Fertil., Suppl.* **6,** 323–337.

Clulow, F. V., and Mallory, F. F. (1970). Oestrus and induced ovulation in the meadow vole, *Microtus pennsylvanicus. J. Reprod. Fertil.* **23,** 341–343.

Conaway, C. H., and Koford, C. B. (1964). Estrous cycle and mating behavior in a free-ranging band of rhesus monkeys. *J. Mammal.* **45,** 577–588.

Conaway, C. H., and Sorenson, M. W. (1966). Reproduction in tree-shrews. *Symp. Zool. Soc. London* **15,** 471–492.

Craig, A. M. (1964). Histology of reproduction and the estrous cycle in the female fur seal *Callorhinus ursinus. J. Fish. Res. Board Can.* **21,** 773–812.

Craighead, J. J., Hornocker, M. G., and Craighead, F. C. J. (1969). Reproductive biology of young female grizzly bears. *J. Reprod. Fertil., Suppl.* **6,** 447–475.

Crichton, E. G. (1969). Reproduction in the pseudomyine rodent *Mesembriomys gouldii* (Gray) (Muridae). *Aust. J. Zool.* **17,** 785–795.

Crisp, T. M. (1964). Studies of reproduction in the female ovoviviparous lizard, *Sceloporus cyanogenys* (Cope). *Tex. J. Sci.* **16,** 481.

Deanesly, R. (1943). Delayed implantation in the stoat (*Mustela erminea*). *Nature (London)* **151,** 365.

Deanesly, R. (1966). Observations on reproduction in the mole, *Talpa europaea. Symp. Zool. Soc. London* **15,** 387–402.

Delost, P. (1955). Etude de la biologie sexuelle du campagnol des champs (*Microtus arvalis* P.). *Arch. Anat. Microsc. Morphol. Exp.* **44,** 150–190.

Dobrowolski, W., Stupnicka, E., and Dománski, E. (1968). Progesterone levels in ovarian venous blood during the oestrous cycle of the cow. *J. Reprod. Fertil.* **15**, 409–414.

Dodd, J. M., Evennett, P. J., and Goddard, C. K. (1960). Reproductive endocrinology in cyclostomes and elasmobranchs. *Symp. Zool. Soc. London* **1**, 77–103.

Dominic, C. J. (1960). A study of the post-ovulatory follicle in the ovary of the domestic pigeon, *Columba livia. J. Zool. Soc. India* **12**, 27–33.

Dryden, G. L. (1969). Reproduction in *Suncus murinus. J. Reprod. Fertil., Suppl.* **6**, 377–396.

Duke, K. L. (1966). Histological observations on the ovary of the white-tailed mole, *Parascaptor leucurus. Anat. Rec.* **154**, 526–532.

Duke, K. L., and Luckett, W. P. (1965). Histological observations on the ovary of several species of tree shrews (*Tupaiidae*). *Anat. Rec.* **151**, 450.

Dunham, D., and Clapp, R. (1962). The effects of exogenous gonadotrophins and egg removal on clutch size in the domesticated canary *Serinus canarius. Auk* **79**, 458–462.

Dunmore, R., and Davis, D. E. (1963). Reproductive condition of feral pigeons in winter. *Auk* **80**, 374.

Dunn, R. S., and Tyler, A. V. (1969). Aspects of the anatomy of the winter flounder ovary with hypotheses on oocyte maturation time. *J. Fish. Res. Board Can.* **26**, 1943–1947.

Eadie, W. R., and Hamilton, W. J., Jr. (1958). Reproduction in the fisher in New York. *N.Y. Fish Game J.* **5**, 77–83.

Egoscue, H.-J. (1970). Laboratory colony of the Polynesian rat, *Rattus exulans. J. Mammal.* **51**, 261–266.

El-Sheikh, A. S., Sakla, F. B., and Amin, S. O. (1969). Changes in the density and progesterone content of luteal tissue in the Egyptian buffalo during pregnancy. *J. Endocrinol.* **43**, 1–8.

Enders, A. C. (1966). The reproductive cycle of the nine-banded armadillo (*Dasypus novemcinctus*). *Symp. Zool. Soc. London* **15**, 295–310.

Enders, R. K., and Enders, A. C. (1963). Morphology of the female reproductive tract during delayed implantation in the mink. *In* "Delayed Implantation" (A. C. Enders, ed.), pp. 129–139. Univ. of Chicago Press, Chicago, Illinois.

England, B. G., Foote, W. C., Matthews, D. H., Cardozo, A. G., and Riera, S. (1969). Ovulation and corpus luteum function in the llama (*Lama glama*). *J. Endocrinol.* **45**, 505–513.

Erickson, B. H. (1966). Development and senescence of the post-natal bovine ovary. *J. Anim. Sci.* **25**, 800–805.

Erickson, C. J., and Lehrman, D. S. (1964). The effect of castration of male ring doves upon ovarian activity of females. *J. Comp. Physiol. Psychol.* **58**, 164–166.

Erpino, M. J. (1969). Seasonal cycle of reproductive physiology in the black-billed magpie. *Condor* **71**, 267–279.

Evans, C. S., and Goy, R. W. (1968). Social behaviour and reproductive cycles in captive ring-tailed lemurs (*Lemur catta*). *J. Zool.* **156**, 181–197.

Evennett, P. J., and Dodd, J. M. (1963). The pituitary gland and reproduction in the lamprey (*Lampetra fluviatilis* L.). *J. Endocrinol.* **26**, xiv–xv.

Fairall, N. (1970). Research on the reproduction of wild ungulates. *Proc. S. Afr. Soc. Anim. Prod.* **9**, 57–61.

Farner, D. S., Follett, B. K., King, J. R., and Morton, M. L. (1966). A quantitative examination of ovarian growth in the white-crowned sparrow. *Biol. Bull.* (*Woods Hole, Mass.*) **130**, 67–75.

Farris, D. A. (1963). Reproductive periodicity in the sardine (*Sardinops caerulea*) and the Jack mackerel (*Trachurus symmetricus*) on the Pacific coast of North America. *Copeia* pp. 182–184.

Fitch, H. S., and Greene, H. W. (1965). The breeding cycle in the ground skink, *Lygosoma laterale. Publ. Mus. Nat. Hist. Univ. Kans.* **15**, 565–575.

Folley, S. J., and Malpress, F. H. (1944). The response of the bovine ovary to pregnant mares' serum and horse pituitary extract. *Proc. R. Soc. London, Ser. B* **132**, 164–188.

Foster, M. A. (1934). The reproductive cycle in the female ground squirrel *Citellus tridecemlineatus* (Mitchell). *Am. J. Anat.* **54**, 487–511.

Gambell, R. (1968). Seasonal cycles and reproduction in Sei whales of the southern hemisphere. *"Discovery" Rep.* **35**, 31–134.

Gambell, R. (1973). Some effects of exploitation on reproduction in whales. *J. Reprod. Fertil., Suppl.* **19**, 533–553.

Gibbons, J. W. (1968). Reproductive potential, activity, and cycles in the painted turtle, *Chrysemys picta. Ecology* **49**, 399–408.

Gilbert, A. B. (1968). Observations on the ultrastructure of the post-ovulatory follicle of the domestic hen. *Proc. Int. Congr. Anim. Reprod. Artif. Insem., 6th, 1968 Paris* Vol. 2, pp. 1629–1631.

Gilbert, A. B. (1971). The ovary. *In* "Physiology and Biochemistry of the Domestic Fowl" (D. J. Bell and B. M. Freeman, eds.), Vol. 3, pp. 1163–1202. Academic Press, New York.

Godfrey, G. K. (1969). Reproduction in a laboratory colony of the marsupial mouse, *Sminthopsis larapinta* (Marsupialia: Dasyuridae). *Aust. J. Zool.* **17**, 637–654.

Goldberg, S. R. (1970). Seasonal ovarian histology of the ovoviviparous iguanid lizard *Sceloporus jarrovi* Cope. *J. Morphol.* **132**, 265–276.

Gomes, W. R., Herschler, R. C., and Erb, R. E. (1965). Progesterone levels in ovarian venous effluent of the non-pregnant sow. *J. Anim. Sci.* **24**, 722–725.

Gopalakrishna, A. (1964). Post-partum pregnancy in the Indian fruit bat *Rousettus leschenaulti* (Desm). *Curr. Sci.* **33**, 558–559.

Gopalakrishna, A. (1969). Unusual persistence of the corpus luteum in the Indian fruit bat, *Rousettus leschenaulti* (Desmaret). *Curr. Sci.* **38**, 388–389.

Govan, A. D. T. (1968). Human ovary in early pregnancy. *J. Endocrinol.* **40**, 421–428.

Greenwald, G. S. (1956). The reproductive cycle of the field mouse *Microtus californicus. J. Mammal.* **37**, 213–222.

Greenwald, G. S. (1968). Failure of hypophysectomy to affect regression of cyclic hamster corpus luteum. *J. Reprod. Fertil.* **16**, 495–497.

Greer, A. E. (1966). Viviparity and oviparity in the snake genera *Conopsis, Toluca, Gyalopion* and *Ficimia* with comments on *Tomodon* and *Helicops. Copeia* pp. 371–373.

Greer, A. E. (1968a). Mode of reproduction in the squamate faunas of three altitudinally correlated life zones in East Africa. *Herpetologica* **24**, 229–232.

Greer, A. E. (1968b). The predominance of oviparity as a mode of reproduction in scincid lizards. *Copeia* p. 171.

Griffiths, M., McIntosh, D. L., and Coles, R. E. A. (1969). The mammary gland of the echidna, *Tachyglossus aculeatus,* with observations on the incubation of the egg and on the newly-hatched young. *J. Zool.* **158,** 371–386.

Guinness, F., Lincoln, G. A., and Short, R. V. (1971). The reproductive cycle of the female red deer, *Cervus elaphus* L. *J. Reprod. Fertil.* **27,** 427–438.

Gulamhusein, A. P. (1973). Studies of reproduction in the Mustelidae. Ph.D. thesis, University of London, London.

Hafs, H. D., and Armstrong, D. T. (1968). Corpus luteum growth and progesterone synthesis during the bovine estrous cycle. *J. Anim. Sci.* **27,** 134–141.

Haller, J. (1968). Hormonal data and regulation of the menstrual cycle. *In* "Normal Human Endometrium" (H. Schmidt-Matthiesen, ed.), pp. 1–23. McGraw-Hill, New York.

Hamilton, W. J., Jr., and Eadie, W. R. (1964). Reproduction in the otter, *Lutra canadensis. J. Mammal.* **45,** 242–252.

Hanks, J., and Short, R. V. (1972). The formation and function of the corpus luteum in the African elephant, *Loxodonta africana. J. Reprod. Fertil.* **29,** 79–89.

Harris, M. P. (1969). Breeding seasons of sea-birds in the Galapagos Islands. *J. Zool.* **159,** 145–165.

Harrison, R. J. (1969). Reproduction and reproductive organs. *In* "The Biology of Marine Mammals" (H. Anderson, ed.), Chapter 8, pp. 253–348. Academic Press, New York.

Harrison, R. J., and Ridgway, S. H. (1971). Gonadal activity in some bottle-nose dolphins (*Tursiops truncatus*). *J. Zool.* **165,** 355–366.

Hartman, C. G. (1921). Breeding habits, development and birth of the opossum. *Smithson. Inst., Annu. Rep.* pp. 347–363.

Hartman, C. G. (1923). The oestrous cycle of the opossum. *Am. J. Anat.* **35,** 353–421.

Henderson, N. E. (1963). Influence of light and temperature on the reproductive cycle of the eastern brook trout, *Salvelinus fontinalis* (Mitchill). *J. Fish. Res. Board Can.* **20,** 899–908.

Hendrickx, A. G., and Kraemer, D. C. (1969). Observations on the menstrual cycle, optimal mating time and pre-implantation embryos of the baboon, *Papio anubis* and *Papio cynocephalus. J. Reprod. Fertil., Suppl.* **6,** 119–128.

Henry, V. G. (1968). Length of estrous cycle and gestation in European wild hogs. *J. Wildl. Manage.* **32,** 406–408.

Herlant, M. (1968). Cycle sexuel chez les Chiroptères des régions tempérées. *In* "Cycles Génitaux des Mammifères Sauvages" (R. Canivenc, ed.), pp. 111–113. Masson, Paris.

Hisaoka, K. K., and Firlit, C. F. (1962). Ovarian cycle and egg production in the zebra fish, *Brachydanio rerio. Copeia* pp. 788–792.

Hoar, W. S. (1955). Reproduction in teleost fish. *Mem. Soc. Endocrinol.* **4,** 5–22.

Hoar, W. S. (1965). Comparative physiology: hormones and reproduction in fishes. *Annu. Rev. Physiol.* **27,** 51–70.

Holmes, R. T. (1966). Breeding ecology and annual cycle adaptations of the red-backed sandpiper (*Calidris alpina*) in northern Alaska. *Condor* **68,** 3–46.

Hyder, M. (1970). Gonadal and reproductive patterns in *Tilapia leucosticta* (Teleostei: Cichlidae) in an equatorial lake, Lake Naivasha (Kenya). *J. Zool.* **162,** 179–196.

Inger, R. F., and Greenberg, B. (1963). Annual reproductive pattern of the frog *Rana erythraea* in Sarawak. *Physiol. Zool.* **36**, 21–33.

Jainudeen, M. R., Eisenberg, J. F., and Tilakeratne, N. (1971). Oestrous cycle of the Asiatic elephant, *Elephus maximus,* in captivity. *J. Reprod. Fertil.* **27**, 321–328.

Jastrzebski, M. (1968). Morphological changes in the reproductive organs in the female marsh frog (*Rana arvalis* Nilss.) in the yearly cycle. *Acta Biol. Cracov., Ser. Zool.* **11**, 9–12.

Johansson, E. D. B. (1969). Progesterone levels in peripheral plasma during the luteal phase of the normal human menstrual cycle measured by a rapid competitive protein binding technique. *Acta Endocrinol. (Copenhagen)* **61**, 592–606.

Johansson, E. D. B., and Wide, L. (1969). Periovulatory levels of plasma progesterone and luteinizing hormones in women. *Acta Endocrinol. (Copenhagen)* **62**, 88–97.

John, K. R. (1963). Effect of torrential rains on the reproductive cycle of *Rhinichthys osculus* in the Chiricahua Mountains, Arizona. *Copeia* pp. 286–291.

Johnson, R. P. (1963). Life history and ecology of bigmouth buffalo *Ictiobus cyprinellus* (Valenciennes). *J. Fish. Res. Board Can.* **20**, 1397–1429.

Kayanja, F. I. B. (1969). The ovary of the impala, *Aepyceros melampus* (Lichtenstein, 1812). *J. Reprod. Fertil., Suppl.* **6**, 311–317.

Kayanja, F. I. B., and Blankenship, L. A. (1973). The ovary of the giraffe. *J. Reprod. Fertil.,* **35**, 305–313.

Kellas, L. M., van Lennep, E. W., and Amoroso, E. C. (1958). Ovaries of some foetal and prepubertal giraffes (*Giraffa camelopardalis* Linnaeus). *Nature (London)* **181**, 487–488.

Kern, M. D. (1972). Seasonal changes in the reproductive system of the female white-crowned sparrow, *Zonotrichia leucophrys gambelii,* in captivity and in the field. 1. The ovary. *Z. Zellforsch. Mikrosk. Anat.* **126**, 297–319.

Kessel, R. G., and Panje, W. R. (1968). Organisation and activity in the pre- and post-ovulatory follicle of *Necturus maculosus. J. Cell Biol.* **39**, 1–34.

King, J. R., Follett, B. K., Farner, D. S., and Morton, M. L. (1966). Annual gonadal cycles and pituitary gonadotropins in *Zonotrichia leucophrys gambelii. Condor* **68**, 476–487.

Koering, M. J. (1969). Cyclic changes in ovarian morphology during the menstrual cycle of *Macaca mulatta. Am. J. Anat.* **126**, 73–101.

Koering, M., Resko, J. A., Phoenix, C. H., and Goy, R. W. (1968). Ovarian morphology and progesterone levels throughout the menstrual cycle in *Macaca mulatta. Anat. Rec.* **160**, 378.

Kopstein, F. (1938). Ein Beitrag zur Eierkunde und zur Fortpflanzung der malaiischen Reptilien. *Bull. Raffles Mus.* **14**, 81–167.

Labhsetwar, A. P., and Enders, A. C. (1968). Progesterone in the corpus luteum and placenta of the armadillo, *Dasypus novemcinctus. J. Reprod. Fertil.* **16**, 381–387.

Lagios, M. D. (1965). Seasonal changes in the cytology of the adenohypophysis, testes and ovaries of the black surfperch, *Embiotoca jacksoni,* a viviparous percomorph fish. *Gen. Comp. Endocrinol.* **5**, 207–221.

Lambert, J. G. D., and Van Oordt, P. G. W. J. (1965). Preovulatory corpora lutea atretica in the guppy, *Poecilia reticulata.* A histological and histochemical study. *Gen. Comp. Endocrinol.* **5**, 693.

Lamotte, M., Rey, P., and Vogeli, M. (1964). Recherches sur l'ovaire de *Nectophrynoides orientalis*. Batracien, Anoure vivipare. *Arch. Anat. Microsc. Morphol. Exp.* **53**, 179–224.

Lance, V., and Callard, I. P. (1969). A histochemical study of ovarian function in the ovoviviparous elasmobranch, *Squalus acanthius*. *Gen. Comp. Endocrinol.* **13**, 255–267.

Laws, R. M. (1961). Reproduction, growth and age of southern fin whales. *"Discovery" Rep.* **31**, 327–486.

Laws, R. M. (1969). Aspects of reproduction in the African elephant, *Loxodonta africana. J. Reprod. Fertil., Suppl.* **6**, 193–218.

Laws, R. M., and Clough, G. (1966). Observations on reproduction in the hippopotamus, *Hippopotamus amphibius* Linn. *Symp. Zool. Soc. London* **15**, 117–140.

Lehri, G. K. (1968). Cyclical changes in the ovary of the catfish, *Clarias batrachus* (Linn.). *Acta Anat.* **69**, 105–124.

Lewis, J. C., and McMillan, D. B. (1965). Development of the ovary of the sea lamprey (*Petromyzon marinus* L.) *J. Morphol.* **117**, 425–466.

Loeb, L. (1911). The cyclic changes in the ovary of the guinea-pig. *J. Morphol.* **22**, 37–70.

Lofts, B., Murton, R. K., and Westwood, N. J. (1966). Gonadal cycles and the evolution of breeding seasons in British Columbidae. *J. Zool.* **150**, 249–272.

McCoy, C. J. (1968). Reproductive cycles and viviparity in Guatemalan *Corythophanes percarinatus* (Reptilia: Iguanidae). *Herpetologica* **24**, 175–178.

McCoy, C. J., and Hoddenbach, G. A. (1966). Geographic variation in ovarian cycles and clutch size in *Cnemidophorus tigris* (Teiidae). *Science* **154**, 1671–1672.

McKeever, S. (1966). Reproduction in *Citellus beldingi* and *Citellus lateralis* in north eastern California. *Symp. Zool. Soc. London* **15**, 365–385.

Mackie, R. J., and Buechner, H. K. (1963). The reproductive cycle of the chukar. *J. Wildl. Manage.* **27**, 246–260.

Magnin, E. (1966). Quelques données biologiques sur la reproduction des esturgeons *Acipenser fulvescens* Raf. de la rivière Nottaway, tributaire de la baie James. *Can. J. Zool.* **44**, 257–263.

Manley, G. H. (1966). Reproduction in lorisoid primates. *Symp. Zool. Soc. London* **15**, 493–509.

Marden, W. G. R. (1953). The hormone control of ovulation in the calf. *J. Agric. Sci.* **43**, 381–406.

Marshall, A. J., and Hook, R. (1960). The breeding biology of equatorial vertebrates: reproduction of the lizard *Agama agama lionotus* Boulenger at lat. 0° 01′ N. *Proc. Zool. Soc. London* **134**, 197–205.

Marza, V. D. (1938). "Histophysiologie de l'ovogenèse." Hermann, Paris.

Matthews, R. G., Ropiha, R. T., and Butterfield, R. M. (1967). The phenomenon of foal heat in mares. *Aust. Vet. J.* **43**, 579–582.

Matthews, S. A. (1938). The seasonal cycle in the gonads of *Fundulus. Biol. Bull.* (*Woods Hole, Mass.*) **75**, 66–74.

Mayhew, W. W. (1963). Reproduction in the granite spiny lizard, *Sceloporus orcutti. Copeia* pp. 144–152.

Mayhew, W. W. (1965). Reproduction in the sand-dwelling lizard *Uma inornata. Herpetologica* **21**, 39–55.

Mayhew, W. W. (1966). Reproduction in the psammophilous lizard *Uma scoparia. Copeia* pp. 114–122.

Mayhew, W. W. (1971). Reproduction in the desert lizard, *Dipsosaurus dorsalis. Herpetologica* **27**, 57–77.

Mead, R. A., and Eik-Nes, K. B. (1969). Seasonal variation in plasma levels of progesterone in western forms of the spotted skunk. *J. Reprod. Fertil., Suppl.* **6**, 397–403.

Medway, Lord. (1971). Observations of social and reproductive biology of the bent-winged bat *Miniopterus australis* in northern Borneo. *J. Zool.* **165**, 261–273.

Mendoza, G. (1962). The reproductive cycles of three viviparous teleosts *Alloophorus robustus, Goodea luitpoldii* and *Neoophorus diazi. Biol. Bull. (Woods Hole, Mass.)* **123**, 351–365.

Miller, M. R. (1948). The seasonal histological changes occurring in the ovary, corpus luteum and testis of the viviparous lizard *Xantusia vigilis. Univ. Calif., Berkeley, Publ. Zool.* **47**, 197–223.

Miller, W. W., and Wiggins, E. L. (1964). Ovarian activity and fertility in lactating ewes. *J. Anim. Sci.* **23**, 981–983.

Moore, N. W., Barrett, S., Brown, J. G., Schindler, I., Smith, M. A., and Smyth, B. (1969). Oestrogen and progesterone content of ovarian vein blood in the ewe during the oestrous cycle. *J. Endocrinol.* **44**, 55–62.

Morel, G., and Morel, M. Y. (1962). La reproduction des oiseaux dans une région semi-aride du vallée du Sénégal. *Alauda* **30**, 241–269.

Morrison, J. A. (1971). Morphology of corpora lutea in the Uganda kob antelope, *Adenota kob thomasi* (Neumann). *J. Reprod. Fertil.* **26**, 297–305.

Morrison, J. A., and Buechner, H. K. (1971). Reproductive phenomena during the *post partum*-preconception interval in the Uganda kob. *J. Reprod. Fertil.* **26**, 307–317.

Morrow, D. A. (1969). Estrous behavior and ovarian activity in prepuberal and postpuberal dairy heifers. *J. Dairy Sci.* **52**, 224–227.

Moser, H. G. (1967). Seasonal histological changes in the gonads of *Sebastodes paucispinis*, an ovoviviparous teleost. *J. Morphol.* **123**, 329–354.

Mossman, H. W., and Judas, I. (1949). Accessory corpora lutea, lutein cell origin and the ovarian cycle in the Canadian porcupine. *Am. J. Anat.* **85**, 1–39.

Murton, R. K., Isaacson, A. J., and Westwood, N. J. (1963). The food and growth of nestling wood-pigeons in relation to the breeding season. *Proc. Zool. Soc. London* **141**, 747–781.

Nair, P. V. (1963). Ovular atresia and the formation of the so-called 'corpus luteum' in the ovary of the Indian catfish *Heteropneustes fossilis* (Bloch). *Proc. Zool. Soc. (Calcutta)* **16**, 51–61.

Nawar, M. M., and Hafez, E. S. E. (1972). The reproductive cycle of the crab-eating macaque (*Macaca fascicularis*). *Primates* **13**, 43–56.

Nawito, H. F., Shalash, M. R., Hoppe, R., and Rakha, A. M. (1967). Reproduction in the female camel. *Bull. Anim. Sci. Res. Inst., Cairo* **2**, 1–88.

Neill, W. T. (1962). The reproductive cycle of snakes in a tropical region, British Honduras. *Q. J. Fla. Acad. Sci.* **25**, 234–253.

Neill, W. T. (1964). Viviparity in snakes: some ecological and zoogeographical considerations. *Am. Nat.* **98**, 35–55.

New, J. G. (1966). Reproductive behaviour of the shield darter, *Percina peltata,* in New York. *Copeia* pp. 20–28.

Newson, R. M. (1966). Reproduction in the feral coypu (*Myocastor coypus*). *Symp. Zool. Soc. London* **15**, 323–334.

Novoa, C. (1970). Reproduction in Camelidae. *J. Reprod. Fertil.* **22**, 3–20.

Ohsumi, S. (1964). Comparison of maturity and accumulation rate of corpora albicantia between the left and right ovaries in Cetacea. *Sci. Rep. Whales Res. Inst.* **18**, 123–148.

Øritsland, T. (1964). Klappmysshunneus forplantningsbiologi. *Fisken. Havet* **1**, 1–15.

Ortavant, R., Mauléon, P., and Thibault, C. (1964). Photoperiodic control of gonadal and hypophyseal activity in domestic animals. *Ann. N.Y. Acad. Sci.* **117**, 157–193.

Payne, R. B. (1966). The post-ovulatory follicles of blackbirds (*Agelaius*). *J. Morphol.* **118**, 331–352.

Pearson, O. P. (1944). Reproduction in the shrew (*Blarina brevicauda* Say). *Am. J. Anat.* **75**, 39–93.

Pearson, O. P. (1949). Reproduction of a South American rodent, the mountain viscacha. *Am. J. Anat.* **84**, 143–174.

Perry, J. S. (1953). The reproduction of the African elephant, *Loxodonta africana*. *Philos. Trans. R. Soc. London, Ser. B* **237**, 93–149.

Perry, J. S. (1971). "The Ovarian Cycle of Mammals." Oliver & Boyd, Edinburgh.

Perry, J. S., and Rowlands, I. W. (1961). Effect of hysterectomy on the ovarian cycle of the rat. *J. Reprod. Fertil.* **2**, 332–340.

Perry, J. S., and Rowlands, I. W. (1962). The ovarian cycle in vertebrates. *In* "The Ovary" (S. Zuckerman, ed.), Vol. 1, Chapter 5, pp. 275–309. Academic Press, New York.

Peters, H., and Levy, E. (1966). Cell dynamics of the ovarian cycle. *J. Reprod. Fertil.* **11**, 227–236.

Petter-Rousseaux, A. (1969). Cycles génitaux saisonniers des lémuriens malagaches. *In* "Cycles génitaux saisonniers de mammifères sauvages" (R. Canivenc, ed.), pp. 11–22. Masson, Paris.

Peyre, A. (1969). Cycles génitaux et corrélations hypophyse-génitales chez trois Insectivores européens. *In* "Cycles génitaux saisonniers de mammifères sauvages" (R. Canivenc, ed.), pp. 133–149. Masson, Paris.

Phillips, C. L. (1887). Egg-laying extraordinary in *Colaptes auratus*. *Auk* **4**, 346.

Phoenix, C. H., Goy, R. W., Resko, J. A., and Koering, M. J. (1968). Probability of mating during various stages of the ovarian cycle in *Macaca mulatta*. *Anat. Rec.* **160**, 490.

Pilton, P. E., and Sharman, G. B. (1962). Reproduction in the marsupial *Trichosurus vulpecula*. *J. Endocrinol.* **25**, 119–136.

Plasse, D., Warnick, A. C., and Koger, M. (1970). Reproductive behaviour of *Bos indicus* females in a subtropical environment. IV. Length of estrous cycle, duration of estrus, time of ovulation, fertilization and embryo survival in grade Brahman heifers. *J. Anim. Sci.* **30**, 63–72.

Polder, J. J. W. (1964). On the occurrence and significance of atretic follicles (preovulatory corpora lutea) in ovaries of the bitterling, *Rhodeus amarus* (Bloch). *Proc. K. Ned. Akad. Wet. Ser. C* **67**, 218–222.

Racey, P. A. (1972). Aspects of reproduction in some heterothermic bats. Ph.D. thesis, University of London, London.

Racey, P. A., and Kleiman, D. G. (1970). Maintenance and breeding in captivity of some Vespertilionid bats, with special reference to the noctule. *Int. Zoo Yearb.* **10,** 65–70.

Rahn, H. (1939). Structure and function of placenta and corpus luteum in viviparous snakes. *Proc. Soc. Exp. Biol. Med.* **40,** 381–382.

Rahn, H. (1942). The reproductive cycle of the prairie rattler. *Copeia* pp. 233–240.

Rajakowski, E. (1960). The ovarian follicular system in sexually mature heifers with special reference to seasonal, cyclical and left-right variations. *Acta Endocrinol. (Copenhagen), Suppl.* **52,** 1–68.

Rakha, A. M., Igboeli, G., and Hale, D. (1970). The oestrous cycle of Zebu and Sanga breeds of cattle in Central Africa. *J. Reprod. Fertil.* **23,** 411–414.

Ramaswamy, K. R. (1961). Studies on the sex-cycle of the Indian vampire bat, *Megaderma (Lyroderma) lyra lyra* (Geoffroy). I. Breeding habits. *Proc. Natl. Inst. Sci. India, Part B* **27,** 287–307.

Rasweiler, J. J., IV. (1972). Reproduction in the long-tongued bat, *Glossophaga soricina.* I. Preimplantation development and histology of the oviduct. *J. Reprod. Fertil.* **31,** 249–262.

Reeves, J. J., and Ellington, E. F. (1967). Pituitary gonadotropic activity of the ewe during the breeding and anestrous seasons. *J. Anim. Sci.* **26,** 950.

Richmond, M., and Conaway, C. H. (1969). Induced ovulation and oestrus in *Microtus ochrogaster. J. Reprod. Fertil., Suppl.* **6,** 357–376.

Robinson, E. J., and Rugh, R. (1943). Reproductive processes of the fishes, "*Oryzias latipes.*" *Biol. Bull. (Woods Hole, Mass.)* **84,** 115–125.

Robinson, T. J. (1951). Reproduction in the ewe. *Biol. Rev. Cambridge Philos. Soc.* **26,** 121–157.

Rood, J. P., and Weir, B. J. (1970). Reproduction in female wild guinea-pigs. *J. Reprod. Fertil.* **23,** 393–409.

Rousson, G. (1957). Some considerations concerning sturgeon spawning periodicity. *J. Fish. Res. Board Can.* **14,** 553–572.

Rowan, W. (1929). Experiments in bird migration. 1. Manipulation of the reproductive cycle—seasonal histological changes in the gonads. *Proc. Boston Soc. Nat. Hist.* **39,** 151–208.

Rowell, T. E. (1969). Effect of social environment on the menstrual cycle of baboons: preliminary report. *J. Reprod. Fertil., Suppl.* **6,** 117–118.

Rowlands, I. W. (1957). The ferret (*Mustela putorius furo* L.). *In* "The UFAW Handbook on the Care and Management of Laboratory Animals" (A. N. Worden and W. Lane-Petter, eds.), 2nd ed., Chapter 48, pp. 557–567. UFAW, London.

Rowlands, I. W. (1964). Levels of gonadotropins in tissues and fluids with emphasis on domestic animals. *In* "Gonadotropins" (H. H. Cole, ed.), pp. 74–107. Freeman, San Francisco, California.

Rowlands, I. W., and Heap, R. B. (1966). Histological observations on the ovary and progesterone levels in the coypu, *Myocastor coypus. Symp. Zool. Soc. London* **15,** 335–352.

Rowlands, I. W., and Parkes, A. S. (1935). The reproductive processes of certain mammals. VIII. Reproduction in foxes (*Vulpes* spp.). *Proc. Zool. Soc. London* Part 4, 823–841.

Rowlands, I. W., and Short, R. V. (1959). The progesterone content of the guinea-pig corpus luteum during the reproductive cycle and after hysterectomy. *J. Endocrinol.* **19,** 81–86.

Rowlands, I. W., Tam, W. H., and Kleiman, D. G. (1970). Histological and biochemical studies on the ovary and of progesterone levels in the systemic blood of the green acouchi (*Myoprocta pratti*). *J. Reprod. Fertil.* **22**, 533–545.

Roy, D. J., and Mullick, D. N. (1964). Endocrine function of corpus luteum of buffaloes during estrous cycle. *Endocrinology* **75**, 284–287.

Salzmann, R. C. (1963). Beitrage zur Fortpflangungsbiologie von *Meriones shawi* (Mammalia, Rodentia). *Rev. Suisse Zool.* **70**, 343–452.

San-Martin, M., Copaira, M., Zuniga, J., Rodreguez, R., Bustinza, G., and Acosta, L. (1968). Aspects of reproduction in the alpaca. *J. Reprod. Fertil.* **16**, 395–399.

Sanyal, M. K., and Prasad, M. R. N. (1967). Reproductive cycle of the Indian house lizard *Hemidactylus flaviviridis* Ruppell. *Copeia* pp. 627–633.

Sathyanesan, A. G. (1961). Spawning periodicities as revealed by the seasonal histological changes of the gonads in *Barbus stigma* (Cuv. and Val.). *Proc. Zool. Soc. India* **14**, 15–26.

Sathyanesan, A. G. (1962). The ovarian cycle in the catfish *Mystus seenghala* (Sykes). *Proc. Natl. Acad. Sci. India, Sect. B* **28**, 497–506.

Seth, P., and Prasad, M. R. N. (1969). Reproductive cycle of the female five-striped Indian palm squirrel *Funambulus pennanti* (Wroughton). *J. Reprod. Fertil.* **20**, 211–222.

Shalash, M. R., and Nawito, M. (1964). Some reproductive aspects in the female camel. *Proc. Int. Congr. Anim. Reprod. Artif. Insem., Trento 5th, 1964* Vol. 2, pp. 263–273.

Sharman, G. B. (1955a). Studies on marsupial reproduction. II. The oestrous cycle of *Setonix brachyurus*. *Aust. J. Zool.* **3**, 44–55.

Sharman, G. B. (1955b). Studies on marsupial reproduction. III. Normal and delayed pregnancy in *Setonix brachyurus*. *Aust. J. Zool.* **3**, 56–70.

Sharman, G. B. (1970). Reproductive physiology of marsupials. *Science* **167**, 1221–1228.

Sharman, G. B., and Calaby, J. H. (1964). Reproductive behaviour in the red kangaroo, *Megaleia rufa*, in captivity. *CSIRO Wildl. Res.* **9**, 58–85.

Sharman, G. B., Calaby, J. H., and Poole, W. E. (1966). Patterns of reproduction in female diprotodont marsupials. *Symp. Zool. Soc. London* **15**, 205–232.

Short, R. V. (1966). Oestrous behaviour, ovulation and the formation of the corpus luteum in the African elephant, *Loxodonta africana*. *East Afr. Wildl. J.* **4**, 56–58.

Short, R. V., and Buss, I. O. (1965). Biochemical and histological observations on the corpora lutea of the African elephant, *Loxodonta africana*. *J. Reprod. Fertil.* **9**, 61–67.

Short, R. V., and Hay, M. F. (1966). Delayed implantation in the roe deer, *Capreolus capreolus*. *Symp. Zool. Soc. London* **15**, 173–194.

Simkin, D. W. (1965). Reproduction and productivity of moose in northwestern Ontario. *J. Wildl. Manage* **29**, 740–750.

Simpson, G. G. (1945). Principles of classification and a classification of mammals. *Bull. Am. Mus. Nat. Hist.* **85**, 1–350.

Simpson, T. H., Wright, R. S., and Hunt, S. V. (1968). Steroid biosynthesis *in vitro* by the component tissues of the ovary of the dogfish (*Scyliorhinus caniculus* L.). *J. Endocrinol.* **42**, 519–527.

Sinha, A. A., and Conaway, C. H. (1968). The ovary of the sea otter. *Anat. Rec.* **160**, 795–806.

Sinha, B. M., and Rastogi, R. K. (1967). Cyclical changes in the ovarian activity of the freshwater mud-eel *Amphipnous cuchia. Proc. Natl. Acad. Sci. India, Sect. B* **37**, 175–188.

Slater, P. J. B. (1969). The stimulus to egg-laying in the Bengalese finch. *J. Zool.* **158**, 427–440.

Smeaton, T. C., and Robertson, H. A. (1971). Studies on the growth and atresia of the Graafian follicles in the ovary of the sheep. *J. Reprod. Fertil.* **25**, 243–252.

Smith, J. G., Hanks, J., and Short, R. V. (1969). Biochemical observations on the corpora lutea of the African elephant, *Loxodonta africana. J. Reprod. Fertil.* **20**, 111–117.

Smith, R. F. C. (1969). Studies on the marsupial glider, *Schoinobates volans* (Kerr). I. Reproduction. *Aust. J. Zool.* **17**, 625–636.

Snow, D. W., and Snow, B. K. (1964). Breeding seasons and annual cycles of Trinidad land-birds. *Zoologica (N.Y.)* **49**, 1–39.

Sowls, L. K. (1966). Reproduction in the collared peccary (*Tayassu tajacu*). *Symp. Zool. Soc. London* **15**, 155–172.

Spinage, C. A. (1969). Reproduction in the Uganda defassa waterbuck, *Kobus defassa ugandae* Neumann. *J. Reprod. Fertil.* **18**, 445–457.

Srivastava, P. N., and Rathi, S. K. (1970). Effect of radiation on the reproductive system of the Indian catfish *Heteropneustes fossilis*. I. Annual cycle in the development of ovarian eggs. *Acta Anat.* **75**, 114–125.

Stabenfeldt, G. H., Ewing, L. L., and McDonald, L. E. (1969a). Peripheral plasma progesterone levels during the bovine oestrous cycle. *J. Reprod. Fertil.* **19**, 433–442.

Stabenfeldt, G. H., Holt, J. A., and Ewing, L. L. (1969b). Peripheral plasma progesterone levels during the ovine estrous cycle. *Endocrinology* **85**, 11–15.

Stockard, C. R., and Papanicolaou, G. N. (1917). The existence of a typical oestrous cycle in the guinea-pig, with a study of its histological and physiological changes. *Am. J. Anat.* **22**, 225–283.

Stockard, C. R., and Papanicolaou, G. N. (1919). The vaginal closure membrane, copulation and the vaginal plug in the guinea-pig, with further consideration of the oestrous rhythm. *Biol. Bull. (Woods Hole, Mass.)* **37**, 225–245.

Sundararaj, B. I., and Sehgal, A. (1970). Effects of a long or an increasing photoperiod on the initiation of ovarian recrudescence during the preparatory period in the catfish, *Heteropneustes fossilis* (Bloch). *Biol. Reprod.* **2**, 413–424.

Talmage, R. V., Buchanan, G. D., Kraintz, F. W., Lazo-Wasem, E. A., and Zarrow, M. X. (1954). The presence of a functional corpus luteum during delayed implantation in the armadillo. *J. Endocrinol.* **11**, 44–49.

Tam, W. H. (1970). The function of the accessory corpora lutea in the hystricomorph rodents. *J. Endocrinol.* **48**, 54–55.

Tam, W. H. (1971). The production of hormonal steroids by ovarian tissues of the chinchilla (*Chinchilla laniger*). *J. Endocrinol.* **50**, 267–279.

Tam, W. H. (1973). Progesterone levels during the oestrous cycle and pregnancy in the cuis, *Galea musteloides. J. Reprod. Fertil.* **35**, 105–114.

Telford, S. R., Jr. (1969). The ovarian cycle, reproductive potential, and structure in a population of the Japanese lacertid (*Takydromus tachydromoides*). *Copeia* pp. 548–567.

Tinkle, D. W. (1961). Population structure and reproduction in the lizard *Uta stansburiana stejnegeri. Am. Midl. Nat.* **66**, 206–234.

Tinkle, D. W. (1962). Reproductive potential and cycles in female *Crotalis atrox* from north western Texas. *Copeia* pp. 306–313.

Tinkle, D. W. (1969). The concept of reproductive effort and its relation to the evolution of life history of lizards. *Am. Nat.* **103**, 501–516.

Tomich, P. Q. (1962). The annual cycle of the Californian ground squirrel. *Univ. Calif., Berkeley, Publ. Zool.* **65**, 213–282.

Tripp, H. R. H. (1971). Reproduction in elephant shrews (Macroscelididae) with special reference to ovulation and implantation. *J. Reprod. Fertil.* **26**, 149–159.

Tsusaka, A., and Nakano, E. (1965). The metabolic pattern during oogenesis in the fish, *Oryzias latipes. Acta Embryol. Morphol. Exp.* **8**, 1–11.

Turner, C. L. (1937). Reproductive cycles and superfetation in Poeciliid fishes. *Biol. Bull. (Woods Hole, Mass.)* **72**, 145–164.

Turner, C. L. (1938). The reproductive cycle of *Brachyraphis episcopi,* an ovoviviparous Poeciliid fish, in the natural tropical habitat. *Biol. Bull. (Woods Hole, Mass.)* **75**, 56–65.

Tyndale-Biscoe, C. H. (1963a). Effects of ovariectomy in the marsupial *Setonix brachyurus. J. Reprod. Fertil.* **6**, 25–40.

Tyndale-Biscoe, C. H. (1963b). The role of the corpus luteum in the delayed implantation of marsupials. *In* "Delayed Implanation" (A. C. Enders, ed.), pp. 15–32. Univ. of Chicago Press, Chicago, Illinois.

Vandenbergh, J. G., and Vessey, S. (1968). Seasonal breeding of free-ranging rhesus monkeys and related ecological factors. *J. Reprod. Fertil.* **15**, 71–79.

van der Horst, C. J., and Gillman, J. (1940). Ovulation and corpus luteum formation in *Elephantulus. S. Afr. J. Med. Sci.* **5**, 73–91.

van Rensburg, S. J., and van Niekerk, C. H. (1968). Ovarian function, follicular oestradiol-17β, and luteal progesterone and 20α-hydroxy-preg-4-en-3-one in cycling and pregnant equines. *Onderstepoort J. Vet. Res.* **35**, 301–318.

Varma, S. K. (1970). Morphology of ovarian changes in the garden lizard *Calotes versicolor. J. Morphol.* **131**, 195–210.

Vilter, V., and Vilter, A. (1963). Mise en évidence d'un cycle reproducteur biennial chez le Triton alpestre de montagne. *C. R. Soc. Biol.* **157**, 464–469.

Vilter, V., and Vilter, A. (1964). Sur l'évolution des corps jaunes ovariens chez *Salamandra atra* Laur. des Alpes vaudoises. *C. R. Soc. Biol.* **158**, 457–461.

Watson, R. M. (1969). Reproduction of wildebeest, *Connochaetes taurinus albojubatus* Thomas, in the Serengeti region, and its significance to conservation. *J. Reprod. Fertil., Suppl.* **6**, 287–310.

Weekes, H. C. (1933). On the distribution, habitat and reproductive habits of certain European and Australian snakes and lizards, with particular reference to their adoption of viviparity. *Proc. Linn. Soc. N.S.W.* **58**, 270–274.

Weir, B. J. (1971a). The reproductive organs of the female plains viscacha, *Lagostomus maximus. J. Reprod. Fertil.* **25**, 365–373.

Weir, B. J. (1971b). The evocation of oestrus in the cuis, *Galea musteloides. J. Reprod. Fertil.* **26**, 405–408.

Weir, B. J. (1973a). Another hysticomorph rodent: keeping casiragua (*Proechimys guairae*) in captivity. *Lab. Anim.* **7**, 125–134.

Weir, B. J. (1973b). The rôle of the male in the evocation of oestrus in the cuis, *Galea musteloides* (Rodentia: Hystricomorpha). *J. Reprod. Fertil., Suppl.* **19**, 421–432.

Weir, B. J. (1974). Reproductive characteristics of hystricomorph rodents. *Symp. Zool. Soc. London* **34**, 265–301.

Wiebe, J. P. (1968). The reproductive cycle of the viviparous sea-perch, *Cymatogaster aggregata* Gibbons. *Can. J. Zool.* **46,** 1221–1234.

Wilhoft, D. C. (1963). Gonadal histology and seasonal changes in the tropical Australian lizard, *Leiolopisma rhomboidalis. J. Morphol.* **113,** 185–204.

Wilson, D. E., and Findley, J. S. (1970). Reproductive cycle of a neotropical insectivorous bat, *Myotis nigricans. Nature (London)* **225,** 1155.

Wimsatt, W. A. (1960). Some problems of reproduction in relation to hibernation in bats. *Bull. Mus. Comp. Zool.* **124,** 249–267.

Wimsatt, W. A. (1963). Delayed implantation in the Ursidae, with particular reference to the black bear (*Ursus americanus* Pallas). *In* "Delayed Implantation" (A. C. Enders, ed.), pp. 49–76. Univ. of Chicago Press, Chicago, Illinois.

Wimsatt, W. A., and Parks, H. F. (1966). Ultrastructure of the surviving follicle of hibernation and of the ovum-follicle cell relationship in the vespertilionid bat, *Myotis lucifugus. Symp. Zool. Soc. London* **15,** 419–454.

Wolfson, A. (1966). Environmental and neuroendocrine regulation of annual gonadal cycles and migratory behavior in birds. *Recent Prog. Horm. Res.* **22,** 177–244.

Woolley, P. (1966). Reproduction in *Antechinus* spp. and other dasyurid marsupials. *Symp. Zool. Soc. London* **15,** 281–294.

Worden, A. N., and Lane-Petter, W., eds. (1957). "The UFAW Handbook on the Care and Management of Laboratory Animals," 2nd ed. UFAW, London.

Wright, P. L. (1966). Observations on the reproductive cycle of the American badger (*Taxidea taxus*). *Symp. Zool. Soc. London* **15,** 27–45.

Xavier, F. (1969). Corps jaunes de post-ovulation actifs chez les femelles non fécondées de *Nectophrynoides occidentalis* (Amphibien anoure vivipare). *Gen. Comp. Endocrinol.* **13,** 542.

Xavier, F. (1970). Analyse du rôle des corpora lutea dans le montien de le gestation chez *Nectophrynoides occidentalis* (Angel). *C. R. Acad. Sci. Ser. D* **270,** 2018–2020.

Yamamoto, K. (1956). Studies on the development of fish eggs. 1. Annual cycle in the development of ovarian eggs in the flounder, *Liopsetta obscura. J. Fac. Sci., Hokkaido Univ., Ser. 6,* **12,** 362–373.

Yamamoto, K., and Yamazaki, F. (1961). Rhythm of development in the oocyte of goldfish, *Carassius auratus. Bull. Fac. Fish. Hokkaido Univ.* **12,** 93–110.

Zaki, K., Soliman, F. A., and Ramsees, G. (1963). Reproductive pattern of the Egyptian buffalo. *J. Arab Vet. Med. Assoc.* **23,** 361–374.

Zuckerman, S. (1937). The duration and phases of the menstrual cycle in primates. *Proc. Zool. Soc. London* pp. 315–329.

Zuckerman, S., and Parkes, A. S. (1932). The menstrual cycle of the primates. V. The cycle of the baboon. *Proc. Zool. Soc. London* pp. 139–191.

7

Endocrine Activities of the Ovary

P. Eckstein

The two basic and complementary tasks of the mammalian ovary are to produce fertilizable eggs and a variety of steroid hormones. The following account is concerned only with the latter function, and more specifically with the cellular site of origin of the ovarian steroids. In addition, it deals with ovarian endocrine activities before the attainment and after the end of reproductive maturity, as well as with the relations between the ovary and both the adrenal and the thyroid gland.

I. THE CELLULAR ORIGIN OF THE OVARIAN STEROIDS

A. Biochemical and Structural Aspects of Ovarian Steroidogenesis

The ovary elaborates at least three types of steroids: estrogens, progestins, and androgens. It may also, in some species and special circumstances, secrete corticoids and contribute to the production of a nonsteroidal polypeptide hormone, relaxin. Although it is fully established that the most important of these compounds, estrogens and progestins, are secreted mainly, if not exclusively, by the ovary, their ultimate site of production remains one of the basic problems of ovarian physiology.

Proof of the origin of a particular hormone from the ovary has been obtained from many sources and by diverse methods. Among them were early hormone replacement studies after bilateral ovariectomy and the histochemical demonstration of cholesterol-containing lipids in special cell sites, from which their capacity to form steroid hormones was inferred (see Corner, 1938; White *et al.,* 1951; Young, 1961; Jacoby, 1962; Parkes and Deanesly, 1966a; and also Volume I, Chapter 7). This type of approach has been superseded by the isolation of steroids from specific ovarian structures, after either perfusion of ovarian tissue *in vivo* or incubation *in vitro* and, most directly and convincingly, by qualitative and quantitative identification in ovarian effluent blood, in comparison with peripheral blood (see Volume III, Chapter 5). Work of this kind has been greatly facilitated by the development of modern, highly sensitive and accurate methods for the determination of steroids in tissues and body fluids.

Elucidation of the cellular origin of the ovarian hormones is, however, complicated by several factors. Ovarian steroidogenesis, unlike the formation and release of eggs, involves several structural units, and their respective contributions to steroid secretion must be evaluated separately. Moreover, the cellular elements concerned are not static but undergo periodic changes, in phase with the ovarian cycle. As a result, the ovary is made up of constantly varying types and proportions of tissue throughout the stages of the cycle. During the proliferative phase, it contains the granulosa and theca interna and externa cells of the follicles, in addition to interstitial and regressing luteal tissue. During the secretory phase, freshly formed granulosa lutein and theca lutein cells appear. There is thus not only great morphological diversity throughout the entire ovarian cycle but also, because of the multiplicity of steroids involved and their interrelated biosynthetic pathways, considerable overlapping in hormone production. For this reason, more than one type of steroid is usually formed by the intact ovary at any given stage of the cycle. The principal biochemical and structural components of ovarian steroidogenesis are briefly examined below.

1. Biosynthetic Pathways

The pathways involved in the formation of ovarian steroids have been frequently described (see Ryan and Smith, 1965; Savard *et al.,* 1965; Savard, 1969; McKerns, 1969; Abraham and Tait, 1971; Fotherby, 1972;

Fotherby and James, 1972; Besch and Buttram, 1972; Engel, 1973; and also Volume III, Chapters 5 and 6); they are only outlined here.

The ovary, like other steroid-producing glands, readily converts acetate to cholesterol, the principal precursor of all steroid hormones. The side chain of cholesterol is split by hydroxylation between C-20 and C-22, yielding pregnenolone (3β-hydroxypregn-5-en-20-one) which serves as precursor of all other steroids. It is a C_{21} steroid and can be converted to androstenedione, a C_{19} compound, along one of two, the Δ^5 and Δ^4, pathways. The former is used by compounds with the Δ^5-3β-hydroxy structure such as 17α-hydroxypregnenolone and dehydroepiandrosterone (DHA). The Δ^4 pathway, characteristic of steroids containing the Δ^4-3-oxo group, involves the enzyme 3β-hydroxysteroid dehydrogenase which will convert pregnenolone to progesterone. This, in turn, can be hydroxylated at C-17, so yielding 17α-hydroxyprogesterone and, after cleavage of the side chain, androstenedione (androst-4-ene-3,17-dione). It is believed that the theca interna and granulosa cells of human Graafian follicles follow different biosynthetic pathways, the theca interna favoring the Δ^5 route and the granulosa and lutein cells the Δ^4 route (Ryan and Petro, 1966; Fotherby, 1972).

Estrogens are derived from androgens by aromatization, the mechanism whereby a C_{19} steroid is converted into a C_{18} steroid, or estrogen, by removal of C-19 with simultaneous or consequential aromatization of the A-ring; the process is not yet completely understood (see Abraham and Tait, 1971). The prevalent estrogen formed from androstenedione is estradiol-17β. Androstenedione and testosterone are interconvertible (by the enzyme 17β-hydroxysteroid dehydrogenase), and either will serve as substrates for the aromatizing system.

It is believed that these biosynthetic pathways are common to the adrenal, testis, and ovary (see Ryan and Smith, 1965), but that all three glands show quantitative and qualitative differences in the steroid hormones secreted by them. These differences are probably related to their different cell populations and enzyme makeup.

2. Morphological Compartments of the Ovary

In view of the difficulties of assigning steroidogenic activities to a particular cellular constituent of the ovary (see Section I,A), it is useful to consider the three main ovarian tissue compartments or subunits effectively as separate steroid-forming glands. They are: (a) the Graafian follicle (granulosa and theca interna cells); (b) the corpus luteum (theca lutein and granulosa lutein cells); and (c) the stroma (interstitial cells). The hormone production and physiological role of the special cell aggregates near the ovarian hilus ("hilus cells"; see Volume I, Chapter 4) remain uncertain.

Some of the more important structural and histochemical features of these compartments are summarized below. Full descriptions have been given by Christensen and Gillim (1969) and Guraya (1971), and some of the variations found in different species are discussed in Volume I, Chapter 4.

a. The Ovarian Follicle. *i. The theca interna.* Basically, the ovarian follicle consists of two layers of cells, the membrana granulosa and theca interna (surrounded by the mainly fibrous theca externa) arranged in roughly concentric fashion around the ovum. The two types of cell become differentiated when the developing follicle acquires an antrum, and those lying outside the granulosa hypertrophy into large secretory cells with vesicular nuclei to constitute the theca interna. The thecal cells gradually become vascularized, unlike the granulosa cells which remain separated from them by a conspicuous basement membrane (Fig. 1). Ultrastructurally, the thecal cells possess the features which characterize steroidogenic tissue in general, that is, a tubular smooth endoplasmic reticulum, mitochondria with mainly tubular cristae, and lipid bodies (Enders 1962; Christensen and Gillim, 1969; Guraya, 1971; Bjersing *et al.,* 1972; Fig. 2; also Volume I, Chapter 4, Fig. 11). The presence of 17-hydroxylase and other enzymes known to play a role in the biosynthesis of ovarian hormones has been demonstrated in the theca interna of sows and other mammals (Ryan and Petro, 1966; Bjersing, 1967b; Guraya, 1971).

ii. The membrana granulosa. The granulosa cells multiply by mitosis during growth of the follicle, and form the multilayered membrana granulosa (Brambell, 1956). Their cytoplasm contains a well developed Golgi zone, a predominantly granular endoplasmic reticulum, and lipid bodies unlike those which characterize the theca interna and other steroid-forming cells. Enzyme reactions indicative of steroidogenic activities occur in granulosa cells, but vary considerably between different species and are generally less well developed than in thecal cells (Bjersing, 1967a; Bjersing and Carstensen, 1967; Guraya, 1971).

Short (1962, 1964) and YoungLai and Short (1970) have pointed out that the (essentially avascular) granulosa cells probably secrete into the follicular fluid which bathes them rather than across the basement membrane, although those in immediate contact with the membrane may do so. The secretions of the well-vascularized theca interna cells, however, pass directly into the tissue fluid and capillaries surrounding them; some thecal tissue fluid may also cross the basement membrane and so reach the granulosa cells and follicular fluid (see also Section I,E).

b. The Corpus Luteum. The histological features, development, histogenesis, and life cycle of the mammalian corpus luteum have been

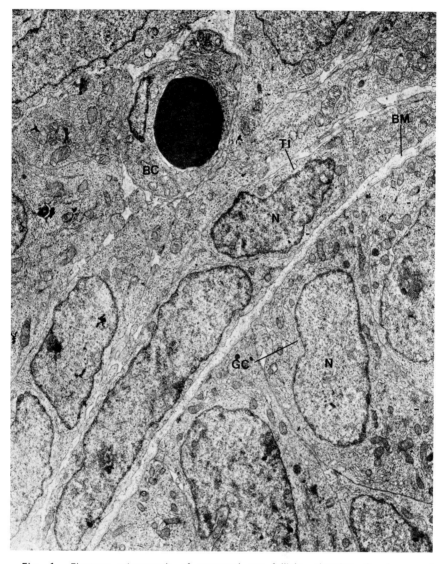

Fig. 1. Electron micrograph of preantral rat follicle, showing the basement membrane (BM) intervening between granulosa cells (GC), below, and theca interna cells (TI) above. BC, blood capillary; N, nucleus (×5300). (Micrograph provided by Dr. L. L. Franchi.)

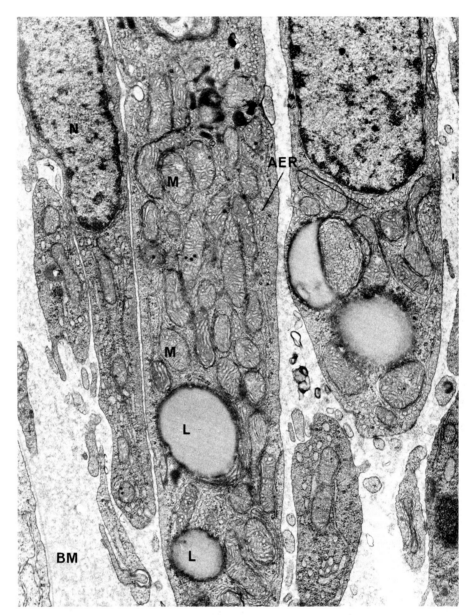

Fig. 2.   Electron micrograph of theca interna from vesicular follicle of rhesus monkey, shortly before ovulation. Note the extensive agranular endoplasmic reticulum cisternae (AER), mitochondria with villiform or tubular cristae (M), and lipid inclusions (L) characteristic of active steroidogenic cells. BM, basement membrane; N, nucleus ($\times$16,250). (Micrograph provided by Dr. Thomas M. Crisp.)

described many times (see Corner, 1945; Brambell, 1956; Harrison, 1948, 1962; and also Volume I, Chapter 4). The relationship between the ultrastructure and secretory activity of the corpus luteum has been studied by Blanchette (1966), Bjersing (1967b), Christensen and Gillim (1969), and Crisp *et al.* (1970).

It is believed that in most mammals both the granulosa and thecal cells are involved in forming the corpus luteum, but their respective contributions vary greatly in different species (see Brambell, 1956; and Volume I, Chapter 4). For instance, in women, rhesus monkeys, and probably cows, the corpus luteum has a dual origin (Corner, 1945; Brambell, 1956; Harrison, 1962; Donaldson and Hansel, 1965; Adams and Hertig, 1969), but in the hamster, mouse, and rabbit it appears to develop entirely from the granulosa (see Perry, 1971). In the monkey and cow, the theca lutein cells regress and lose their identity soon after formation of the corpus luteum (Corner, 1945; Brambell, 1956; Donaldson and Hansel, 1965).

There are equally marked species differences in steroid content and biosynthesis of the corpus luteum. According to Savard *et al.* (1965), mammalian corpora lutea fall into two categories, those which produce only progestins, such as progesterone and 20α- or 20β-hydroxypregnenone, and those which secrete both progestins and estrogens, as well as 17α-hydroxyprogesterone and androstenedione. The cow (Savard *et al.*, 1965), mare (Short, 1964), and probably the monkey (Hotchkiss *et al.*, 1971) belong to the first group, and women to the second (Huang and Pearlman, 1963; Hammerstein *et al.*, 1964).

It has been clearly demonstrated that granulosa cells, maintained in tissue culture and examined electron microscopically, luteinize in a manner comparable to that observed *in vivo*, and may thus serve as a model in the study of the luteinization process (see Christensen and Gillim, 1969; Channing, 1969d,e, 1970b; Crisp and Channing, 1972; see also Section I,D).

Recent electron microscopical observations suggest that the cells of the regressing corpus luteum at the end of pregnancy in women and rabbits may be involved in the synthesis and storage of the polypeptide hormone, relaxin (Crisp *et al.* 1970; Belt *et al.*, 1971).

c. **The Ovarian Stroma (Interstitial Cells).** The interstitial tissue forms an important but frequently inconspicuous and variable component of the mammalian ovary which is thought to participate in its steroidogenic activity (see Section I,B,3).

The constituent cells are large and polyhedral, with the cytological characteristics of glandular elements. Their similarity to the Leydig cells of the testis prompted the early histologists to name them as the ovarian interstitial cells (Tourneux, 1879) and to recognize them as one of the two endocrine glands of the ovary, the second one being the corpus luteum

(Bouin, 1902). Their histological and ultrastructural characteristics, origin, specific distribution, and homologies have been extensively studied (see Brambell, 1956; Harrison, 1962; Christensen and Gillim, 1969; Mossman and Duke, 1973; Volume I, Chapter 4; and Section I,B,3, below).

B. Steroid Biosynthesis by Ovarian Compartments

There is a consensus that the developing Graafian follicle is the principal site of steroid formation during the proliferative phase of the cycle and that the corpus luteum assumes this function after ovulation. It is equally accepted that the preovulatory follicle elaborates mainly androstenedione and estrogens while the corpus luteum is responsible, for instance in women, for the secretion of progestins and estrogens. What is not clearly understood and remains controversial is (1) which cells are responsible for the synthesis of which hormones, (2) how the changeover from estrogen to progesterone secretion is brought about, and (3) the exact role of the interstitial cells in ovarian steroidogenesis. This and the following sections attempt to assemble the more important information that bears on these issues.

1. Follicular Tissue

The close temporal association between the various manifestations of estrus and the presence of mature follicles suggested that the Graafian follicle or some of its cellular components must be involved, and, in 1938, Corner concluded that the most probable site of estrogen production was the theca interna of ripe and developing follicles. Since then, both estradiol-17β and estrone have been chemically identified in human Graafian follicles (see Zander et al., 1959), and estradiol in those of the cow (Velle, 1959), while Short (1964) has identified both of these estrogens, as well as progesterone and several other steroids, in follicular fluid of the mare.

Corner's view (1938) is, however, now known to be an oversimplification. For instance, it is well established that the secretion of estrogen by the ovary does not depend on the presence of structurally organized follicles. Thus some estrogen is formed by the ovaries after complete destruction by X irradiation of the Graafian follicles. This was first demonstrated in mice some 50 years ago (Parkes, 1926, 1929; see Eckstein, 1962), and has since been confirmed in both mice and rats by Mandl and Zuckerman (1956a,b). The experiments of these workers indicate that estrogen may be produced by ovarian cells other than those constituting the normal follicle, even though the nature of the cellular elements involved, and persisting after irradiation, has not been established (see also Section I,B,3).

While these studies make it clear that, at least in rodents, the structural integrity of Graafian follicles is no prerequisite for estrogen formation, they are in no way inconsistent with the view that, in the intact gonad, estrogen-producing cells are present in the follicular wall, and are, in fact, responsible for the normal output of estrogenic hormone by the ovary.

The first attempt to ascertain the steroidogenic capacities of specific ovarian cell elements or "compartments" was made by Falck (1959). He autotransplanted minute single-cell aggregates, representing granulosa, theca interna, interstitial, and luteal cells, either alone or in combination, into the anterior eye chamber of spayed female rats, and used small pieces of vagina placed alongside the ovarian grafts as indicators of estrogenic activity. He showed that estrogen formation required the joint action between (1) theca interna or interstitial cells and (2) granulosa or corpus luteum cells; neither granulosa nor thecal cells alone were able to secrete estrogens. In this way, he introduced the concept of a functional interaction between two different cellular elements, the theca interna and granulosa cells, as the basis of the "estrogen-producing" cell system of the ovary. In Falck's study, estrogen production was assessed indirectly from cornification of the adjacent vaginal graft, but was not biochemically confirmed.

Direct enquiry into the biosynthetic capacities of specific follicular tissues began with the work of Ryan and Smith (1961a,b,c,d, 1965) on the isolated follicle wall from gonadotropin-stimulated human ovaries. These workers used the resulting large cystic follicles, dissected free of other ovarian tissue, and showed that their cellular lining possesses high biosynthetic potencies. Its incubation with labeled precursors, such as acetate, yielded estradiol-17β and estrone as the major radioactive products, and, in order of decreasing activity, 4-androstenedione, dehydroepiandrosterone (DHA), 17α-hydroxyprogesterone, 17α-hydroxypregnenolone, progesterone, and pregnenolone; no testosterone was identified (Fig. 3). It must be realized that in Ryan and Smith's studies the follicular tissue used contained both granulosa and luteinized theca interna elements, and therefore could provide no direct information on the ultimate cellular site of steroid formation. These observations are consistent with those of Short (1964), who determined the steroids present in the follicular fluid of the intact ovary of mares, and found that estradiol-17β was present in highest amounts; he attributed its formation to the theca interna.

It has since been shown that preparations of pure equine granulosa cells can readily convert androstenedione and testosterone into estrogens during short-term incubation or in tissue culture (Ryan and Short, 1965; Channing, 1969b,d). This *in vitro* activity contrasts, however, with the much weaker aromatizing capacity of granulosa cells in the intact follicle of the mare. While no ultimate conclusions can be drawn at present, it is possible that

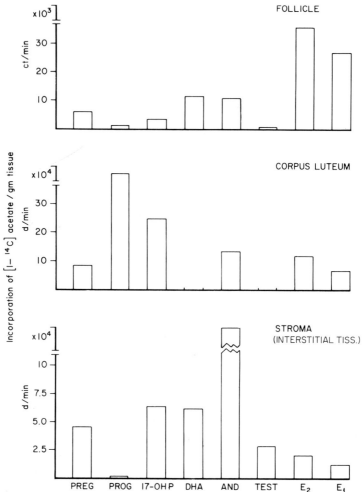

Fig. 3. Patterns of radioactive steroids obtained from isolated compartments of the human ovary incubated with [1-^{14}C]acetate. PREG, 3β-hydroxypregn-5-en-20-one; PROG, progesterone; 17-OHP, 17α-hydroxyprogesterone; DHA, dehydroepiandrosterone; AND, androst-4-ene-3,17-dione; TEST, testosterone; E$_2$, estradiol-17β; E$_1$, estrone. (After Savard *et al.,* 1965.)

steroids produced in the theca interna are altered by the granulosa cells while passing into the follicular fluid or, conversely, that granulosa-derived steroids are modified by the thecal cells before reaching the ovarian venous blood. This potential interaction between the two types of cell, and the extent to which the passage of steroids across the follicle wall is affected by the basement membrane, remain to be elucidated.

2. Luteal Tissue

The corpus luteum has been investigated longer and more thoroughly than all other cellular compartments of the ovary, but many aspects of luteal function pertinent in the present context remain obscure. The unique and striking changes which accompany the transformation of the vesicular follicle into a compact corpus luteum, and alter its principal hormonal output from estrogen to progestin within a matter of days, have been fully characterized anatomically but not yet biochemically. Also, the factors responsible for the process of luteinization are unknown. While gonadotropic hormones are undoubtedly involved in ovulation, it is uncertain whether, and to what extent, they control the formation and subsequent function of the corpus luteum. For instance, the human corpus luteum reaches its steroidogenic peak toward the end of the first week after its formation, yet there appears to be no corresponding increase in the secretion of gonadotropin. Little more is known about the decline and eventual "death," both structural and functional, of the corpus luteum. Although this is believed to be brought about by a uterine hormone, luteolysin, in ruminants and several other species, there is good evidence that such a mechanism does not operate in women and lower primates (see Anderson *et al.,* 1969; and Volume III, Chapter 3).

Steroidogenesis. This has been intensively studied by Savard and his associates, using, as in the case of follicular tissue, slices from fresh specimens incubated with labeled precursors such as acetate and pregnenolone (see Savard *et al.,* 1965, 1969; Savard, 1969). It was found that in such an *in vitro* system human corpora lutea, obtained either during the luteal phase of the cycle or from cases of ectopic pregnancy, synthesized 5-pregnenolone, progesterone, 20α-hydroxypregn-4-en-3-one (20α-hydroxypregnenone), 17-hydroxyprogesterone, androstenedione, estrone, and estradiol-17β, but no testosterone or DHA (Fig. 3). By contrast, bovine corpora lutea synthesize only two major steroids, progesterone and 20α-hydroxypregnenone, but no estrogen. The fact that human luteal tissue produces estrone and estradiol *in vitro* reflects and may account for the capacity of the ovary *in vivo* to secrete both estrogen and progestin during the luteal phase of the menstrual cycle. It is believed that 17α-hydroxyprogesterone not only acts as a precursor of other steroids but is also actively secreted by the human corpus luteum (Ross *et al.,* 1970).

The variable ability of the corpora lutea of different species to secrete estrogen, apart from progestins, may depend on their content of functional theca lutein cells. The human corpus luteum is a mixed glandular structure in which theca lutein (or "paraluteal") elements are clearly recognizable and separate from the granulosa lutein cells during the greater part of the luteal phase. In the monkey, the theca lutein cells are only recognizable during the first few days after formation of the corpus luteum (Corner, 1945; Harrison, 1962), and little if any estrogen is formed after ovulation (Hotchkiss *et al.*, 1971). In the cow, the mature corpus luteum consists entirely of granulosa lutein cells, and appears to produce only progestins, both *in vitro* and *in vivo*.

3. The Stromal (Interstitial) Tissue

It has been thought for many years that the interstitial tissue of the ovary possesses endocrine properties. This was originally suspected on purely histological grounds, and expressed by Bouin as long ago as 1902 when he referred to "the two glands of internal secretion" in the ovary of the rabbit: the interstitial gland and the corpus luteum.

The evidence that the ovarian stroma constitutes an endocrine gland is, however, indirect and frequently controversial, and the biochemical nature of its hormonal output remains unknown. It has been held responsible for the estrogen production that maintains rudimentary estrous cycles in mice and rats after destruction of their follicular apparatus by X irradiation (Parkes, 1926; Zuckerman, 1956; Mandl and Zuckerman, 1956a,b). Equally, the paradoxical androgen production of ovaries grafted to the ears of castrated male mice (Hill, 1937; Hill and Strong, 1940) has been related to the ovarian stromal elements, but has never been fully determined, while in rats androgen secretion by the ovaries appears to be associated with luteinized thecal cells present in long-term ovarian grafts (Deanesly, 1938, 1966). By contrast, in rabbits the interstitial tissue is exceptionally well developed, and is believed to be the source of progestins (Hilliard *et al.*, 1963; Simmer *et al.*, 1963). Both progesterone and 20α-hydroxypregn-4-en-3-one have been identified in ovarian venous blood from intact ovaries and from ovaries with all follicles destroyed by cauterization.

More recently, Scully and Cohen (1964) reported that the stroma of the normal human ovary contains a large number of cells resembling luteinized thecal elements and possessing high glucose-6-phosphate dehydrogenase activity. These authors suggested that the enzymatically active stromal cells might be steroidogenic, but could not establish what hormones were secreted by them. Since then, it has been shown by Rice and Savard (1966) that, unlike human follicular and luteal tissue, normal ovarian stroma produces substantial amounts of testosterone *in vitro*. Among the compounds identified were androstenedione, DHA, and testosterone. In addition to these three androgens, smaller amounts of estrogens and 17α-hydroxyprogesterone were found, but only traces of progesterone (Fig. 3).

Since the formation of testosterone has also been demonstrated in stromal tissue slices (Hammerstein *et al.,* 1964) but not, or only in minute amounts, in human follicular and luteal tissue (see Savard, 1969), it seems that the interstitial cells are the most likely source of plasma testosterone in normal women. The further observation that androgens are present in appreciable concentration in ovarian venous effluent proves that the ovary not only possesses the biosynthetic capacity to produce these steroids, but also secretes them actively. Thus, the levels of testosterone and androstenedione in ovarian venous blood are several times higher than in peripheral blood (Mikhail, 1967; Lloyd *et al.,* 1971; Osborn and Yannone, 1971). The possible physiological role of androgens in conditioning female sexual receptivity has been recognized (Herbert, 1970), although in women and rhesus monkeys receptivity appears to depend more on adrenal androgens than on those of ovarian origin (Everitt and Herbert, 1971; Everitt *et al.,* 1972). The capacity for androgen formation of the ovarian stroma implies, of course, a potential for virilization. This may be slight, for instance in postmenopausal women (see Section III) or marked, as in certain ovarian neoplasms (see Parkes, 1950).

It may be concluded that the stromal compartment of the human ovary constitutes a steroidogenic unit distinct from both the follicle and the corpus luteum. Although the ultimate cellular origin of the steroids formed in this compartment is not clear, it is thought that they are similar to those produced by the Leydig cells of the testis (Savard, 1969).

C. Steroid Hormones in Ovarian and Peripheral Venous Blood

Direct measurement of the steroid content of the ovarian venous plasma represents the best, if technically the most difficult, way of assessing the steroidogenic capabilities of the intact ovary. It has usually been carried out with the organ *in situ* but, more recently, on ovaries autotransplanted to a more accessible site in the neck (for instance, in sheep: Goding *et al.,* 1967; Baird *et al.,* 1968a; McCracken and Baird, 1969). This topic and the various factors affecting the results of such studies, are fully discussed in Volume III, Chapter 5.

Early work (Zander, 1958) demonstrated the presence of the following four steroids in human ovarian tissue: progesterone, 17α-hydroxyprogesterone, 20α-hydroxypregn-4-en-3-one, and androstenedione. Since the same compounds have also been identified in pooled ovarian venous blood (Mikhail, 1967), it must be assumed that they are secreted by the ovary. Table I is based on assays of single samples of ovarian and peripheral blood

TABLE I

Concentrations of Plasma Steroids (in μg/100 ml) in Three Parous Women[a]

Steroids	Subject 1 (day 12 of cycle, preovulatory)			Subject 2 (days 3 to 4, postovulatory)			Subject 3 (4 years postmenopausal)		
	Peripheral blood	Ovarian blood		Peripheral blood	Ovarian blood		Peripheral blood	Ovarian blood	
		Rt	Lt[b]		Rt[c]	Lt		Rt	Lt
Progesterone	0.15	0.39	1.55	2.13	47.10	2.57	0.07	0.18	0.14
17α-Hydroxy-progesterone	<2.00	<2.00	4.43	<2.00	4.05	<2.00	<1.00	<1.00	<1.00
Androstenedione	0.68	8.52	8.85	0.53	1.76	1.71	0.21	0.56	0.35
Testosterone	0.08	0.19	0.24	—	—	—	0.11	0.14	0.10
Estrone	0.04	0.07	0.17	0.04	0.06	0.08	0.82	0.64	0.54
Estradiol-17β	0.12	0.36	1.76	0.06	0.42	0.19	0.04	0.09	—

[a] Slightly modified from Mikhail (1967).
[b] Ovary containing ripe Graafian follicle.
[c] Ovary containing recent corpus luteum.

obtained at laparotomy from three parous women examined, respectively, just before midcycle, shortly after ovulation, and several years postmenopausally (Mikhail, 1967). The table shows that, as a rule, steroid concentrations are considerably higher in ovarian effluent than in peripheral blood, and that androstenedione is the principal ovarian steroid product, except during the luteal phase when more progesterone is secreted. It is also clear that substantial amounts of progesterone can be released before ovulation, and hence from the ripe Graafian follicle (see Table I, subject 1), and that much more estradiol and progesterone are being secreted by the ovary containing either an active follicle or the corpus luteum (Table I, subjects 1 and 2) than from the inactive ovaries. By contrast, concentrations of most steroids in the venous blood of the nonactive ovary generally resembled those in peripheral blood. In the postmenopausal subject (Table I, subject 3) the blood concentrations were mostly far lower than those determined during active reproductive life, and few if any of the steroids present in ovarian blood appear to be secreted by the ovary itself. It may be concluded that all three major types of steroid hormones are being produced by the adult ovary, and more than one group of steroids at any given time. Also, the bulk of steroids is derived from the ovary containing the mature Graafian follicle or a recent corpus luteum.

Essentially similar findings have been reported by Lloyd *et al.* (1971) in a series of eighteen women with normal cycles. These workers stress the wide range in steroid concentrations observed in the women studied by them, and suggest that ovarian hormone production is related to the cyclic growth and regression of the Graafian follicle and corpus luteum.

Observations on the nature and concentration of steroids in the ovarian veins have been reported for sheep (Edgar and Ronaldson, 1958; Short *et al.*, 1963; Lindner *et al.*, 1964; McCracken and Baird, 1969), sows (Gomes *et al.*, 1965), mares (Short, 1964), rabbits (Hilliard *et al.*, 1963; Simmer *et al.*, 1963; Hilliard and Eaton, 1971), and other animals (see Volume III, Chapter 5).

Peripheral Conversion of Endogenous Androgens to Estrogens

It is now recognized that sex steroids are not exclusively produced by the gonads and adrenal glands, but also arise by peripheral conversion of their biogenetic precursors in plasma (see Siiteri and MacDonald, 1973). These precursors (or "prehormones") possess little intrinsic hormonal potency, but when metabolized peripherally they are transformed into more active compounds and can contribute significantly to overall biological activity (Baird *et al.*, 1968b, 1969; see also Volume III, Chapter 5).

In women, estrogens are produced by aromatization of circulating C_{19} precursors by the peripheral tissues, and androstenedione serves as precursor for most of the physiologically active estrone formed in this way. Androstenedione is produced in substantial amounts (approx. 3 mg/day), most of it being derived from the ovaries (Table I) and the rest from the adrenals (see Tagatz and Gurpide, 1973).

Plasma androstenedione is efficiently (ca. 1.2%) converted into estrone, and may contribute from 10 to 50% of the total estrogen production during the cycle in premenopausal women (Siiteri and MacDonald, 1973).

D. Tissue Culture of Ovarian Cells

The studies by Channing and her associates have shown convincingly that tissue culture techniques are suitable for investigating the steroidogenic capacities of the cells making up the various ovarian compartments (Channing, 1969a,b,c,d,e, 1970a,b, 1973; Channing and Grieves, 1969).

Granulosa, thecal, and stromal cells from mares and women have been grown successfully in such preparations for periods up to several weeks and their hormone secretion patterns determined qualitatively and quantitatively. It was shown that cultures of thecal cells obtained by microdissection of follicles from women not treated with gonadotropin can secrete estrogen in the absence of granulosa cells. Thus, in tissue culture conditions, and perhaps also *in vivo* and *in situ,* interaction between these two types of cell in estrogen biosynthesis does not appear to be essential. The thecal elements secreted more 17α-hydroxyprogesterone, androstenedione, and estrogens relative to progesterone than did granulosa cells, and may therefore possess more 17-hydroxylase, 17,20-desmolase, and aromatizing activities than the latter (Channing, 1969e). By contrast, granulosa cells produced chiefly progesterone and only relatively minute amounts of estrogen and androstenedione.

Granulosa cells obtained from women after gonadotropic stimulation, or late in the proliferative phase of the cycle, tended to secrete more progesterone than cells from unstimulated ovaries, or cells procured at other stages of the cycle. The steroid excretion pattern of cultured human granulosa cells resembled that inferred from the concentrations of steroids in the ovarian venous blood of women during the luteal phase (Mikhail, 1967; Lloyd *et al.,* 1971; Section I,C).

It is important to realize that granulosa cells from Graafian follicles of women and estrous mares, when cultured, soon undergo spontaneous hyperplasia and other changes characteristic of "luteinization," and will then resemble luteal cells histologically and steroidogenically (Channing

1969d,e). For instance, they are able to synthesize progesterone from small molecular substrates, and their rate of progestin secretion approximates that of intact luteal tissue *in situ*. Luteinization appears to occur when equine granulosa cells are taken from large, vascular follicles at estrus and grown in tissue culture. Those harvested during the luteal phase from small follicles do not become luteinized in culture and produce far less progesterone. Thus, the *in vivo* environment of these cells within the follicle appears to influence their subsequent behavior in culture.

E. Conclusions

It is generally agreed that the ovarian progestins are formed by the corpus luteum and by luteinized granulosa cells both *in vivo* and *in vitro*. Only the site of origin of estrogen remains doubtful and somewhat controversial.

The concept of a functional interrelationship between the theca interna and granulosa cells of the follicle in the biosynthesis of estrogen stems from the work of Falck (1959) in the rat. He observed that small autografts of pure theca interna cells were unable to secrete estrogen unless some granulosa cells were also present. Falck, therefore, concluded that in the rat both types of cells are essential, and that estrogen formation depends on an "interplay" between them. He relied, however, wholly on histological criteria, and did not define the nature of this interaction. A different explanation, the so-called "two-cell type" theory, was advanced by Short (1962, 1964) to account for the pattern of steroid secretion observed in the ovary of the mare. In his view, the theca interna cells possess all the enzyme systems needed for converting progesterone to androgens and aromatizing them, but lack 20-reductase activity. The other type, the luteinized granulosa cells, have an active 20-reductase system, but are relatively deficient in 17-hydroxylase and 17-desmolase activity, and therefore are relatively inefficient in converting pregnenolone or progesterone into other steroids. Aromatization of androgens may be performed by the theca cells or, after transport across the basement membrane, by the granulosa cells.

According to Short's concept, the theca interna and granulosa cells differ in their steroidogenic potentialities, and the dramatic change from an androgen/estrogen type of secretion before ovulation to one predominantly of progestin after ovulation can be attributed to the acquisition of a blood supply by the previously avascular membrana granulosa, possibly because this permits essential nutrients to reach the follicular fluid (see Short, 1967).

More recent research, however, suggests that the "two-cell type" concept does not provide an adequate explanation of ovarian steroidogenesis in all the species studied so far, and requires modification. The extensive tissue culture studies of Channing have demonstrated that isolated theca cells, without granulosa cells, from normal human and equine follicles, secrete estrogen (Channing, 1969d,e; Channing and Grieves, 1969). Granulosa cells uncontaminated by thecal tissue can also secrete estrogen, but the amount formed by them is much less than that produced by theca cells (Channing, 1969e, 1970b). The differences between the two types of cell, at least in the equine and human ovary, may thus be more quantitative than qualitative (but see Section I,E, p. 293).

This conclusion is indirectly supported by several other observations. Electron microscopical studies of monkey and sheep ovaries actively secreting estrogen have revealed ultrastructural evidence of steroidogenesis in the thecal cells of the growing follicle, but little or none in the granulosa cells of the same follicle (Bjersing *et al.,* 1972; C. P. Channing, personal communication). It was previously reported by Short (1964) that in the mare estrogen secretion by the mature Graafian follicle into the ovarian vein did not change significantly after aspiration of the follicular fluid, which left the stratum granulosum collapsed and dry. Since it is difficult to see how such collapsed granulosa cells can participate in estrogen synthesis (as required by Falck's theory), it appears unlikely that they contribute materially to follicular estrogen biosynthesis. More recent experiments by YoungLai and Short (1970) on mares also support the view that at estrus the bulk of the secreted estrogen is formed outside the follicle itself, most probably in the theca interna cells. These authors showed that only a very small proportion of androstenedione injected into the follicular fluid *in vivo* is converted locally into estrogen by the granulosa cells. In Short's view (1967), the latter are relatively inactive until after rupture of the follicle when they become vascularized and begin to secrete progesterone freely, as well as smaller amounts of other steroids, depending on the species.

In spite of the new insight gained through tissue culture experiments, the problem of the cellular site of estrogen biosynthesis in the follicle remains complex. For example, the granulosa cells of the mare during short-term incubation readily transform androstenedione and testosterone into estrone and estradiol-17β (Ryan and Short, 1965; Channing, 1969b,d), while granulosa cells maintained *in vitro* for longer periods produce more progestin than estrogen. It is, however, well established that during long-term culture, granulosa cells undergo spontaneous luteinization (see Section I,D), and this may account for the observed difference in steroid production. Moreover, there is always doubt about the extent to which *in vitro* studies reflect the physiological (*in vivo*) condition and function of the same

group of cells. For instance, in the intact ovary, after ovulation, both follicular and freshly luteinized elements coexist in close biosynthetic and metabolic association, so that both estrogenic and progestational steroids and
their metabolites occur in adjacent ovarian compartments. Again, the lack of
a blood supply in tissue culture conditions precludes the immediate removal
of newly formed steroids from their site of origin, as in the ovary *in situ.*

There is no reasonable doubt that the theca interna cells, whether consti-
tuted in the isolated follicle or in single-cell tissue culture, secrete chiefly
estrogen as well as 17α-hydroxyprogesterone and androstenedione, and
granulosa cells mainly, if not exclusively, progestins. It seems equally well
established that either type of cell can produce its particular steroids independently of the other, even though there is a marked quantitative difference between their principal and what might be termed secondary steroid
output. Further progress in the elucidation of the old and difficult issue of
the ultimate cytological origin of ovarian estrogen will depend on additional,
preferably combined, *in vivo* and *in vitro* studies (see Short, 1964; YoungLai
and Short, 1970; Bjersing *et al.,* 1972).

The steroidogenic properties of the corpus luteum appear to be related to
its histogenesis and cell population, and consequently show marked species
variability. The human corpus luteum is composed of both theca lutein and
granulosa lutein cells and secretes a mixture of progestins and estrogens,
presumably derived from both types of luteal cell. In the rat, these cells cannot be clearly distinguished (see Crisp *et al.,* 1970), and in the mare and cow
the luteal tissue is thought to be formed mainly from the granulosa elements
(see Brambell, 1956; Harrison, 1962); in these three species the luteal tissue
produces virtually only progestins (see Short, 1967; Crisp *et al.,* 1970). In
sheep, studied by Bjersing *et al.* (1972), the estrogen concentration in the
ovarian veins appears to be largely unaffected by the presence or absence of
a corpus luteum in the ovary, suggesting that in this species the corpus
luteum makes no substantial contribution to the secretion of ovarian
estrogen. Again, in rhesus monkeys the concentration of blood estrogens
rarely shows the conspicuous postovulatory peak which characterizes the
human menstrual cycle (Hotchkiss *et al.,* 1971). This may be related to the
early disappearance of the theca lutein cells as separate elements from the
corpus luteum of the monkey (Corner, 1945; Volume I, Chapter 4). It
seems, therefore, that in respect of steroidogenesis by the corpus luteum
there exist true, qualitative species differences, although these may be
largely explained by the variable presence or persistence of theca lutein cells
during the functional life of the corpus luteum (see also Sections I,B,2 and
I,E).

II. THE OVARY BEFORE SEXUAL MATURITY

Anatomically, the immature ovary differs from the sexually mature one mainly, apart from size, by the absence of large Graafian follicles and corpora lutea (see Potter, 1963; Pryse-Davies and Dewhurst, 1971). Functionally, it neither releases eggs nor secretes progesterone, but is sensitive to exogenous hormonal stimulation and is not wholly devoid of endocrine activities of its own.

The ovaries of the fetal horse, seal, and African elephant during advanced gestation show marked hypertrophy of the interstitial cells, and those of the foal may approach or exceed the maternal ovaries in bulk (Cole *et al.*, 1933; Amoroso *et al.*, 1951; Volume I, Chapters 1 and 4 and Volume II, Chapter 8). In the giraffe, advanced follicular development and the presence of corpus luteum-like structures have been observed in the ovaries during late gestation and after birth (Kellas *et al.*, 1958; Amoroso and Finn, 1962; Kayanja and Blankenship, 1973), and it has been inferred that in this species the fetal ovaries may exert both hormonal and gametogenic functions (see Fig. 4). The hypertrophy of the interstitial cells in the gonads of the fetal horse (Volume II, Chapter 8) has been ascribed to the (presumably indirect) effects of maternal estrogens. This is supported by the disintegration of these elements, which occurs shortly before birth, and coincides with a reduced output of estrogen from the mother (Amoroso and Rowlands, 1951). It is generally assumed, though not yet proved, that transplacental passage of maternal hormones is also involved in other species in which the fetal or neonatal gonads display similar hypertrophy or precocity (for the condition in man see Section II,A).

It is, however, well established that the ovary of infantile mice and rats will respond to suitable gonadotropic stimulation with follicular growth, ovulation, and corpus luteum formation (Rowlands, 1944; Sasamoto, 1969). Following this stage, the ovaries pass through a phase of relative insensitivity to hormones, the length of which varies with the species (Price and Ortiz, 1944; Ortiz, 1947; Picon, 1956; Young, 1961), but which ends some time before puberty itself.

A. Hormonal Activities

With regard to endogenous estrogen, the amounts produced by the neonatal and prepubertal ovary are clearly subthreshold, that is, insufficient to induce opening of the vaginal membrane in rodents, cornification in the vagina, and an active endometrium.

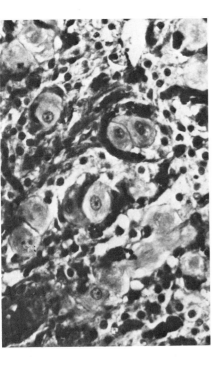

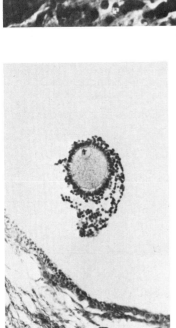

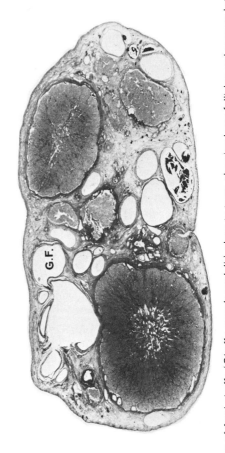

G.F.

Fig. 4. Bottom, ovary of fetal giraffe (*Giraffa camelopardalis*), close to term, showing large follicles and corpora lutea (×6). Left, mature oocyte in Graafian follicle (×184); right, luteal cells (×184). (Section provided by Dr. F. I. B. Kayanja.).

The fetal ovaries appear to exert little influence on development and differentiation of the gonoducts, and it has therefore been concluded that they have no significant endocrine function (Jost, 1953, 1965; Volume I, Chapter 3). This contrasts markedly with the situation in the male in which sexual differentiation definitely depends on the presence of functional testes (Price and Williams-Ashman, 1961).

The endocrine activities of the fetal ovaries may, however, not be wholly negligible. The human ovary acquires the capacity to convert labeled progesterone into 20α-hydroxypregnenone after about the fifth month of intrauterine life (Bloch, 1964). In the rat, so-called castration changes have been reported in the basophils of the anterior pituitary lobe in 1- to 5-day-old females (Presl *et al.,* 1965); ovariectomy in infantile animals is also followed by enhanced secretion of gonadotropin (see Greep and Chester Jones, 1950; Chester Jones and Ball, 1962). There was, however, little direct proof of estrogen production by the fetal or neonatal ovary (see Ramirez and McCann, 1963), although steroid 3β-ol-dehydrogenase, which is known to be involved in steroid biosynthesis, had been demonstrated in the theca interna of 9- to 10-day-old rats (Taylor, 1961; Presl *et al.,* 1965). It is now known that the appearance of this enzyme is accompanied by a sudden rise in total plasma estrogen (from a mean value of 0.13 μg/100 ml at the age of 5 days to 0.38 μg/100 ml at 10 days), and by intensive growth of the uterus (Presl *et al.,* 1969). The occurrence of active estrogen biosynthesis so soon after birth in rats can account for the establishment of a negative feedback relationship between the ovary and the diencephalohypophysial system in this species.

The ovary of near-term fetuses and young children not infrequently contains well developed and occasionally luteinized follicles (see Potter, 1963; Pryse-Davies and Dewhurst, 1971; Baker, 1972). It is thought that few of them rupture and that most degenerate soon after being formed. It has, however, been suggested that they may become transformed into interstitial tissue, and so augment the potential capacity of the prepubertal ovary to form estrogen (Sturgis, 1961; Baker, 1972). The development and luteinization of these follicles may be related to the action of maternal gonadotropins, and low concentrations of HCG have been demonstrated in both the blood and urine of newborn children (Lauritzen and Lehmann, 1965).

Small quantities of urinary estrogens can usually be found in young girls below the age of 10 (Dewhurst, 1963; Pennington and Dewhurst, 1969). The amount increases rapidly after the onset of puberty and reaches adult levels about the time of the menarche. The concentration of testosterone in the blood plasma of prepubertal girls is about one-half that of adult women (August *et al.,* 1969). The ultimate source of all gonadal types of hormone secreted before puberty may well be the adrenal cortex rather than the ovary (see Wilkins, 1965; Liddle, 1974).

The event which both characterizes and most immediately determines puberty is the maturation of follicles to an "ovulable" size. As ripe Graafian follicles develop and subsequently rupture, the concentrations of both estrogenic and, later, progestational hormones in the blood and urine begin to approach adult levels, and eventually become cyclic. The female experiences her first estrous or menstrual cycle, and at the same time undergoes the series of external changes associated with sexual maturation (see Tanner, 1962, 1967).

B. The Influence of Neonatal Androgens on Ovarian Function

Many studies have confirmed early observations that mature female rats do not ovulate and display persistent vaginal cornification if they receive a testicular transplant (Pfeiffer, 1936) or are injected with androgenic hormone (Bradbury, 1941) soon after birth. This subject, and the larger one of the role of the CNS in determining sexual behavior, has been repeatedly reviewed, and reference is made to several comprehensive and expert accounts dealing with it (Barraclough, 1966, 1968; Everett, 1964; Harris, 1964; Donovan and van der Werff ten Bosch, 1965; Bogdanove, 1972; Dörner, 1972). Only the effects on gonadal function, but not those on sexual behavior, will be summarized below.

The effects of postnatal administration of testosterone (usually given as the propionate, "TP") on ovarian function depend both on the dose and the time of administration (see Gorski, 1968). A single injection of 1 mg TP given on day 5 after birth will cause complete sterility, at maturity, in both rats and mice (Barraclough, 1961, 1966; Barraclough and Leathem, 1954). A dose of 10 μg induces sterility in 70% of rats (Barraclough, 1961, 1968), while a mere 1.2 to 1.6 μg, implanted intracranially, will do so in 100% of animals (Nadler, 1968). Similar results have been reported in hamsters given an injection of 300 μg TP on day 2 (Swanson, 1966).

The ovaries of such animals contain large vesicular follicles and hypertrophied interstitial tissue, but no corpora lutea ("anovulatory sterility"). Rats treated with doses of 5 to 10 μg will, however, exhibit regular vaginal cycles and ovulation for several weeks after reaching puberty, but then develop anovulatory cycles and polyfollicular ovaries. During the immediate postpubertal phase such animals may conceive, deliver, and rear litters (Swanson and van der Werff ten Bosch, 1964a,b). Similarly, androgen administration on day 10, postnatally, causes less ovarian damage than between days 1 and

5, and at 20 days none (Barraclough, 1966). Thus the female rat is most steroid-sensitive during the first 5 days of life, and this appears to be a critical phase in the development of the CNS, when exposure to androgen for probably no more than 12 to 24 hours (following a single injection) is sufficient to produce permanent, if not immediate, sterility (see Arai and Gorski, 1968). Similar effects, as well as masculinization of the external genitalia, can be produced by administering TP directly to the fetus *in utero* a few days before delivery (Swanson and van der Werff ten Bosch, 1965). More recently, Dörner and Fatschel (1970) have reported subfertility rather than complete sterility in rats given an injection of 20 μg TP on day 3 of life; the animals conceived but then absorbed their fetuses. This effect could be prevented by the antiandrogen, cyproterone acetate, administered for 3 days immediately after the injection of TP.

It appears that the presence of androgen during the neonatal period in female rodents exerts its sterilizing action not by a direct effect on the ovary or pituitary gland, but by inhibiting the hypothalamic release mechanism, which induces the normal rhythmic pattern of secretion and discharge of gonadotropic hormones from the anterior pituitary. When grafted into intact cyclic females, androgen-sterilized ovaries will ovulate and form corpora lutea (Bradbury, 1941; see Donovan, 1970). Also, transplantation of the pituitary gland from an androgen-sterilized rat to a hypophysectomized female restored normal vaginal cycles and ovulation, but not *vice versa* (Segal and Johnson, 1959).

III. THE OVARY AT THE END OF REPRODUCTIVE LIFE

The length of reproductive life and its relation to total life-span vary greatly in different species of mammals, from 8 to 9 months in hamsters (Peczenik, 1942) to over 30 years in women and probably no less in the elephant, horse, and bear (see Perry, 1953; Eckstein, 1955; Talbert, 1968). The most characteristic feature of the ovary during the fertile phase is the continuous depletion of the number of oocytes, a process which is not significantly checked by the intervention of pregnancy (Block, 1952; Baker, 1963, 1972; Jones, 1970). There are marked differences in the rate of atresia of oocytes between species and even between different strains of the same species (e.g., mice), and hence in reproductive potential (Fekete, 1953; Thung *et al.*, 1956; Jones and Krohn, 1961). In women, the decline and ultimate cessation of fertility can be directly attributed to the progressive exhaustion of the oocyte stock; this is virtually complete at the menopause, which occurs well before the end of the average life span. In rodents, however, the reduction in oocyte numbers is more gradual, and the last stages may not be reached before the end of life itself. The human menopause is characterized by the cessation of cyclic changes in the ovary and accessary reproductive organs. Follicular maturation and ovulation stop, and the postmenopausal ovary is usually composed exclusively of stromal tissue and blood vessels.

It is generally assumed that after the menopause the human ovaries make no significant contribution to the formation of endogenous estrogen or its precursors, and that the residual output of hormones from the ovary must be attributed to the stromal compartment. The human ovarian stroma has only been recognized comparatively recently as a source of steroids, although the anatomical and probable physiological role of the interstitial tissue in lower mammals has long been known (see Mossman *et al.,* 1964). The work of Rice and Savard (1966) has revealed the ovarian stroma of postmenopausal women as a distinct endocrine tissue which elaborates predominantly androgen. On incubation with labeled substrates, it displays little aromatizing activity and yields mainly androstenedione, DHA, and testosterone, only trace amounts of estrogen and no progesterone (Rice and Savard, 1966; Plotz, 1969; Schenker *et al.,* 1971).

There is some evidence that the ovarian stroma from women and other mammals in the reproductive phase of life can synthesize estrogen independently of the follicular apparatus (Wotiz *et al.,* 1956; Falck, 1959; see also Section I,E). It cannot be assumed, however, that the stromal compartment of postmenopausal ovaries is biochemically equivalent to the stromal tissue that surrounds the normal follicle during the reproductive phase. It must also be recognized that the results of incubation studies do not necessarily bear a direct relationship to the activity of the postmenopausal ovary *in vivo.* In particular, they do not indicate whether the secretion of androgen is high enough to cause virilization, although the frequent tendency towards hirsutism in postmenopausal women suggests that it may be. This issue can only be settled by direct measurement of the steroid content of the venous blood from aged ovaries.

Clinical observations are, however, generally consistent with findings *in vitro.* The plasma concentrations of estrone and estradiol-17β are substantially lower after the menopause than before it (Tulchinsky and Korenman, 1970; Dufau *et al.,* 1970), while the total urinary excretion of estrogen of postmenopausal women rarely exceeds 10 μg/24 hours (Procope and Adlercreutz, 1969). This is approximately one-third of the level in normally cyclic women, and is comparable to that observed after ovariectomy. Since it was also established that ovariectomy in postmenopausal subjects with breast cancer does not affect estrogen excretion (Barlow *et al.,* 1969), it has been concluded that the operation, undertaken with the object of lowering estrogen concentration, cannot be justified unless accompanied by adrenalectomy.

Such estrogen as persists in the blood and urine after the menopause is thought to be derived from the adrenal gland or by extragonadal conversion from the plasma pool (see Section I,C). In either case, it is extremely low, and clearly insufficient to prevent the atrophy of the reproductive tract and vulva which is a frequent characteristic of postmenopausal women.

IV. THYRO-OVARIAN RELATIONSHIPS

The influence of thyroid secretions on gonadal activity and reproductive function in general is well known, but its clinical and experimental investigation has usually yielded variable or controversial findings, and the exact relationships between the thyroid and ovary remain far from clear (Maqsood, 1952; Leathem, 1959, 1972; Eckstein, 1962).

Myxedema in women is frequently associated with menstrual irregularities and relative infertility, which can be improved by thyroid medication; few women with cretinism or untreated myxedema become pregnant (Parkin and Greene, 1943; Leathem, 1959, 1972). Similarly, experimentally induced hypothyroidism in laboratory animals tends to disturb or inhibit both estrous and menstrual cycles (Engle, 1944; Krohn and White, 1950; Maqsood, 1952), probably by a direct effect upon the ovary. Thus, in some species such as the mouse, hypothyroidism interferes with normal development and onset of function of the reproductive organs, and may also cause ovarian degeneration in the mature female (Morris *et al.,* 1946). In poultry, hypothyroidism lowers egg production and egg weight (see Maqsood, 1952). There are, however, marked species differences in response, and similar effects have been variously produced by both deficiency and excess of thyroid hormone (Ershoff, 1945; Maqsood, 1952; Leathem, 1959, 1972). For instance, in man an association between precocious puberty and juvenile hypothyroidism has been reported (Jenkins, 1965); in this case, the ultimate cause may have been a lesion in the hypothalamus, the hypothyroidism being due to a deficiency of TSH. It is also worth noting that manipulation of the protein or amino acid content of the diet may produce results comparable to those of hypothyroidism. Estrous cycles will become irregular or cease altogether when adult rats are fed on a diet containing only 3.5–5% protein; a 6% casein ration will sustain normal cyclicity, whereas 6% gelatin does not (Leathem, 1959).

Thyroidectomy or feeding with antithyroid drugs, while rarely abolishing reproductive capacity altogether, usually results in lowered fecundity and other adverse effects, such as abortion and fetal resorption (Chu, 1944; Krohn and White, 1950; see Leathem, 1959, 1972). Conversely, treatment of intact animals with thyroxine may lead to litters of greater than normal size (Peterson et al., 1952; Schultze and Noonan, 1970).

Cyclic ovarian activity, in turn, is believed to affect the thyroid gland. An increase in thyroid activity at the time of spontaneous ovulation occurs in some rodents, guinea pigs, and sheep, but not in the rabbit, in which ovulation is induced by mating (Brown-Grant, 1968).

Although it seems clear that thyroid activity is closely connected with normal gonadal function and reproduction in general, its influence probably involves a complex series of interactions. Thus it is not known whether any of the above-mentioned effects are mediated through the anterior pituitary, by an altered output of, or ovarian sensitivity to, gonadotropins, or by a more generalized influence on tissue metabolism. Leathem's studies (1958a,b, 1959) on the connection between hypothyroidism and the formation of ovarian cysts in experimental animals have provided more specific information about these poorly understood interrelationships. His finding that following thiouracil treatment the response of the rat's ovary to gonadotropin is altered so that luteinizing gonadotropin becomes predominantly follicle stimulating and induces cyst formation is relevant in this context.

Thyroid function during human pregnancy has been discussed by Stuart Mason (1970) and Leathem (1972).

V. ADRENO-OVARIAN RELATIONSHIPS

The existence of close functional relationships between the adrenal gland and the ovary is firmly established and based on many clinical and experimental observations (see Parkes, 1945; Courrier et al., 1953; Chester Jones, 1957; Young, 1961; Parkes and Deanesly, 1966b).

The ovaries and the adrenal cortex have a common embryological origin and elaborate their respective steroidal secretions along similar biosynthetic pathways. Each gland can also substitute to some extent for the other. Thus low levels of pregnanediol persist in the urine after ovariectomy, and its excretion in normal human subjects is raised by corticotropin (Klopper et al., 1957; Nabarro and Moxham, 1957). Since, in addition, progesterone and its metabolites have been identified in adrenal venous blood of man and other mammals the adrenocortical origin of these compounds must be accepted (Balfour et al., 1959; Short, 1960). Similarly, ACTH after ovariectomy causes a rapid rise in urinary estrogen, especially estriol, but not after ovariectomy combined with adrenalectomy (Brown et al., 1959; Barlow et *al.*, 1969). The production of estrogen by the normal adrenal gland, however, appears too low to be of biological significance, and the atrophy of the female genital tract that follows bilateral ovariectomy supports this view.

In certain strains of mice the adrenals are capable of maintaining the accessory reproductive organs after neonatal excision of the ovaries (Fekete *et al.*, 1941; Hill, 1962). Conversely, in adrenalectomized ground squirrels adrenal cortex-like cells appear in the ovarian stroma, and the ovary as a whole assumes adrenocortical functions (Groat, 1944). In rats, pseudopregnancy and injected progesterone can prolong survival after bilateral adrenalectomy (Hartman and Brownell, 1949), as will ovaries grafted into the ear of adrenalectomized mice (Hill, 1962).

The influence of the adrenals on reproductive processes is illustrated by a variety of observations (see Parkes and Deanesly, 1966b). After adrenalectomy or during chronic cortical deficiency estrous and menstrual cycles are disturbed or abolished, but can be restored by corticosteroids and, in rats, even by sodium chloride alone (Kutz *et al.*, 1934). Again, in laboratory rodents as well as in seasonally breeding species, such as the mole and ground squirrel, the size of the adrenals varies during the cycle, generally increasing at estrus or ovulation (Andersen and Kennedy, 1932; Bourne and Zuckerman, 1941a). In rats this is due to enlargement of the cortex, itself brought about by hypertrophy of the cells in the zona fasciculata (Bourne and Zuckerman, 1941b). It has also been shown that artificial "threshold cycles" may continue in spayed rats and monkeys injected with constant low doses of estrogen (Zuckerman, 1941; Bourne and Zuckerman, 1941b). This suggests that the occurrence of such cycles depends on the presence of the adrenals, even though the exact nature of their influence has still to be explained.

As already mentioned (Section I,B,3), recent work suggests that androgens of adrenal origin are involved in regulating sexual receptivity in female primates. Responsiveness is greatly reduced or abolished by adrenalectomy in rhesus monkeys, but can be restored by small doses of an adrenal androgen such as androstenedione (Herbert, 1970; Everitt *et al.*, 1972).

In view of the marked overlap in adrenocortical and ovarian physiology referred to above, it is not surprising that abnormalities in human adrenal function are frequently associated with deficient ovarian activities and vice versa (see Goldzieher, 1968; Kistner, 1972).

The chief differences in the biogenesis of the adrenal and ovarian steroids can be summarized as follows. The adrenal cortex possesses potent 11- and 21-hydroxylating enzymes which result in the formation of glucocorticoids such as cortisol; conversion of dehydroepiandrosterone yields androgen, and hydroxylation at C-18 aldosterone. The ovary, by contrast, possesses no or only minimal 20- and 11-hydroxylating activity. Instead, it splits 17-hydroxypregnenolone and 17-hydroxyprogesterone to C_{19} steroids such as progesterone and its 20-hydroxy derivatives, and then converts androstenedione to estrone by hydroxylation at C-19 and subsequent aromatization of ring A.

A pathological crossing of these normally separate biosynthetic pathways results in the "adrenogenital syndrome." In its most common form, characterized by congenital adrenal hyperplasia, there is a lack of 21-hydroxylase. This results in a deficiency of cortisol and an excessive production of androgens, with consequent virilization of the external genitalia of the genetic, that is, sex chromatin-positive, female. In the absence of treatment, complete virilization, anovulation, and amenorrhea will follow. With appropriate treatment, however, the condition reverts, and normal cycles and even fertility may be restored (Wilkins, 1965; Hall *et al.,* 1969, Liddle, 1974).

ACKNOWLEDGMENTS

The author wishes to express his gratitude to Miss M. Howdle and Mr. A. Greenman for assistance with the illustrations and to Mrs. G. Macbeth for comprehensive secretarial help.

REFERENCES

Abraham, G. E., and Tait, A. D. (1971). Ovarian metabolism of steroid hormones. *Res. Reprod.* **3,** No. 5.

Adams, E. C., and Hertig, A. T. (1969). Studies on the human corpus luteum. I. Observations on the ultrastructure of development and regression of the luteal cells during the menstrual cycle. *J. Cell Biol.* **41,** 696–715.

Amoroso, E. C., and Finn, C. A. (1962). Ovarian activity during gestation, ovum transport and implantation. *In* "The Ovary" (S. Zuckerman, ed.), 1st ed., Vol. 1, Chapter 9, pp. 451–537. Academic Press, New York.

Amoroso, E. C., and Rowlands, I. W. (1951). Hormonal effects in the pregnant mare and foetal foal. *J. Endocrinol.* **7,** l–liii.

Amoroso, E. C., Harrison, R. J., Matthews, L. H., and Rowlands, I. W. (1951). Reproductive organs of near-term and new born seals. *Nature (London)* **168,** 771–772.

Andersen, D. H., and Kennedy, H. S. (1932). Studies on the physiology of reproduction. IV. Changes in the adrenal gland of the female rat associated with the oestrous cycle. *J. Physiol. (London)* **76,** 247–260.

Anderson, L. L., Bland, K. P., and Melampy, R. M. (1969). Comparative aspects of uterine-luteal relationships. *Recent Prog. Horm. Res.* **25,** 57–99.

Arai, Y., and Gorski, R. A. (1968). Critical exposure time for androgenization of the developing hypothalamus in the female rat. *Endocrinology* **82,** 1010–1014.

August, G. B., Tkachuk, M., and Grumbach, M. M. (1969). Plasma testosterone-binding affinity and testosterone in umbilical cord plasma, late pregnancy, prepubertal children and adults. *J. Clin. Endocrinol. Metab.* **29,** 891–899.

Baird, D. T., Goding, J. R., Ichikawa, Y., and McCracken, J. A. (1968a). The secretion of steroids from the autotransplanted ovary in the ewe spontaneously and in response to systemic gonadotrophin. *J. Endocrinol.* **42,** 283–299.

Baird, D. T., Horton, R., Longcope, C., and Tait, J. F. (1968b). Steroid prehormones. *Perspect. Biol. Med.* **11,** 348–421.

Baird, D. T., Horton, R., Longcope, C., and Tait, J. F. (1969). Steroid dynamics under steady-state conditions. *Recent Prog. Horm. Res.* **25,** 611–664.

Baker, T. G. (1963). A quantitative and cytological study of germ cells in human ovaries. *Proc. R. Soc. London, Ser. B* **158,** 417–433.

Baker, T. G. (1972). Oogenesis and ovarian development. *In* "Reproductive Biology" (H. Balin and S. Glasser, eds.), pp. 398–437. Excerpta Med. Found., Amsterdam.

Balfour, W. E., Comline, R. S., and Short, R. V. (1959). Changes in the secretion of 20α-hydroxy-pregn-4-en-3-one by the adrenal gland of young calves. *Nature (London)* **183,** 467–468.

Barlow, J. J., Emerson, K., Jr., and Saxena, B. N. (1969). Estradiol production after ovariectomy for carcinoma of the breast. *N. Engl. J. Med.* **280,** 633–637.

Barraclough, C. A. (1961). Production of anovulatory, sterile rats by single injections of testosterone propionate. *Endocrinology* **68,** 62–67.

Barraclough, C. A. (1966). Modifications in the CNS regulation of reproduction after exposure of prepubertal rats to steroid hormones. *Recent Prog. Horm. Res.* **22,** 503–539.

Barraclough, C. A. (1968). Alterations in reproductive function following prenatal and early postnatal exposure to hormones. *Adv. Reprod. Physiol.* **3,** 81–112.

Barraclough, C. A., and Leathem, J. H. (1954). Infertility induced in mice by a single injection of testosterone propionate. *Proc. Soc. Exp. Biol. Med.* **85,** 673–674.

Belt, W. D., Anderson, L. L., Cavazos, L. F., and Melampy, R. M. (1971). Cytoplasmic granules and relaxin levels in porcine corpora lutea. *Endocrinology* **89,** 1–10.

Besch, P. K., and Buttram, V. C. B. (1972). Steroidogenesis in the human ovary. *In* "Reproductive Biology" (H. Balin and S. Glasser, eds.), pp. 552–571. Excerpta Med. Found., Amsterdam.

Bjersing, L. (1967a). On the ultrastructure of follicles and isolated follicular granulosa cells of porcine ovary. *Z. Zellforsch. Mikrosk. Anat.* **82,** 173–186.

Bjersing, L. (1967b). On the ultrastructure of granulosa lutein cells in porcine corpus luteum. *Z. Zellforsch. Mikrosk. Anat.* **82,** 187–211.

Bjersing, L., and Carstensen, H. (1967). Biosynthesis of steroids by granulosa cells of the porcine ovary *in vitro. J. Reprod. Fertil.* **14,** 101–111.

Bjersing, L., Hay, M. F., Kann, G., Moor, R. M., Naftolin, F., Scaramuzzi, R. J., Short, R. V., and YoungLai, E. V. (1972). Changes in gonadotrophins, ovarian steroids and follicular morphology in sheep at oestrus. *J. Endocrinol.* **52,** 465–479.

Blanchette, E. J. (1966). Ovarian steroid cells. II. The lutein cell. *J. Cell Biol.* **31,** 517–542.

Bloch, E. (1964). Metabolism of 4-^{14}C-progesterone by human fetal testes and ovaries. *Endocrinology* **74,** 833–845.

Block, E. (1952). Quantitative morphological investigations of the follicular system in women. *Acta Anat.* **14,** 108–123.

Bogdanove, E. M. (1972). Hypothalamic-hypophyseal interrelationships: basic aspects. *In* "Reproductive Biology" (H. Balin and S. Glasser, eds.), pp. 5–70. Excerpta Med., Found., Amsterdam.

Bouin, P. (1902). Les deux glandes à sécrétion interne de l'ovaire; la glande interstitielle et le corps jaune. *Rev. Med. Est* **34,** 465–472.

Bourne, G., and Zuckerman, S. (1941a). The influence of the adrenals on cyclical changes in the accessory reproductive organs of female rats. *J. Endocrinol.* **2,** 268–282.

Bourne, G., and Zuckerman, S. (1941b). Changes in the adrenals in relation to the normal and artificial threshold oestrous cycle in the rat. *J. Endocrinol.* **2,** 283–310.

Bradbury, J. T. (1941). Permanent after-effects following masculinization of the infantile female rat. *Endocrinology* **28,** 101–106.

Brambell, F. W. R. (1956). Ovarian changes. *In* "Marshall's Physiology of Reproduction" (A. S. Parkes, ed.), 3rd ed., Vol. I, Part 1, pp. 397–542. Longmans Green, London and New York.

Brown, J. B., Falconer, C. W. A., and Strong, J. A. (1959). Urinary oestrogens of adrenal origin in women with breast cancer. *J. Endocrinol.* **19,** 52–63.

Brown-Grant, K. (1968). Ovulation and thyroid function in the rabbit. *J. Endocrinol.* **41,** 85–89.

Channing, C. P. (1969a). Tissue culture of equine ovarian cell types: culture methods and morphology. *J. Endocrinol.* **43,** 381–390.

Channing, C. P. (1969b). Studies on tissue culture of equine ovarian cell types: pathways of steroidogenesis. *J. Endocrinol.* **43,** 403–414.

Channing, C. P. (1969c). Studies on tissue culture of equine ovarian cell types: effect of gonadotrophins and stage of cycle on steroidogenesis. *J. Endocrinol.* **43,** 415–425.

Channing, C. P. (1969d). The use of tissue culture of granulosa cells as a method of studying the mechanism of luteinization. *In* "The Gonads" (K. W. McKerns, ed.), pp. 245–275. North-Holland Publ., Amsterdam.

Channing, C. P. (1969e). Steroidogenesis and morphology of human ovarian cell types in tissue culture. *J. Endocrinol.* **45,** 297–308.

Channing, C. P. (1970a). Effects of stage of the menstrual cycle and gonadotrophins on luteinization of rhesus monkey granulosa cells in culture. *Endocrinology* **87,** 49–60.

Channing, C. P. (1970b). Influences of the *in vivo* and *in vitro* hormonal environment upon luteinization of granulosa cells in tissue culture. *Recent Prog. Horm. Res.* **26,** 589–622.

Channing, C. P. (1973). Regulation of luteinization granulosa cell cultures. *In* "The Regulation of Mammalian Reproduction" (S. J. Segal *et al.,* eds.), pp. 505–518. Thomas, Springfield, Illinois.

Channing, C. P., and Grieves, S. A. (1969). Studies on tissue culture of equine ovarian cell types: steroidogenesis. *J. Endocrinol.* **43,** 391–402.

Chester Jones, I. (1957). "The Adrenal Cortex." Cambridge Univ. Press, London and New York.

Chester Jones, I., and Ball, J. N. (1962). Ovarian-pituitary relationships. *In* "The Ovary" (S. Zuckerman, ed.), 1st ed., Vol. 1, Chapter 7, pp. 361–434. Academic Press, New York.

Christensen, A. K., and Gillim, S. W. (1969). The correlation of fine structure and function in steroid secreting cells, with emphasis on those of the gonads. *In* "The Gonads" (K. W. McKerns, ed.), pp. 415–488. North-Holland Publ., Amsterdam.

Chu, J. P. (1944). Influence of the thyroid on pregnancy and parturition in the rabbit. *J. Endocrinol.* **4**, 109–114.

Cole, H. H., Hart, G. H., Lyons, W. R., and Catchpole, H. R. (1933). The development and hormonal content of foetal horse gonads. *Anat. Rec.* **56**, 275–289.

Corner, G. W. (1938). Sites of formation of oestrogenic substances. *Physiol. Rev.* **18**, 154–172.

Corner, G. W. (1945). Development, organization and breakdown of the corpus luteum in the rhesus monkey. *Contrib. Embryol. Carnegie Inst.* **31**, 117–146.

Courrier, R., Baclesse, M., and Marois, M. (1953). Rapports de la cortico-surrénale et de la sexualité. *J. Physiol. Pathol. Gen.* **45**, 327–374.

Crisp, T. M., and Channing, C. P. (1972). Fine structural events correlated with progestin secretion during luteinization of rhesus monkey granulosa cells in culture. *Biol. Reprod.* **7**, 55–72.

Crisp, T. M., Dessouky, A. D., and Denys, F. R. (1970). The fine structure of the human corpus luteum of early pregnancy and during the progestational phase of the menstrual cycle. *Am. J. Anat.* **127**, 37–70.

Deanesly, R. (1938). The androgenic activity of ovarian grafts in castrated male rats. *Proc. R. Soc. London, Ser. B* **126**, 122–135.

Deanesly, R. (1966). The endocrinology of pregnancy and foetal life. *In* "Marshall's Physiology of Reproduction" (A. S. Parkes, ed.), 3rd ed., Vol. 3, pp. 891–1063. Longmans Green, London and New York.

Dewhurst, C. J. (1963). "Gynaecological Disorders of Infants and Children." Cassell, London.

Donaldson, L., and Hansel, W. (1965). Histological study of bovine corpora lutea. *J. Dairy Sci.* **48**, 905–909.

Donovan, B. T. (1970). "Mammalian Neuroendocrinology." McGraw-Hill, New York.

Donovan, B. T., and van der Werff ten Bosch, J. J. (1965). "Physiology of Puberty." Arnold, London.

Dörner, G. (1972). "Sexualhormonabhängige Gehirndifferenzierung und Sexualität." Fischer, Jena.

Dörner, G., and Fatschel, J. (1970). Wirkungen neonatal verabreichter Androgene und Antiandrogene auf Sexualverhalten und Fertilität von Rattenweibchen. *Endokrinologie* **56**, 29–48.

Dufau, M. L., Dulmanis, A., Catt, K. J., and Hudson, B. (1970). Measurement of plasma estradiol-17β by competitive binding assay employing pregnancy plasma. *J. Clin. Endocrinol. Metab.* **30**, 351–356.

Eckstein, P. (1955). The duration of reproductive life in the female. *Rep. 3rd. Congr. Int. Assoc. Gerontol., London, 1954* pp. 190–199.

Eckstein, P. (1962). Ovarian physiology in the non-pregnant female. *In* "The Ovary" (S. Zuckerman, ed.), 1st ed., Vol. I, pp. 311–359. Academic Press, New York.

Edgar, D. G., and Ronaldson, J. W. (1958). Blood levels of progesterone in the ewe. *J. Endocrinol.* **16**, 378–384.

Enders, A. C. (1962). Observations on the fine structure of lutein cells. *J. Cell Biol.* **12**, 101–113.

Engel, L. L. (1973). The biosynthesis of estrogens. *In* "Handbook of Physiology" (Am. Physiol. Soc., J. Field, ed.), Sect. 7, Vol. II, pp. 467–483. Williams & Wilkins, Baltimore, Maryland.

Engle, E. T. (1944). The effect of hypothyroidism on menstruation in adult rhesus monkeys. *Yale J. Biol. Med.* **17**, 59–66.

Ershoff, B. H. (1945). Effects of thyroid feeding on ovarian development in the rat. *Endocrinology* **37**, 218–220.

Everett, J. W. (1964). Central neural control of reproductive functions of the adeno-hypophysis. *Physiol. Rev.* **44**, 373–431.

Everitt, B. J., and Herbert, J. (1971). The effects of dexamethasone and androgens on sexual receptivity of female rhesus monkeys. *J. Endocrinol.* **51**, 575–588.

Everitt, B. J., Herbert, J., and Hamer, J. D. (1972). Sexual receptivity of bilaterally adrenalectomised female rhesus monkeys. *Physiol. Behav.* **8**, 409–415.

Falck, B. (1959). Site of production of oestrogen in rat ovary as studied in micro-transplants. *Acta Physiol. Scand.* **47**, *Suppl.* 1–101.

Fekete, E. (1953). A morphological study of the ovaries of virgin mice of eight inbred strains showing quantitative differences in their hormone producing components. *Anat. Rec.* **117**, 93–113.

Fekete, E., Woolley, G. W., and Little, C. C. (1941). Histological changes following ovariectomy in mice. I. Dba high tumor strain. *J. Exp. Med.* **74**, 1–8.

Fotherby, K. (1972). Biochimie de l'ovaire normal. *Rev. Med. (Paris)* **24**, 1597–1603.

Fotherby, K., and James, F. (1972). Biochemistry of steroids in normal subjects. *In* "Endocrine Therapy in Malignant Disease" (B. A. Stoll, ed.), pp. 3–23. Saunders, Philadelphia, Pennsylvania.

Goding, J. R., McCracken, J. A., and Baird, D. T. (1967). The study of ovarian function in the ewe by means of a vascular autotransplantation technique. *J. Endocrinol.* **39**, 37–52.

Goldzieher, J. W. (1968). The interplay of adrenocortical and ovarian function. *In* "The Ovary" (H. C. Mack, ed.), pp. 106–129. Thomas, Springfield, Illinois.

Gomes, W. R., Herschler, R. C., and Erb, R. E. (1965). Progesterone levels in ovarian venous effluent of the non-pregnant sow. *J. Anim. Sci.* **24**, 722–725.

Gorski, R. A. (1968). Influence of age on the response to paranatal administration of a low dose of androgen. *Endocrinology* **82**, 1001–1004.

Greep, R. O., and Chester Jones, I. (1950). Steroid control of pituitary function. *Recent Prog. Horm. Res.* **5**, 197–254.

Groat, R. A. (1944). Formation and growth of adrenocortical-like tissue in ovaries of the adrenalectomized ground squirrel. *Anat. Rec.* **89**, 33–41.

Guraya, S. S. (1971). Morphology, histochemistry and biochemistry of human ovarian compartments and steroid hormone synthesis. *Physiol. Rev.* **51**, 785–807.

Hall, R., Anderson, J., and Smart, G. A. (1969). "Fundamentals of Clinical Endocrinology." Pitman, London.

Hammerstein, J., Rice, B. F., and Savard, K. (1964). Steroid hormone formation in the human ovary. I. Identification of steroids formed *in vitro* from acetate-1-^{14}C in the corpus luteum. *J. Clin. Endocrinol. Metab.* **24**, 597–605.

Harris, G. W. (1964). Sex hormones, brain development and brain function. *Endocrinology* **75**, 624–648.

Harrison, R. J. (1948). The development and fate of the corpus luteum in the vertebrate series. *Biol. Rev., Cambridge Philos. Soc.* **23**, 296–331.

Harrison, R. J. (1962). The structure of the ovary. *In* "The Ovary" (S. Zuckerman, ed.), 1st ed., Vol. 1, pp. 143–187. Academic Press, New York.

Hartman, F. A., and Brownell, K. A. (1949). "The Adrenal Gland." Kimpton, London.

Herbert, J. (1970). Hormones and reproductive behaviour in rhesus and talapoin monkeys. *J. Reprod. Fertil., Suppl.* **11**, 119–140.

Hill, R. T. (1937). Ovaries secrete male hormone. I. Restoration of the castrate type of seminal vesicle and prostate glands to normal by grafts of ovaries in mice. *Endocrinology* **21**, 495–502.

Hill, R. T. (1962). Paradoxical effects of ovarian secretions. *In* "The Ovary" (S. Zuckerman, ed.), 1st ed., Vol. 2, Chapter 16, pp. 231–261. Academic Press, New York.

Hill, R. T., and Strong, M. T. (1940). Ovaries secrete male hormone. V. A comparison of some synthetic androgens with naturally occurring ovarian androgen in mice. *Endocrinology* **27**, 79–82.

Hilliard, J., and Eaton, L. W., Jr. (1971). Estradiol-17β, progesterone and 20α-hydroxypregn-4-en-3-one in rabbit ovarian venous plasma. II. From mating through implantation. *Endocrinology* **89**, 522–527.

Hilliard, J., Archibald, D., and Sawyer, C. H. (1963). Gonadotropic activation of preovulatory synthesis and release of progestin in the rabbit. *Endocrinology* **72**, 59–66.

Hotchkiss, J., Atkinson, L. E., and Knobil, E. (1971). Time course of serum estrogen and luteinizing hormone (LH) concentrations during the menstrual cycle of the rhesus monkey. *Endocrinology* **89**, 177–183.

Huang, W. Y., and Pearlman, W. H. (1963). The corpus luteum and steroid hormone formation. II. Studies on the human corpus luteum *in vitro*. *J. Biol. Chem.* **238**, 1308–1315.

Jacoby, F. (1962). Ovarian histochemistry. *In* "The Ovary" (S. Zuckerman, ed.), 1st ed., Vol. 1, pp. 189–245. Academic Press, New York.

Jenkins, M. E. (1965). Precocious menstruation in hypothyroidism. *Am. J. Dis. Child.* **109**, 252–254.

Jones, E. C. (1970). The ageing ovary and its influence on reproductive capacity. *J. Reprod. Fertil., Suppl.* **12**, 17–30.

Jones, E. C., and Krohn, P. L. (1961). The relationships between age, numbers of oocytes and fertility in virgin and multiparous mice. *J. Endocrinol.* **21**, 469–495.

Jost, A. (1953). Problems of fetal endocrinology: the gonadal and hypophyseal hormones. *Recent Prog. Horm. Res.* **8**, 379–418.

Jost, A. (1965). Gonadal hormones in the sex differentiation of the mammalian fetus. *In* "Organogenesis" (R. L. De Haan and H. Ursprung, eds.), pp. 611–628. Holt, New York.

Kayanja, F. I. B., and Blankenship, L. H. (1973). The ovary of the giraffe, *Giraffa camelopardalis*. *J. Reprod. Fertil.* **34**, 305–313.

Kellas, L. M., van Lennep, E. W., and Amoroso, E. C. (1958). Ovaries of some foetal and prepubertal giraffes [*Giraffa camelopardalis* (Linnaeus)]. *Nature (London)* **181**, 487–488.

Kistner, R. W. (1972). Ovulation: clinical aspects. *In* "Reproductive Biology" (H. Balin and S. Glasser, eds.), pp. 477–538. Excerpta Med. Found., Amsterdam.

Klopper, A., Strong, J. A., and Cook, L. R. (1957). The excretion of pregnanediol and adrenocortical activity. *J. Endocrinol.* **15**, 180–189.

Krohn, P. L., and White, H. C. (1950). The effect of hypothyroidism on reproduction in the female albino rat. *J. Endocrinol.* **6**, 375–385.

Kutz, R. L., McKeown, T., and Selye, H. (1934). Effect of salt treatment on certain changes following adrenalectomy. *Proc. Soc. Exp. Biol. Med.* **32**, 331–332.

Lauritzen, C., and Lehmann, W. D. (1965). Choriongonadotropin im Blut und Harn von Neugeborenen. *Arch. Gynaekol.* **200**, 578–589.

Leathem, J. H. (1958a). Hormones and protein nutrition. *Recent Prog. Horm. Res.* **14**, 141–182.

Leathem, J. H. (1958b). Hormonal influences on the gonadotrophin sensitive hypothyroid rat ovary. *Anat. Rec.* **131**, 487–497.

Leathem, J. H. (1959). Extragonadal factors in reproduction. *In* "Recent Progress in the Endocrinology of Reproduction" (C. W. Lloyd, ed.), pp. 179–203. Academic Press, New York.

Leathem, J. H. (1972). Role of the thyroid. *In* "Reproductive Biology" (H. Balin and S. Glasser, eds.), pp. 856–876. Excerpta Med. Found., Amsterdam.

Liddle, G. W. (1974). The adrenals. *In* "Textbook of Endocrinology" (R. H. Williams, ed.), 5th ed., pp. 233–283. Saunders, Philadelphia, Pennsylvania.

Lindner, H. R., Sass, M. B., and Morris, B. (1964). Steroids in the ovarian lymph and blood of conscious ewes. *J. Endocrinol.* **30**, 361–376.

Lloyd, C. W., Lobotsky, J., Baird, D. T., McCracken, J. A., and Weisz, J. (1971). Concentration of unconjugated estrogens, androgens and gestagens in ovarian and peripheral venous plasma of women: the normal menstrual cycle. *J. Clin. Endocrinol. Metab.* **32**, 155–166.

McCracken, J. A., and Baird, D. T. (1969). The study of ovarian function by means of transplantation of the ovary in the ewe. *In* "The Gonads" (K. W. McKerns, ed.), pp. 176–209. North-Holland Publ., Amsterdam.

McKerns, K. W., ed. (1969). "The Gonads." North-Holland Publ., Amsterdam.

Mandl, A. M., and Zuckerman, S. (1956a). The reactivity of the X-irradiated ovary. *J. Endocrinol.* **13**, 243–261.

Mandl, A. M., and Zuckerman, S. (1956b). Changes in the mouse after X-ray sterilization. *J. Endocrinol.* **13**, 262–268.

Maqsood, M. (1952). Thyroid function in relation to reproduction of mammals and birds. *Biol. Rev., Cambridge Philos. Soc.* **27**, 281–319.

Mikhail, G. (1967). Sex steroids in blood. *Clin. Obstet. Gynecol.* **10**, 29–39.

Morris, H. P., Dubnik, C. S., and Dalton, A. J. (1946). Effect of prolonged ingestion of thiourea on mammary glands and the appearance of mammary tumours in adult C_3H mice. *J. Natl. Cancer Inst.* **7**, 159–169.

Mossman, H. W., and Duke, K. L. (1973). Some comparative aspects of the mammalian ovary. *In* "Handbook of Physiology" (Am. Physiol. Soc., J. Field, ed.), Sect. 7, Vol. II, pp. 389–402. Williams & Wilkins, Baltimore, Maryland.

Mossman, H. W., Koering, M. J., and Ferry, D., Jr. (1964). Cyclic changes of interstitial gland tissue of the human ovary. *Am. J. Anat.* **115**, 235–256.

Nabarro, J. D. N., and Moxham, A. (1957). Effect of corticotrophin on urinary excretion of pregnanediol and pregnanetriol. *Lancet* **2**, 624–625.

Nadler, R. D. (1968). Masculinization of female rats by intracranial implantation of androgen in infancy. *J. Comp. Physiol. Psychol.* **66**, 157–167.

Ortiz, E. (1947). Postnatal development of the reproductive system of the golden hamster (*Cricetus auratus*) and its reactivity to hormones. *Physiol. Zool.* **20**, 45–66.

Osborn, R. H., and Yannone, M. E. (1971). Plasma androgens in the normal and androgenic female. *Obstet. & Gynecol. Surv.* **26**, 195–228.

Parkes, A. S. (1926). On the occurrence of the oestrous cycle after X-ray sterilization. I. Irradiation of mice at three weeks old. *Proc. R. Soc. London, Ser. B* **100**, 172–199.

Parkes, A. S. (1929). "The Internal Secretions of the Ovary." Longmans Green, London and New York.

Parkes, A. S. (1945). The adrenal-gonad relationship. *Physiol. Rev.* **25**, 203–254.

Parkes, A. S. (1950). Androgenic activity of the ovary. *Recent Prog. Horm. Res.* **5**, 101–114.

Parkes, A. S., and Deanesly, R. (1966a). The ovarian hormones. *In* "Marshall's Physiology of Reproduction" (A. S. Parkes, ed.), 3rd ed., Vol. 3, pp. 570–828. Longmans Green, London and New York.

Parkes, A. S., and Deanesly, R. (1966b). Relation between the gonads and the adrenal glands. *In* "Marshall's Physiology of Reproduction" (A. S. Parkes, ed.), 3rd ed., Vol. 3, pp. 1064–1111. Longmans Green, London and New York.

Parkin, G., and Greene, J. A. (1943). Pregnancy occurring in cretinism and in juvenile and adult myxoedema. *J. Clin. Endocrinol. Metab.* **3**, 466–468.

Peczenik, O. (1942). Actions of sex hormones on oestrous cycle and reproduction of the golden hamster. *J. Endocrinol.* **3**, 157–167.

Pennington, G. W., and Dewhurst, C. J. (1969). Hormone excretion in premenarcheal girls. *Arch. Dis. Child.* **44**, 629–636.

Perry, J. S. (1953). The reproduction of the African elephant, *Loxodonta africana. Philos. Trans. R. Soc. London, Ser. B* **237**, 93–149.

Perry, J. S. (1971). "The Ovarian Cycle of Mammals." Oliver & Boyd, Edinburgh.

Peterson, R. R., Webster, R. C., Rayner, B., and Young, W. C. (1952). The thyroid and reproductive performance in the adult female guinea pig. *Endocrinology* **51**, 405–518.

Pfeiffer, C. A. (1936). Sexual differences of the hypophyses and their determination by the gonads. *Am. J. Anat.* **58**, 195–225.

Picon, L. (1956). Sur le role de l'age dans la sensibilité de l'ovaire à l'hormone gonadotrophe chez le rat. *Arch. Anat. Microsc. Morphol. Exp.* **45**, 311–341.

Plotz, J. (1969). In vitro synthesis of steroids in ovaries of post-menopausal patients with and without endometrial carcinoma. *Adv. Biosci.* **3**, 123–128.

Potter, E. L. (1963). The ovary in infancy and childhood. *In* "The Ovary" (H. G. Grady and D. E. Smith, eds.), pp. 11–23. Williams & Wilkins, Baltimore, Maryland.

Presl, J., Jirasek, J., Horsky, J., and Henzl, M. (1965). Observations on steroid-3β-ol dehydrogenase activity in the ovary during early postnatal development in the rat. *J. Endocrinol.* **31**, 293–294.

Presl, J., Herzmann, J., and Horsky, J. (1969). Oestrogen concentrations in blood of developing rats. *J. Endocrinol.* **45**, 611–612.

Price, D., and Ortiz, E. (1944). The relation of age to reactivity in the reproductive system of the rat. *Endocrinology* **34**, 215–239.

Price, D., and Williams-Ashman, H. G. (1961). The accessory reproductive glands of mammals. *In* "Sex and Internal Secretions" (W. C. Young, ed.), 3rd ed., Vol. 1, pp. 366–448. Williams and Wilkins, Baltimore, Maryland.

Procope, B.-J., and Adlercreutz, H. (1969). Studies on the influence of age on oestrogens in post-menopausal women with atrophic endometrium and normal liver function. *Acta Endocrinol. (Copenhagen)* **62**, 461–467.

Pryse-Davies, J., and Dewhurst, C. J. (1971). The development of the ovary and uterus in the foetus, newborn and infant: a morphological and enzyme histochemical study. *J. Pathol.* **103**, 5–25.

Ramirez, D. V., and McCann, S. M. (1963). Comparison of the regulation of luteinizing hormone (LH) secretion in immature and adult rats. *Endocrinology,* **72**, 452–464.

Rice, B. F., and Savard, K. (1966). Steroid hormone formation in the human ovary. IV. Ovarian stromal compartment; formation of radioactive steroids from acetate-1-^{14}C and action of gonadotropins. *J. Clin. Endocrinol. Metab.* **26**, 593–609.

Ross, G. T., Cargille, C. M., Lipsett, M. B., Rayford, P. L., Marshall, J. R., Strott, C. A., and Rodbard, D. (1970). Pituitary and gonadal hormones in women during spontaneous and induced ovulatory cycles. *Recent Prog. Horm. Res.* **26**, 1–48.

Rowlands, I. W. (1944). The production of ovulation in the immature rat. *J. Endocrinol.* **3**, 384–391.

Ryan, K. J., and Petro, Z. (1966). Steroid biosynthesis by human ovarian granulosa and thecal cells. *J. Clin. Endocrinol. Metab.* **26**, 46–52.

Ryan, K. J., and Short, R. V. (1965). Formation of estradiol by granulosa and theca cells of the equine ovarian follicle. *Endocrinology* **76**, 108–114.

Ryan, K. J., and Smith, O. W. (1961a). Biogenesis of estrogens by the human ovary. I. Conversion of acetate 1-C^{14} to estrone and estradiol. *J. Biol. Chem.* **236**, 705–709.

Ryan, K. J., and Smith, O. W. (1961b). Biogenesis of estrogens by the human ovary. II. Conversion of progesterone-4-C^{14} to estrone and estradiol. *J. Biol. Chem.* **236**, 710–714.

Ryan, K. J., and Smith, O. W. (1961c). Biogenesis of estrogens by the human ovary. III. Conversion of cholesterol-4-C^{14} to estrone. *J. Biol. Chem.* **236**, 2204–2206.

Ryan, K. J., and Smith, O. W. (1961d). Biogenesis of estrogens by the human ovary. IV. Formation of neutral steroid intermediates. *J. Biol. Chem.* **236**, 2207–2212.

Ryan, K. J., and Smith, O. W. (1965). Biogenesis of steroid hormones in the human ovary. *Recent Prog. Horm. Res.* **21**, 367–403.

Sasamoto, S. (1969). Inhibition of HCG-induced ovulation by anti-HCG serum in immature mice pretreated with PMSG. *J. Reprod. Fertil.* **20**, 271–277.

Savard, K. (1969). Biogenesis of steroids in the human ovary. *In* "The Ovary" (W. Ingiulla and R. B. Greenblatt, eds.), pp. 264–273. Thomas, Springfield, Illinois.

Savard, K., Marsh, J. M., and Rice, B. F. (1965). Gonadotropins and ovarian steroidogenesis. *Recent Prog. Horm. Res.* **21**, 285–356.

Savard, K., Le Maire, W., and Kumari, L. (1969). Progesterone synthesis from labelled precursors in the corpus luteum. *In* "The Gonads" (K. W. McKerns, ed.), pp. 119–136. North-Holland Publ., Amsterdam.

Schenker, J. G., Polishuk, W. Z., and Eckstein, B. (1971). Pathways in the biosynthesis of androgens in the post-menopausal ovary *in vitro. Acta Endocrinol. (Copenhagen)* **66**, 325–332.

Schultze, A. B., and Noonan, S. (1970). Thyroxine administration and reproduction in rats. *J. Anim. Sci.* **30**, 774–776.

Scully, R. E., and Cohen, R. B. (1964). Oxidative enzyme activity in normal and pathologic human ovaries. *Obstet. Gynecol.* **24**, 667–681.

Segal, S. J., and Johnson, D. C. (1959). Inductive influence of steroid hormones on the neural system. Ovulation controlling mechanisms. *Arch. Anat. Microsc. Morphol. Exp.* **48**, 261–274.

Short, R. V. (1960). Steroids present in the follicular fluid of the mare. *J. Endocrinol.* **20**, 147–156.

Short, R. V. (1962). Steroids in the follicular fluid and the corpus luteum of the mare. A "two cell type" theory of ovarian steroid synthesis. *J. Endocrinol.* **24**, 59–63.

Short, R. V. (1964). Ovarian steroid synthesis and secretion *in vivo. Recent Prog. Horm. Res.* **20**, 303–333.

Short, R. V. (1967). Reproduction. *Annu. Rev. Physiol.* **29**, 373–400.

Short, R. V., McDonald, M. F., and Rowson, L. E. A. (1963). Steroids in the ovarian venous blood of ewes before and after gonadotrophic stimulation. *J. Endocrinol.* **26**, 155–169.

Siiteri, P. K., and MacDonald, P. G. (1973). Role of extraglandular estrogen in human endocrinology. *In* "Handbook of Physiology" (Am. Physiol. Soc., J. Field, ed.), Sect. 7, Vol. II, pp. 615–629. Williams & Wilkins, Baltimore, Maryland.

Simmer, H. H., Hilliard, J., and Archibald, D. (1963). Isolation and identification of progesterone and 20α-hydroxypreg-4-en-3-one in ovarian venous blood of rabbits. *Endocrinology* **72**, 67–70.

Stuart Mason, A. (1970). The thyroid. *In* "Scientific Foundations of Obstetrics and Gynaecology" (E. E. Philipp, J. Barnes, and M. Newton, eds.), pp. 531–534. Heinemann, London.

Sturgis, S. H. (1961). Factors influencing ovulation and atresia of ovarian follicles. *In* "Control of Ovulation" (C. L. Villee, ed.), pp. 213–221. Pergamon, Oxford.

Swanson, H. H. (1966). Modification of the reproductive tract of hamsters of both sexes by neonatal administration of androgen. *J. Endocrinol.* **36**, 327–328.

Swanson, H. E., and van der Werff ten Bosch, J. J. (1964a). The "early-androgen" syndrome; its development and the response to hemi-spaying. *Acta Endocrinol. (Copenhagen)* **45**, 1–12.

Swanson, H. E., and van der Werff ten Bosch, J. J. (1964b). The "early-androgen" syndrome; differences in response to pre-natal and post-natal administration of various doses of testosterone propionate in female and male rats. *Acta Endocrinol. (Copenhagen)* **47**, 37–50.

Swanson, H. E., and van der Werff ten Bosch, J. J. (1965). The "early-androgen" syndrome; effects of pre-natal testosterone propionate. *Acta Endocrinol. (Copenhagen)* **50**, 379–390.

Tagatz, G. E., and Gurpide, E. (1973). Hormone secretion by the normal human ovary. *In* "Handbook of Physiology" (Am. Physiol. Soc., J. Field, ed.), Sect. 7, Vol. II, pp. 603–613. Williams & Wilkins, Baltimore, Maryland.

Talbert, G. B. (1968). Effect of maternal age on reproductive capacity. *Am. J. Obstet. Gynecol.* **102**, 451–477.

Tanner, J. M. (1962). "Growth at Adolescence," 2nd ed. Blackwell, Oxford.

Tanner, J. M. (1967). Puberty. *Adv. Reprod. Physiol.* **2**, 311–347.

Taylor, F. B. (1961). Histochemical changes in the ovaries of normal and experimentally treated rats. *Acta Endocrinol. (Copenhagen)* **36**, 361–374.

Thung, P. J., Boot, L. M., and Mühlbock, O. (1956). Senile changes in the oestrous cycle and in ovarian structure in some inbred strains of mice. *Acta Endocrinol. (Copenhagen)* **23**, 8–32.

Tourneaux, F. (1879). Des cellules interstitielles du testicule. *J. Anat. (Paris)* **15,** 305–328.

Tulchinsky, D., and Korenman, S. G. (1970). A radio-ligand assay for plasma estrone; normal values and variations during the menstrual cycle. *J. Clin. Endocrinol. Metab.* **31,** 76–80.

Velle, W. (1959). Thesis, Norges Veterinaerhogskole, Oslo. (cited by Eckstein, 1962).

White, R. F., Hertig, A. T., Rock, J., and Adams, E. C. (1951). Histological and histochemical observations on the corpus luteum of human pregnancy. *Contrib. Embryol. Carnegie Inst.* **34,** 57–74.

Wilkins, L. (1965). "The Diagnosis and Treatment of Endocrine Disorders in Childhood and Adolescence." Thomas, Springfield, Illinois.

Wotiz, H. H., Davis, J. W., Lemon, H. M., and Gut, M. (1956). Studies in steroid metabolism. *J. Biol. Chem.* **222,** 487–495.

Young, W. C. (1961). The mammalian ovary. *In* "Sex and Internal Secretions" (W. C. Young, ed.), 3rd ed., Vol. 1, pp. 449–496. Williams and Wilkins, Baltimore, Maryland.

YoungLai, E. V., and Short, R. V. (1970). Pathways of steroid biosynthesis in the intact Graafian follicle of mares in oestrus. *J. Endocrinol.* **47,** 321–331.

Zander, J. (1958). Steroids in the human ovary. *J. Biol. Chem.* **232,** 117–122.

Zander, J., Brendle, E., von Münstermann, A. M., Diczfalusy, E., Martinsen, B., and Tillinger, K.-G. (1959). Identification and estimation of oestradiol-17β and oestrone in human ovaries. *Acta Obstet. Gynecol. Scand.* **38,** 724–736.

Zuckerman, S. (1941). Periodic uterine bleeding in spayed rhesus monkeys injected daily with a constant threshold dose of oestrone. *J. Endocrinol.* **2,** 263–267.

Zuckerman, S. (1956). The regenerative capacity of ovarian tissue. *Ciba Found. Colloq. Ageing* **2,** 31–54.

8

Ovarian Activity during Gestation

E. C. Amoroso and J. S. Perry

315

I. INTRODUCTION

The purpose of this chapter is to give an account of the part played by the ovary in the initiation, maintenance, and termination of pregnancy in viviparous vertebrates. Our primary concern is with the mammals, particularly the eutherian orders, and with the modification of the estrous or menstrual cycle that gestation involves. Having formed and discharged the egg which is its contribution to a future generation, the ovary reverts to an intermediary role, so that a consideration of its function falls naturally into two parts: the factors that influence or control ovarian activity, and the effects of that activity upon other organs. At what point the course of events in pregnancy diverges from that of the "normal" cycle is not always clear and, in some species at least, embryos reside in the uterus for some time before ovarian activity is modified. In the sheep, for example, ovulation recurs after the normal 16-day interval unless the uterus has been occupied by a viable embryo during the preceding 3 or 4 days (Moor and Rowson, 1964, 1966a,b).

The delay in the recurrence of estrus is associated with the persistence of the corpus luteum beyond its duration in the normal cycle, and the relation between this organ and the maintenance of pregnancy was first appreciated in the early years of this century. The existence of such a relationship, and the idea that the corpus luteum is a functional organ of pregnancy and not the fortuitous result of an emptied follicle, is attributed to Gustav Born, of Breslau. The chain of evidence establishing the fact of the secretory function of the corpus luteum is detailed by Corner (1947), who credits Fraenkel with the vindication of Born's hypothesis. Although John Beard (1897) and Louis Augustus Prenant (1898) had earlier suggested that the corpora lutea might exercise an inhibitory influence on ovulation, the first experimental evidence that these bodies are necessary for implantation and maintenance of embryos in the uterus was the discontinuance of pregnancy, following their removal (Fraenkel and Cohn, 1901; Fraenkel, 1927). The evidence was derived from the results of experiments on pregnant rabbits in which the ovaries were removed or the corpora lutea destroyed; in each case the pregnancy was brought to an end. Control experiments proved that the effects were not merely due to the operation itself.

Very similar experiments were being carried out about the same time by Vilhelm Magnus, in Scandinavia. He too was inspired by Born, who, he records, "developed an hypothesis that the corpus luteum serves to maintain pregnancy by secreting a substance beneficial to the fetus, to which it is carried by the blood." Magnus's experiments were described in 1901. He first used rats, but could not determine the time of mating and encountered difficulty in cauterizing the corpora lutea without other damage to the ovary. He therefore used rabbits, as did Fraenkel, and showed that bilateral ovariectomy, performed between day 4 and day 19 of pregnancy, led to the loss of the embryos and an "atrophic" uterus. Cautery of the corpora lutea before the 21st day of pregnancy also led to the loss of the embryos but the uterus was not rendered atrophic. When this operation was performed later than day 21, living young were delivered at the normal time. Magnus also performed unilateral ovariectomy, and found that a single ovary prevented atrophy of both uterine horns. He considered this to be best explained by an internal secretion from the ovary, and quoted experiments by Rubinstein, and by Halban, both using rabbits, who had shown by ovarian autotransplantation that the influence of the ovary on the uterus was independent of its nerve supply.

Both Fraenkel and Magnus supposed that the luteal bodies elaborated a substance which was carried in the blood and which somehow stimulated the uterine mucosa and assisted in the attachment of the fertilized ovum and the maintenance of its nutrition during the first half of pregnancy. This opinion was further strengthened by Loeb's (1907) discovery that the corpora lutea specifically induce a decidual response in the uterus of the guinea pig so that implantation becomes possible. He found that mechanical irritation of the endometrium in the nonpregnant uterus from the second to the eighth or ninth day after ovulation elicits the formation of tumor-like decidual growths (deciduomata) at the injured sites in much the same way as the embryo produces a decidual reaction during its interstitial implantation. He showed, moreover, that deciduomata form in transplanted portions of the uterus and here, as in the normal uterus, the reaction depends upon the integrity of the corpus luteum.

The early experiments on ovariectomy and luteal ablation, as well as those of Loeb (1907) on the production of deciduomata, have been confirmed in several species of mammals and by numerous other investigators (see Amoroso and Finn, 1962), but it was not until Bouin and Ancel (1910) had shown that the corpus luteum exerts a comparable influence on the rabbit's uterus in pseudopregnancy that the general conclusions regarding the functions of that organ were placed on a completely firm foundation.

In recent years, techniques have been developed that make it possible to measure very small amounts of steroid compounds, and these methods have been applied to the study of the role of estrogen and progesterone, as will be described in a later section. The origin of the name "progesterone" was reported by W. M. Allen (1970).

II. OVARIAN CHANGES IN PREGNANCY AND THE FORMATION OF THE CORPUS LUTEUM IN EUTHERIAN MAMMALS

The formation and structure of the mammalian corpus luteum is described in Volume I, Chapter 4 and a detailed review of the earlier literature is given by Hett (1933) and by Brambell (1956).

A. Inhibition of Ovulation and Cyclic Ovarian Activity

Corpora lutea are normally formed from all ruptured follicles at ovulation and, in the majority of spontaneously ovulating species at least, their cyclical persistence is sufficient to permit the developing egg to reach the stage at which implantation is possible and to maintain uterine sensitivity long enough for implantation to occur. If fertilization and implantation do occur, the corpora lutea are transformed into corpora lutea of pregnancy. In addition, ovarian follicles which would have been expected to undergo maturation with the onset of another cycle are inhibited and with them ovulation and estrus itself. This inhibition of estrus and ovulation in pregnancy, which is usually attributed to the secretions of the corpus luteum—estrogen and progesterone—may be complete or, more rarely, the ovaries may continue to undergo rhythmical changes, especially in early pregnancy (see Amoroso and Finn, 1962).

Experimental proof of the importance of the corpus luteum in suppressing ovulation was provided by Loeb (1923), by Drummond-Robinson and Asdell (1926), and by Hammond (1927) when they showed that removal of this organ in the guinea pig, goat, and cow, respectively, was soon followed by ovulation regardless of whether the products of conception were expelled immediately or retained temporarily. Additional evidence that the premature occurrence of ovulation following removal of the corpora lutea is due to the withdrawal of the active principle progesterone was provided by Parkes and Bellerby (1927) and by Selye *et al.* (1936).

That there is a temporal relationship in the rabbit between mating, release of gonadotropin, secretion of progesterone, and ovulation was first demonstrated by Fee and Parkes (1929) and Deanesly *et al.* (1930).

B. Growth and Life Span of the Corpus Luteum

The pattern of growth and survival of the corpus luteum of pregnancy varies very widely between species. In some it grows larger as well as surviving longer than the corpus luteum of the cycle, in many it survives longer but does not grow larger, and in a few eutherian species, as in most marsupials, the corpus luteum of pregnancy is like that of the cycle in size and persistence.

In many species, such as the giraffe, the corpus luteum of pregnancy does not regress appreciably before parturition. In some other species it does not persist throughout gestation; thus, in two African bats, *Nycteris luteola* and *Triaenops afer,* Matthews (1941) found no trace of a corpus luteum in the ovary after an early stage of pregnancy. The same is true of the black rhinoceros and the mare (see below). There are, moreover, species differences in the relative importance of apparently degenerating corpora lutea and long-lasting ones, as shown by a comparison of the changes in the luteal bodies of the rabbit, cat, rat, and guinea pig. In the rabbit, regressive changes can be detected in the corpora lutea toward the end of pregnancy, yet only occasionally can the ovaries be dispensed with and then not unless the fetuses are exteriorized (Courrier, 1941) or eliminated altogether, leaving the placentas *in situ* (Klein, 1933). By contrast, in the rat, in which the ovaries are likewise necessary throughout pregnancy, no similar regressive changes are noted. In the cat, luteal regression is well under way by the sixth week of pregnancy, but removal of both ovaries at 49 days or later was followed by normal parturition at term (64 days), although spaying at 46 days or earlier regularly interrupted pregnancy (Gros, 1936). The decline of the corpora lutea of pregnancy in the cat is gradual, and in some cases they remain large until very near term. After parturition, if the newborn kittens are removed, and a further ovulation and mating ensue, these surviving corpora lutea may undergo a temporary recrudescence under the influence of the endocrine factors of a new pregnancy. A similar phenomenon has been observed in wild Norway rats, where fertile mating frequently occurs at the immediately postpartum estrus (personal observation, J. S. Perry). In the guinea pig there is no histological evidence of degeneration in the corpora lutea until the end of pregnancy, although gestation may not be interrupted by removal of the ovaries a considerable time before parturition (see Table I). It is therefore clear that "persistence" of the corpora lutea and lack of "degeneration" have a different significance. The evidence summarized above, which has been substantiated by direct measurements of ovarian steroid output, cautions against the assumption that the endocrine activity of the gland can be adjudged from its histological appearance or its size. On the other hand, because lutein cells, in common with most steroid-producing tissues, release their product immediately, and do not store it, the concentration of progesterone in the corpus luteum usually reflects the secretory activity of the gland (Gomes and Erb, 1965; Masuda *et al.,* 1967).

TABLE I

Effects of Ovariectomy or Ablation of Corpora Lutea on the Course of Gestation in Mammals

Species and gestation length	Authorities	Operation	Effect on gestation
1. Animals in which implantation follows ovariectomy in the absence of exogenous steroids			
Guinea pig (ca. 66 days)	Loeb and Hesselberg (1917); Deanesly (1960)	Corpora lutea (CL) cauterized or ovaries removed	Ovariectomy at $3\frac{1}{2}$ to 5 days postcoitum did not prevent implantation which normally occurs $6\frac{1}{2}$ to $7\frac{1}{2}$ days postcoitum. Development continues up to about day 14, a few embryos remaining viable as late as day 17. For normal growth, progesterone is needed from about day 12 to day 18 or 20, after which the placenta alone can maintain pregnancy
Armadillo (variable: delayed implantation)	Buchanan et al. (1956); Enders and Buchanan (1959)	Bilateral ovariectomy or removal of ovary containing corpora lutea	Implantation could occur 24–30 days after ovariectomy provided the latter was carried out about the middle of the 4-month period of delayed implantation. If ovary with corpora lutea is removed, progesterone maintains blastocysts. After bilateral ovariectomy, fewer blastocysts are recovered, even with progesterone treatment
2. Animals in which experimental evidence indicates that pregnancy may continue following ovariectomy after a certain stage has been reached			
Brush possum (17.5 days)	Sharman (1965a)	Luteal ablation	Embryos developed to term
Quokka (27 days)	Tyndale-Biscoe (1963a)	Ovariectomy	After day 7, embryos developed normally; parturition failed
Tammar wallaby (29 days)	Tyndale-Biscoe (1970)	Ovariectomy	After day 7, embryos developed normally

	Removal of ovary containing corpora lutea		
Armadillo (variable: delayed implantation)		Buchanan et al. (1956)	Pregnancy continued if operation performed later than the first one-third of the postimplantation period
Guinea pig (ca. 66 days)	Ovariectomy	Daels (1908); Loeb (1923); Herrick (1928); Benazzi (1933); Nelson (1934); Artunkal and Colonge (1949); Deanesly (1963)	At 14 to 27 days postcoitum and sometimes up to 40 days pregnancy terminated. Then may continue normally
Cat (63 days)	Ovariectomy	Gros (1936); Courrier and Gros (1935); Amoroso (1955a)	Up to 46 days postcoitum abortion results. At 49 days or later young are carried to term
Sheep (136 days)	Ovariectomy	Casida and Warwick (1945); Dutt and Casida (1948); Neher and Zarrow (1954); Denamur and Martinet (1955); Foote et al. (1957)	Abortion or resorption follows removal before 55th day of pregnancy; thereafter the chances that pregnancy will go to term appear to increase the later the operation is performed. Daily treatment with 1 mg progesterone plus 0.25 µg estrone (ratio 4000 : 1) will cause embryos to implant and maintain pregnancy
Horse (350 days)	Ovariectomy	E. C. Amoroso, J. A. Laing and P. Messervy (unpublished)	The ovaries may be removed between 170 and 270 days after breeding without interruption of the pregnancy. After the operation, estrogen excretion falls, but soon rises again to normal level and is maintained within the normal range for the remainder of pregnancy
Rhesus monkey (ca. 164 days)	Ovariectomy	Hartman (1941); Hartman and Corner (1947); Tullner and Hertz (1966)	May be performed as early as 25th day of pregnancy without causing abortion. The bright red color of the sexual skin is maintained throughout gestation and lactation, thus giving further proof that estrogens are secreted by extragonadal sources or accessory ovarian tissue (see Hartman and Corner, 1947)

(continued)

321

TABLE I (continued)

Species and gestation length	Authorities	Operation	Effect on gestation
Women (267 days)	Blair-Bell (1920); Ask-Upmark (1926); Asdell (1928); Douglass (1931); Corbet (1932); Melinkoff (1950); Kulseng-Hanssen (1951); Glasser (1952); Tulsky and Koff (1957); Venning (1957); Diczfalusy and Borell (1961); see also Deanesly (1966)	Ovariectomy	Ovaries have been removed or corpora lutea destroyed as early as the 40th day without terminating pregnancy

3. Animals in which experimental evidence indicates that pregnancy cannot be maintained in the absence of the ovaries

Species and gestation length	Authorities	Operation	Effect on gestation
Virginian opossum (13 days)	Hartman (1925)	Ovariectomy	Death of embryos at all times
Spermophile (28 days)	Drips (1919); Johnson and Challans (1932)	Ovariectomy	Terminated at all times but could be maintained with luteal extracts
Hamster (16 days)	Klein (1938); Orsini and Meyer (1959)	Ovariectomy	Early (2 to 3 days postcoitum) prevents implantation; late (11–13 days) results in abortion. Eggs made to implant with progesterone alone. 4 mg daily
Mouse (19 days)	Harris (1927); Parkes (1928); Robson (1936)	Ovariectomy or one ovary sterilized with X rays, fertile ovary removed	At any time during pregnancy causes abortion or resorption of the embryos. Daily injections of 1 to 1.5 mg progesterone required to repair the effects of oophorectomy
Chinchilla (111 days)	B. J. Weir (unpublished)	Ovariectomy	Bilateral ovariectomy causes abortion or resorption of embryos. Pregnancy maintained if only one ovary removed or with >1 mg progesterone daily

Rabbit (32 days)	Fraenkel and Cohn (1901); Magnus (1901); Fraenkel (1903, 1910); McIlroy (1912); Hammond (1925); Corner (1928); Pincus and Werthessen (1938); Courrier and Kehl (1938); Allen and Heckel (1939)	Ovariectomy or ablation of corpora lutea	Terminated at all times. In ovariectomized does complete maintenance of pregnancy is possible with a dosage of 1 mg progesterone/day during first 10–12 days and 2 to 4 mg/day thereafter. When fetuses within their amniotic sacs are exteriorized, they may survive in oophorectomized animals
Goat (112 days)	Drummond-Robinson and Asdell (1926); Meites et al. (1951)	Ovariectomy or dissection of corpora lutea	At any time during pregnancy causes abortion or resorption of embryos. With all corpora lutea removed, 15–25 mg of progesterone daily are required to maintain pregnancy
Sow (112 days)	du Mesnil du Buisson and Dauzier (1957); Day et al. (1959); Spies et al. (1960)	Ovariectomy	As early as 4 days postcoitum or as late as 89 to 106 days resulted in abortion within 3 days; effects not due to trauma. Treatment with 1.2 mg progesterone plus 0.3 μg estrone caused embryos to implant but many embryos failed to survive

4. Animals in which the evidence is inconclusive

| Rat (21 days) | Johnson and Challans (1930); Hain (1934); Selye et al. (1935a); Haterius (1936); McKeown and Zuckerman (1938); Kirsch (1938); Haterius and Kempner (1939); Zeiner (1943); Frazer and Alexander (1954) | Ovariectomy or cauterization of corpora lutea | Most reports state that early removal leads to abortion. After 16th day, pregnancy is not speedily terminated, especially if ovaries are removed in two stages. Pregnancy maintained in absence of ovaries, provided all fetuses but one are removed and all placentas are allowed to remain. Pregnancy will continue in ovariectomized hypophysectomized rats given 1 μg estrone plus 2 mg progesterone |

(continued)

TABLE I (*continued*)

Species and gestation length	Authorities	Operation	Effect on gestation
Dog (60–63 days)	Marshall and Jolly (1905, 1906)	Ovariectomy	At 3, 10, 14, 21, and 28 days caused abortion in all cases. No animals ovariectomized after the first month
Cow (ca. 280 days)	Hess (1920); Schmaltz (1921); Hammond (1927); Raeside and Turner (1950); McDonald et al. (1952, 1953); Venable and McDonald (1958); Estergreen et al. (1967)	Ovariectomy or squeezing out corpus luteum	Abortion before days 165 to 180, thereafter variable

C. Accessory Luteal Structures

In the horse (Amoroso *et al.*, 1948) the corpus luteum of pregnancy persists only for about the same length of time as the normal cyclical corpus luteum. A type of gonadotropic activity then follows, which is attributed to the placenta (see Allen and Moor, 1972), and new corpora lutea are formed, partly by direct luteinization and partly by the rupture of follicles (Amoroso *et al.*, 1948). Some of them appear to be formed by the luteinization of the theca interna cells with degeneration of the ovum and granulosa (Cole *et al.*, 1931). These accessory structures, which are a rich source of progesterone, have a longer life span than the ones they replace, and may be a factor of safety ensuring implantation (Amoroso and Rowlands, 1951; Amoroso, 1959). The accessory corpora lutea degenerate completely by 150 days, and from then until the end of gestation the ovaries contain neither corpora lutea nor large follicles. The accessory structures cannot, therefore, be regarded as providing a continuing source of progesterone in this species. In the black rhinoceros, the ovaries during late pregnancy are similarly devoid of corpora lutea or large follicles, but their condition during earlier stages of gestation has not yet been described. In the African elephant, on the other hand, Perry (1953) recorded a number of accessory corpora lutea at all stages of pregnancy, some of which persist until term. Laws (1969) concluded that they were formed by the luteinization of unovulated follicles early in pregnancy. The occurrence of luteal tissue during pregnancy, other than that of the corpus luteum of conception, has also been recorded in monotremes (Bretschneider and de Wit, 1947) and in the ovaries of the plains viscacha (Weir, 1971a), the armadillo (Newman and Paterson, 1910), the human female (Meyer, 1911), and the rhesus monkey (Corner *et al.*, 1936; Corner, 1940). In these, the accessory luteal structures are formed at the same time as those resulting from an ovulation. In others, including the peccary (Wislocki, 1931), baboon (Zuckerman and Parkes, 1932), bank vole (Brambell and Rowlands, 1936), mountain viscacha (Pearson, 1949; Weir, 1971b), chinchilla (Weir, 1966), Canadian porcupine (Mossman and Judas, 1949), Norway rat (Hall, 1952), water shrew (Price, 1953), nilgai (Amoroso, 1955a), fin whale (Laws, 1958), sow (du Mesnil du Buisson and Dauzier, 1959), and cat, accessory corpora lutea may arise during the course of pregnancy.

In the Canadian porcupine, *Erethizon dorsatum,* the accessory corpora lutea are formed in large numbers at estrus and in early pregnancy as a result of luteinization of atretic follicles. They develop in both ovaries, but persist only on the side carrying the normal corpus luteum, an observation which led Mossman and Judas (1949) to suggest that the latter was concerned in their maintenance, although the presence of an empty uterine horn may be sufficient to account for the asymmetry.

In contrast to the Canadian porcupine, a feature of the ovarian cycle of the mountain viscacha, the chinchilla, the peccary, and the nilgai is the appearance in midpregnancy of large numbers of accessory corpora lutea, probably formed by the luteinization of unruptured follicles. In the chinchilla, peccary, and nilgai, they develop in both ovaries, but are found only on the right side in the mountain viscacha where, in addition to the primary and apparently functional corpus luteum, twelve or more accessory corpora lutea may be present at the time of parturition. In the ovaries of a pregnant nilgai bearing twin fetuses, as many as thirty-four accessory corpora lutea have been recorded, and in a chinchilla with only two fetuses, ninety-three luteal bodies were present (Weir, 1966; Fig. 1).

The ovulation of supernumerary follicles in the African insectivore *Elephantulus myurus* (van der Horst and Gillman, 1940, 1946) was for many years the most bizarre example known of multiple ovulation and egg wastage, as many as 60 ova being shed from each ovary, to produce only two implanted embryos. An even greater prodigality has recently been described in the plains viscacha, *Lagostomus maximus* (Weir, 1971a), which

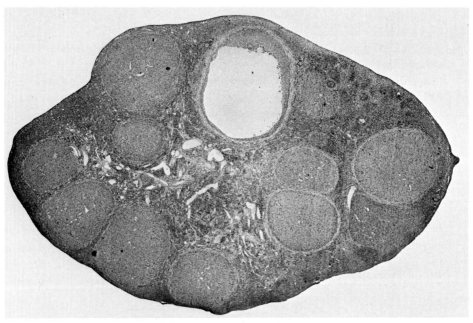

Fig. 1. Ovary of a 90-day pregnant chinchilla with two fetuses showing eleven corpora lutea and a luteinizing follicle. There were twenty-one corpora lutea in the pair of ovaries but those of the pregnancy were not distinguishable from the accessory corpora. × 20. (By courtesy of B. J. Weir.)

may shed hundreds of ova from follicles which form as many small corpora lutea, although only about seven ova are fertilized and implant, and only two fetuses normally survive to term.

A feature of the ovary of the gibbon is the intensive luteinization of all follicular elements during the luteal phase of the cycle, pregnancy, and lactation (Dempsey, 1940). This constitutes a distinctive feature in which the gibbon resembles the New-World monkeys (Dempsey, 1939) and differs from man and the Old-World monkeys, in whom luteinization of the theca interna takes place only during pregnancy.

D. Fetal Ovarian Activity

It has become evident in recent years that the feto–placental unit exercises a considerable degree of autonomy and that the fetus itself contributes to the hormonal maintenance of pregnancy. The extent of this activity is beyond the scope of this chapter (see Volume II, Chapter 7 and Volume III, Chapter 3) but mention must be made here of morphological changes in the fetal gonads that are a prominent feature of gestation in several species. The hypertrophy of the testicular and ovarian elements described below is not, however, clearly related to the important part which the fetal gonads, or at least the fetal testes, appear to play in the process of sex differentiation of the accessory reproductive organs (see Raynaud, 1942; Jost, 1970; Price, 1970).

1. Luteal Structures

Widespread luteinization of the follicular elements is seen in the ovaries of the human full-term fetus (see Potter, 1953) and in those of the fetal giraffe during late gestation (Kellas *et al.,* 1958). Whether these changes reflect a genuine development of ovarian sensitivity to the gonad-stimulating substances which appear in the urine of pregnant women (Zondek, 1928) and the pregnant giraffe (Wilkinson and de Fremery, 1940), or may be referred to some other cause, is not yet clear. Nevertheless, to judge from the histological appearance of the fetal gonad as well as the development and luteinization of the follicles, it would appear that the ovary of the giraffe may be capable of secretory activity and ovum maturation long before it is normally called upon to exercise its cyclical functions.

2. Interstitial Elements

In some other species, such as the horse (Fig. 2), the fetal gonads are strikingly enlarged by an enormous development of the interstitial cells. The available evidence points to the existence of an association between the

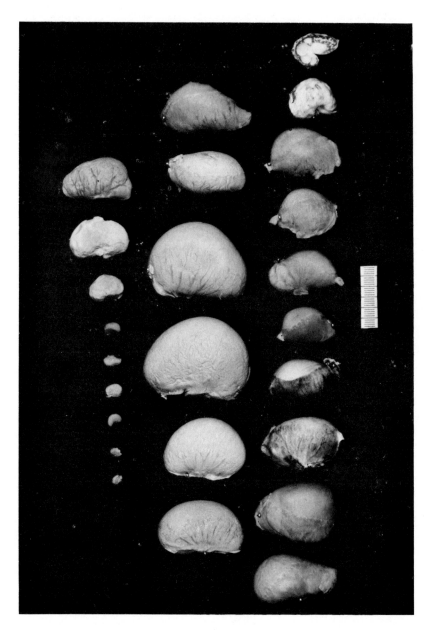

Fig. 2. Stages in the development and regression of fetal mare ovaries from 45 days postcoitum (top left). The gonad attains maximum size between 150 and 180 days and regresses thereafter. As the two ovaries (bottom right) from a neonatal filly (5 days postpartum) show, there is a striking decline after birth. ×0.5. (from Deanesly, 1966.)

concentration of urinary estrogens in the pregnant mare and the quantitative development and regression of the fetal gonads (Amoroso and Rowlands, 1951). They reach their greatest size (Fig. 3) at a time when the urinary excretion of equilin, a steroid peculiar to the feto–placental unit of the horse, is maximal. But, although similar qualitative and quantitative changes have been observed in the gonads of the fetal elephant (Perry, 1953) and of the grey and elephant seals (Amoroso et al., 1965; Bonner, 1955), no correlations between urinary estrogens and gonadal development have, as yet, been reported.

III. THE INITIATION AND MAINTENANCE OF LUTEAL FUNCTION

We have already emphasized the fact that the outstanding ovarian change involved in pregnancy is the increased functional survival of the corpus luteum and, as a consequence, the predominance of progesterone in the ovarian steroid output. Progesterone has been described as "the hormone of pregnancy" for without a source of progesterone no pregnancy, as far as is

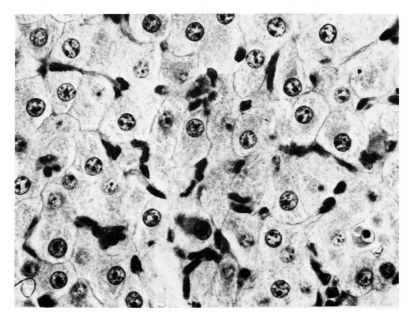

Fig. 3. Photomicrograph of the interstitial tissue that constitutes the mass of the fetal horse ovary, 120 days postcoitum. The polyhedral cells bear a striking resemblance to lutein cells and are abundantly supplied with vessels engorged with blood. × 650.

known, can be maintained (see, however, Short, 1966; Hanks and Short, 1972). Yet the mechanism controlling the persistence of the corpus luteum, and the factors that at first maintain it, and those that later cause its regression, remain controversial.

A. The Role of the Pituitary Gland

The original discovery that luteal activity may follow the injection of pituitary extracts in normal rats was made by Long and Evans (1922). The formation of a corpus luteum appears to follow automatically on the rupture of a follicle under the influence of pituitary LH. Its subsequent persistence and function is also controlled, at least initially, by the pituitary. This luteotropic function depends chiefly on prolactin in the rat (see Deanesly, 1966), and perhaps chiefly on LH in the cow (Hansel and Seifart, 1967). Both these hormones are probably involved, in different proportions, in the initial maintenance of luteal function in most, if not all, species. Evidence that LH has a luteotropic function in the rat, for example, is provided by the work of Moudgal *et al.* (1969), who were able to terminate pregnancy in rats by administering a specific rabbit antiserum to purified ovine LH on any day between days 8 and 11 of pregnancy. This effect of the antiserum could be antagonized by daily administration of 4 mg of progesterone.

This situation—functional and necessary corpora lutea maintained by pituitary gonadotropin—persists throughout gestation in some species (rabbit). In others (rat and mouse), the corpora lutea continue to secrete progesterone and the ovaries continue to be essential to the maintenance of pregnancy, but the luteotropic function is assumed by the placenta and pituitary luteotropin can be dispensed with. In yet another group (primates) the placenta itself secretes progesterone and so assumes the original role of the ovaries, which can then be removed without affecting the course of pregnancy. The role of the pituitary has been investigated by hypophysectomy in a wide variety of species (Table II) and the results have been found to vary with the species and with the stage of gestation at the time of operation.

The mechanism responsible for the interruption of pregnancy as a result of hypophysectomy, on the one hand, and that responsible for its prolongation, on the other, is revealed by consideration of the effects of ovariectomy in conjunction with those of hypophysectomy (see Tables I and II). In the rabbit, in which the corpora lutea promptly degenerate after hypophysectomy, the maintenance of pregnancy is seen to depend upon ovarian secretions produced under hypophysial stimulation. In rats and mice, removal of the ovaries invariably terminates pregnancy although hypophysectomy does not (see above). Hence the continuation of pregnancy after hypophysectomy during the second half of gestation in these species must depend on the functional activity of the corpora lutea and this activity must be maintained by stimulation other than from the pituitary. The balance of evidence at present available suggests that the placenta is the new source of stimulation (Amoroso, 1955a; Diczfalusy and Troen, 1961; Amoroso and Porter, 1966, 1970). The data also support the earlier observation of Astwood and Greep (1938) and probably explain the findings of Selye *et al.* (1933a), and those of Alloiteau *et al.* (1963), that, after hypophysectomy in the rat during the latter half of gestation, the corpora lutea of pregnancy can survive beyond their normal life span.

In a third group of animals, as outlined above, pregnancy may continue in the absence of both ovaries and pituitary. In these, the placenta has been shown to secrete progesterone, and so to assume the direct role of the ovary, rather than the gonadotropic role of the pituitary gland.

B. The Nature of the Luteotropic Complex

As already indicated, luteal function is normally controlled by a combination of at least two, often more, hormones, so that it is proper to discuss the nature of the luteotropic complex, rather than to attempt to identify "the luteotropin," in any given species. In the rat, in which the pituitary–ovarian relationship was first and most extensively studied, the principal pituitary luteotropin (LTH) is prolactin. According to one recent concept, pituitary control in this species is exerted by changes in the extent to which progesterone is metabolized to the much less active 20α-dihydro-progesterone (Hashimoto and Wiest, 1969; Wiest *et al.*, 1968). The enzyme responsible for the conversion, 20α-hydroxysteroid dehydrogenase (20α-OH-SDH) is inhibited during pregnancy, as it is by hypophysectomy or by the stimulus of suckling. LH treatment stimulates the activity of this enzyme and promotes luteal regression, although it stimulates steroidogenesis in surviving corpora lutea—a property called into play when it acts together with prolactin, which directly antagonizes 20α-OH-SDH (Hashimoto and Wiest, 1969) and at the same time promotes cholesterol synthesis (Armstrong *et al.*, 1969).

Prolactin is similarly important in the pituitary control of the corpora lutea in the hamster and mouse. In the hamster it appears to act synergistically with FSH, prolactin maintaining luteal survival and FSH stimulating the steroidogenic activity of the lutein cells (Greenwald, 1967; Greenwald

TABLE II

Effects of Hypophysectomy on Gestation

Species and gestation length	Authorities	Stage of pregnancy at operation	Effects
1. Fishes, amphibians, and reptiles			
Mustelus canis	Hisaw and Abramowitz (1938, 1939); Hisaw (1959)	Very early and soon after eggs had entered uterus	No effect on development of young or of corpora atretica for at least 5 months of a 10-month gestation. Effect on late gestation not reported
Salamandra salamandra	Joly and Picheral (1972)	Throughout pregnancy	No effect on embryos and parturition normal
Natrix cyclopion; N. sipedon confluens; Thamnophis butleri; Stereia dekayi	Clausen (1940)	Early	Fetal resorption
		Late	Abortion of dead young
Natrix sipedon	Bragdon (1951, 1952)	From second week after ovulation to late pregnancy	No effect on embryos, but young retained *in utero* past term and parturition protracted
Zootoca vivipara	Panigel (1956)		No effect on embryos, but young retained in eggs within body of parent beyond normal limits
2. Mammals			
a. Species in which pregnancy cannot be maintained in the absence of the hypophysis			
Rabbit (32 days)	(1) Smith and White (1931) (2) White (1933) (3) Firor (1933) (4) Westman and Jacobsohn (1937) (5) Robson (1937a,b)	Early (1, 3, 4, 5)[a]	Atrophy of corpora lutea within 4 days. Pregnancy maintained with estrogen and progesterone after hypophysectomy on day 3 (5)[a]
		Late (2, 5)[a]	Abortion within 24 to 48 hours, except within 2–4 days of term, when both living and dead young were born at term (2)[a]

Species	Reference	Time	Result
Bitch (60–63 days)	Aschner (1912)	Day 14	Fetuses resorbed
		Day 24, 25, 26	Resorption or abortion
		Later (?)	Two aborted within 3–4 days, one produced a litter 12 days later
	Houssay (1935)	Probably near term	Two produced living pup
	Votquenne (1936)	Between 5th and 7th week	Seven aborted within 2 days; five resorbed fetuses or aborted as revealed by laparotomy
Pig (112 days)	Anderson et al. (1967)	Day 1 to 2	Embryos died before day 20. Failure apparently due to regression of corpora lutea
Goat (150 days)	Cowie and Tindal (1960); Cowie et al. (1963)	38 to 120 days	Abortion 3 to 9 days after operation. Pituitary stalk section gave similar result.

b. Species in which the evidence is inconclusive

Species	Reference	Time	Result
Ferret (ca. 42 days)	McPhail (1935a)	Day 21	Four animals: two died 3–4 days after operation; two resorbed or aborted fetuses 12 days after operation
		Day 35	Six animals: three delivered live and dead young on day 43; three delivered young prematurely 3 to 4 days after operation
	E. C. Amoroso and A. Galil (unpublished)	Day 34	Four animals: two died after 2 to 3 days, with no sign of fetal resorption; one delivered of 3 live and 1 dead pup by Caesarean section on day 43; one gave birth to 7 live and 1 dead pup 9 days after operation
Cat (63 days)	McPhail (1935b)	Second half	Four animals; one gave birth to full-term kittens 2 to 3 days after operation; three aborted 2 to 3 days after operation

(continued)

TABLE II (continued)

Species and gestation length	Authorities	Stage of pregnancy at operation	Effects
c. Species in which pregnancy may continue following hypophysectomy after a certain stage of gestation			
Guinea pig (ca. 66 days)	Pencharz and Lyons (1934); Desclin (1932, 1939); Nelson (1935); Rowlands and Perry (1964)	Day 34 to 36	Fetuses resorbed within 2 days
		Very near term	Three animals: one died, two delivered successfully. When fetuses implanted (Section d below) pregnancy continued to term in some cases, but parturition did not occur (pubic symphysis did relax)
Mouse (19 days)	Selye et al. (1933b, 1935a); Newton and Beck (1939)	Second half	No abortions; dead fetuses produced at term
		Day 12 or 13	Fetuses were removed at time of hypophysectomy; placentas retained until animals killed on day 19
	Newton and Richardson (1940); Gardner and Allen (1942)	Day 10	Fetuses carried to term. Parturition protracted
Rat (21 days)	Pencharz and Long (1931); (1933)	Day 4	Failure to implant ⎫ Total
		Day 7–10	Resorption 1–3 days ⎬ hypophysectomy after operation ⎭
	Leonard (1945)	Day 11–20	Fetuses unaffected and carried to term. Parturition greatly protracted
	Cutuly (1941a,b, 1942)	Day 1–5	Implantation occurred if prolactin administered
	Selye et al. (1933a)	Day 10–14	Resorption variable; parturition normal if fetus carried to term
	Fortgang and Simpson (1953)	Preimplantation or day 12	Ablation of posterior and intermediate lobe: no effect. Ablation of anterior lobe: effect as total hypophysectomy

Species	Authorities[a]	Stage	Results
Rhesus monkey (ca. 164 days)	Agate (1952); Smith (1946, 1954)	Between days 27 to 156	Thirty-three monkeys: sixteen came to term 20 to 131 days after operation; 12 live babies, 5 by Caesarean section
Women (267 days)	Della-Beffa (1951); (see also Williams, 1952); Little et al. (1958);	26th week	No effect on gestation. Pituitary destroyed by endocranial tumor / Delivery induced in 35th week. Urinary estrogen and pregnanediol within normal limits. Pituitary removed for neoplasia of breast
	Kaplan (1961)	12th week	Normal pregnancy and parturition. Hypophysectomy for neoplasia. Hydrocortisone, thyroid extract, and vasopressin administered
Sheep (136 days)	Denamur and Martinet (1961)	Day 50–134 / Day 70 (6 animals)	Fetuses carried to term. Parturition normal / Healthy fetuses removed on day 148

d. Species in which implantation may follow hypophysectomy without administration of exogenous steroids

Species	Authorities[a]	Stage	Results
Guinea pig (ca. 66 days)	Heap et al. (1967)	Days 1–5 (preimplantation)	Twenty-seven animals. Implantation occurred in 20, normal conceptuses found in 16 killed after implantation

[a] Numbers in parentheses refer to references listed in column "authorities."

and Rothchild, 1968). In the mouse, the supporting role involves LH rather than FSH: prolactin alone will not maintain pregnancy after hypophysectomy on day 7 (Jaitly *et al.*, 1966).

The guinea pig is unusual in that pregnancy can proceed even when the pituitary is removed within the first 3 days postcoitum (Heap *et al.*, 1967). It has been suggested that in this species "the growth and function of the corpus luteum after its early formation in pregnancy is probably more dependent upon the suppression of a uterine (luteolytic) factor which otherwise inhibits it than upon a continuous pituitary secretion" (Heap *et al.*, 1972), since hysterectomy very markedly promotes both growth and persistence of the corpus luteum. In this respect the guinea pig differs from the mouse, rat, and rabbit. In the guinea pig, as in the rat, the corpora lutea of pregnancy grow to approximately twice the volume attained in the estrous cycle of the nonpregnant animal, but in the former species this growth also occurs in hysterectomized animals, irrespective of the presence or absence of the pituitary, and in hypophysectomized pregnant animals where one or more embryos survive (Heap *et al.*, 1967). However, hypophysial support is necessary for maximal embryonic survival, and it has recently been shown that prolactin can exert a luteotropic effect in this species (Illingworth and Perry, 1971). Luteal growth, which is slowed in nonpregnant animals hypophysectomized early in the cycle, was restored when prolactin was administered, and it was not impaired by hypophysial stalk section, an operation assumed to allow prolactin secretion in the absence of FSH or LH release.

In recent years an increasing amount of research has been carried out on the problem of the control of luteal function in farm animals. Luteal maintenance, and its corollary luteal regression, have been extensively studied in sheep, pigs, and cattle, especially since the demonstration by du Mesnil du Buisson (1961) of a "local" effect of unilateral hysterectomy on the corpora lutea of the pig. Among the three species, prolactin appears to predominate in the luteotropic complex of the sheep (Denamur *et al.*, 1966; Denamur, 1968) and LH in that of the cow (Hansel and Seifart, 1967), whereas in the pig, the requirements for luteal maintenance appear to change during the course of gestation (du Mesnil du Buisson and Denamur, 1968). In the sheep, prolactin will maintain the function of the corpus luteum in the hypophysectomized nonpregnant animal, provided it is also hysterectomized to eliminate the overriding luteolytic influence of the uterus (see Section III,C). The luteotropic effect, however, as judged by the secretory function of the corpus luteum, approximates more closely to the normal situation when LH is administered with prolactin (Denamur *et al.*, 1966; Denamur, 1968).

The observation that active corpora lutea can survive in the hysterectomized cow, after stalk section, suggests that prolactin is important in the luteotropic hormone complex in this species also, but there is a wealth of evidence to suggest that the dominant factor here is LH. Estrogen is luteolytic in the cow, and its effect can be reversed by LH administration (Denamur, 1968). Similarly, following the discovery that oxytocin, administered from the second to the sixth day, inhibits normal luteal function in the cow (by a mechanism not yet explained), this reaction has been used extensively by Hansel and his colleagues in studies of luteal maintenance that demonstrate the luteotropic activity of LH (Hansel and Seifart, 1967).

The corpora lutea of the pig survive for about 2 weeks, unaffected by either hypophysectomy or pituitary stalk section, in either pregnant or nonpregnant animals (du Mesnil du Buisson and Denamur, 1968). Beyond this time, pituitary hormones are required for the maintenance of the corpora lutea during pregnancy. Up to about day 40, LH seems to predominate in the luteotropic complex, and thereafter the role of prolactin becomes increasingly important (du Mesnil du Buisson and Denamur, 1968). The corpora lutea survive pituitary stalk section, but not hypophysectomy, after about day 70. Luteal function can also be enhanced by estrogen administration in the last half of gestation and this effect, it has been suggested, is due to stimulation of prolactin secretion, an effect recorded in a number of species (Pasteels, 1970). It may be significant that large amounts of estrogen are secreted by the pig placenta in late pregnancy (Lunaas, 1962; Rombauts, 1962; Raeside, 1963); reference is made in the following section (III,C) to the chorionic production of estrogen early in pregnancy.

In the human female, the pituitary, or the ovaries, can be dispensed with relatively early in gestation (see Tables I and II). In nonpregnant women the corpus luteum begins to regress about 14 days after its formation. Its maintenance beyond that time during pregnancy appears to depend chiefly on human chorionic gonadotropin (HCG), which appears in the urine about the tenth day and can be detected in the blood by the fifteenth day (Goldstein et al., 1968). Another chorionic factor, believed to synergize with HCG, is human (chorionic) placental lactogen (mammotropin, HPL), which also appears early in pregnancy (see Section III,D). It is puzzling, however, that the secretory activity of the corpus luteum declines from about day 50 (Yoshimi et al., 1969) and luteal regression begins soon afterward, whereas both HCG and HPL concentrations are still rising at that time. The ovaries can be removed, without interrupting pregnancy, somewhat earlier; the operation has been performed at least as early as the thirty-fifth day (see Table I). On the other hand, the corpus luteum, although markedly regressed, does not cease to secrete progesterone before the end of pregnancy (LeMaire et al., 1970).

C. The Role of the Conceptus

By transforming the cyclic corpus luteum into the corpus luteum of pregnancy, it is apparent that the gravid uterus exerts a sustaining influence on the corpus luteum during pregnancy which it does not exert during the cycle of the nonpregnant animal. Alternatively, one may suppose that during pregnancy the uterus is restrained from exerting a luteolytic effect that would otherwise cause luteal regression. This latter interpretation seems appropriate in the case of species in which luteal persistence is markedly prolonged by hysterectomy, as in the pig and guinea pig, already referred to (Section III,B).

The effect of the embryos on the persistence of the corpus luteum has been studied, principally in sheep, in experiments involving the transfer of an embryo before implantation to the uterus of a nonpregnant animal at the equivalent stage of the ovarian cycle (Moor and Rowson, 1966a,c; Moor, 1968). It was found that when the transfer was made on or before the twelfth day of the cycle the embryos would survive and the recipient ewe's corpus luteum would persist for the normal duration of pregnancy. The presence of the embryo apparently prevents or overrides the luteolytic effect which the uterus normally exerts in the nonpregnant ewe. Conversely, if the embryo is removed from a pregnant ewe by the twelfth day postcoitum luteal regression follows and ovulation recurs at the normal time (Moor and Rowson, 1966b). In a further series of experiments, in which embryos were transferred to surgically isolated segments of the uterus (Moor and Rowson, 1966c; Niswender and Dziuk, 1966), corpora lutea were maintained only in the ovary adjacent to the isolated gravid uterine horn; embryos confined to the contralateral horn were entirely without effect on the life-span of the luteal bodies. This unilateral effect can only be demonstrated when the embryo is confined to one uterine horn in such a way as to prevent any tissue or fluid from passing directly into the nongravid horn.

The luteolytic effect of the uterus thus demonstrated in the sheep has also been attributed to several other species, including the pig, cow, and guinea pig (see Anderson *et al.*, 1969). In all these species, the effect appears to involve a local pathway, as yet undetermined. The nature of the luteolytic substance is still open to doubt, but its role has been attributed to prostaglandin $F_{2\alpha}$ in the sheep (see Goding, 1973). Gleeson, Thorburn, and Cox (1974) have demonstrated a marked rise in the concentration of prostaglandin F in the plasma of nonpregnant sows from the eleventh to the sixteenth day after ovulation, coinciding with the decline of progesterone output by the corpora lutea. It is not yet known whether the presence of blastocysts suppresses the production or the release of the uterine prostaglandin or counteracts its effect on the corpus luteum, but it has been shown that the trophoblast has pronounced enzymatic capacity for steroidogenesis, particularly the synthesis of estrogens, at this time (Huff and Eik-Nes, 1966; Perry et al., 1973). It has also been shown that the uterine endometrium and myometrium of the pig are capable of conjugating the estrogens so produced and that the luteal tissue can hydrolyze the conjugated steroids (Perry et al., 1975). The latter authors suggested that this series of steroid transformations, demonstrated in vitro, may be involved in the prolongation of luteal activity in the pregnant pig, bearing in mind that exogenous estrogen had earlier been shown to exert a luteotropic action in this species at this time, both in vivo (Gardner et al., 1963) and in vitro (Goldenberg et al., 1970). Histochemical evidence of steroidogenic activity in preimplantation stages of development in the rat has been described by Dickmann and Dey (1974), and here too it is tempting to suppose that it may be concerned in the "maternal recognition of pregnancy" (see Heap and Perry, 1974), especially as estrogen has long been known to be luteotropic in the rat, at least in certain conditions (see Section F, below). Thus in the pig the early conceptus may have both an "antiluteolytic" effect, in suppressing the production or the release of $PGF_{2\alpha}$, and a luteotropic effect, through the direct action of estrogen on the ovary. Antiluteolytic and positive luteotropic effects have been more clearly distinguished in the guinea pig (Heap et al., 1973).

Loeb (1907) was the first to demonstrate that deciduomata prolong the life of the corpus luteum in the guinea pig and the same has been shown to be true for the rat (Ershoff and Deuel, 1943) but not for the mouse (Kamell and Atkinson, 1948). Furthermore, since a quantitative relationship has been shown to exist between the amount of decidual tissue present and the length of prolongation of pseudopregnancy in the rat (Velardo et al., 1953), it might be supposed that the products of conception prolong the life-span of the corpus luteum by distending the uterus (Bourg and Spehl, 1948). However, the secretory functions of the corpora lutea are not affected when the fetuses are exteriorized (rabbit: Courrier, 1941, 1945; cat: Amoroso, 1956) or when the majority of the fetuses are eliminated altogether, thereby reducing the distension of the uterus (see Amoroso and Finn, 1962, p. 468; Fig. 4). Supravaginal ligation of one uterine horn of the rat shortly before the expected date of delivery (thereby prolonging the distension of the uterus) was not effective in delaying parturition or in maintaining the corpora lutea beyond normal term (Klein and Mayer, 1942b). On the other hand, Selye (1934) was able to maintain corpora lutea of pregnancy and prolong "gestation in rats by substituting paraffin for the products of conception." The paraffin was said to stimulate the anterior pituitary gland,

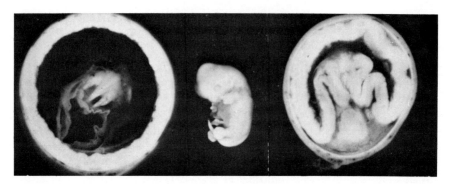

Fig. 4. The conceptus of the cat, 30 days postcoitum. On the left, a section through a normal specimen; on the right, a section through one from which the fetus had been removed at 21 days, showing the continuing growth of the placenta. The volume of the placental band was similar in the control and experimental specimens. A normal, 30-day fetus is shown (center). × 1.5.

causing persistence and function of the luteal tissue. The link between the stimulus arising in the uterus and the adenohypophysis was assumed to be neural (Moore and Nalbandov, 1953; Nalbandov *et al.*, 1955).

These observations suggest an interrelationship between the uterus and the pituitary gland, and inasmuch as deciduoma formation is invariably induced by mechanical or electrical stimulation it is possible that a reflex mechanism, not unlike that associated with the postcoital discharge of ovulating hormones, might be involved.

Although there is considerable evidence for neural mediation in the production of luteotropic substances by the adenohypophysis (cow: Hansel and Trimberger, 1951; rat: Everett, 1952; rabbit: Markee *et al.*, 1952), hypophysectomized mice (Newton and Beck, 1939; Deanesly and Newton, 1940; Newton and Richardson, 1940) and rats (Pencharz and Long, 1931; Gardner and Allen, 1942) remain pregnant when the operation is performed during the second half of pregnancy. It is therefore unlikely that the pituitary plays any considerable role as a central reflex pathway for maintaining corpora lutea in the later stages of gestation in these species. Hence, by exclusion, it may be inferred that the organ chiefly responsible for furthering luteal function is the placenta.

D. Placental Luteotropins

Astwood and Greep (1938) prepared luteotropic extracts from rat placentas, and their original findings have been confirmed. Lyons (1951) and his collaborators (Averill *et al.*, 1950; Ray *et al.*, 1955) were able, in ovariectomized–hypophysectomized animals, to elicit mammotropic, lacto-genic, and pigeon crop-stimulating activity with injections of extracts of 12-day rat placentas. They also reported that, when compared with pituitary mammotropin (prolactin), the rat placenta is potent with regard to its luteotropic, mammotropic, and lactogenic properties, but weak in its crop-stimulating activity.

Heap and Perry (1974) discussed the role of human chorionic gonadotropin (HCG) in relation to the maternal recognition of pregnancy. Although this hormone is generally regarded as being involved in maintain-ing the corpus luteum in early pregnancy, it is not certain that it is responsi-ble for arresting luteal regression, which occurs about 10 days after ovula-tion in the nonpregnant woman. HCG is first detectable in plasma by radioimmunoassay at a time which coincides with, rather than precedes, that when plasma progesterone concentrations in pregnancy diverge from those of the menstrual cycle. In the rhesus monkey, however, the temporal relations between CG production and changes in luteal function have been more precisely defined (Hodgen et al., 1972; Knobil, 1973) and it has been shown that the first detectable increase in plasma or urinary CG coincides with the "rescue" of a declining corpus luteum. That this is not the sole function of CG in these species is suggested by the fact that, in both of them, the concentration of placental hormone in the blood continues to rise, and reaches very high levels, after luteal progesterone secretion has reached its maximum (see Amoroso and Perry, 1975).

A polypeptide hormone chemically and immunologically similar to human pituitary GH, with both prolactin- and GH-like activity, was demonstrated in the human placenta by Ito and Higashi (1953), who extracted it by the method employed by Lyons (1937) to extract prolactin from sheep pituitary glands. Ito and Higashi assayed the substance by its effect on the pigeon crop gland and showed that it was present in the "mature" placenta and most conspicuous during the second to third month of pregnancy. They subsequently devised an improved method of extraction (Ito and Higashi, 1961; Higashi, 1962) and they credited Ehrhardt (1936) with being the first to suggest the presence of a prolactinlike substance in the human placenta. This hormone, now usually referred to as human placental lactogen (HPL), or human chorionic somatomammotropin (HCS), was isolated from the serum as well as from the placenta by Josimo-vich and MacLaren (1962), Josimovich et al. (1963), and Josimovich and Atwood (1964), who emphasized its luteotropic and lactogenic activity in the rabbit, rat, and pigeon. It can be demonstrated by the twelfth week of pregnancy (Sciarra et al., 1963) and, like HCG, it is present in the syncytial trophoblast (de Ikonicoff and Cédard, 1973). HPL has now been extensively studied (Grumbach and Kaplan, 1964; Kaplan and Grumbach, 1965a,b;

Grumbach *et al.,* 1968) and it has been suggested (Josimovich *et al.,* 1969) that this placental polypeptide, rather than pituitary GH, functions as an important metabolic growth hormone in late pregnancy.

A comparable hormone has been identified in the monkey placenta (Kaplan and Grumbach, 1964). The simian hormone bears an even closer immunochemical relationship to the human placental hormone than either has to the respective pituitary GH and, like its human counterpart, it may act as a metabolic hormone during late stages of pregnancy (Kaplan and Grumbach, 1964). Friesen *et al.* (1969) state that monkey placental lactogen (MPL) is more closely allied to GH than to HPL, and is likewise more potent than HPL in stimulating growth in hypophysectomized rats.

E. The Influence of the Fetus in Late Pregnancy

The fetus is so obviously a "parasite" that the manifestations of pregnancy at any time might be suspected as being the direct result of the presence of the fetus itself. However, the removal of the fetuses, leaving the placentas, from a series of monkeys, mice, rabbits, cats (Fig. 4), or rats (see Amoroso and Finn, 1962, p. 469) did not alter the course of pregnancy as judged by the maintenance of the corpora lutea, the continued growth of the retained placentas, the maintenance of body weight, the development of the interpubic ligament, and the occurrence of parturition at the normal time. In short, the animals in these experiments remained physiologically pregnant despite the absence of the fetuses. It would be a mistake, however, to imagine that the fetus is altogether a passenger. Even in the mouse, in which placental pregnancy so closely resembles the normal, the act of parturition may be sluggish (Newton, 1935), mammary development is sometimes impaired, and lactation occasionally fails, while pubic separation does not occur (Newton and Lits, 1938). In rats, only an occasional interference with parturition was reported, although pregnancy was usually prolonged (Selye *et al.,* 1935a). The role of the ovary in the initiation of parturition is discussed more fully in Section IV,I, below.

F. Steroid Control of Luteal Function in Pregnancy

It is well known that estrogens can maintain the activity of the corpus luteum in the rabbit, mouse, and rat in the absence of the pituitary (see Amoroso and Finn, 1962). That such luteal tissue exerts its normal secretory activity is shown by the fact that in nonpregnant animals it produces its typical effects on the endometrium and in pregnant animals it maintains normal gestation. Estrogen treatment has likewise been shown to have a beneficial effect on the corpora lutea of unmated rats and gilts causing follicular luteinization when injected during the cycle (see, however, Chu and You, 1945). However, Kidder (1954), Greenstein *et al.* (1958), and Loy *et al.* (1960) were unable to demonstrate this effect in the cow, though Hammond and Day (1947) had earlier reported that corpora lutea persisted for prolonged periods in heifers implanted with stilbestrol.

The stimulating effect of injected estrogen on the growth and survival of the corpora lutea of pseudopregnant and of hysterectomized rabbits has also been reported. That estrogen-maintained corpora lutea are functional when their life-span is prolonged in pseudopregnant rabbits, is shown by uterine insensitivity to pituitrin (Heckel and Allen, 1936), by uterine histology (Westman and Jacobsohn, 1937), and also by the maintenance of the mammary glands beyond the normal duration of pseudopregnancy (Hammond, 1956). That they are functional after hypophysectomy is shown by maintenance of the uterine glands (Robson, 1937a). The work of Heckel and Allen (1936, 1939) on the prolongation of pregnancy in the rabbit with continuous estrogen treatment is of importance, primarily in showing the effect of estrogen on the corpora lutea of pregnancy, since the conceptus was dislodged, though not expelled, as soon as the injections were started.

The belief that estrogen may be essential in the control of luteal persistence and function in the rabbit has been confirmed. Keyes and Nalbandov (1967) and Keyes and Armstrong (1968, 1969) showed that pregnancy was maintained by the injection of estradiol (2–4 μg/day) into rabbits in which one ovary had been X-irradiated, thus leaving only corpora lutea and interstitial tissue, and the other ovary had been excised. Withdrawal of the estrogen caused the cessation of progesterone production and led to abortion. Scott and Rennie (1971) have described the presence of an estrogen receptor in the lutein cells of the rabbit.

The evidence in rats and mice, in contrast to that in rabbits, is equivocal. Klein and Mayer showed that appropriate doses of estradiol benzoate prolonged the life span of corpora lutea in pseudopregnant (1944) and in pregnant (1942a) rats. The treatment maintained luteal function, as indicated by vaginal mucification, and it was concluded that estrogen plays an important part in these conditions. Hain (1935) was able to prolong pregnancy in the rat for 3 to 5 days with estrogen treatment, and some living fetuses were born. Selye *et al.* (1935b) also recorded the prolongation of gestation, and luteal function, in rats injected in late pregnancy with daily doses of 500 μg estrone in oil. In mice, however, Dewar (1957) found no evidence of luteotropic activity of estrogen. Whereas exogenous estrogen has usually been found to prolong the life-span of the corpus luteum in the species so far investigated, the effects of progesterone administration have proved quite variable (see Amoroso and Finn, 1962).

Two different mechanisms of luteal maintenance have been postulated on the basis of experiments with hormones administered to intact or hypophysectomized animals. In one, the gonadotropic hormones are envisaged as exerting a direct effect on the corpus luteum (Lindner *et al.*, 1951; Desclin, 1952). According to the alternative hypothesis (Robson, 1947), the role of the gonadotropic hormone is indirect, its function being to stimulate the ovary to produce estrogen, which in turn acts directly, even locally, upon the luteal tissue (Hammond and Robson, 1951; Hammond, 1952, 1956). Thus, ovarian estrogen has a luteotropic function in addition to its effect on the accessory organs and its feedback action on the hypothalamus. According to this hypothesis, moreover, a luteotropic function is the primary role of estrogen during pregnancy, when the accessory organs are dominated by the influence of progesterone.

The work already quoted (and see also Hilliard *et al.*, 1969), provides convincing evidence that, in the rabbit at least, naturally secreted estrogen has an actual luteotropic role in maintaining pregnancy. There is little convincing evidence that the same is true of other species, even though exogenous estrogen, in sufficient doses, can be shown to be luteotropic. Indeed, the fact that both the rabbit and rat experience a postpartum estrus is strong evidence that the terminal failure of the corpus luteum occurs in the presence of increasing amounts of estrogen. Moreover, it is well established that the placentas of several rodents play an important role in maintaining luteal activity (see above, Section II,A), but there is no evidence to show that this effect is mediated by estrogen; in fact, the latter substance has not yet been found in extracts of rodent placenta, though it is abundantly present in the placenta of the sow, cow, mare, monkey, and human female (for references, see Corner, 1938; Newton, 1938; Mayer and Klein, 1955).

IV. ENDOCRINE FUNCTIONS OF THE OVARY DURING PREGNANCY

A. The Secretory Activity of the Oviduct

The mammalian oviduct is known to undergo cyclical histological changes in phase with the reproductive cycle. In general, cell height and secretory activity are maximal during estrus, intermediate during pregnancy, and minimal after ovariectomy. Likewise, demonstrations by cytochemical methods indicate cyclical fluctuations in such components as alkaline phosphatase, mucopolysaccharides, lipid, and glycogen in the tubal epithelium of a number of mammals, including the hamster, rat, cat, dog, ewe, and man (Hadek, 1955; Fredricsson, 1959; Fredricsson and Bjorkman, 1962). In the rat, the decline after ovariectomy can be reversed by the administration of estrogen (Robboy and Kahn, 1964).

Both the magnitude and timing of cyclic changes in oviductal secretion vary with the species (Greenwald, 1969). In rabbits, the most pronounced changes occur after ovulation, when the eggs acquire their remarkable mucin coat during passage through the oviduct. In women, there appears to be more secretory activity during the follicular than during the luteal phase of the cycle, and there seems to be little cyclic variation in the secretion rate in rats. Biochemical studies on the tubal fluid of rabbits indicate that the fluid contains certain metabolic substrates, notably lactate and phospholipids, which may serve the energy requirements of the spermatozoa and eggs (Bishop, 1956c, 1957), and that these, as well as the oxygen tension and pH concentration, vary in response to the estrogen and progesterone domination of the animal (Bishop, 1956a,b). It has been shown that exogenous progesterone results in a twofold increase in the glycine and serine content of the tubal fluid (Gregoire et al., 1961). The same authors (1962) also demonstrated an increase in the inositol content of the fluid and suggested that it might serve as a reserve of carbohydrate. It must be pointed out, however, that although the secretory function of the oviduct is conditioned by the ovarian hormones, little has been learned of the precise manner in which this is brought about, or how the secretions may modify, positively or negatively, the natural processes of gamete survival, fertilization, ovum transport, and blastocyst development.

B. The Transport of Eggs through the Oviduct

The tendency of the ovum to pass more rapidly through the upper portion of the oviduct and more slowly through the lower portion has been observed in several species of mammals (see Amoroso and Finn, 1962; Blandau, 1969). Sobotta (1895) long ago emphasized that the rather constant 3- to 4-day tubal sojourn of the ovum in the majority of species is independent of the length and caliber of the tube, the length of gestation, and the ultimate size or degree of development attained. The opossum (Hartman, 1939), the cat (Gros, 1936; Amoroso, 1952), and the bitch (Bonnet, 1907; Griffiths and Amoroso, 1939) are apparently exceptions, the eggs reaching the uterus 1, 5, and 7 days, respectively, after ovulation.

Direct observation of the movements of rat ova (within the periovarial space and in the oviduct) suggests that entrance of the egg into the oviduct is effected primarily by the action of cilia, the motility of which has been shown to be stimulated by estrogens. However, while ciliary action is the primary factor in effecting entrance of the egg into and through the ampulla, the action of the tubal musculature controls further advancement (Alden, 1942b). Muscular action produces a hesitant advance of the tubal contents, the so-called "Pendulum Bewegung" observed by von Mikulicz-Radecki and Nahmmacher (1925) and von Mikulicz-Radecki (1930).

That the tubal musculature may be subject to hormonal action is indicated by various kinds of evidence. Thus the tube-locking of mouse ova with estrogen injections of proper dosage and the acceleration of passage in cases of superovulation in the rabbit suggest that ovarian hormones control egg transport (see Deanesly, 1966); small amounts of estrogenic substances inhibit egg passage, while progesterone accelerates it (see, however, Alden, 1942a).

Additional support for the view that the loss of transportational function is due to lack of hormones normally secreted by the corpus luteum is provided by the observations of Long and Evans (1922) and those of Burdick *et al.* (1940). The former noted that tubal eggs of unmated rats (animals without functional corpora lutea) never reach the uterus, while Burdick *et al.* (1940) have shown that testosterone propionate causes the retention of blastocysts in the oviducts of mice injected with this hormone. Acceleration of ova through the tubes occurs in pregnant and pseudopregnant rabbits in which a second ovulation has been experimentally induced; ova of the second or induced ovulation reach the uterus in less than $2\frac{1}{2}$ days following injection of extract of pregnancy urine (Wislocki and Snyder, 1933). More recently, Chang and his collaborators have shown that administration of progesterone or synthetic progestins daily to female rabbits for 3 days before insemination and the induction of ovulation caused not only rapid egg transport from the tube to the uterus (Chang, 1966; Chang and Hunt, 1970) but also inhibition of fertilization (Chang, 1967, 1969; Chang and Hunt, 1970). Boling (1969) remarked that, in attempting to elucidate the role of estrogen and progesterone by hormonal treatment of ovariectomized animals, it is difficult so to gauge the dosage as to simulate the normal endocrine balance. Reviewing the many experiments in which one or both of the ovarian steroids have been so administered, he concluded that "near the time of ovulation they work together in the induction of muscular contractility."

Further evidence that the corpora lutea have a decisive role in the nutrition and transport of ova during the period preceding implantation is provided by experiments involving their extirpation. Thus the passage of ova through the tubes, which normally requires 3 days in the rabbit and mouse, may be delayed by removing the ovaries (rabbit: Corner, 1928; Ancel and Vintemberger, 1929; mouse: Whitney and Burdick, 1939), or the corpora lutea alone (Corner, 1928). Corner studied the rate of transport of the rabbit ovum after partial or total ovariectomy and concluded that ". . . the uterine proliferation is necessary not only for implantation, but for the nutrition and protection of the free blastocysts during 3 or 4 days between arrival in the uterus and implantation" (Corner, 1928, p. 81).

In the rat, it would seem that although ovarian hormones may, and apparently do, affect the activity of the tubal musculature, ovariectomy following ovulation neither impedes development nor prevents delivery into the uterus, at essentially the proper time, of normally cleaved ova. Moreover, since blastocysts occur, in the rabbit and in the rat, both in the uterus and in the oviduct following bilateral removal of the ovaries, it appears that the dependence of the developing embryo upon ovarian secretions may not become critical until immediately preceding implantation, when the absence of progestational proliferation, and perhaps full blastocyst expansion, precludes implantation and further development (Davies and Hesseldahl, 1971).

Fertilization usually appears to occur in the ampullary region of the Fallopian tube, so that, before the zygote is conveyed toward the uterus, the oviduct has to provide for the transport of spermatozoa in the opposite direction. At this time, or earlier, the female gamete has to be transported from the surface of the ovary through the fimbria of the oviduct and into the ampulla. Oviduct function is, therefore, manifestly complex, and although it is clear that it is largely under the control of ovarian hormones, the mechanism of the control remains obscure. For a detailed review of the available evidence the reader should refer to the volume edited by Hafez and Blandau (1969).

C. Ovum Implantation

Ovarian hormones stimulate endometrial proliferation in preparation for the reception of the embryo. Only after the arrival of the embryo is there, in most species, appreciable further differentiation within the uterine mucosa. In those where the trophoblast is invasive, the degree of endometrial growth and cellular differentiation is considerable, and involves the development of such specialized regions as the metrial gland and deciduoma of rodents (Mahoney, 1969; Finn, 1971). It is evident that estrogen and progesterone act synergistically, rather than antagonistically, in controlling these changes (Deanesly, 1966).

Implantation is a crucial stage in the development of the young individual, and it depends heavily upon the pituitary and ovaries in all viviparous mammals. The pituitary maintains the activity of the corpus luteum up to this stage, and the steroid hormones of the ovary—estrogen and progesterone—are essential for the continued growth of the blastocysts and for the full differentiation of the endometrium. Removal of either the pituitary or the ovaries before this condition has been established prevents normal implantation and placentation (see Table I). It must be pointed out, however, that Loeb and Hesselberg (1917), Loeb (1923), and Deanesly (1960, 1963, 1971, 1972) have shown that, provided the uterus has received a minimum of progestational hormone, the whole process of ovo-implantation can occur in the guinea pig in the absence of ovaries or of exogenous hormones, and that the embryos will continue to develop normally for several days. Their subsequent failure appears to be due to inadequate differentiation of the uterine tissues accompanied by extensive hemorrhages rather than to intrinsic factors in the embryos themselves (Deanesly, 1971, 1972). The only other records of implantation following ovariectomy are those of Buchanan *et al.* (1956), Enders and Buchanan (1959), and Enders (1966) who have reported that implantation can occur in the armadillo 1 month after ovariectomy if the operation is carried out during the 4-month period when the eggs are normally delayed in the uterus.

Implantation can be accomplished only during the brief period when both the uterus and the blastocysts have reached, but have not proceeded beyond, the appropriate stage of modification or development. This precise timing is controlled by the ovarian hormones. Chambon (1949a,b) showed that in rats ovariectomized 2 days after ovulation, free uterine blastocysts could be maintained in a viable condition, without implanting, by treatment with progesterone. This observation was followed up in investigations in which estrogen was shown to be necessary for implantation in the rat. By varying the time of ovariectomy, it was shown that the necessary "estrogen surge" (Shelesnyak, 1960) occurs over a brief, and precisely timed, period which varies according to the diurnal light cycle (Psychoyos, 1967). The concept of such a surge implies a dependence on cyclic or varying LH release from the anterior pituitary, and it has been suggested that the surge coincides with the time when such a gonadotropin release would occur in the course of the normal cycle. It seems probable that the delay of implantation induced by suckling a large litter (see below) is due to the suppression of LH release by the effect of the suckling stimulus on the hypothalamus (Rothchild, 1960). Zeilmaker (1964) has shown that reducing the number of young suckled, before the 4th day but not later, allows implantation without delay.

D. Delayed Implantation

The interval between mating and implantation is 5–6 days in the rat, 9 days in the monkey, 12–17 days in the cat (Amoroso, 1952), and 18 days in the plains viscacha (Roberts and Weir, 1973). In some other species there is normally a long period of "delayed implantation" during which the embryos lie dormant in the uterus until favorable uterine conditions for implantation have developed. This phenomenon of delayed implantation was first observed in deer (see Short and Hay, 1965), and has since been recorded as a normal occurrence in a wide variety of mammals (see Enders, 1963; Wright, 1966). Excluding the marsupials (which are dealt with in a subsequent section), Deanesly (1966) lists some eighteen species of which this is true, and the phenomenon has been widely studied in recent years (Enders, 1963).

Among eutherian mammals, the condition appears to be widespread among members of the Mustelidae. Experiments on the pine marten and mink suggest that artificial illumination at the end of the hours of daylight can hasten implantation, and so shorten the total duration of pregnancy (see Eckstein *et al.,* 1959; Canivenc, 1960; Amoroso and Marshall, 1960). There seems little reason to doubt that the effect of light is mediated via the pituitary on the corpora lutea, and that the uterus needs to be conditioned in some special way, not yet understood, before implantation may proceed (Mayer, 1959a). If we assume this much, the question still remains as to the hormone or hormones responsible.

Hammond (1951) has shown that implantation in the mink is not simply a matter of the supply of progesterone. Accompanying implantation in this species, he postulates the presence of an extraovarian factor acting upon the uterus and, clearly enough, its production or activity seems to be regulated by light. In the opinion of Yeates (1954), the existence of some such special uterotropic hormone is suggested by, and probably accounts for, the fact that, whereas implantation is satisfactory in ewes whose sexual season has been reversed by light treatment (Yeates, 1949), it is defective in animals in which ovulation and estrus have been induced by gonadotropin injection during the anestrus (Robinson, 1951).

In some species, estrus recurs very soon after parturition, and lactating mothers may become pregnant while still nursing young litters. This does not affect the course or duration of pregnancy in the rabbit, nor does it in the European hare (Martinet *et al.,* 1970) or the South American casiragua, an hystricomorph rodent (Weir, 1974); in both these species estrus and mating may actually precede parturition. In rats and mice, pregnancies resulting from (immediately) postpartum mating are usually prolonged because implantation is delayed. Such species may be said to show facultative embryonic diapause, unlike the obligate diapause of those species in which there is always a considerable period of delay. Gestation is not prolonged in mice unless more than two young are suckled, and in rats (Brambell, 1937) implantation is not always delayed unless the female is suckling five or more young; in both species there is a degree of correlation between the number of young being suckled and the duration of gestation. There is no evidence that the number of embryos in the uterus has any influence on the delay of implantation. A similar prolongation of gestation during lactation has also been described in some other families of rodents and in some insectivores (see Eckstein and Zuckerman, 1956).

This delay of implantation in rats and mice cannot be due to the functional failure of the corpora lutea but it is highly probable that normal implantation in the rat depends upon the delicately balanced activities of several hormones, chief among which are the steroid hormones of the ovary and the gonadotropic hormones of the pituitary. Thus, it has been shown by Krehbiel (1941a) that in nursing rats, made unilaterally pregnant by ligation of one uterine tube, deciduomata could be produced in the sterile horn at a time when unimplanted blastocysts were present in the fertile horn. Moreover, the injection of small doses of estrogens into such animals permits the blastocysts to implant at the normal time (Krehbiel, 1941b; Weichert, 1942), suggesting that something more than progestational proliferation is required to produce the physiological conditions favorable for implantation. Canivenc and Laffargue (1956), Cochrane and Meyer (1957), and Whitten (1958) all believe that the prolongation of gestation in lactating mice is due to an inhibition of the blastocyst by progesterone not balanced by estrogen, and Amoroso and Finn (1962) indicated some of the supporting evidence. Estrogens are effective at low doses and are believed to produce their effects either by acting synergistically with endogenous progesterone or by reinforcing the activity of the corpus luteum. Although the exact means by which estrogen activates the corpus luteum is unknown, there is evidence that in the rat it may do so through its effect on the pituitary by causing the release of prolactin (see Meites and Turner, 1948a,b).

E. The Role of the Corpus Luteum after Implantation

The effect of loss of luteal tissue or loss of all ovarian tissue on the course of gestation has been studied in a number of vertebrate species (see Table I and references in Courrier, 1945; Mayer, 1953; Amoroso, 1955a; Deanesly, 1966), and the importance of the corpus luteum secretion during implantation has been stressed in these and other reviews on the endocrinology of early pregnancy (Kehl, 1950; Courrier and Baclesse, 1955; Mayer, 1959a,b; Eckstein *et al.*, 1959; Amoroso, 1960a; Amoroso and Finn, 1962; Amoroso and Porter, 1970).

The indispensibility of the ovaries during early pregnancy has already been emphasized. Removal of the ovaries or the corpora lutea during the later stages of pregnancy does not disturb the course of gestation in some species, notably the human female, monkey, horse, guinea pig, sheep, pig, and cat. In others, including the hamster, rabbit, mouse, and goat, abortion or resorption invariably occurs following such operations (see Table I). In the latter group, evacuation of the uterus is not, however, always immediate (see Amoroso and Finn, 1962, p. 483).

Whether or not the ovaries can be dispensed with in the later stages of pregnancy depends on the production of progesterone from an alternative source which, as far as is known at present, is invariably the placenta. The only other source which has been suggested is the adrenal gland (Balfour *et al.*, 1957). For example, in the guinea pig, placental production of progesterone was indicated by the high levels of the steroid in uterine venous plasma, compared with those in systemic plasma, in ovariectomized pregnant animals (Heap and Deanesly, 1966). In the human female (Venning, 1957) and in the mare (Suzuki, 1947), the life span of the placenta coincides with the excretion of large quantities of pregnanediol, indicative of progesterone metabolism. These aspects are considered elsewhere in this volume and also in the reviews by Diczfalusy and Troen (1961), Amoroso and Porter (1970), and Fuchs and Klopper (1971). The earlier reviews can also be approached in the account given by Amoroso and Finn (1962).

F. Ovarian Hormones and Mammary Growth

It is generally recognized that the mammary duct system present at birth grows at the same rate as the rest of the body (isometric growth) until about the time of puberty, when it becomes significantly more rapid (allometric growth) under the influence of ovarian hormones (Nelson, 1936). The pioneer work of Turner (1939) and his co-workers led to the belief that estrogens are mainly responsible for duct growth, and that estrogen and progesterone together are necessary for normal lobule–alveolar development, but later work has demonstrated that in some species (goat, cow, sheep, though not the bitch) considerable, but usually abnormal, growth can be produced with estrogen alone (see Cowie, 1971). Mixner and Turner (1943) and Cowie *et al.* (1952) reported that in the goat estrogen alone caused the development of udders often characterized by abnormally large, even cystic, alveoli, whereas combinations of estrogen and progesterone produced growth more nearly comparable to that seen in normal udders during gestation (Sykes and Wrenn, 1951; Benson *et al.*, 1955).

Although the guinea pig was previously regarded as one of those relatively few species whose mammary glands respond to estrogen with complete growth of both the lobule–alveolar and duct components (Turner, 1939; Folley and Malpress, 1948), Benson *et al.* (1957) have amassed evidence indicating that progesterone acts synergistically with estrogen in this animal. Utilizing both objective and subjective methods of assessment, they reported that over a wide range of estrogen dosage greater mammary development was elicited in the spayed guinea pig by estrone plus progesterone than by estrone alone. They also noted that complete mammary development did not occur with estrogen alone as previously reported (Turner and Gomez, 1934; Nelson, 1937) although considerable growth of lobules with normal structure was obtained (Folley, 1952; Cowie, 1957). They concluded that the absolute doses of the two hormones supplied are more important than their ratios. The estrogen:progesterone ratio had no significance per se, since administration of the same ratios, but different doses, of estrogen and progesterone resulted in differences in mammary responses which varied both quantitatively and qualitatively (see Cowie, 1971). Small doses of progesterone injected simultaneously with estrogens produce optimal lobule-alveolar growth of the mammary gland in the mouse and rabbit. Relatively large doses of progesterone are required to elicit a comparable reaction in the rat, though combinations of estradiol and progesterone which do not produce lobulation of the mammary parenchyma of the ovariectomized rat will do so when ovarian extracts containing relaxin are added to the treatment. Relaxin has also been shown to influence mammary growth in other species of laboratory animals. Relaxin may slightly increase the percentage of mice showing lobule–alveolar responses to simultaneously administered estrogen and progesterone, and it potentiates the action of estrogen on the alveolar system of the guinea pig. There is insufficient evidence at present to indicate that it may also affect duct growth (see, however, Steinetz *et al.*, 1959).

There is some conflict of evidence as to whether or not ovarian hormones or other steroids will promote mammary growth in the absence of the pituitary (see Folley, 1952; Lyons *et al.*, 1958). Asdell and Seidenstein (1935) and Fredikson (1939) reported growth of mammary ducts and alveoli in completely hypophysectomized rabbits receiving estrogen and progesterone, whereas it has been shown that these hormones did not induce extensive proliferation of the mammary gland in the hypophysectomized mouse (Trentin and Turner, 1948; Nandi, 1958, 1959) or rat (Uyldert *et al.*, 1940; Trentin and Turner, 1948). Leonard (1943) evoked limited mammary growth in hypophysectomized rats receiving estrogen and testosterone, singly or in combination, but only if the treatment began immediately after the operation. The importance of pituitary mediation is also suggested by the experiments of Cowie *et al*. (1968), who showed that mammary growth, and lactation, could be induced in ovariectomized goatlings by regular "milking" but that this response could not be evoked after pituitary stalk section.

1. Pregnancy

Bouin and Ancel (1909) and Ancel and Bouin (1911) were the first to demonstrate that the corpora lutea were responsible for mammary development during pregnancy and pseudopregnancy in the rabbit. They showed that removal of the corpora lutea by excision or cauterization prevented mammary development, and these observations have since been confirmed in a number of species by various workers (Folley, 1952). The initial growth changes of the mammary gland during early pregnancy depend, to a large extent, upon the degree of lobule–alveolar development attained during prior sex cycles. Sod-Moriah and Schmidt (1968), using a microphotometric method of measuring the DNA content of individual cells, have shown that proliferative activity is high during the first half of pregnancy in the rabbit, and decreases thereafter until a day before parturition, when there is a sudden increase which continues during parturition, followed by a decrease on the first day of lactation.

Despite the evidence of Ray *et al*. (1955) and of Lyons *et al*. (1958) indicating a complex of endocrine substances in the control of the growth of the mammary parenchyma, it remains a basic postulate that growth results from hormonal stimuli set in train by one or both of the ovarian hormones. Numerous experiments on several species of animals show, however, that the continuing presence of the ovary is not essential for full differentiation of the mammary gland. It would be natural to suppose that the pituitary played some part in the process (see Leonard, 1945), but, as shown below, full mammary development can be obtained during pregnancy after hypophysectomy. Thus, in circumstances in which the pregnant state is terminated through lack of progesterone (and possibly of estrogen) the presence of living placental tissue is sufficient to determine mammary growth (Newton, 1949; Mayer and Canivenc, 1950).

The effect of hypophysectomy or section of the pituitary stalk upon the development of the mammary gland during pregnancy has been described for several species, including cat, rat, guinea pig, mouse, rhesus monkey (see Amoroso and Finn, 1962 for references), sheep, and goats (Cowie and Tindal, 1960; Cowie *et al*., 1963, 1964a,b). Hypophysectomy during pregnancy in rats and mice results in involution of the mammary gland only when the placenta is removed at the same time, but lack of the anterior pituitary promptly makes itself felt after parturition, when lactation fails to occur, or at the most, makes a transient appearance. It is thus clear that the placenta, in rats and mice (and probably in the monkey and goat), plays a special part in mammary development which is largely independent of both ovaries and pituitary. It is perhaps significant that these are species in which there is evidence of the production of a prolactinlike or mammotropic hormone by the placenta (see Section III,D); lactogenic activity has been detected in the blood of goats in the late stages of pregnancy by radioimmunoassay (Bryant et al., 1968) and by organ culture assay procedures (Brumby and Forsyth, 1969). In sheep, although a number of animals continued to term and delivered their lambs spontaneously after pituitary stalk transection, "there was no palpable development of the udder, and no milk could be expressed from the teat" (Cowie et al., 1963).

The question as to why lactation is not normally initiated during pregnancy still remains unanswered. The concept which has probably gained widest acceptance is that the ovarian hormones, particularly estrogens, inhibit the secretion of prolactin by the pituitary (Nelson, 1936). This view is held despite the fact that no direct evidence has yet demonstrated that the ovarian hormones can reduce prolactin either in the pituitary or in the circulation. On the contrary, such studies have consistently shown that ovarian hormones, particularly estrogens, increase the prolactin content of the pituitary (Meites and Turner, 1948a,b). Kuhn (1969a,b) has shown that the natural lactogenic stimulus occurs some 24–30 hours before parturition in the rat, and results from the conversion of progesterone to 20α-dihydroprogesterone. Meites and Sgouris (1953) concluded that estrone and progesterone together can effectively inhibit the milk-secreting action of prolactin on the mammary gland of the rabbit, whereas progesterone alone is ineffective and estrone alone only slightly effective in this respect (see, however, Ray et al., 1955; Lyons et al., 1958).

2. Pseudopregnancy

The duration of pseudopregnancy (Ancel and Bouin, 1911) is roughly half that of gestation in the rabbit, rat, and cat and extensive growth of the mammary alveolar system has been observed during this period in these species. In the ferret (Hammond and Marshall, 1930) pseudopregnancy is equivalent to pregnancy in duration and the mammary changes are similar in both conditions.

There is no active luteal phase in the absence of copulation in the foregoing species, but in the bitch and in the vixen, and in the two marsupials, *Dasyurus viverrinus* and *Didelphis virginiana*, ovulation is spontaneous and in the absence of conception it is followed by a luteal phase (pseudopregnancy) which lasts as long as gestation, whether or not copulation occurs. During the prolonged metestrus in the bitch, the vixen, the marsupial cat *Dasyurus*, and the opossum *Didelphis* the mammary glands proliferate in a fashion equivalent to their development during pregnancy. In the case of the bitch, lactation which allows the suckling of a litter may occur at the end of pseudopregnancy.

G. Ovarian Activity in Adrenalectomized Pregnant Animals

It has long been known that pregnancy prolongs the survival time of adrenalectomized animals and temporarily relieves the symptoms of Addison's disease in the human female, and that the weight of the fetal adrenals is increased by adrenalectomy of the mother (see Amoroso and Finn, 1962 for references). Firor and Grollman (1933) and Jost *et al.* (1955) in the rat, and Billmann and Engel (1939) in the bitch, suggested that the enlarged adrenals of the fetus can secrete sufficient amounts of cortical hormones to maintain the life of the adrenalectomized mother during pregnancy. Christianson and Jones (1957), however, found no evidence that fetal adrenal secretions affect the maternal organism in the rat. They concluded that the enlargement of the fetal adrenals. after adrenalectomy of the mother, was due to activity of the fetal pituitary. There is considerable evidence in support of the alternative hypothesis, that adrenalectomy is better tolerated during pregnancy than at other times because of the activity of the (maternal) ovaries (Parkes and Deanesly, 1966). Pseudopregnancy and administration of progesterone have been shown to maintain adrenalectomized animals in good health and it has also been found that intensive stimulation and luteinization of the ovaries by gonadotropins increases the survival period in the rat. Moreover, in view of the placental secretion of a luteotropic substance (RPL) by the rat's placenta, already alluded to (Section III), it is probable that the sustaining effect of pregnancy in adrenalectomized rats is chiefly due to the luteotropic action of the placenta on the ovary.

H. The Role of the Ovary in Pelvic Relaxation

Skeletal adjustments in pelvic architecture which facilitate the birth of young and which do not involve any endocrine mechanism are found in a number of animals (see Amoroso and Finn, 1962). Pelvic adjustments, which are definitely known to be conditioned by hormones and reversed after parturition, have been described only in the guinea pig and mouse. In the female pocket gopher, resorption of the pubic symphysis occurs at the first estrus, apparently under the influence of ovarian (but not luteal) hormone(s). The process is not reversible, and so leads to a sexual dimorphism in this respect among adult animals (Hisaw, 1925; Hisaw and Zarrow, 1950),

It is now well established (Hall, 1960; Deanesly, 1966) that the reversible effects seen in guinea pigs and mice are largely attributable to the action of the polypeptide hormone relaxin present in the ovaries (especially the corpora lutea) and the placenta of many mammals. Although pelvic relaxation is the most conspicuous effect of relaxin in species where this phenomenon occurs, the hormone has been shown to exert many other effects in a wide range of species. It is capable, for example, of inhibiting spontaneous uterine motility, *in vivo* and *in vitro* (Felton *et al.*, 1953; Miller *et al.*, 1957; Bloom *et al.*, 1958; Kroc *et al.*, 1958), and of inducing the delivery of living young in prepubertal or adult pregnant mice, spayed and maintained on progesterone (Smithberg and Runner, 1956; Hall, 1956, 1957). Dallenbach-Hellweg *et al.* (1966) and Hisaw *et al.* (1967) have described the effects of relaxin on the endothelium of endometrial blood vessels in rhesus monkeys where it was found to act synergistically with progesterone, and in conjunction with estrogen, in the process leading to implantation and probably the formation of blood spaces in the establishment of the placenta.

I. The Role of the Ovary at Parturition

The extent to which gonadal hormones participate in the mechanism of labor in various species is still far from settled (see Knight and Fitzsimons, 1976). The "progesterone block" theory of Csapo developed as a result of work with rabbits. This theory (Csapo, 1956, 1969; Schofield, 1957; Caldeyro-Barcia, 1965) derived from the observation that the pattern of myometrial activity varies according to whether the dominating hormone is estrogen or progesterone; the progesterone-dominated myometrium exhibits a negative correlation between frequency of stimulation and tension (negative staircase effects) whereas the reverse is true of the estrogen-dominated myometrium (positive staircase effect). Using this test on pregnant rabbits, it was found that estrogen fully dominated the uterus at the time of mating, but progesterone became the dominant hormone within 20 hours postcoitum, and remained so until about 24 hours before parturition. The uterus then became progressively more estrogen-dominated until maximal myometrial contractile power was developed at the time of labor (Corner and Csapo, 1953; Csapo, 1956; Schofield, 1957). The development of this line of research has recently been reviewed by Csapo and Wood (1968; see also Wolstenholme and Knight, 1969).

In contrast to the observations on the rabbit, Porter (1970) showed that progesterone had little effect on the spontaneous or oxytocin-induced activity of the myometrium of the guinea pig. In women too, it would appear that progesterone withdrawal may not be central to the initiation or completion of parturition, for plasma progesterone levels, which are very high during pregnancy (as they are in the guinea pig), do not clearly decline before term (Csapo *et al.*, 1971) and may even reach maximal values at the time of parturition (Fuchs, 1971).

More or less coincident with the preparturient fall in circulating progesterone levels recorded in a number of species, there is an equally or even more dramatic rise in estrogen levels. Estrogens reach their highest levels at parturition in women (Roy and Mackay, 1962), sheep (Challis, 1971), goats (Challis and Linzell, 1971), and rats (Yoshinaga *et al.*, 1969) and just before its onset in guinea pigs (Challis *et al.*, 1971) and cows (Robinson *et al.*, 1970). This steroid production is ovarian rather than placental in the rat and the goat, but mainly or exclusively placental in the other species mentioned. Species differences in the mode of action of various steroids are unrelated, however, to their site of production, except in so far as those formed within the placenta may have a local action and may, therefore, exert a greater effect than might be expected from comparisons of circulating or peripheral levels.

In recent years, work on the problem of the initiation of parturition has largely focused on the sheep, mainly as a result of the demonstration that hypophysectomy or pituitary stalk section of the fetal lamb prolonged gestation (Liggins *et al.*, 1967), and the relation of this finding to the dramatic rise in fetal adrenocortical activity in intact animals near term (see Liggins, 1969; Bassett and Thorburn, 1969). Later, it was found possible to induce parturition by the injection of synthetic corticotropin into the hypophysectomized lamb *in utero*, while premature delivery could be induced in the pregnant ewe by infusing cortisol or dexamethasone into the intact fetus *in utero* (Liggins, 1969). The evidence summarized here points strongly to the conclusion that, in sheep at least, there is a causal relationship between the fetal corticosteroids and the onset of parturition. Even in sheep, and in other species in which estrogen and progesterone production in late pregnancy is mainly or entirely placental, the ovaries are not necessarily without any function during parturition. Their role, however, is more prominent in species such as the rat and goat, where circulating progesterone levels are high during pregnancy and fall sharply during the last 2 days of gestation (rat: Wiest and Kidwell, 1969; goat: Thorburn and Schneider, 1972). Notwithstanding the fact that progesterone production is principally ovarian in the goat (Linzell and Heap, 1968) in contrast to the sheep, the fetal pituitary–adrenal component of the mechanism initiating parturition at the end of gestation is similar in the two species (Thorburn *et al.*, 1972). If the fall in progesterone concentration at the onset of parturition in the goat is a consequence of increasing fetal corticosteroid levels, as has been suggested in the sheep, it would appear that regression of the maternal corpus luteum must be controlled by changes in fetal adrenal function.

Interest in the mechanism of parturition in the sheep has been further stimulated by the evidence that fetal adrenal activity, probably under pituitary control, influences myometrial contractility in women (Turnbull and Anderson, 1969a,b; Anderson *et al.*, 1972). In the first place, gestation is often prolonged in cases of anencephaly, provided there is no hydramnios. Anencephaly is almost invariably associated with adrenocortical hypoplasia, attributed to pituitary insufficiency. Conversely, it has been shown that premature labor may be significantly associated with fetal adrenal hypertrophy. Further evidence that the fetus may play an active part in the timing of parturition in the human lies in the correlation between maternal estrogen excretion and gestation length, together with the fact that much of this estrogen is known to be of fetal origin. The pattern of hormonal activity in women during gestation and at parturition, however, is very different from that in sheep, and it is not to be expected that the latter will serve as a model for the analysis of events in the former, except in very general terms.

V. THE ROLE OF THE CORPUS LUTEUM IN THE MAINTENANCE OF GESTATION IN MONOTREMES AND MARSUPIALS

In the monotremes, regression of the corpus luteum begins shortly before the eggs are laid (Hill and Gatenby, 1926). As yet, however, their function remains ill-defined although they may be concerned with the transport of the egg, the development of the mammary gland, and the initiation of lactation (Allen *et al.*, 1939). It has been shown that the intrauterine phase of embryonic development is between 12 and 28 days (Asdell, 1964) and the endometrial glands apparently secrete a nutritive substance which is absorbed by the egg for some time before the outer shell is applied to it (Hill, 1933, 1941).

It may not be entirely fanciful to think of the endocrine function of the corpus luteum as one of facilitating the old reptilian custom of carrying ova in the uterus until the most propitious time for oviposition. Indeed, it might be supposed that, in the monotremes, this has developed into a timing device for laying the egg in coordination with the involution of the corpus luteum. Thus, in the duckbill platypus, regression of the corpus luteum begins before the egg is ready for laying, and in this species, and in the spiny anteater, intrauterine development proceeds to a stage approximately equivalent to that of the chick embryo after 40 hours incubation.

In all marsupial species so far studied, ovulation is spontaneous and the corpora lutea of the nonmated female induce changes in the reproductive tract and mammary glands of similar degree and duration to those that occur in the pregnant female (see Chapter 6). That the anatomical and histological similarities indicate an endocrinological equivalence of the pregnant and nonpregnant states is suggested by various lines of reasoning and experiment. In the quokka, *Setonix brachyurus* (Sharman, 1955b), the Tasmanian rat-kangaroo, *Potorous tridactylus* (Hughes, 1962a), the brush possum, *Trichosurus vulpecula* (Pilton and Sharman, 1962), and the red kangaroo, *Megaleia rufa* (Sharman and Calaby, 1964), no constant differences could be detected between the postestrous vaginal smears of pregnant and nonmated females. In addition, equivalent postovulatory mammary gland development occurs in pregnant and nonmated marsupials. This was first demonstrated by O'Donoghue (1911) in the Australian native cat, *Dasyurus viverrinus*, and the glands of virgin and nonmated females have since been shown to become functional at a postestrous stage equivalent to that at which parturition occurs in the Virginian opossum, *Didelphis virginiana* (Hartman, 1922, 1923), and in the brush possum (Sharman, 1962).

The functional condition of the mammary gland in the nonpregnant brush possum and the red kangaroo was demonstrated by transferring newborn young to the pouch of nonmated, nonlactating, parous or virgin females of the same or other species at the end of the uterine luteal phase. The young attached to teats, lactation was initiated, and the young were reared to maturity (Sharman, 1962; Sharman and Calaby, 1964; Merchant and Sharman, 1966). Furthermore, the normal development of early embryos transferred to the uteri of nonpregnant females demonstrated the physiological equivalence of pregnant and nonpregnant states in the quokka, *Setonix* (Tyndale-Biscoe, 1963b). Finally, compared to their pregnant counterparts, nonmated females of *Trichosurus* and *Setonix* have equivalent concentrations of pregnanediol in their urine (Sharman, 1955a; Tyndale-Biscoe, 1955; Pilton and Sharman, 1962), and Lemon (cited by Sharman, 1970b, p. 1225) has shown that progesterone concentrations are similar in the blood of pregnant and nonpregnant animals at equivalent postovulatory stages.

These experiments, while they do not prove that there is no enhanced hormone secretion during gestation in marsupials, do suggest that this is the case. Sharman (1956) concluded that pregnancy in the quokka is the equivalent of a specialized estrous cycle during which all the hormone-induced changes which normally follow cyclic ovulation are exploited by the developing embryo. He states, moreover, that in *Trichosurus* and *Setonix* the embryos may develop to term following ablation of the corpora lutea and, in *Setonix*, at least, the same may occur after complete removal of the ovaries after the seventh day of gestation (Tyndale-Biscoe, 1963a). Similarly, Tyndale-Biscoe (1970) has shown that gestation in the tammar wallaby (*Macropus eugenii*) can proceed normally to term in the absence of the ovaries if these are removed after day 8. The situation is similar in the American opossum, *Didelphis*; Hartman (1925) reported that "double ovariectomy early in pregnancy invariably causes the death of the embryos" and Renfree (1974) has shown that "the effects of ovariectomy in the opossum are indistinguishable from those in the quokka, the tammar and the brush possum, the only other marsupials so far studied."

Fetal or placental participation in the maintenance of pregnancy can probably be discounted in marsupials, since the uterine luteal phase persists in nonpregnant ovariectomized quokkas for the same time as in pregnant animals (Tyndale-Biscoe, 1963a). Among eutherian mammals, only the armadillo (Buchanan *et al.*, 1956) can dispense with the ovaries as completely as the marsupials so far investigated and the hypothesis most in accord with these results, in the marsupials particularly, is that the endometrial transformation induced by secreted or injected progesterone usually persists in the absence of corpora lutea once it has been initiated (Sharman, 1965a,b; Tyndale-Biscoe, 1963b). In marsupials, pregnancy neither extends the interval between successive estrous periods nor prolongs the uterine luteal phase. The latter is known to be induced and maintained by the secretion of progesterone from the corpora lutea (Sharman, 1965a,b). Thus, with the possible exception of the gray kangaroo, *Macropus canguru* (*M. giganteus*; *M. major* - Pilton, 1961; Poole and Pilton, 1964), and the swamp wallaby, *Protemnodon bicolor* (Sharman *et al.*, 1966), the gestation periods of all marsupials so far studied are shorter than the length of an estrous cycle (see Sharman, 1959a, 1965a), varying between 9 days in the native cat, *Dasyurus viverrinus* (Tyndale-Biscoe, 1973) and about 38 days in the rat kangaroo, *Potorous tridactylus* (Hughes, 1962a).

In the Australian native cat, *Dasyurus*, the American opossum, *Didelphis* (Reynolds, 1952), and the brush-tail possum, *Trichosurus*, the young are born at the end of the luteal phase (Table III), and parturition is coincident with the beginning of the degenerative changes in the uterine mucosa which themselves may be attributed to cessation of secretory activity by the cor-

TABLE III

Cyclic Reproductive Processes in Some Marsupials[a]

Species	Luteal phase	Estrus interval	Gestation period	RPY[b] to estrus	Pouch life	Remarks
Didelphis virginiana	13[c]	29	13			
Antechinus flavipes	32		32			Monestrous species. No postpartum or subsequent mating until following year
Dasyurus viverrinus	8–20		8–20			
Isoodon macrourus			ca. 13			
Perameles nasuta	26		ca. 12			
Trichosurus vulpecula	17	26	17.5	8	112	
Setonix brachyurus	20	28	27	26	175	
Protemnodon eugenii		30	29	28	250	Postpartum mating, leading to delayed implantation (embryonic diapause) while young in pouch
Protemnodon rufogrisea		30–31	30	28	284	
Megaleia rufa	22	35	33	33	235	
Potorous tridactylus		42	38		126	
Bettongia leseuri		20–23	22	17–21	120	
Macropus canguru		32–55	29–38	9	280–328	Fertile mating may occur when pouch young about 235 days old, and implantation may be delayed until pouch young is weaned
Protemnodon bicolor		32	35	26	256	Prepartum mating. Implantation probably delayed while young in pouch

[a] Based on data from Sharman (1964, 1965a) and Sharman et al. (1966).
[b] RPY = removal of pouch young.
[c] Time intervals given in days.

pora lutea. The history of the corpora lutea of the ring-tailed possum, *Pseudocheirus*, agrees closely with observations on *Trichosurus* (Sharman, 1959b; Hughes *et al.*, 1965). In the quokka, *Setonix brachyurus*, on the other hand, the luteal phase is of longer duration and degenerative changes in the corpora lutea of pregnant quokkas begin 20 days after estrus as in nonpregnant females (Sharman, 1955b). Thus *Setonix* differs from *Didelphis* and *Dasyurus* in that pregnancy outlasts the uterine luteal phase, and the terminal stages of pregnancy are accompanied by maturation of an ovarian follicle and proestrous changes in the uterus. Parturition in *Setonix* occurs 27 days after insemination and is followed by a postpartum estrus and ovulation, which occur at the time expected, had pregnancy not taken place (i.e., 28 days after the preceding estrus and ovulation) (Sharman, 1955a,b). In the bandicoots, *Perameles nasuta*, and *Isoodon macrourus*, the corpora lutea do not appear to degenerate until late in lactation (Hughes, 1962b) and the same appears to be true of the koala, *Phascolarctos cinereus*, and the wombat, *Phascolomis mitchelli* (O'Donoghue, 1916).

In *Didelphis, Trichosurus,* and *Perameles* the formation of a patent median vaginal canal to form the birth canal (see Tyndale-Biscoe, 1966) and its enlargement with the approach of parturition, appears to depend upon progesterone secreted by the corpus luteum, still large at this time (Tyndale-Biscoe, 1966). In some other marsupials, however, including *Setonix*, as indicated in Table III, the corpora lutea regress much earlier, but here too the ovaries are necessary for parturition (Tyndale-Biscoe, 1963a, 1973). In this connection it is tempting to speculate upon the possible role of relaxin in marsupial parturition, for birth occurred in two of nine ovariectomized quokkas that were given a relaxin preparation, whereas no success was obtained with estrogen or oxytocin (Tyndale-Biscoe, 1963a). The preparation used was "Lutrexin," an extract of pregnant sows' ovaries. In *Trichosurus vulpecula* (Table III), Tyndale-Biscoe (1966) found "a considerable amount of relaxin-like activity" in the ovary 11 days after mating but it appeared to be confined to the corpus luteum. This author remarked, in connection with the work on the quokka already referred to, that "It will be particularly interesting if relaxin, secreted by the ovaries, is shown to facilitate parturition in marsupials although acting upon different structures from those by which it facilitates birth in eutherian species." It would, as he suggests, support the idea, perhaps first formulated by Danforth (1939) and more recently expressed by Medawar (1953), in the aphorism that "'endocrine evolution' is not an evolution of hormones but an evolution of the uses to which they are put; an evolution not, to put it crudely, of chemical formulae but of reactivities, reaction patterns and tissue competences."

The influence of the corpus luteum of postpartum ovulation (Tyndale-Biscoe, 1963c) or the corpus luteum of lactation (Sharman, 1955b, 1963) on ovulation has been studied in several marsupials. Sharman (1955b) at first suggested that the corpus luteum of lactation inhibited ovulation in the lactating quokka, *Setonix brachyurus*, but later (Sharman, 1963) concluded that it was the stimulus of suckling which was important in withholding ovulation. This latter viewpoint was supported by Tyndale-Biscoe (1963b) who reported that eleven of 268 lactating quokkas did not have a corpus luteum of postpartum ovulation and showed no evidence of follicular growth during lactation. Likewise, Newsome (1964) found that some lactating red kangaroos, *Megaleia rufa*, with a pouch young, lacked a corpus luteum of lactation and hence could be presumed not to have had a postpartum ovulation; the same appears to be true for the brush possum, *Trichosurus vulpecula* (Pilton and Sharman, 1962). In the grey kangaroo, *Macropus canguru*, also, lactation apparently inhibits ovulation during at least the initial period of pouch suckling (Poole and Pilton, 1964). In contrast to the foregoing, however, observations on the red kangaroo, *Megaleia rufa*, following ablation of corpora lutea during lactation, suggest that the luteal bodies in this marsupial exert an ovulation-inhibiting effect and that ovulation in this species, unlike that in some other marsupials, is not inhibited by suckling (Sharman and Clark, 1967).

Several marsupial genera, perhaps including all the Macropodidae except *Megaleia rufa*, are characterized by the phenomenon known as "embryonic diapause" or "delayed implantation" (Table III). In these forms, estrus follows soon after parturition and is normally accompanied by fertilization. The embryo enters the uterus, but does not implant until the pouch young ("joey") is weaned or removed or accidentally lost. Tyndale-Biscoe (1963a,b) showed that the corpus luteum is essential for the resumption of blastocyst growth, and for the endometrial activity necessary for implantation, in the quokka. These functions could also be induced by exogenous estrogen and progesterone. Following ovariectomy, the dormant blastocysts remained viable for a considerable time (Tyndale-Biscoe, 1963c). Similarly, Berger and Sharman (1969a) showed that complete absence of the ovaries for periods of up to 30 days had no effect on the retention of normal dormant blastocysts in the tammar wallaby; development was resumed when progesterone was given to ovariectomized animals after the blastocysts had lain dormant in the uterus for 3 weeks.

It seems evident that lactational delay of implantation in marsupials has many features in common with the facultative embryonic diapause of suckling rats and mice (see Section IV,D). The proximate factor in both appears to be the suckling stimulus (Tyndale-Biscoe, 1963c; Sharman and Berger,

1969; Sharman, 1970a,b) and in both the delay of implantation is apparently associated with a reduced secretion of pituitary gonadotropins other than prolactin. The effect on the ovaries is different in the two groups because prolactin is not luteotropic in the marsupials (see Tyndale-Biscoe, 1963c) as it is in the rat.

In a number of eutherian species, implantation delay is a normal feature of gestation (Section IV,D). This obligatory embryonic diapause is usually associated with a specialized breeding habit and bears an apparent relation to the ecology of the species, as in animals that hibernate or in seals that come ashore only once a year. It has been argued (Ealey, 1963) that delayed implantation is an advantage to those marsupials in which it occurs, in that the presence of a viable embryo, ready to implant immediately if the pouch young should be lost, reduces the time that must elapse before replacement is possible after such an accident. A similar advantage can be postulated for the facultative diapause in nonsuckling tammar wallabies, described by Berger and Sharman (1969b).

VI. THE CORPUS LUTEUM AND THE MAINTENANCE OF GESTATION IN NONMAMMALIAN VERTEBRATES

Although the corpus luteum has generally been regarded as a mammalian organ, structures of similar origin and appearance have been found in representatives of other vertebrate classes and certain protochordates. It is, nevertheless, a matter of opinion whether the term "corpus luteum" can properly be applied to the discharged follicles in all forms on grounds of analogy with those of mammals, the endocrine functions of which alone are beyond dispute.

Among lower vertebrates, "postovulation" corpora lutea (Bretschneider and de Wit, 1947), formed by proliferation of cells of the emptied follicle, are not restricted to vivipara, but are present also in ovipara and ovovivipara. Their functional significance is, however, not clear, though they are regarded by some as being less significant as endocrine organs than the "preovulation" corpora lutea or corpora atretica produced by glandular transformation of immature follicles (Bretschneider and de Wit, 1947; Hoar, 1955). Nevertheless, as will presently be shown, these postovulatory corpora lutea are, in some species at least, capable of steroidogenic activity.

In ascidians the corpus luteum is of the preovulatory type, and Carlisle (1955) provides evidence that in *Ciona* it has a secretory function. In ovoviviparous ascidians the corpora lutea persist throughout retention of embryos in the brood pouch (Amoroso, 1955b).

A. Elasmobranchs

Oviparity, ovoviviparity, and viviparity are all encountered in elasmobranchs, and corpora lutea resulting from the transformation of follicles before or after ovulation have been described in all species so far examined (see Amoroso, 1960b; Dodd *et al.*, 1960). Opinion is, however, divided as to the part these so-called corpora lutea may play as glands of internal secretion. Indeed, evidence of their endocrine function is, as yet, so sparse that their comparability with the mammalian corpus luteum must be accepted with reservation, at least for the time being. Nevertheless, in many cases, morphological correspondence can be recognized.

Te Winkel (1950), on the basis of his own and other morphological studies, concluded that ovarian hormones present in selachians at the time of ovulation or slightly preceding it, may stimulate the secretion of the egg case by the oviducal glands. Similarly, Matthews (1955), on indirect evidence and by analogy with mammals, has suggested that luteal hormones may stimulate the hypertrophy of the uterine mucosa, thus producing the complex "trophonemata" (uterine folds) seen in ovoviviparous species (e.g., *Squalus acanthias, Torpedo marmorata,* and *Pteroplatea micrura*) as well as the more elaborate uterine changes in such viviparous forms as *Mustelus canis* and *Scoliodon* spp. (Amoroso, 1952).

These morphological observations alone are certainly not sufficient to identify the structures responsible for the biosynthesis of steroid hormones. Better substantiating evidence is provided by Chieffi's (1961, 1962) observations of a correlation between the lengthening of the trophonemata and the increasing numbers of preovulatory corpora lutea in *Torpedo marmorata*, and by the more recent studies of Lupo di Prisco (1968) indicating the production of steroid hormones by these luteal structures *in vitro*. Furthermore, Della Corte and Chieffi (1961) observed cytological changes in the hypophysis of *T. marmorata* during pregnancy, and steroid hormones, including estrone, estriol, and progesterone, as well as corticosteroids and testosterone, have been found in the circulating plasma of this species (Lupo di Prisco *et al.*, 1967). Estradiol-17β, estriol, and progesterone were found in ovarian extracts (Chieffi and Lupo, 1963; Chieffi, 1967). Lupo di Prisco *et al.* (1967) made the further interesting discovery that total corticosteroid concentration was lower in the plasma during pregnancy than at other times—the reverse of the situation in mammals.

Hisaw and Hisaw (1959) showed the mode of formation of corpora lutea to be essentially similar in oviparous, ovoviviparous, and viviparous selachians, and Hisaw and Abramowitz (1938, 1939) observed that the hypertrophy of the granulosa was associated with the ingestion of yolk in preovulatory corpora lutea and of its remnants in postovulatory follicles. Moreover, when the pituitary was removed soon after the eggs had entered the uterus in *Mustelus canis* the embryos continued to develop normally for at least $3\frac{1}{2}$ months. Because of the difficulty of ovariectomy in selachians, this operation has not been successfully performed in these fishes; hence, it has not been possible to follow the function of the corpus luteum throughout the several phases of the reproductive cycle and, in any event, their action on target organs may be different from those so far studied in mammals. The experiments of Hisaw and his collaborators nevertheless imply that pituitary stimulation is not involved in the formation of the corpora lutea and these authors suggest that the luteal bodies are not essential for the maintenance of gestation in these fishes. However, since the yolk sac of *Mustelus canis* does not fuse with the uterine wall before the fourth month of pregnancy (Chieffi and Botte, 1970), the latter conclusion must be open to doubt.

B. Teleosts

In teleosts the situation is again that corpora lutea are formed as the result of two different processes, ovulation and follicular atresia, and are present in both oviparous and viviparous species (e.g., *Sebastes*: Williamson, 1911; *Anableps*: Turner, 1938). Among teleosts, with rare exceptions, the postovulatory follicles disappear soon after ovulation. In many, atretic follicles transform into luteal structures. It has been suggested that "oestrogen synthesis was one of the responsibilities of the granulosa from the earliest stages of vertebrate phylogeny" (Hoar, 1969).

It has been suggested that the so-called preovulatory corpora lutea, which are universally present in fish ovaries, form the main mass of endocrine tissue of the teleost ovary (Bretschneider and de Wit, 1947; Hoar, 1955, 1969; Ball, 1960; Bara, 1965; Lambert, 1966; Rubin, 1968). Various conflicting statements have been made about the chemical nature and site of production of the hormone or hormones in the ovaries of viviparous teleosts (Pickford and Atz, 1957) and there is as yet no answer to the question whether the luteal bodies have an endocrine function in controlling gestation. Indeed, there is a strong case for its being unanswerable, since ablation of the corpora lutea is unlikely to be successfully performed on a pregnant teleost, and ovariectomy would in any case terminate pregnancy since the embryos are lodged within the ovary. Recently, however, Lambert (1966, 1970) has implicated the granulosa cells as a possible site of synthesis of steroid hormone or hormones prior to follicular atresia in the viviparous teleost *Poecilia reticulata*. It would appear that postovulatory corpora lutea are not formed in this species. The atretic follicles appear to have no influence upon normal gestation, and indeed they may not be necessary, since they were entirely absent in 67% of pregnant fish (Lambert and van Oordt, 1965; Lambert, 1970).

C. Amphibians

There is no doubt that structures previously called corpora atretica, but described by Bretschneider and de Wit (1947) as preovulatory corpora lutea, regularly occur in the amphibian ovary. But, contrary to the opinion of the Dutch workers, the morphological formation of the corpus atreticum is independent of pituitary stimulation (Smith, 1955). Whether the formation of steroid hormones in these structures is under the dominance of the pituitary remains to be demonstrated. Such a function is implied by the evidence of gonadotropic control of the secondary sexual characters in amphibians and in other lower vertebrates with similar ovarian histology (Tuchmann-Duplessis, 1945; Houssay, 1949, 1952).

Postovulatory changes in the follicles have been described by Burns (1932) in *Ambystoma tigrinum* and by Rugh (1935) in *Rana pipiens*, both of which are oviparous; Burns stated that there is no evidence for any functional importance for this structure. However, Galli-Mainini (1950) obtained evidence for an ovarian hormone in *Bufo arenarum* that brought about the secretion of jelly by the oviductal glands; although it was not chemically identified, other experiments by Houssay (1947, 1949, 1952), indicated the presence of a progesterone-like agent, which he believed to be necessary for the functioning of the oviducts during ovulation. It was observed that this substance appeared in the ovary before ovulation, but its site of production remained unknown (Smith, 1955).

It is now very generally assumed that postovulatory corpora lutea in these oviparous forms are rapidly resorbed. However, certain histochemical characteristics of the granulosa of the luteal structures in *Rana esculenta* and *Triturus oristatus carnifex*, such as their richness in lipids and cholesterol and the presence of some steroidogenic enzymes (Botte and Cottino, 1964; Chieffi, 1967; Nandi, 1967; Guraya, 1968, 1969; Lance and Callard, 1969) and in some cases also their ultrastructural features, e.g., the occurrence of a smooth endoplasmic reticulum (Chieffi and Botte, 1970, p. 86) suggest secretory activity. Callard and Leathem (1966) have shown that the steroid synthetic pathways in the ovaries of *Rana pipiens* and *Necturus maculosus* are similar to each other and to those known for mammals and recently demonstrated also in reptilian and elasmobranch ovaries (Callard and Leathem, 1965). Furthermore, estrogen and progesterone have been demonstrated in the ovaries of *Rana vulgaris* (Chieffi and Lupo, 1963) and estrogens have been identified in the ovaries of *Xenopus laevis* (Gallien and Chalumeau-Le Foulgoc, 1960) and in the blood of *Rana temporaria* (Cédard and Ozon, 1962). More recently, Thornton (1971) has demonstrated by bioassay the presence of a progesteronelike factor in the blood of the toad *Bufo bufo* just prior to ovulation.

Among anurans, the transformation of the ruptured follicle has been studied in the viviparous species *Nectophrynoides occidentalis* (Lamotte and Rey, 1954; Gallien, 1959; Vilter and Lugand, 1959a,b; Lamotte *et al.*, 1964) and in the ovoviviparous marsupial frog *Gastrotheca marsupiata* (Amoroso *et al.*, 1957). The history of the postovulatory corpora lutea has also been described in *Salamandra salamandra*, a viviparous urodele (Vilter and Vilter, 1960; Joly *et al.*, 1969). These luteal structures assume simple morphological characteristics and persist throughout gestation in the various forms (9 months in *N. occidentalis*; 3–4 months in *G. marsupiata*; 3–4 years in *S. salamandra*), but they regress rapidly after the young are born. Histological investigations, carried out especially on the corpora lutea of *S. salamandra* (Joly, 1964, 1965; Joly *et al.*, 1969) and on those of pregnant and pseudopregnant females of *N. occidentalis* (Xavier, 1970), have demonstrated the presence of cholesterol and moderate steroid 3β-ol-dehydrogenase activity. These cells also exhibit the typical ultrastructural characteristics of steroid-secreting cells (Joly *et al.*, 1969), and hence the potential synthesis of steroid hormones can be assumed.

The evidence here summarized notwithstanding, the function of these luteal bodies in viviparous amphibians remains doubtful. Thus, Lamotte and Rey (1954) and Vilter and Vilter (1960) credit the corpora lutea with the inhibition of ovulation and the control of oviductal secretions in *N. occidentalis* and the black salamander, respectively, and Lamotte and Rey conclude that the corpora lutea control gestation in the former species. On the other hand, Joly *et al.* (1969) are doubtful of the homology between the corpora lutea of the salamander and those of mammals, and have shown that castration of *S. salamandra* at the beginning of pregnancy does not interfere with the development of the larvae; Xavier (1970) found that ovariectomy terminated gestation in *Nectophrynoides* only when it was carried out at the beginning of pregnancy in primigravid females.

D. Reptiles

The discharged follicles of reptiles have long been known to give rise to distinct corpora lutea and such postovulatory corpora lutea, of essentially mammalian structure, have been described in several species of oviparous and viviparous lizards and snakes (see Weekes, 1934; Betz, 1963; Eyeson, 1971). Such luteal bodies have been found in at least seven species of the genera *Storeria, Potamophis, Thamnophis,* and *Natrix*. Corpora lutea have also been described in the snapping turtle (Rahn, 1938) and the box turtle (Altland, 1951).

The timing of the successive events that occur in the ruptured follicle has been studied most extensively in lizards, and there appears to be a definite correlation between the longevity of the corpus luteum and the egg-laying or egg-retaining habit of the species. In oviparous species, luteal regression is rapid and is completed shortly after oviposition (Cunningham and Smart, 1934). In vivipara, the corpora lutea remain well developed for most, if not the whole, of gestation which, in some species, lasts 2 to 3 months or longer (Weekes, 1934; Rahn, 1939), but regress completely within 2 weeks post partum (Cummingham and Smart, 1934; Weekes, 1934; Rahn, 1939; Bragdon, 1952).

The specific function of the reptilian corpus luteum is by no means clear, although its presence in both viviparous and oviparous species seems to indicate that it is not primarily related to viviparity (see elasmobranchs). Cunningham and Smart (1934), chiefly on indirect evidence, credit the corpora lutea of viviparous lizards with the inhibition of ovulation, and dismiss the possibility of the corpora lutea causing progestational changes in the uterus. An important chance observation by Boyd (1942) in relation to viviparity (later confirmed by Miller, 1948) shows, however, that the progestational changes of the uterine mucosa are not wholly the consequence of direct stimulation by the embryos. For instance, in certain viviparous lizards, e.g., *Hoplodactylus maculatus* and *Xanthusia vigilis*, if only one oviduct is occupied by developing embryos, the other will remain in a state characteristic of pregnancy throughout the entire period of gestation, indicating stimulation by steroid hormones.

Conflicting evidence has, likewise, been presented as to the necessity of the corpus luteum for the maintenance of gestation in viviparous snakes. It is possible that early pregnancy depends on the continued presence of the ovaries or corpora lutea, because ovariectomy (Clausen, 1935, 1940) or deluteinization (Fraenkel *et al.*, 1940a,b) in late pregnancy is not invariably followed by abortion; in these experiments the hypophysis was necessary at all stages of gestation. Hence these authors assumed that the corpora lutea had the same function as the similar structures in mammals (that is, to protect pregnancy until a well advanced period). Significantly, exogenous progesterone failed to repair the effects of ovariectomy in early pregnancy (Clausen, 1940), although progestins have been demonstrated in extracts of ovaries containing corpora lutea (Porto, 1942; Valle and Valle, 1943) and in the plasma of pregnant viviparous snakes (Bragdon *et al.*, 1954). More

recently, pregnenolone, 17α-hydroxypregnenolone, dehydroxyepiandros-
terone, progesterone, 17α-hydroxyprogesterone, androstenedione, estrone,
and estradiol have been demonstrated in the ovaries of the lizard *Lacerta
sicula* (Lupo di Prisco *et al.*, 1968). However, the experiments of Rahn
(1939) and Bragdon (1951) on the viviparous garter snake, *Thamnophis*,
and the water snake, *Natrix*, and those of Panigel (1956) on the ovoviviparous lizard *Zootoca*, indicate that ovariectomy or hypophysectomy during
gestation (even when performed as early as the first week) will not usually
lead to abortion, but will interfere with the process of parturition,
manifested by retention of the young *in utero* past full term. This
interference with parturition was more complete after hypophysectomy than
after ovariectomy. That the neurohypophysis might be involved is suggested
by the observation of Clausen (1935) that injection of posterior lobe extract
in early or midgestation was without effect on the course of pregnancy in
three species of water snakes and two species of garter snakes, whereas its
administration late in gestation was followed immediately by birth of the
young. It should, of course, be remembered that, even in those cases in
which experimental observations seem to deny ovarian participation in the
maintenance of gestation, there is abundant evidence that the ovaries are
capable of producing steroid hormones; their action on target organs may
be different from those so far studied in mammals.

VII. CONCLUDING REMARKS

From the standpoint of comparative endocrinology there is now a
considerable body of evidence for the suggestion of Hisaw and Hisaw (1959)
that the follicular phase of the mammalian estrous cycle is comparable in
most respects to the oviparous cycle of lower vertebrates, and that the control of events leading up to and including ovulation is similar throughout. In
essence, the pituitary secretes follicle-stimulating and luteinizing hormones
and these, in turn, bring about the production of estrogen by the ovaries,
among which estradiol-17β and estrone are invariably present. From this
point on, the innovation of a luteal phase in the estrous cycle of mammals
clearly sets them apart from other vertebrates.

In mammals, a corpus luteum is formed by the action of pituitary luteinizing hormone and it secretes progesterone in response to a pituitary
luteotropic hormone. The corpus luteum has thus acquired the capacity to
respond to two pituitary hormones as well as that of secreting postovulatory
progesterone—features as peculiarly mammalian as hair and mammary
glands.

In nonmammalian vertebrates, however, including some viviparous forms, the pituitary is perhaps not involved in postovulatory ovarian function. Thus, although progesterone is present in the selachian ovary, its production by the corpus luteum seems not to be functionally integrated in the maintenance of gestation, and the primitive function of the granulosa cells after ovulation, in elasmobranchs and in all vertebrates with yolk-laden eggs, is to "dispose of moribund ova and to tidy up empty follicles" (Hisaw and Hisaw, 1959).

Viviparity as we know it today is the product of evolution along widely separate paths, having arisen independently many times in different lines of descent, and with the exception of birds is a recurring feature of all living vertebrate groups from fish to man. In fact, viviparity, in one form or another, is of common occurrence in the animal kingdom, among invertebrates as well as vertebrates.

Within the Eutheria, as we have seen, there exists a bewildering variety of postovulatory control systems superimposed on a pattern of preovulatory activity that is essentially similar in all vertebrates. It may be that this diversity survives among living species because they are phylogenetically recent and their viviparous (and suckling) habit, coupled with their advanced thermoregulatory systems, has so far conferred on them a high degree of immunity from the competition of lower forms. Be that as it may, it must be borne in mind that reproductive processes were carried out successfully and in coordinated fashion long before the advent of a pituitary gland. The situation as it applies to vertebrates, particularly to mammals, must represent the culmination of a long series of adaptive mechanisms, in which the pituitary gonadotropins may be regarded as a link in the informational chain between the central nervous system and the gonads, and the pituitary–gonadal interactions as a means of regulating gonadal activities which, more primitively, would respond directly to environmental stimuli.

REFERENCES

Agate, F. J., Jr. (1952). The growth and secretory activity of the mammary glands of the pregnant rhesus monkey (*Macaca mulatta*) following hypophysectomy. *Am. J. Anat.* **90**, 257–276.

Alden, R. H. (1942a). Aspects of the egg-ovary-oviduct relationship in the albino rat. I. Egg passage and development following ovariectomy. *J. Exp. Zool.* **90**, 159–179.

Alden, R. H. (1942b). The oviduct and egg transport in the albino rat. *Anat. Rec.* **84**, 137–161.

Allen, E., Hisaw, F. L., and Gardner, W. U. (1939). The endocrine functions of the ovaries. *In* "Sex and Internal Secretions" (E. Allen, ed.), 2nd ed., pp. 452–629. Williams & Wilkins, Baltimore, Maryland.

Allen, W. M. (1970). Progesterone: How did the name originate? *South. Med. J.* **63**, 1151–1155.

Allen, W. M., and Heckel, G. P. (1939). Maintenance of pregnancy by progesterone in rabbits castrated on the 11th day. *Am. J. Physiol.* **125**, 31–35.

Allen, W. R., and Moor, R. M. (1972). The origin of the equine endometrial cups. I. Production of PMSG by fetal trophoblast cells. *J. Reprod. Fertil.* **29**, 313–316.

Alloiteau, J. J., Psychoyos, A., and Acker, G. (1963). Durée de la vie fontionelle du corps jaune gestatif chez la ratte hypophysectomisée. *C. R. Acad. Sci. (Paris)* **256**, 4284–4287.

Altland, P. D. (1951). Observations on the structure of the reproductive organs of the box turtle. *J. Morphol.* **89**, 599–616.

Amoroso, E. C. (1952). Placentation. *In* "Marshall's Physiology of Reproduction" (A. S. Parkes, ed.), 3rd ed., Vol. 2, pp. 127–311. Longmans Green, London and New York.

Amoroso, E. C. (1955a). Endocrinology of pregnancy. *Br. Med. Bull.* **11**, 117–125.

Amoroso, E. C. (1955b). The comparative anatomy and histology of the placental barrier. *In* "Gestation" (L. B. Flexner, ed.), Vol. I, pp. 119–224. Josiah Macy, Jr. Found., New York.

Amoroso, E. C. (1956). The endocrine environment of the foetus. *Proc. Int. Congr. Anim. Reprod., 3rd, 1956* Vol. 1, pp. 25–32.

Amoroso, E. C. (1959). The biology of the placenta. *In* "Gestation" (C. A. Villee, ed.), Vol. V, pp. 13–76. Josiah Macy, Jr. Found., New York.

Amoroso, E. C. (1960a). Comparative aspects of the hormonal functions. *In* "The Placenta and Fetal Membranes" (C. A. Villee, ed.), pp. 3–28. Williams & Wilkins, Baltimore, Maryland.

Amoroso, E. C. (1960b). Viviparity in fishes. *Symp. Zool. Soc. London* **1**, 153–181.

Amoroso, E. C., and Finn, C. A. (1962). Ovarian activity during gestation, ovum transport and implantation. *In* "The Ovary" (S. Zuckerman, ed.), 1st ed., Vol. 1, pp. 451–537. Academic Press, New York.

Amoroso, E. C., and Marshall, F. H. A. (1960). External factors in sexual periodicity. *In* "Marshall's Physiology of Reproduction" (A. S. Parkes, ed.), 3rd ed., Vol. 1, Part 2, pp. 707–831. Longmans Green, London and New York.

Amoroso, E. C., and Perry, J. S. (1975). The existence during gestation of an immunological buffer zone at the interface between maternal and fetal tissues. *Philos. Trans. R. Soc. London, Ser.* B **271**, 343–361.

Amoroso, E. C., and Porter, D. G. (1966). Anterior pituitary function in pregnancy. *In* "The Pituitary Gland" (G. W. Harris and B. T. Donovan, eds.), Vol. 2, pp. 364–411. Butterworth, London.

Amoroso, E. C., and Porter D. G. (1970). The endocrine functions of the placenta. *In* "Scientific Foundations of Obstetrics and Gynaecology" (E. E. Philipp, J. Barnes, and M. Newton, eds.), pp. 556–586. Heinemann, London.

Amoroso, E. C., and Rowlands, I. W. (1951). Hormonal effects in the pregnant mare and foetal foal. *J. Endocrinol.* **7**, 1–1iii.

Amoroso, E. C., Hancock, J. L., and Rowlands, I. W. (1948). Ovarian activity in the pregnant mare. *Nature (London)* **161**, 355–356.

Amoroso, E. C., Austin, J., Goffin, A., and Langford, E. (1957). Breeding habits of an amphibian (*Gastrotheca marsupiatum*). *J. Physiol. (London)* **135**, 38 P. (Film.)

Amoroso, E. C., Bourne, G. H., Harrison, R. J., Matthews, L. H., Rowlands, I. W., and Sloper, J. C. (1965). Reproductive and endocrine organs of foetal, newborn and adult seals. *J. Zool.* **147,** 430–486.

Ancel, P., and Bouin, P. (1911). Recherches sur les fonctions du corps jaune gestatif. II. Sur le déterminisme du développement de la glande mammaire au cours de la gestation. *J. Physiol. Pathol. Gen.* **13,** 31–41.

Ancel, P., and Vintemberger, P. (1929). De l'action du corps jaune sur l'évolution des oeufs chez la lapine. *C. R. Soc. Biol.* **100,** 852–857.

Anderson, A. B. M., Pierrepoint, C. G., Griffiths, K., and Turnbull, A. C. (1972). Steroid metabolism in the adrenals of fetal sheep in relation to natural and corticotrophin-induced parturition. *J. Reprod. Fertil., Suppl.* **16,** 25–37.

Anderson, L. L., Dyck, G. W., Mori, H., Henricks, D. M., and Melampy, R. M. (1967). Ovarian function in pigs following hypophysial stalk transection or hypophysectomy. *Am. J. Physiol.* **212,** 1188–1194.

Anderson, L. L., Bland, K. P., and Melampy, R. M. (1969). Comparative aspects of uterine-luteal relationships. *Recent Prog. Horm. Res.* **25,** 57–104.

Armstrong, D. T., Jackanicz, T. M., and Keyes, P. L. (1969). Regulation of steroidogenesis in the rabbit ovary. *In* "The Gonads" (K. W. McKerns, ed.), pp. 3–25. Appleton, New York.

Artunkal, T., and Colonge, R. A. (1949). Action de l'ovariectomie sur la gestation du cobaye. *C. R. Soc. Biol.* **143,** 1590–1592.

Aschner, I. (1912). Über die Funktion der Hypophyse. *Pfluegers Arch. Gesamte Physiol. Menschen Tiere* **146,** 1–146.

Asdell, S. A. (1928). The growth and function of the corpus luteum. *Physiol. Rev.* **8,** 313–345.

Asdell, S. A. (1964). "Patterns of Mammalian Reproduction," 2nd ed. Cornell Univ. Press, Ithaca, New York.

Asdell, S. A., and Seidenstein, H. R. (1935). Theelin and progestin injections on uterus and mammary glands of ovariectomized and hypophysectomized rabbits. *Proc. Soc. Exp. Biol. Med.* **32,** 931–933.

Ask-Upmark, M. E. (1926). Le corps jaune est-il necessaire pour l'accomplissement physiologique de la gravidité humaine? *Acta Obstet. Gynecol. Scand.* **5,** 211–229.

Astwood, E. B., and Greep, R. O. (1938). A corpus luteum stimulating substance in the rat placenta. *Proc. Soc. Exp. Biol. Med.* **38,** 713–716.

Averill, S. C., Ray, E. W., and Lyons, W. R. (1950). Maintenance of pregnancy in hypophysectomised rats with placental implants. *Proc. Soc. Exp. Biol. Med.* **75,** 3–9.

Balfour, W. E., Comline, R. S., and Short, R. V. (1957). Secretion of progesterone by the adrenal gland. *Nature (London)* **180,** 1480–1481.

Ball, J. N. (1960). Reproduction in female bony fishes. *Symp. Zool. Soc. London* **1,** 105–135.

Bara, G. (1965). Histochemical localization of Δ^5-3β-hydroxysteroid dehydrogenase in the ovaries of a teleost fish, *Scomber scomber*, L. *Gen. Comp. Endocrinol.* **5,** 284–286.

Bassett, J. M., and Thorburn, G. D. (1969). Foetal plasma corticosteroids and the initiation of parturition in sheep. *J. Endocrinol.* **44,** 285–286.

Beard, J. (1897). "The Span of Gestation and the Cause of Birth." Jena.

Benazzi, M. (1933). Sulla funzione ovarica in gravidanza. *Arch. Sci. Biol. (Bologna)* **18,** 409–419.

Benson, G. K., Cowie, A. T., Cox, C. P., Flux, D. S., and Folley, S. J. (1955). Studies on the hormonal induction of mammary growth and lactation in the goat. II. Functional and morphological studies of hormonally developed udders with special reference to the effect of 'triggering' doses of oestrogen. *J. Endocrinol.* **13**, 46–58.

Benson, G. K., Cowie, A. T., Cox, C. P., and Goldzveig, S. M. (1957). Effects of oestrone and progesterone on mammary development in the guinea-pig. *J. Endocrinol.* **15**, 126–144.

Berger, P. J., and Sharman, G. B. (1969a). Progesterone-induced development of dormant blastocysts in the Tammar wallaby, *Macropus eugenii* Desmarest; Marsupialia. *J. Reprod. Fertil.* **20**, 201–210.

Berger, P. J., and Sharman, G. B. (1969b). Embryonic diapause initiated without the suckling stimulus in the wallaby, *Macropus eugenii. J. Mammal.* **50**, 630–632.

Betz, T. W. (1963). The ovarian histology of the diamond-backed watersnake, *Natrix rhombifera. J. Morphol.* **113**, 245–260.

Billmann, F., and Engel, R. (1939). Vikariierender Einsatz fetaler Nebennieren in der Schwangerschaft beim nebennierenlosen Hund. *Klin. Wochenschr.* **18**, 599–600.

Bishop, D. W. (1956a). Oxygen concentration in the rabbit genital tract. *Proc. Int. Congr. Anim. Reprod., 3rd, 1956* Part 1, pp. 53–55.

Bishop, D. W. (1956b). The tubal secretions of the rabbit oviduct. *Anat. Rec.* **125**, 631.

Bishop, D. W. (1956c). Active secretion in the rabbit oviduct. *Am. J. Physiol.* **187**, 347–352.

Bishop, D. W. (1957). Metabolic conditions within the oviduct of the rabbit. *Int. J. Fertil.* **2**, 11–22.

Blair-Bell, W. (1920). "The Sex Complex," 2nd ed. Baillière, London.

Blandau, R. J. (1969). Gamete transport; comparative aspects. *In* "The Mammalian Oviduct. Comparative Biology and Methodology" (E. S. E. Hafez and R. J. Blandau, eds.), pp. 129–162. Univ. of Chicago Press, Chicago, Illinois.

Bloom, G., Paul, K.-G., and Wiquist, N. (1958). A uterine-relaxing factor in the pregnant rat. *Acta Endocrinol. (Copenhagen)* **28**, 112–118.

Boling, J. L. (1969). Endocrinology of oviducal musculature. *In* "The Mammalian Oviduct. Comparative Biology and Methodology" (E. S. E. Hafez and R. J. Blandau, eds.), pp. 163–181. Univ. of Chicago Press, Chicago, Illinois.

Bonner, W. N. (1955). Reproductive organs of foetal and juvenile elephant seals. *Nature (London)* **176**, 982–983.

Bonnet, R. (1907). "Lehrbuch der Entwicklungsgeschichte." Berlin.

Botte, V., and Cottino, E. (1964). Ricerche istochimiche sulla distribuzione del colesterolo e di alcuni enzimi della steroidogenesi nei follicoli atresici e post-ovulatori di *Rana esculenta* e *Triturus cristatus. Boll. Zool.* **31**, 491–500.

Bouin, P., and Ancel, P. (1909). Le développement de la glande mammaire pendant la gestation est determiné par le corps jaune. *C. R. Soc. Biol.* **67**, 466–467.

Bouin, P., and Ancel, P. (1910). Recherches sur les fonctions du corps jaune gestatif. 1. Sur le déterminisme de la préparation de l'utérus à la fixation de l'oeuf. *J. Physiol. Pathol. Gen.* **12**, 1–16.

Bourg, R., and Spehl, E. (1948). Etude histophysiologique des consequences de la ligature des cornes utérines chez la rate adulte. Pseudo-gestation, hyperplasie glandulo-kystique utérine (H.G.K.) et métaplasie degénérative kystique (M.D.K.). *Bull. Acad. R. Med. Belg.* **13,** 118–169.

Boyd, M. M. M. (1942). The oviduct, foetal membranes and placentation in *Hoplodactylus maculatus. Proc. Zool. Soc. London* **A112,** 65–104.

Bragdon, D. E. (1951). The non-essentiality of the corpora lutea for the maintenance of gestation in certain live-bearing snakes. *J. Exp. Zool.* **118,** 419–435.

Bragdon, D. E. (1952). Corpus luteum formation and follicular atresia in the common garter snake, *Thamnophis sirtalis. J. Morphol.* **91,** 413–437.

Bragdon, D. E., Lazo-Wasem, E. A., Zarrow, M. X., and Hisaw, F. L. (1954). Progesterone-like activity in the plasma of ovoviviparous snakes. *Proc. Soc. Exp. Biol. Med.* **86,** 477–480.

Brambell, F. W. R. (1937). The influence of lactation on the implantation of the mammalian embryo. *Am. J. Obstet. Gynecol.* **33,** 942–953.

Brambell, F. W. R. (1956). Ovarian changes. *In* "Marshall's Physiology of Reproduction" (A. S. Parkes, ed.), 3rd ed., Vol. 1, Part 1, pp. 397–542. Longmans Green, London and New York.

Brambell, F. W. R., and Rowlands, I. W. (1936). Reproduction in the bank vole (*Evotomys glareolus* Schreber). *Philos. Trans. R. Soc. London, Ser. B* **226,** 71–97.

Bretschneider, L. H., and de Wit, J. J. D. (1947). "Sexual Endocrinology of Non-mammalian Vertebrates" (Monographs on the Progress of Research in Holland during the War, No. 11). Elsevier, Amsterdam.

Brumby, H. I., and Forsyth, I. A. (1969). Bioassay of prolactin in the blood of goats at parturition. *J. Endocrinol.* **43,** xxiii–xxiv.

Bryant, G. D., Greenwood, F. C., and Linzell, J. L. (1968). Plasma prolactin levels in the goat: physiological and experimental modification. *J. Endocrinol.* **40,** iv–v.

Buchanan, G. D., Enders, A. C., and Talmage, R. V. (1956). Implantation in armadillos ovariectomized during the period of delayed implantation. *J. Endocrinol.* **14,** 121–128.

Burdick, H. O., Emerson, B. B., and Whitney, R. (1940). Effects of testosterone propionate on pregnancy and on passage of ova through the oviducts of mice. *Endocrinology* **26,** 1081–1086.

Burns, R. K. (1932). Follicular atresia in the ovaries of hypophysectomized salamanders in relation to yolk formation. *J. Exp. Zool.* **63,** 309–322.

Caldeyro-Barcia, R. (1965). Regulation of myometrial activity in pregnancy. *In* "Muscle" (W. M. Paul *et al.,* eds.), pp. 317–347. Pergamon, Oxford.

Callard, I. P., and Leathem, J. H. (1965). *In vitro* steroid synthesis by the ovaries of elasmobranchs and snakes. *Arch. Microsc. Morphol. Exp.* **54,** 35–48.

Callard, I. P., and Leathem, J. H. (1966). Steroid synthesis by amphibian ovarian tissue. *Gen. Comp. Endocrinol.* **7,** 80–84.

Canivenc, R. (1960). L'ovo-implantation différée des animaux sauvages. *In* "Les Fonctions de Nidation utérine et leurs Troubles," pp. 33–86. Masson, Paris.

Canivenc, R., and Laffargue, M. (1956). Survie prolongée d'oeufs féconds non implantés dans l'utérus de rattes castrées et injectées de progestérone. *C. R. Acad. Sci. (Paris)* **242,** 2857–2860.

Carlisle, D. B. (1955). Discussion on "The hormones of sex and reproduction and their effects in fish and lower chordates." *Mem. Soc. Endocrinol.* **4,** 185.

Casida, L. E., and Warwick, E. J. (1945). The necessity of the corpus luteum for maintenance of pregnancy in the ewe. *J. Anim. Sci.* **4,** 34–36.

Cédard, L., and Ozon, R. (1962). Teneur en oestrogènes du sang de la Grenouille mousse (*Rana temporaria*, L.). *C. R. Soc. Biol.* **156**, 1805–1806.

Challis, J. R. G. (1971). Sharp increase in free circulating oestrogens immediately before parturition in sheep. *Nature (London)* **229**, 208.

Challis, J. R. G., and Linzell, J. L. (1971). The concentration of total unconjugated oestrogens in the plasma of pregnant goats. *J. Reprod. Fertil.* **26**, 401–404.

Challis, J. R. G., Heap, R. B., and Illingworth, D. V. (1971). Concentrations of oestrogen and progesterone in the plasma of non-pregnant, pregnant and lactating guinea-pigs. *J. Endocrinol.* **37**, 333–345.

Chambon, Y. (1949a). Absence d'influence sur l'implantation de fortes doses de progéstérone chez la ratte. *C. R. Soc. Biol.* **143**, 753–756.

Chambon, Y. (1949b). Réalisation du retard de l'implantation par les faibles doses de progéstérone chez la ratte. *C. R. Soc. Biol.* **143**, 756–758.

Chang, M. C. (1966). Effects of oral administration of medroxyprogesterone acetate and ethinyloestradiol on the transportation and development of rabbit eggs. *Endocrinology* **79**, 939–948.

Chang, M. C. (1967). Effects of progesterone and related compounds on fertilization, transportation and development of rabbit eggs. *Endocrinology* **81**, 1251–1260.

Chang, M. C. (1969). Fertilization, transportation and degeneration of eggs in pseudopregnant or progesterone-treated rabbits. *Endocrinology* **84**, 356–361.

Chang, M. C., and Hunt, D. M. (1970). Effects of various progestins and estrogen on gamete transport and fertilization in the rabbit. *Fertil. Steril.* **21**, 683–686.

Chieffi, G. (1961). La luteogenesi nei selaci ovovivipari. Ricerche istologiche ed istochimiche in *Torpedo marmorata* e *Torpedo ocellata*. *Pubbl. Stn. Zool. Napoli* **32**, 145–166.

Chieffi, G. (1962). Aspetti endocrinologici della riproduzione nei pesci. *Boll. Zool.* **29**, 150–186.

Chieffi, G. (1967). Occurrence of steroids in gonads of nonmammalian vertebrates and sites of their biosynthesis. *Proc. Int. Congr. Horm. Steroids, 2nd, 1966* Excerpta Med. Found. Int. Congr. Ser. No. 132, pp. 1047–1057.

Chieffi, G., and Botte, V. (1970). The problem of "luteogenesis" in non-mammalian vertebrates. *Boll. Zool.* **37**, 85–102.

Chieffi, G., and Lupo, C. (1963). Identification of sex hormones in the ovarian extracts of *Torpedo marmorata* and *Bufo vulgaris*. *Gen. Comp. Endocrinol.* **3**, 149–152.

Christianson, M., and Jones, I. C. (1957). The inter-relationship of the adrenal glands of mother and foetus in the rat. *J. Endocrinol.* **15**, 17–42.

Chu, J. P., and You, S. S. (1945). The role of the thyroid gland and oestrogen on the regulation of gonadotrophic activity of the anterior pituitary. *J. Endocrinol.* **4**, 115–124.

Clausen, H. J. (1935). The effects of ovariectomy and hypophysectomy on parturition in snakes. *Anat. Rec.* **64**, Suppl. 1, 88.

Clausen, H. J. (1940). Studies on the effects of ovariectomy and hypophysectomy on gestation in snakes. *Endocrinology* **27**, 700–704.

Cochrane, R. L., and Meyer, R. L. (1957). Delayed nidation in the rat induced by progesterone. *Proc. Soc. Exp. Biol. Med.* **96**, 155–159.

Cole, H. H., Howell, C. E., and Hart, G. H. (1931). The changes occurring in the ovary of the mare during pregnancy. *Anat. Rec.* **49**, 199–200.

Corbet, R. M. (1932). A case of the removal of the corpus luteum of pregnancy with persistence of the pregnancy. *Ir. J. Med. Sci.* [6], 520–521.

Corner, G. W. (1928). Physiology of the corpus luteum. 1. The effect of very early ablation of the corpus luteum upon embryos and uterus. *Am. J. Physiol.* **86**, 74–81.

Corner, G. W. (1938). The sites of formation of estrogenic substances in the animal body. *Physiol. Rev.* **18**, 154–172.

Corner, G. W. (1940). Accessory corpora lutea in the ovary of the monkey, *Macaca rhesus. An. Fac. Med. Univ. Repub., Montevideo* **25**, 553–560.

Corner, G. W. (1947). "The Hormones in Human Reproduction." Princeton Univ. Press, Princeton, New Jersey.

Corner, G. W., and Csapo, A. (1953). Action of the ovarian hormones on uterine muscle. *Br. Med. J.* **1**, 687–691.

Corner, G. W., Bartelmez, G. W., and Hartman, C. G. (1936). On the normal and aberrant corpora lutea of the rhesus monkey. *Am. J. Anat.* **59**, 433–443.

Courrier, R. (1941). Evolution de la grossesse extra-utérine chez la lapine castrée. *C. R. Soc. Biol.* **135**, 820–822.

Courrier, R. (1945). "Endocrinologie de la Gestation," Chapter VII, pp. 70–93. Masson, Paris.

Courrier, R., and Baclesse, M. (1955). L'équilibre hormonal au cours de la gestation. *Rapp. Reun. Endocrinol. Langue Fr., 3rd, 1955* 1–16.

Courrier, R., and Gros, G. (1935). Contribution à l'endocrinologie de la grossesse chez la chatte. *C. R. Soc. Biol.* **120**, 5–7.

Courrier, R., and Kehl, R. (1938). Sur le besoin hormonal quantitatif chez la lapine gestante castrée. *C. R. Soc. Biol.* **128**, 188–191.

Cowie, A. T. (1957). Mammary development and lactation. *In* "Progress in the Physiology of Farm Animals" (J. Hammond, ed.), pp. 907–961. Butterworth, London.

Cowie, A. T. (1971). Influence of hormones on mammary growth and milk secretion. *In* "Lactation" (I. R. Falconer, ed.), pp. 123–140. Butterworth, London.

Cowie, A. T., and Tindal, J. S. (1960). Effects of hypophysectomy of the pregnant and lactating goat. *Advan. Abstr. Int. Congr. Endocrinol., 1st, 1960* pp. 679–680.

Cowie, A. T., Folley, S. J., Malpress, F. H., and Richardson, K. C. (1952). Studies on the hormonal induction of mammary growth and lactation in the goat. *J. Endocrinol.* **8**, 64–88.

Cowie, A. T., Daniel, P. M., Prichard, M. M. L., and Tindal, J. S. (1963). Hypophysectomy in pregnant goats, and section of the pituitary stalk in pregnant goats and sheep. *J. Endocrinol.* **28**, 93–102.

Cowie, A. T., Daniel, P. M., Knaggs, G. S., Prichard, M. M. L., and Tindal, J. S. (1964a). Lactation in the goat after section of the pituitary stalk. *J. Endocrinol.* **28**, 253–265.

Cowie, A. T., Knaggs, G. S., and Tindal, J. S. (1964b). Complete restoration of lactation in the goat after hypophysectomy. *J. Endocrinol.* **28**, 267–279.

Cowie, A. T., Knaggs, G. S., Tindal, J. S., and Turvey, A. (1968). The milking stimulus and mammary growth in the goat. *J. Endocrinol.* **40**, 243–252.

Csapo, A. (1956). Progesterone "block." *Am. J. Anat.* **98**, 273–291.

Csapo, A. (1969). The four direct regulatory factors of myometrial function. *Ciba Found. Study Group* **34**, 13–55.

Csapo, A., and Wood, C. (1968). The endocrine control of the initiation of labour in the human. *In* "Recent Advances in Endocrinology" (V. H. T. James, ed.), 8th ed., pp. 207–239. Churchill, London.

Csapo, A., Knobil, E., van der Molen, H. J., and Wiest, W. G. (1971). Peripheral plasma progesterone levels during human pregnancy and labor. *Am. J. Obstet. Gynecol.* **110**, 630–632.

Cunningham, J. T., and Smart, W. A. M. (1934). The structure and origin of corpora lutea in some of the lower vertebrates. *Proc. R. Soc. London, Ser. B* **116**, 258–281.

Cutuly, E. (1941a). Maintenance of pregnancy in hypophysectomized rats. *Proc. Soc. Exp. Biol. Med.* **47**, 126–128.

Cutuly, E. (1941b). Implantation following mating in hypophysectomized rats injected with lactogenic hormone. *Proc. Soc. Exp. Biol. Med.* **48**, 315–318.

Cutuly, E. (1942). Effects of lactogenic and gonadotrophic hormones on hypophysectomized pregnant rats. *Endocrinology* **31**, 13–22.

Daels, F. (1908). On the relations between the ovaries and the uterus. *Surg., Gynecol. Obstet.* **6**, 153–159.

Dallenbach-Hellweg, G., Dawson, A. B., and Hisaw, F. L. (1966). The effect of relaxin on the endometrium of monkeys. Histological and histochemical studies. *Am. J. Anat.* **119**, 61–78.

Danforth, C. H. (1939). Relation of genic and endocrine factors in sex. *In* "Sex and Internal Secretions" (E. Allen, ed.), 2nd ed., p. 337. Williams & Wilkins, Baltimore, Maryland.

Davies, J., and Hesseldahl, H. (1971). Comparative embryology of mammalian blastocysts. *In* "The Biology of the Blastocyst" (R. J. Blandau, ed.), pp. 27–48. Univ. of Chicago Press, Chicago, Illinois.

Day, B. N., Anderson, L. L., Emmerson, M. A., Hazel, L. N., and Melampy, R. M. (1959). Effect of estrogen and progesterone on early embryonic mortality in ovariectomized gilts. *J. Anim. Sci.* **18**, 607–613.

Deanesly, R. (1960). Implantation and early pregnancy in ovariectomized guinea-pigs. *J. Reprod. Fertil.* **1**, 242–248.

Deanesly, R. (1963). Early embryonic growth and progestagen function in ovariectomized guinea-pigs. *J. Reprod. Fertil.* **6**, 143–152.

Deanesly, R. (1966). The endocrinology of pregnancy and foetal life. *In* "Marshall's Physiology of Reproduction" (A. S. Parkes, ed.), 3rd ed., Vol. 3, pp. 891–1063. Longmans Green, London and New York.

Deanesly, R. (1971). The differentiation of the decidua at ovo-implantation in the guinea-pig contrasted with that of the traumatic deciduoma. *J. Reprod. Fertil.* **26**, 91–97.

Deanesly, R. (1972). Retarded embryonic development and pregnancy termination in ovariectomized guinea-pigs; progesterone deficiency and decidual collapse. *J. Reprod. Fertil.* **28**, 241–247.

Deanesly, R., and Newton, W. H. (1940). The influence of the placenta on the corpus luteum of pregnancy in the mouse. *J. Endocrinol.* **2**, 317–321.

Deanesly, R., Fee, A. R., and Parkes, A. S. (1930). Studies on ovulation. II. The effect of hypophysectomy on the formation of the corpus luteum. *J. Physiol. (London)* **70**, 38–44.

de Ikonicoff, L. K., and Cédard, L. (1973). Localization of human chorionic gonadotropic and somatomammotropic hormones by the peroxidase immunohistoenzymologic method in villi and amniotic epithelium of human placentas (from six weeks to term). *Am. J. Obstet. Gynecol.* **116**, 1124–1132.

Della-Beffa, A. (1951). Gravidanza, parto ed allattamento ad ipofisi distrutta de neoplasia endocranica. *Folia Endocrinol.* **4**, 119–132.

Della Corte, F., and Chieffi, G. (1961). Modificazioni citologiche della ipofisi di *Torpedo marmorata* Risso, furante la gravidanza. *Boll. Zool.* **28**, 219–225.

Dempsey, E. W. (1939). The reproductive cycle of New World monkeys. *Am. J. Anat.* **64**, 381–401.

Dempsey, E. W. (1940). The structure of the reproductive tract in the female gibbon. *Am. J. Anat.* **67**, 229–249.

Denamur, R. (1968). Formation and maintenance of corpora lutea in domestic animals. *J. Anim. Sci.* **27**, Suppl. 1, 163–180.

Denamur, R., and Martinet, J. (1955). Effets de l'ovariectomie chez la brebis pendant la gestation. *C. R. Soc. Biol.* **149**, 2105–2107.

Denamur, R., and Martinet, J. (1961). Effets de l'hypophysectomie et de la section de la tige pituitaire sur la brebis. *Ann. Endocrinol.* **22**, 768–776.

Denamur, R., Martinet, J., and Short, R. V. (1966). Sécrétion de la progestérone par les corps jaunes de la brebis après hypophysectomie, section de la tige pituitaire et hystérectomie. *Acta Endocrinol. (Copenhagen)* **52**, 72–90.

Desclin, L. (1932). A propos des interactions entre l'utérus et le corps jaune au cours de la grossesse chez le cobaye. *C. R. Soc. Biol.* **109**, 972–973.

Desclin, L. (1939). Influence de l'hypophysectomie sur la glande mammaire du Cobaye gravide. *C. R. Soc. Biol.* **131**, 837–840.

Desclin, L. (1952). Recherches sur le déterminisme des phénomènes de sécrétion dans la glande mammaire du rat. *Ann. Endocrinol.* **13**, 120–136.

Dewar, A. D. (1957). The endocrine control of the extra-uterine weight gain of pregnant mice. *J. Endocrinol.* **15**, 216–229.

Dickmann, Z., and Dey, S. K. (1974). Steroidogenesis in the preimplantation rat embryo and its possible influence on morula-blastocyst transformation and implantation. *J. Reprod. Fertil.* **37**, 91–93.

Diczfalusy, E., and Borell, U. (1961). Influence of oophorectomy on steroid excretion in early pregnancy. *J. Clin. Endocrinol. Metab.* **21**, 1119–1126.

Diczfalusy, E., and Troen, P. (1961). Endocrine functions of the human placenta. *Vitam. Horm. (N. Y.)* **19**, 229–311.

Dodd, J. M., Evennett, P. J., and Goddard, C. K. (1960). Reproductive endocrinology in cyclostomes and elasmobranchs. *Symp. Zool. Soc. London* **1**, 77–103.

Douglass, M. (1931). Persistence of pregnancy after excision of the corpus luteum in the early weeks. *Surg., Gynecol. Obstet.* **52**, 52–55.

Drips, D. (1919). Studies on the ovary of the spermophile (*Spermophilus citellus tridecemlineatus*) with special reference to the corpus luteum. *Am. J. Anat.* **25**, 117–184.

Drummond-Robinson, G., and Asdell, S. A. (1926). The relation between the corpus luteum and the mammary gland. *J. Physiol. (London)* **61**, 608–614.

du Mesnil du Buisson, F. (1961). Régression unilatérale des corps jaunes après hystérectomie partielle chez la truie. *Ann. Biol. Anim., Biochim. Biophys.* **1**, 105–112.

du Mesnil du Buisson, F., and Dauzier, L. (1957). Influence de l'ovariectomie chez la truie pendant la gestation. *C. R. Soc. Biol.* **161**, 311–313.

du Mesnil du Buisson, F., and Dauzier, L. (1959). Contrôle mutuel de l'utérus et de l'ovaire chez la truie. *Ann. Zootech.* **8**, Suppl. 147–159.

du Mesnil du Buisson, F., and Denamur, R. (1968). Mécanismes du contrôle de la fonction lutéale chez la truie, la brebis et la vache. *Proc. Int. Congr. Endocrinol., 3rd,* 1968 Excerpta Med. Found. Congr. Ser. No. 157, pp. 927–934.

Dutt, R. G., and Casida, L. E. (1948). Alteration of the estrual cycle in the sheep by use of progesterone and its effect upon subsequent ovulation and fertility. *Endocrinology* **43,** 208–217.

Ealey, E. H. M. (1963). The ecological significance of delayed implantation in a population of the hill kangaroo (*Macropus robustus*). *In* "Delayed Implantation" (A. C. Enders, ed), pp. 33–48. Univ. of Chicago Press, Chicago, Illinois.

Eckstein, P., and Zuckerman, S. (1956). The oestrous cycle in the Mammalia. *In* "Marshall's Physiology of Reproduction" (A. S. Parkes, ed.), 3rd ed., Vol. 1, pp. 226–396. Longmans Green, London and New York.

Eckstein, P., Shelesnyak, M. C., and Amoroso, E. C. (1959). A survey of the physiology of ovum implantation in mammals. *Mem. Soc. Endocrinol.* **6,** 3–12.

Ehrhardt, K. (1936). Ueber das Laktationshormon des Hypophysenvorderlappens. *Muench. Med. Wochenschr.* **83,** 1163–1164.

Enders, A. C., ed. (1963). "Delayed Implantation." Univ. of Chicago Press, Chicago, Illinois.

Enders, A. C. (1966). The reproductive cycle of the nine-banded armadillo (*Dasypus novemcinctus*). *Symp. Zool. Soc. London* **15,** 295–310.

Enders, A. C., and Buchanan, G. D. (1959). Some effects of ovariectomy and injection of ovarian hormones in the armadillo.*J. Endocrinol.* **19,** 251–258.

Ershoff, B. H., and Deuel, H. J. (1943). Prolongation of pseudo-pregnancy by induction of deciduomata in rats. *Proc. Soc. Exp. Biol. Med.* **54,** 167–168.

Estergreen, V. L., Jr., Frost, O. L., Gomes, W. R., Erb, R. E., and Bullard, J. F. (1967). Effect of ovariectomy on pregnancy maintenance and parturition in dairy cows. *J. Dairy Sci.* **50,** 1293–1395.

Everett, J. W. (1952). Presumptive hypothalamic control of spontaneous ovulation. *Ciba Found. Colloq. Endocrinol. [Proc.]* **5,** 167–178.

Eyeson, K. N. (1971). Pituitary control of ovarian activity in the lizard *Agama agama. J. Zool.* **165,** 367–372.

Fee, A. R., and Parkes, A. S. (1929). Studies on ovulation. I. The relation of the anterior pituitary to ovulation in the rabbit. *J. Physiol. (London)* **67,** 383–388.

Felton, L. C., Frieden, E. H., and Bryant, H. H. (1953). The effects of ovarian extracts upon activity of the guinea-pig uterus *in situ. J. Pharmacol. Exp. Ther.* **107,** 160–164.

Finn, C. A. (1971). The biology of decidual cells. *Adv. Reprod. Physiol.* **5,** 1–26.

Firor, W. M. (1933). Hypophysectomy in pregnant rabbits. *Am. J. Physiol.* **104,** 204–215.

Firor, W. M., and Grollman, A. (1933). Studies on the adrenals. I. Adrenalectomy in mammals with particular reference to the white rat (*Mus norvegicus*). *Am. J. Physiol.* **103,** 686–698.

Folley, S. J. (1952). Lactation. *In* "Marshall's Physiology of Reproduction" (A. S. Parkes, ed.), 3rd ed., Vol. 2, pp. 525–647. Longmans Green, London and New York.

Folley, S. J., and Malpress, F. H. (1948). Hormonal control of mammary growth. *In* "The Hormones" (G. Pincus and K. V. Thimann, eds.), Vol. 1, pp. 695–743. Academic Press, New York.

Foote, W. D., Gooch, L. D., Pole, A. L., and Casida, L. E. (1957). The maintenance of early pregnancy in the ovariectomized ewe by injections of ovarian hormones. *J. Anim. Sci.* **16**, 986–989.

Fortgang, A., and Simpson, M. E. (1953). Effects on parturition in the rat of ablation of different lobes of the hypophysis. *Proc. Soc. Exp. Biol. Med.* **84**, 663–666.

Fraenkel, L. (1903). Die Funktion des Corpus luteum. *Arch. Gynaekol.* **68**, 438–535.

Fraenkel, L. (1910). Neue Experimente zur Funktion des Corpus luteum. *Arch. Gynaekol.* **91**, 705–761.

Fraenkel, L. (1927). Structure and function of endocrine glands, particularly ovary. *Am. J. Obstet. Gynecol.* **13**, 606–610.

Fraenkel, L., and Cohn, F. (1901). Experimentelle Untersuchung über den Einfluss des Corpus luteum und die Insertion des Eies. *Anat. Anz.* **20**, 294–300.

Fraenkel, L., Martins, T., and Mello, R. F. (1940a). Observaciones sobre el embarazo de las serpientes viviparas. *Arch. Soc. Biol. Montevideo* **10**, 1–4.

Fraenkel, L., Martins, T., and Mello, R. F. (1940b). Studies on the pregnancy of viviparous snakes. *Endocrinology* **27**, 836–837.

Frazer, J. F. D., and Alexander, D. P. (1954). The effect of spaying in the pregnant rat. *J. Physiol. (London)* **124**, 36–37P.

Fredricsson, B. (1959). Histochemical observations on the epithelium of human fallopian tubes. *Acta Obstet. Gynecol. Scand.* **38**, 109–134.

Fredricsson, B., and Bjorkman, N. (1962). Studies on the ultrastructure of the human oviduct epithelium in different functional states. *Z. Zellforsch. Mikrosk. Anat.* **58**, 387–402.

Fredrikson, H. (1939). Endocrine factors involved in the development and function of the mammary glands of female rabbits. *Acta Obstet. Gynecol. Scand.* **19**, Suppl. 1, 167.

Friesen, H. G., Suwa, S., and Pare, P. (1969). Synthesis and secretion of placental lactogen and other proteins by the placenta. *Recent Prog. Horm. Res.* **25**, 161–205.

Fuchs, F. (1971). Endocrinology of labor. *In* "Endocrinology of Pregnancy" (F. Fuchs and A. Klopper, eds.), pp. 306–327. Harper, New York.

Fuchs, F., and Klopper, A., eds. (1971). "Endocrinology of Pregnancy." Harper, New York.

Gallien, L. (1959). Endocrine basis for reproductive adaptations in Amphibia. *In* "Comparative Endocrinology" (A. Gorbman, ed.), pp. 479–487. Wiley, New York.

Gallien, L., and Chalumeau-Le Foulgoc, M. T. (1960). Mise en évidence de steroides oestrogènes dans l'ovaire juvénile de *Xenopus laevis* (Daudin) et cycle des oestrogènes au cours de la ponte. *C. R. Soc. Biol.* **156**, 1805–1806.

Galli-Mainini, C. (1950). Secreción de oviducto del sapo por el ovario en ovulación. *Rev. Soc. Argent. Biol.* **26**, 166–178.

Gardner, M. L., First, N. L., and Casida, L. E. (1963). Effect of exogenous estrogens on corpus luteum maintenance in gilts. *J. Anim. Sci.* **22**, 132–134.

Gardner, W. U., and Allen, E. (1942). Effects of hypophysectomy at mid-pregnancy in the mouse. *Anat. Rec.* **83**, 75–94.

Glasser, J. W. H. (1952). Early removal of the corpus luteum of pregnancy. *Bull. Margaret Hague Matern. Hosp.* **5**, 112.

Gleeson, A. R., Thorburn, G. D., and Cox, R. G. (1974). Prostaglandin F concentrations in the utero-ovarian venous plasma of the sow during the late luteal phase of the oestrous cycle. *Prostaglandins* **5**, 521–529.

Goding, J. R. (1973). The demonstration that $PGF_{2\alpha}$ is the uterine luteolysin in the ewe. *In* "Le Corps Jaune" (R. Denamur and A. Nettar, eds.), pp. 311–323. Masson et Cie, Paris.

Goldenberg, R. L., Bridson, W. E., and Kohler, P. O. (1970). Estrogen stimulation of progesterone synthesis by porcine granulosa cells in culture. *Biochem. Biophys. Res. Commun.* **48**, 101–107.

Goldstein, D. P., Aono, T., Taymor, M. L., Jochelson, K., Todd, R., and Hines, E. (1968). Radioimmunoassay of serum chorionic gonadotrophin activity in normal pregnancy. *Am. J. Obstet. Gynecol.* **102**, 110–114.

Gomes, W. R., and Erb, R. E. (1965). Progesterone in bovine reproduction: a review. *J. Dairy Sci.* **48**, 314–330.

Greenstein, J. S., Murray, R. W., and Foley, R. C. (1958). Effects of exogenous hormones on the reproductive processes of the cycling dairy heifer. *J. Dairy Sci.* **41**, 1834.

Greenwald, G. S. (1967). Luteotropic complex of the hamster. *Endocrinology* **80**, 118–130.

Greenwald, G. S. (1969). Endocrinology of oviducal secretions. *In* "The Mammalian Oviduct. Comparative Biology and Methodology" (E. S. E. Hafez and R. J. Blandau, eds.), pp. 183–201. Univ. of Chicago Press, Chicago, Illinois.

Greenwald, G. S., and Rothchild, I. (1968). Formation and maintenance of corpora lutea in laboratory animals. *J. Anim. Sci.* **27**, Suppl. 1, 139–162.

Gregoire, A. T., Gongsakdi, D., and Rakoff, A. E. (1961). The free amino acid content of the female rabbit genital tract. *Fertil. Steril.* **12**, 322–327.

Gregoire, A. T., Gongsakdi, D., and Rakoff, A. E. (1962). The presence of inositol in genital tract secretions of the female rabbit. *Fertil. Steril.* **13**, 432–435.

Griffiths, W. F. B., and Amoroso, E. C. (1939). Prooestrus, oestrus and mating in the greyhound bitch. *Vet. Rec.* **51**, 1279–1284.

Gros, G. (1936). "Contribution à l'endocrinologie sexuelle. Le cycle génital de la chatte," No. 21. Thesis, University of Algeria.

Grumbach, M. M., and Kaplan, S. L. (1964). On the placental origin and purification of chorionic 'growth hormone-prolactin' and its immunoassay in pregnancy. *Trans. N. Y. Acad. Sci.* [2] **27**, 167–188.

Grumbach, M. M., Kaplan, S. L., Sciarra, J. J., and Burr, I. M. (1968). Chorionic growth hormone-prolactin (CGP): secretion, disposition, biologic activity in man, and postulated function as the "growth-hormone" of the second half of pregnancy. *Ann. N. Y. Acad. Sci.* **148**, 501–531.

Guraya, S. S. (1968). Histochemical study of granulosa (follicular) cells in the pre-ovulatory and post-ovulatory follicles of amphibian ovary. *Gen. Comp. Endocrinol.* **10**, 138–146.

Guraya, S. S. (1969). Histochemical observations on the corpora atretica of amphibian ovary. *Gen. Comp. Endocrinol.* **12**, 165–180.

Hadek, R. (1955). The secretory processes in the sheep's oviduct. *Anat. Rec.* **121**, 187–201.

Hafez, E. S. E., and Blandau, R. J., eds. (1969). "The Mammalian Oviduct; Comparative Biology and Methodology." Univ. of Chicago Press, Chicago, Illinois.

Hain, A. M. (1934). The physiology of pregnancy in the rat. Further data bearing on the prolongation of pregnancy, with a study of the effects of oophorectomy during pregnancy. *Q. J. Exp. Physiol.* **24**, 101–112.

Hain, A. M. (1935). The physiology of pregnancy in the rat: an hormonal investigation into the mechanism of parturition. Effect on the female rat of the antenatal administration of oestrin to the mother. *Q. J. Exp. Physiol.* **25**, 131–143.

Hall, K. (1956). An evaluation of the roles of oestrogen, progesterone and relaxin in producing relaxation of the symphysis pubis of the ovariectomized mouse, using the technique of metachromatic staining with toluidine blue. *J. Endocrinol.* **13**, 384–393.

Hall, K. (1957). The effect of relaxin extracts, progesterone and oestradiol on maintenance of pregnancy, parturition and rearing of young after ovariectomy in mice. *J. Endocrinol.* **15**, 108–117.

Hall, K. (1960). Relaxin. *J. Reprod. Fertil.* **1**, 368–384.

Hall, O. (1952). Accessory corpora lutea in the wild Norway rat. *Tex. Rep. Biol. Med.* **10**, 32–38.

Hammond, J. (1925). "Reproduction in the Rabbit." Oliver & Boyd, Edinburgh.

Hammond, J. (1927). "The Physiology of Reproduction in the Cow." Cambridge Univ. Press, London and New York.

Hammond, J., and Day, F. T. (1947). Oestrogen treatment of cattle. Induced lactation and other effects. *J. Endocrinol.* **4**, 53–82.

Hammond, J., and Marshall, F. H. A. (1930). Oestrus and pseudopregnancy in the ferret. *Proc. R. Soc. London, Ser. B* **105**, 607–630.

Hammond, J., Jr. (1951). Failure of progesterone treatment to affect delayed implantation in mink. *J. Endocrinol.* **7**, 330–334.

Hammond, J., Jr. (1952). Maintenance of grafted luteal tissue. *Nature (London)* **169**, 330–331.

Hammond, J., Jr. (1956). The rabbit corpus luteum: oestrogen prolongation and accompanying changes in the genitalia. *Acta Endocrinol. (Copenhagen)* **21**, 307–320.

Hammond, J., Jr. and Robson, J. M. (1951). Local maintenance of the rabbit corpus luteum with oestrogen. *Endocrinology* **49**, 384–389.

Hanks, J., and Short, R. V. (1972). The formation and function of the corpus luteum in the African elephant, *Loxodonta africana. J. Reprod. Fertil.* **29**, 79–89.

Hansel, W., and Seifart, K. H. (1967). Maintenance of luteal function in the cow. *J. Dairy Sci.* **50**, 1948–1958.

Hansel, W., and Trimberger, C. W. (1951). Atropine blockade of ovulation in the cow and its possible significance. *J. Anim. Sci.* **10**, 719–724.

Harris, R. G. (1927). Effect of bilateral ovariectomy upon duration of pregnancy in mice. *Anat. Rec.* **37**, 83–93.

Hartman, C. G. (1922). Breeding habits, development and birth of the opossum. *Smithson. Inst., Annu. Rep.* **1**, 347–363.

Hartman, C. G. (1923). The oestrous cycle of the opossum. *Am. J. Anat.* **32**, 353–421.

Hartman, C. G. (1925). Interruption of pregnancy by ovariectomy in the placental opossum. Study in physiology of implantation. *Am. J. Physiol.* **71**, 436–454.

Hartman, C. G. (1939). Ovulation, fertilization and the transport and viability of eggs and spermatozoa. *In* "Sex and Internal Secretions" (E. Allen, ed.), 2nd ed., pp. 647–733. Williams & Wilkins, Baltimore, Maryland.

Hartman, C. G. (1941). Non-effect of ovariectomy on the 25th day of pregnancy in the rhesus monkey. *Proc. Soc. Exp. Biol. Med.* **48**, 221–223.

Hartman, C. G., and Corner, G. W. (1947). Removal of the corpus luteum and of the ovaries of the rhesus monkey during pregnancy; observations and cautions. *Anat. Rec.* **98**, 539–546.

Hashimoto, I., and Wiest, W. G. (1969). Luteotrophic and luteolytic mechanisms in rat corpora lutea. *Endocrinology* **84**, 886–892.

Haterius, H. O. (1936). Reduction of litter size and maintenance of pregnancy in the oophorectomized rat: evidence concerning the endocrine role of the placenta. *Am. J. Physiol.* **114,** 399–406.

Haterius, H. O., and Kempner, M. I. (1939). Uterine distention and maintenance of pregnancy following oophorectomy in the rat. *Proc. Soc. Exp. Biol. Med.* **42,** 322–325.

Heap, R. B., and Deanesly, R. (1966). Progesterone in systemic blood and placentae of intact and ovariectomized pregnant guinea-pigs. *J. Endocrinol.* **34,** 417–423.

Heap, R. B., and Perry, J. S. (1974). The maternal recognition of pregnancy. *Br. J. Hosp. Med.* **12,** 8–14.

Heap, R. B., Perry, J. S., and Rowlands, I. W. (1967). Corpus luteum function in the guinea-pig; arterial and luteal progesterone levels, and the effects of hysterectomy and hypophysectomy. *J. Reprod. Fertil.* **13,** 537–553.

Heap, R. B., Perry, J. S., and Challis, J. R. G. (1972). The hormonal maintenance of pregnancy. *In* "The Handbook of Physiology" (Am. Physiol. Soc., J. Field, ed.), Sec. 7, Vol. II, Part 2, pp. 217–260. Williams & Wilkins, Baltimore, Maryland.

Heap, R. B., Illingworth, D. V., and Perry, J. S. (1973). The secretory activity of the corpus luteum in the guinea-pig and its role in the establishment and maintenance of pregnancy. *In* "Le Corps Jaune" (R. Denamur and A. Netter, eds.), pp. 69–80. Masson, Paris.

Heckel, G. P., and Allen, W. M. (1936). Prolongation of the corpus luteum in the pseudopregnant rabbit. *Science* **84,** 161–162.

Heckel, G. P., and Allen, W. M. (1939). Maintenance of the corpus luteum and inhibition of parturition in the rabbit by injection of oestrogenic hormone. *Endocrinology* **24,** 137–148.

Herrick, E. H. (1928). Duration of pregnancy in guinea-pigs after removal and also after transplantation of ovaries. *Anat. Rec.* **39,** 193–200.

Hess, E. (1920). "Die Sterilität des Rindes." Schaper, Hannover.

Hett, J. (1933). Vergleichende Anatomie der Corpora lutea. *In* "Handbuch der vergleichenden Anatomie der Wirbeltiere" (W. Bolk *et al.,* eds.), Vol. VI, pp. 253–266. Urban & Schwarzenberg, Berlin.

Higashi, K. (1962). Studies on the prolactin-like substance in human placenta. IV. *Endocrinol. Jpn.* **9,** 1–11.

Hill, C. J. (1933). The development of Monotremata. Part I. The histology of the oviduct during gestation. *Trans. Zool. Soc. London* **21,** 413–443.

Hill, C. J. (1941). The development of the Monotremata. Part V. Further observations on the histology and the secretory activities of the oviduct prior to and during gestation. *Trans. Zool. Soc. London* **25,** 1–31.

Hill, J. P., and Gatenby, J. B. (1926). The corpus luteum of the Monotremata. *Proc. Zool. Soc. London* **47,** 715–763.

Hilliard, J., Spies, H. G., and Sawyer, C. H. (1969). Hormonal factors regulating ovarian cholesterol mobilization and progestin secretion in intact and hypophysectomized rabbits. *In* "The Gonads" (K. W. McKerns, ed.), p. 55. Appleton, New York.

Hisaw, F. L. (1925). The influence of the ovary on the resorption of the pubic bones of the pocket gopher, *Geomys bursarius* (Shaw). *J. Exp. Zool.* **42,** 411–433.

Hisaw, F. L. (1959). Endocrine adaptations of the mammalian estrous cycle and gestation. *In* "Comparative Endocrinology" (A. Gorbman, ed.), pp. 533–552. Academic Press, New York.

Hisaw, F. L., and Abramowitz, A. A. (1938). The physiology of reproduction in the dog fish, *Mustelus canis. Rep. Wood's Hole Oceanogr. Inst.*, pp. 21–22.

Hisaw, F. L., and Abramowitz, A. A. (1939). Physiology of reproduction in the dog fish, *Mustelus canis* and *Squalus acanthius. Rep. Wood's Hole Oceanogr. Inst.*, p. 22.

Hisaw, F. L., and Hisaw, F. L., Jr. (1959). Corpora lutea of elasmobranch fishes. *Anat. Rec.* **135,** 269–279.

Hisaw, F. L., and Zarrow, M. X. (1950). The physiology of relaxin. *Vitam. Horm.* (*N. Y.*) **8,** 151–178.

Hisaw, F. L., Hisaw, F. L., Jr., and Dawson, A. B. (1967). Effects of relaxin on the endothelium of endometrial blood vessels in monkeys (*Macaca mulatta*). *Endocrinology* **81,** 375–385.

Hoar, W. S. (1955). Reproduction in bony fishes. *Mem. Soc. Endocrinol.* **4,** 5–24.

Hoar, W. S. (1969). Reproduction. *Fish Physiol.* **3,** 1–72.

Hodgen, G. D., Dufau, M. L., Catt, K. J., and Tullner, W. W. (1972). Estrogens, progesterone and chorionic gonadotropin in pregnant rhesus monkeys. *Endocrinology* **91,** 896–900.

Houssay, B. A. (1935). Action de l'hypophysectomie sur la grossesse et la secretion lactée chez la chienne. *C. R. Soc. Biol.* **120,** 496–497.

Houssay, B. A. (1947). Ovulación y postura del sapo *Bufo arenarum* Hensel. V. Transporte de los ovulos por el oviducto y el utero. *Rev. Soc. Argent. Biol.* **23,** 275–287.

Houssay, B. A. (1949). Hypophyseal functions in the toad *Bufo arenarum* Hensel. *Rev. Biol.* **24,** 1–27.

Houssay, B. A. (1952). La régulation hormonale des fonctions de l'oviducte du crapeau. *Schweiz. Med. Wochenschr.* **39,** 997–999.

Huff, R. L., and Eik-Nes, K. B. (1966). Metabolism *in vitro* of acetate and certain steroids by six-day-old rabbit blastocysts. *J. Reprod. Fertil.* **11,** 57–63.

Hughes, R. L. (1962a). Reproduction in the macropod marsupial *Potorous tridactylus* (Kerr). *Aust. J. Zool.* **10,** 193–224.

Hughes, R. L. (1962b). Role of the corpus luteum in marsupial reproduction. *Nature* (*London*) **194,** 890–891.

Hughes, R. L., Thomson, J. A., and Owen, W. H. (1965). Reproduction in natural populations of the Australian ringtail possum, *Pseudocheirus peregrinus* (Marsupialia: Phalangeridae) in Victoria. *Aust. J. Zool.* **13,** 383–406.

Illingworth, D. V., and Perry, J. S. (1971). The effect of hypophysial stalk-section on the corpus luteum of the guinea-pig. *J. Endocrinol.* **50,** 625–635.

Ito, Y., and Higashi, K. (1953). Lactogenic factors in placenta. *J. Pharm. Soc. Jpn.* **73,** 89–93.

Ito, Y., and Higashi, K. (1961). Studies on the prolactin-like substance in human placenta. II and III. *Endocrinol. Jpn.* **8,** 279–287 and 288–296.

Jaitly, K. D., Robson, J. M., Sullivan, F. M., and Wilson, C. (1966). Hormonal requirements for the maintenance of gestation in hypophysectomized mice. *J. Endocrinol.* **34,** iv.

Johnson, G. E., and Challans, J. S. (1930). Ovariectomy and corpus luteum extract experiments in pregnant rats. *Anat. Rec.* **47,** 300–301.

Johnson, G. E., and Challans, J. S. (1932). Ovariectomy and corpus luteum extract studies on rats and ground squirrels. *Endocrinology* **16,** 278–284.

Joly, J. (1964). La présence de lipides biréfringents dans les corps jaunes post-ovulatoires de la salamandre tachetée. *C. R. Acad. Sci.* (*Paris*) **258,** 3563–3565.

Joly, J. (1965). Mise en évidence histochemique d'une delta-5-3-beta-hydroxystéroide-déshydrogénase dans l'ovaire de l'urodèle *Salamandra salamandra* (L) à différents stades du cycle sexuel. *C. R. Acad. Sci. (Paris)* **261,** 1569–1571.

Joly, J., and Picheral, B. (1972). Ultrastructure histochemie et physiologie du follicle pré-ovulatoire et du corps jaune de l'urodèle ovo-vivipare; *Salamandra salamandra* (L.). *Gen. Comp. Endocrinol.* **18,** 235–259.

Joly, J., Picheral, B., and Boisseau, C. (1969). Lutéinisation du follicule ovarien chez un amphibien ovo-vivipare, *Salamandra salamandra* L. Etude histochemique et ultrastructurale. *Gen. Comp. Endocrinol.* **13,** 510–511.

Josimovich, J. B., and Atwood, B. L. (1964). Human placental lactogen (HPL), a trophoblastic hormone synergizing with a chorionic gonadotropin and potentiating the anabolic effects of pituitary growth hormone. *Am. J. Obstet. Gynecol.* **88,** 867–879.

Josimovich, J. B., and MacLaren, J. A. (1962). Presence in the human placenta and term serum of a highly lactogenic substance immunologically related to pituitary growth hormone. *Endocrinology* **71,** 209–220.

Josimovich, J. B., Atwood, B. L., and Goss, D. A. (1963). Luteotrophic, immunologic and electrophoretic properties of human placental lactogen. *Endocrinology* **73,** 410–420.

Josimovich, J. B., Kosor, B., and Mintz, D. H. (1969). Roles of placental lactogen in foetal-maternal relations. *Foetal Auton., Ciba Found. Symp., 1968* pp. 117–131.

Jost, A. (1970). Hormonal factors in the sex differentiation of the mammalian foetus. *Philos. Trans. R. Soc. London, Ser. B* **259,** 119–132.

Jost, A., Jacquot, R., and Cohen, A. (1955). Sur les interrélations entre les hormones corticosurrénaliennes maternelles et la surrénale du foetus de rat. *C. R. Soc. Biol.* **149,** 1319–1322.

Kamell, S. A., and Atkinson, W. B. (1948). Absence of prolongation of pseudopregnancy by induction of deciduomata in the mouse. *Proc. Soc. Exp. Biol. Med.* **67,** 415–416.

Kaplan, N. M. (1961). Successful pregnancy following hypophysectomy during the twelfth week of gestation. *J. Clin. Endocrinol. Metab.* **21,** 1139–1145.

Kaplan, S. L., and Grumbach, M. M. (1964). Studies of a human and simian placental hormone with growth hormone-like and prolactin-like activities. *J. Clin. Endocrinol. Metab.* **24,** 80–100.

Kaplan, S. L., and Grumbach, M. M. (1965a). Serum chorionic "growth hormone prolactin" and serum pituitary growth hormone in mother and fetus at term. *J. Clin. Endocrinol. Metab.* **25,** 1370–1374.

Kaplan, S. L., and Grumbach, M. M. (1965b). Immunoassay for human chorionic "growth-hormone-prolactin" in serum and urine. *Science* **147,** 751–753.

Kehl, R. (1950). La nidation normale et pathologique. *Bull. Assoc. Gynecol. Obstet. Lang. Fr.* **2,** 68–70.

Kellas, L. M., van Lennep, E. W., and Amoroso, E. C. (1958). Ovaries of some foetal and prepubertal giraffes (*Giraffa camelopardalis,* Linnaeus). *Nature (London)* **181,** 487–488.

Keyes, P. L., and Armstrong, D. T. (1968). Endocrine role of follicles in the regulation of corpus luteum function in the rabbit. *Endocrinology* **83,** 509–515.

Keyes, P. L., and Armstrong, D. T. (1969). Development of corpora lutea from follicles autotransplanted under the kidney capsule in rabbits. *Endocrinology* **85,** 423–427.

Keyes, P. L., and Nalbandov, A. V. (1967). Maintenance and function of corpora lutea in rabbits depend on estrogen. *Endocrinology* **80,** 938-946.

Kidder, H. E. (1954). "Studies on fertilization failure and embryonic death: effects of the male and pre-ovulatory environment of ova." Thesis, University of Wisconsin, Madison.

Kirsch, R. E. (1938). A study in the control of length of gestation in the rat with notes on maintenance and termination of gestation. *Am. J. Physiol.* **122,** 86-93.

Klein, M. (1933). Sur l'ablation des embryons chez la lapine gravide et sur les facteurs qui déterminent le maintien du corps jaune pendant la deuxième partie de la grossesse. *C. R. Soc. Biol.* **113,** 441-443.

Klein, M. (1938). Relation between the uterus and the ovaries in the pregnant hamster. *Proc. R. Soc. London, Ser. B* **125,** 348-364.

Klein, M., and Mayer, G. (1942a). Effets d'injections d'oestrogènes sur l'ovaire gestatif de la rate. Maintien de l'état gestatif de l'ovaire après ablation de l'utérus gravide par des injections d'oestrogènes chez la rate. *Arch. Phys. Biol. Chim.-Phys. Corps Organ.* **16,** Suppl., 40 and 41-42.

Klein, M., and Mayer, G. (1942b). Sur la parturition partielle chez la rate. 1. Position du problème; technique expérimentale; comportement maternal et allaitement. 2. Réactions du tractus génital et de la glande mammaire. *Arch. Phys. Biol. Chim.-Phys. Corps Organ.* **16,** 125 and 127-128.

Klein, M., and Mayer, G. (1944). Corps jaune cyclique et taux d'oestrone chez la rate. Réaction de l'ovaire, du vagin et de la glande mammaire à l'administration d'oestrogène. *Arch. Phys. Biol. Chim.-Phys. Corps Organ.* **18,** 169.

Knight, J., and Fitzsimons, D. W. (eds.) (1976). The Fetus and Birth. Ciba Foundation Symp. (New Series) No. 47, Elsevier, Amsterdam (in press).

Knobil, E. (1973). On the regulation of the primate corpus luteum. *Biol. Reprod.* **8,** 246-258.

Krehbiel, R. H. (1941a). The effects of lactation on the implantation of ova of a concurrent pregnancy in the rat. *Anat. Rec.* **81,** 43-64.

Krehbiel, R. H. (1941b). The effects of theelin on delayed implantation in the pregnant lactating rat. *Anat. Rec.* **81,** 381-392.

Kroc, R. L., Steinetz, B. G., and Beach, V. L. (1958). The effects of estrogens, progestagens and relaxin in pregnant and non-pregnant laboratory rodents. *Ann. N.Y. Acad. Sci.* **75,** 942-980.

Kuhn, N. J. (1969a). Progesterone withdrawal as the lactogenic trigger in the rat. *J. Endocrinol.* **44,** 39-54.

Kuhn, N. J. (1969b). Specificity of progesterone inhibition of lactogenesis. *J. Endocrinol.* **45,** 615-616.

Kulseng-Hanssen, K. (1951). Maintenance of early pregnancy despite extirpation of the corpora lutea. *Acta Obstet. Gynecol. Scand.* **30,** 420-427.

Lambert, J. G. D. (1966). Location of hormone production in the ovary of the guppy, *Poecilia reticulata. Experientia* **22,** 476-477.

Lambert, J. G. D. (1970). The ovary of the guppy, *Poecilia reticulata.* The atretic follicle, a corpus atreticum or a corpus luteum preovulationis? *Z. Zellforsch. Mikrosk. Anat.* **107,** 54-67.

Lambert, J. G. D., and van Oordt, P. G. W. J. (1965). Preovulatory corpora lutea or corpora atretica in the guppy, *Poecilia reticulata.* A histological and histochemical study. *Gen. Comp. Endocrinol.* **5,** 693-694.

Lamotte, M., and Rey, P. (1954). Existence de corpora lutea chez un batracien anoure vivipare, *Nectophrynoides occidentalis* Angel; leur évolution morphologique. *C. R. Acad. Sci. (Paris)* **238,** 393-395.

Lamotte, M., Rey, P., and Vogeli, M. (1964). Recherches sur l'ovaire de *Nectophrynoides occidentalis* batracien anoure vivipare. *Arch. Anat. Microsc. Morphol. Exp.* **53**, 179–224.

Lance, V., and Callard, I. P. (1969). A histochemical study of ovarian function in the ovoviviparous elasmobranch, *Squalus acanthias*. *Gen. Comp. Endocrinol.* **13**, 255–267.

Laws, R. M. (1958). Recent investigations on fin whale ovaries. *Norw. Whaling Gaz.* **5**, 225–254.

Laws, R. M. (1969). Aspects of reproduction in the African elephant, *Loxodonta africana*. *J. Reprod. Fertil., Suppl.* **6**, 193–217.

LeMaire, W. J., Conly, P. W., Moffett, A., and Cleveland, W. M. (1970). Plasma progesterone secretion by the corpus luteum of term pregnancy. *Am. J. Obstet. Gynecol.* **108**, 132–134.

Leonard, S. L. (1943). Stimulation of mammary glands in hypophysectomized rats by estrogen and testosterone. *Endocrinology* **32**, 229–237.

Leonard, S. L. (1945). The relation of the placenta to the growth of the mammary gland of the rat during the last half of pregnancy. *Anat. Rec.* **91**, 65–71.

Liggins, G. C. (1969). The foetal role in the initiation of parturition in the ewe. *Foetal Auton., Ciba Found. Symp., 1968* pp. 218–244.

Liggins, G. C., Kennedy, P. C., and Holm, L. W. (1967). Failure of initiation of parturition after electro-coagulation of the pituitary of the foetal lamb. *Am. J. Obstet. Gynecol.* **98**, 1080–1086.

Lindner, A., Satke, I., and Voelkel, O. (1951). Die inverse Wirkung von Hexöstrol-implantationen auf den Geschlechtszyklus weiblicher Ratten. *Arch. Int. Phamacodyn. Ther.* **86**, 421–433.

Linzell, J. L., and Heap, R. B. (1968). Comparison of progesterone metabolism in the pregnant sheep and goat: sources of production and an estimate of uptake by some target organs. *J. Endocrinol.* **41**, 433–438.

Little, B., Smith, O. W., Kessiman, A. G., Selenkow, H. A., van't Hoff, W., Eglin, J. M., and Moore, F. D. (1958). Hypophysectomy during pregnancy in a patient with cancer of the breast: case report with hormone studies. *J. Clin. Endocrinol. Metab.* **18**, 425–443.

Loeb, L. (1907). Über die experimentelle Erzeugung von Knoten von Deci-duagewebe in dem Uterus des Meerschweinchens nach stattgefundener Copula-tion. *Zentralbl. Allg. Pathol. pathol. Anat.* **18**, 563–565.

Loeb, L. (1923). Mechanism of the sexual cycle with special reference to the corpus luteum. *Am. J. Anat.* **32**, 305–343.

Loeb, L., and Hesselberg, C. (1917). The cyclic changes in the mammary gland under normal and pathological conditions. 1. The changes in the non-pregnant guinea-pig. *J. Exp. Med.* **25**, 285–321.

Long, J. A., and Evans, H. M. (1922). The oestrous cycle in the rat and its associated phenomena. *Mem. Univ. Calif.* **6**, 1–111.

Loy, R. G., Zimbelman, R. G., and Casida, L. E. (1960). Effects of injected ovarian hormones on the corpus luteum of the estrual cycle in cattle. *J. Anim. Sci.* **19**, 175–182.

Lunaas, T. (1962). Urinary oestrogen levels in the sow during oestrous cycle and early pregnancy. *J. Reprod. Fertil.* **4**, 13–20.

Lupo di Prisco, C. (1968). Biosintesi di ormoni steroidi nell'ovario di due specie di selaci: *Scyliorhinus stellaris* (oviparo) e *Torpedo marmorata* (ovoviviparo). *Riv. Biol.* **61**, 113–146.

Lupo di Prisco, C., Vellano, C., and Chieffi, G. (1967). Steroid hormones in the plasma of the elasmobranch *Torpedo marmorata* at various stages of the sexual cycle. *Gen. Comp. Endocrinol.* **8**, 325–331.

Lupo di Prisco, C., Delrio, G., and Chieffi, G. (1968). Sex hormones in the ovaries of the lizard, *Lacerta sicula. Gen. Comp. Endocrinol.* **10**, 292–295.

Lyons, W. R. (1937). Preparation and assay of mammotropic hormone. *Proc. Soc. Exp. Biol. Med.* **35**, 645–648.

Lyons, W. R. (1951). Lobule-alveolar mammary growth in the rat. *Colloq. Int. C.N.R.S.*, 29.

Lyons, W. R., Li, C. H., and Johnson, R. E. (1958). The hormonal control of mammary growth and lactation. *Recent Prog. Horm. Res.* **14**, 219–248.

McDonald, L. E., Nichols, R. E., and McNutt, S. H. (1952). Studies on corpus luteum ablation and progesterone replacement therapy during pregnancy in the cow. *Am. J. Vet. Res.* **13**, 446–451.

McDonald, L. E., Nichols, R. E., and McNutt, S. H. (1953). On the essentiality of the bovine corpus luteum of pregnancy. *Am. J. Vet. Res.* **14**, 539–541.

McIlroy, A. L. (1912). Some experimental work upon the physiological function of the ovary. *J. Obstet. Gynaecol. Br. Emp.* **22**, 19–26.

McKeown, T., and Zuckerman, S. (1938). The suppression of oestrus in the rat during pregnancy and lactation. *Proc. R. Soc. London, Ser. B* **124**, 464–475.

McPhail, M. K. (1935a). Studies on the hypophysectomized ferret. 9. The effect of hypophysectomy on pregnancy and lactation. *Proc. R. Soc. London, Ser. B* **117**, 34–44.

McPhail, M. K. (1935b). Hypophysectomy of the cat. *Proc. R. Soc. London, Ser. B* **117**, 45–63.

Magnus, V. (1901). Ovariets betydning for svangerskabet med saerligt hensyn til corpus luteum. *Nor. Mag. Laegevidensk.* **62**, 1138–1145.

Mahoney, C. J. R. (1969). "The metrial gland of the rodent uterus." M.Sc. Thesis, University of Wales, Bangor.

Markee, J. E., Everett, J. W., and Sawyer, C. H. (1952). The relationship of the nervous system to the release of gonadotrophin in the regulation of the sex cycle. *Recent Prog. Horm. Res.* **7**, 139–163.

Marshall, F. H. A., and Jolly, W. A. (1905). The oestrous cycle in the dog. The ovary as an organ of internal secretion. *Proc. R. Soc. London, Ser. B* **76**, 395–398.

Marshall, F. H. A., and Jolly, W. A. (1906). The oestrous cycle in the dog. II. The ovary as an organ of internal secretion. *Philos. Trans. R. Soc. London, Ser. B* **198**, 123–141.

Martinet, L., Legouis, J-J., and Moret, B. (1970). Quelques observations sur la reproduction du lièvre européen (*Lepus europaeus* Pallas) en captivité. *Ann. Biol. Anim., Biochim., Biophys.* **10**, 195–202.

Masuda, H., Anderson, L. L., Henricks, D. M., and Melampy, R. M. (1967). Progesterone in ovarian venous plasma and corpora lutea of the pig. *Endocrinology* **80**, 240–246.

Matthews, L. H. (1941). The genitalia and reproduction of some African bats. *Proc. Zool. Soc. London B* **111**, 289–346.

Matthews, L. H. (1955). The evolution of viviparity in vertebrates. *Mem. Soc. Endocrinol.* **4**, 129–148.

Mayer, G. (1953). Histophysiologie de l'état gravidique. *C. R. Assoc. Anat.* **40**, 28–78.

Mayer, G. (1959a). L'ovo-implantation et la vie latente de l'oeuf. *Bull. Soc. Belge Gynecol. Obstet.* **29**, 1.

Mayer, G. (1959b). Recent studies on hormonal control of delayed implantation and superimplantation in the rat. *Mem. Soc. Endocrinol.* **6**, 76–83.

Mayer, G., and Canivenc, R. (1950). Placenta et lactogénèse. *In* "Mécanisme Physiologique de la Sécrétion Lactée," p. 125. CNRS, Paris.

Mayer, G., and Klein, M. (1955). Les hormones du placenta. *Rapp. Reun. Endocrinol. Long. Fr., Paris. 3rd, 1900* p. 47.

Medawar, P. B. (1953). Some immunological and endocrinological problems raised by the evolution of viviparity in vertebrates. *Symp. Soc. Exp. Biol.* **7**, 320–338.

Meites, J., and Sgouris, J. T. (1953). Can the ovarian hormones inhibit the mammary response to prolactin? *Endocrinology* **53**, 17–23.

Meites, J., and Turner, C. W. (1948a). Studies concerning the induction and maintenance of lactation. II. The influence of various factors on the lactogen content of the pituitary. *Mo., Agr. Exp. Sta., Res. Bull.* **415.**

Meites, J., and Turner, C. W. (1948b). Studies concerning the induction of lactation. I. The experimental maintenance and experimental inhibition and augmentation of lactation. *Mo., Agr. Exp. Sta., Res. Bull.* **416.**

Meites, J., Webster, H. D., Young, F. W., Thorpe, F. J., and Hatch, R. N. (1951). Effect of corpora lutea ablation and replacement therapy with progesterone on gestation in goats. *J. Anim. Sci.* **10**, 411–416.

Melinkoff, E. (1950). Questionable necessity of the corpus luteum. *Am. J. Obstet. Gynecol.* **60**, 437–439.

Merchant, J. C., and Sharman, G. B. (1966). Observations on the attachment of marsupial pouch young to the teats and on the rearing of pouch young by foster mothers of the same or different species. *Aust. J. Zool.* **14**, 593–609.

Meyer, R. (1911). Ueber Corpus luteum-Bildung beim Menschen. *Arch. Gynaekol.* **93**, 354–405.

Miller, J. W., Kisley, A., and Murray, W. J. (1957). The effects of relaxin-containing ovarian extracts on various types of smooth muscle. *J. Pharmacol. Exp. Ther.* **120**, 426–437.

Miller, M. R. (1948). The seasonal histological changes occurring in the ovary, corpus luteum and testis of the viviparous lizard, *Xantusia vigilis. Univ. Calif., Berkeley, Publ. Zool.* **47**, 197–224.

Mixner, J. P., and Turner, C. W. (1943). The mammogenic hormones of the anterior pituitary. II. The lobule-alveolar growth factor. *Mo., Agr. Exp. Sta., Res. Bull.* **378.**

Moor, R. M. (1968). Effect of embryo on corpus luteum function. *J. Anim. Sci.* **27**, Suppl. 1, 97–118.

Moor, R. M., and Rowson, L. E. A. (1964). Influence of the embryo and uterus on luteal function in the sheep. *Nature (London)* **201**, 522–523.

Moor, R. M., and Rowson, L. E. A. (1966a). The corpus luteum of the sheep: functional relationship between the embryo and the corpus luteum. *J. Endocrinol.* **34**, 233–239.

Moor, R. M., and Rowson, L. E. A. (1966b). The corpus luteum of the sheep: effect of the removal of embryos on luteal function. *J. Endocrinol.* **34**, 497.

Moor, R. M., and Rowson, L. E. A. (1966c). Local maintenance of the corpus luteum in sheep with embryos transferred to various isolated portions of the uterus. *J. Reprod. Fertil.* **12**, 539–550.

Moore, W. W., and Nalbandov, A. V. (1953). Neurogenic effects of uterine disten-sion on the estrous cycle of the ewe. *Endocrinology* **53**, 1–11.

Mossman, H. W., and Judas, I. (1949). Accessory corpora lutea, lutein cell origin, and the ovarian cycle in the Canadian porcupine. *Am. J. Anat.* **85**, 1–39.

Moudgal, N. R., Madhwa Raj, H. G., Jagannadha Rao, A., and Sairam, M. R. (1969). Need of luteinizing hormone for maintaining early pregnancy in rat. *Indian J. Exp. Biol.* **7**, 45–46.

Nalbandov, A. V., Moore, W. W., and Norton, H. W. (1955). Further studies on the neurogenic control of the estrous cycle by uterine distension. *Endocrinology* **56**, 225–231.

Nandi, J. (1967). Comparative endocrinology of steroid hormones in vertebrates. *Am. J. Zool.* **7**, 115–133.

Nandi, S. (1958). Endocrine control of mammary gland development and function, in the C3H/He Crgl mouse. *J. Natl. Cancer Inst.* **21**, 1039–1063.

Nandi, S. (1959). Hormonal control of mammogenesis and lactogenesis in the C3H/He Crgl mouse. *Univ. Calif. Berkeley, Publ. Zool.* **65**, 1–128.

Neher, G. M., and Zarrow, M. X. (1954). Concentration of progestin in the serum of the non-pregnant, pregnant and post partum ewe. *J. Endocrinol.* **11**, 323–330.

Nelson, W. O. (1934). Studies on the physiology of lactation. III. The reciprocal hypophysial-ovarian relationship as a factor in the control of lactation. *Endocrinology* **18**, 33–46.

Nelson, W. O. (1935). The effect of hypophysectomy upon mammary gland develop-ment and function in the guinea pig. *Proc. Soc. Exp. Biol. Med.* **33**, 222–224.

Nelson, W. O. (1936). Endocrine control of the mammary gland. *Physiol. Rev.* **16**, 488–526.

Nelson, W. O. (1937). Studies on the physiology of lactation. VI. The endocrine influences concerned in the development and function of the mammary gland in the guinea pig. *Am. J. Anat.* **60**, 341–360.

Newman, H. H., and Paterson, J. J. T. (1910). Development of the nine-banded armadillo from the primitive streak stage to birth: with special reference to the question of specific polyembryony. *J. Morphol.* **21**, 359–423.

Newsome, A. E. (1964). Anoestrus in the red kangaroo, *Megaleia rufa* (Desmarest). *Aust. J. Zool.* **12**, 9–17.

Newton, W. H. (1935). 'Pseudo-parturition' in the mouse, and the relation of the placenta to the post-partum oestrus. *J. Physiol. (London)* **84**, 196–207.

Newton, W. H. (1938). Hormones and the placenta. *Physiol. Rev.* **18**, 419–446.

Newton, W. H. (1949). "Recent Advances in Physiology." Churchill, London.

Newton, W. H., and Beck, N. (1939). Placental activity in the mouse in the absence of the pituitary gland. *J. Endocrinol.* **1**, 65–75.

Newton, W. H., and Lits, F. J. (1938). Criteria of placental endocrine activity in the mouse. *Anat. Rec.* **72**, 333–348.

Newton, W. H., and Richardson, K. C. (1940). The secretion of milk in hypophysec-tomized pregnant mice. *J. Endocrinol.* **2**, 322–328.

Niswender, G. D., and Dziuk, P. J. (1966). A study of the unilateral relationship between the embryo and the corpus luteum by egg transfer in the ewe. *Anat. Rec.* **154**, 394–395.

O'Donoghue, C. H. (1911). The growth-changes in the mammary apparatus of *Dasyurus* and the relation of the corpora lutea thereto. *Q. J. Microsc. Sci.* **57**, 187–235.

O'Donoghue, C. H. (1916). On the corpora lutea and interstitial tissue of the ovary in the Marsupialia. *Q. J. Microsc. Sci.* **61**, 433–473.

Orsini, M. W., and Meyer, R. K. (1959). Implantation of the castrate hamster in the absence of exogenous oestrogen. *Anat. Rec.* **134**, 619–620.

Panigel, M. (1956). Contribution à l'étude de l'ovoviviparité chez les reptiles: gestation et parturition chez le lézard vivipare, *Zootoca vivipara*. *Ann. Sci. Nat., Zool. Biol. Anim.* [11] **18**, 569–668.

Parkes, A. S. (1928). The role of the corpus luteum in the maintenance of pregnancy. *J. Physiol. (London)* **65**, 341–349.

Parkes, A. S., and Bellerby, C. W. (1927). Studies on the internal secretion of the ovary. III. The effects of injection of oestrin during lactation. *J. Physiol. (London)* **62**, 301–314.

Parkes, A. S., and Deanesly, R. (1966). Relation between the gonads and the adrenal glands. *In* "Marshall's Physiology of Reproduction" (A. S. Parkes, ed.), 3rd ed., Vol. 3, pp. 1064–1111. Longmans Green, London and New York.

Pasteels, J. L. (1970). Control of prolactin secretion. *In* "The Hypothalamus" (L. Martini, M. Motta, and F. Fraschini, eds.), pp. 385–399. Academic Press, New York.

Pearson, O. P. (1949). Reproduction of a South American rodent, the mountain viscacha. *Am. J. Anat.* **84**, 143–173.

Pencharz, R. I., and Long, J. A. (1931). The effect of hypophysectomy on gestation in the rat. *Science* **74**, 206.

Pencharz, R. I., and Long, J. A. (1933). Hypophysectomy in the pregnant rat. *Am. J. Anat.* **53**, 117–139.

Pencharz, R. I., and Lyons, W. R. (1934). Hypophysectomy in the pregnant guinea-pig. *Proc. Soc. Exp. Biol. Med.* **31**, 1131–1132.

Perry, J. S. (1953). The reproduction of the African elephant, *Loxodonta africana*. *Philos. Trans. R. Soc. London, Ser. B* **237**, 93–149.

Perry, J. S., Heap, R. B., and Amoroso, E. C. (1973). Steroid hormone production by pig blastocysts. *Nature (London)* **245**, 45–47.

Perry, J. S., Heap, R. B., and Burton, R. D. (1975). Oestrogen metabolism by trophoblast and endometrium and its probable relation to luteal maintenance in the pig. *J. Endocrinol.* **64**, 40–41P.

Pickford, G. E., and Atz, J. W. (1957). "The Physiology of the Pituitary Gland of Fishes." Zoological Society, New York.

Pilton, P. E. (1961). Reproduction in the great grey kangaroo. *Nature (London)* **189**, 984–985.

Pilton, P. E., and Sharman, G. B. (1962). Reproduction in the marsupial *Trichosurus vulpecula*. *J. Endocrinol.* **25**, 119–136.

Pincus, G., and Werthessen, N. T. (1938). The maintenance of embryo life in ovariectomized rabbits. *Am. J. Physiol.* **124**, 484–490.

Poole, W. E., and Pilton, P. E. (1964). Reproduction in the grey kangaroo, *Macropus canguru*, in captivity. *CSIRO Wildl. Res.* **9**, 218–234.

Porter, D. G. (1970). The failure of progesterone to affect myometrial activity in the guinea-pig. *J. Endocrinol.* **46**, 425–434.

Porto, A. (1942). Sobre a presenca de progesterone no corpo amarelo de serpentes ovovivipares. *Mem. Inst. Butantan Sao Paulo* **15**, 27–30.

Potter, E. L. (1953). "Pathology of the Fetus and Infant." Yearbook Publ., Chicago, Illinois.

Prenant, L. A. (1898). De la valeur morphologique du corps jaune, son action physiologique et thérapeutique possible. *Rev. Gen. Sci.* **9**, 646–650.

Price, D. (1970). *In-vitro* studies on differentiation of the reproductive tract. *Philos. Trans. R. Soc. London, Ser. B* **259**, 133–139.

Price, M. (1953). The reproductive cycle of the water shrew, *Neomys fodiens bicolor* Shaw. *Proc. Zool. Soc. London* **123**, 599–621.

Psychoyos, A. (1967). The hormonal interplay controlling egg-implantation in the rat. *Adv. Reprod. Physiol.* **2**, 257–277.

Raeside, J. I. (1963). Urinary oestrogen excretion in the pig during pregnancy and parturition. *J. Reprod. Fertil.* **6**, 427–431.

Raeside, J. I., and Turner, C. W. (1950). Progesterone in the maintenance of pregnancy in dairy heifers. *J. Anim. Sci.* **9**, 681.

Rahn, H. (1938). The corpus luteum of reptiles. *Anat. Rec.* **72**, Suppl., 55.

Rahn, H. (1939). Structure and function of placenta and corpus luteum in viviparous snakes. *Proc. Soc. Exp. Biol. Med.* **40**, 381–382.

Ray, E. W., Averill, S. C., Lyons, W. R., and Johnson, R. E. (1955). Rat placental hormonal activities corresponding to those of pituitary mammotropin. *Endocrinology* **56**, 359–373.

Raynaud, A. (1942). Modification expérimentale de la différenciation sexuelle des embryons de souris par action des hormones androgènes et oestrogènes. *Actual. Sci. Ind.* Nos. **925–926.**

Renfree, M. B. (1974). Ovariectomy during gestation in the American opossum, *Didelphis marsupialis virginiana. J. Reprod. Fertil.* **39**, 127–130.

Reynolds, H. C. (1952). Studies on reproduction in the opossum, *Didelphis virginiana. Univ. Calif., Berkeley, Publ. Zool.* **52**, 223–283.

Robboy, S. J., and Kahn, R. H. (1964). Electrophoretic separation of hydrolytic enzymes of the female rat reproductive tract. *Endocrinology* **75**, 97–103.

Roberts, C. M., and Weir, B. J. (1973). Implantation in the plains viscacha, *Lagostomus maximus. J. Reprod. Fertil.* **33**, 299–307.

Robinson, R., Baker, R. D., Anastassiadas, P. A., and Common, R. H. (1970). Estrone concentrations in the peripheral blood of pregnant cows. *J. Dairy Sci.* **53**, 1592–1595.

Robinson, T. J. (1951). Reproduction in the ewe. *Biol. Rev. Cambridge Philos. Soc.* **26**, 121–157.

Robson, J. M. (1936). Maintenance of pregnancy in the hypophysectomized rabbit with progestin. *J. Physiol. (London)* **86**, 415–424.

Robson, J. M. (1937a). Maintenance by oestrin of the luteal function in hypophysectomized rabbits. *J. Physiol. (London)* **90**, 435–439.

Robson, J. M. (1937b). Maintenance of pregnancy and of the luteal function in the hypophysectomized rabbit. *J. Physiol. (London)* **90**, 145–166.

Robson, J. M. (1947). "Recent Advances in Sex and Reproductive Physiology." Churchill, London.

Rombauts, P. (1962). Excrétion urinaire d'oestrogènes chez la truie pendant la gestation. *Ann. Biol. Anim., Biochim., Biophys.* **2**, 151–156.

Rothchild, I. (1960). The corpus luteum-pituitary relationship: the association between the cause of luteotrophin secretion and the cause of follicular quiescence during lactation; the basis for a tentative theory of the corpus luteum-pituitary relationship in the rat. *Endocrinology* **67**, 9–41.

Rowlands, I. W., and Perry, J. S. (1964). Luteal function in the guinea-pig. *Bienn. Rep. Inst. Anim. Physiol., Babraham, 1962/1963* pp. 34–35.

Roy, E. J., and Mackay, R. (1962). The concentration of oestrogens in blood during pregnancy. *J. Obstet. Gynaecol. Br. Commonw.* **69**, 13–17.

Rubin, B. L. (1968). Assay of Δ^5-3β-hydroxysteroid dehydrogenase in fish ovaries. *Gen. Comp. Endocrinol.* **11**, 251–253.

Rugh, R. (1935). Ovulation in the frog. 2. Follicular rupture to fertilization. *J. Exp. Zool.* **71**, 163–189.

Schmaltz, R. (1921). "Das Geschlechtsleben der Haussaugetiere." Richarts Schoets, Berlin.

Schofield, B. M. (1957). The hormonal control of myometrial function during pregnancy. *J. Physiol. (London)* **138**, 1–10.

Sciarra, J. J., Kaplan, S. L., and Grumbach, M. M. (1963). Localization of antihuman growth hormone within the human placenta: evidence for a human chorionic "growth hormone-prolactin." *Nature (London)* **199**, 1005–1006.

Scott, R. S., and Rennie, P. I. C. (1971). An oestrogen receptor in the corpora lutea of the rabbit. *Abstr. Asia Oceania Congr. Endocrinol., 4th, New Zealand* p. 85.

Selye, H. (1934). Influence of the uterus on the ovary and mammary gland. *Proc. Soc. Exp. Biol. Med.* **31**, 488–490.

Selye, H., Collip, J. B., and Thomson, D. L. (1933a). Effect of hypophysectomy upon pregnancy and lactation. *Proc. Soc. Exp. Biol. Med.* **30**, 589–590.

Selye, H., Collip, J. B., and Thomson, D. L. (1933b). Effect of hypophysectomy upon pregnancy and lactation in mice. *Proc. Soc. Exp. Biol. Med.* **31**, 82–83.

Selye, H., Collip, J. B., and Thomson, D. L. (1935a). Endocrine interrelationships during pregnancy. *Endocrinology* **19**, 151–159.

Selye, H., Collip, J. B., and Thomson, D. L. (1935b). Effects of oestrin on ovaries and adrenals. *Proc. Soc. Exp. Biol. Med.* **32**, 1377–1381.

Selye, H., Browne, J. S. L., and Collip, J. B. (1936). Effect of combined administration of oestrone and progesterone in adult ovariectomized rats. *Proc. Soc. Exp. Biol. Med.* **34**, 198–200.

Sharman, G. B. (1955a). Studies on marsupial reproduction. II. The oestrous cycle of *Setonix brachyurus*. *Aust. J. Zool.* **3**, 44–55.

Sharman, G. B. (1955b). Studies on marsupial reproduction. III. Normal and delayed pregnancy in *Setonix brachyurus*. *Aust. J. Zool.* **3**, 56–70.

Sharman, G. B. (1956). Some aspects of marsupial reproduction. *Proc. Zool. Soc. London* **127**, 141–143.

Sharman, G. B. (1959a). Marsupial reproduction. *In* "Biogeography and Ecology in Australia" (W. Junk, ed.), Vol. VIII, pp. 332–368. Monographiae Biologicae, The Hague.

Sharman, G. B. (1959b). Evolution of marsupials. *Aust. J. Sci.* **22**, 40–45.

Sharman, G. B. (1962). The initiation and maintenance of lactation in the marsupial *Trichosurus vulpecula*. *J. Endocrinol.* **25**, 375–385.

Sharman, G. B. (1963). Delayed implantation in marsupials. *In* "Delayed Implantation" (A. C. Enders, ed.), pp. 3–14. Univ. of Chicago Press, Chicago, Illinois.

Sharman, G. B. (1964). The female reproductive system of the red kangaroo, *Megaleia rufa*. *CSIRO Wildl. Res.* **9**, 50–57.

Sharman, G. B. (1965a). Marsupials and the evolution of viviparity. *Viewpoints Biol.* **4**, 1–28.

Sharman, G. B. (1965b). The effects of the suckling stimulus and oxytocin injection on the corpus luteum of delayed implantation in the red kangaroo. *Proc. Int. Congr. Endocrinol., 2nd, 1964* Excerpta Med. Found. Int. Congr. Ser. No. 83, p. 452–457.

Sharman, G. B. (1970a). The biology of sex in marsupials. *Aust. J. Sci.* **32**, 307–314.

Sharman, G. B. (1970b). Reproductive physiology of marsupials. *Science* **167**, 1221–1228.

Sharman, G. B., and Berger, P. J. (1969). Embryonic diapause in marsupials. *Adv. Reprod. Physiol.* **4**, 211–240.

Sharman, G. B., and Calaby, J. H. (1964). Reproductive behaviour in the red kangaroo, *Megaleia rufa* in captivity. *CSIRO Wildl. Rev.* **9**, 58–85.

Sharman, G. B., and Clark, M. J. (1967). Inhibition of ovulation by the corpus luteum in the red kangaroo, *Megaleia rufa. J. Reprod. Fertil.* **14**, 129–137.

Sharman, G. B., Calaby, J. H., and Poole, W. E. (1966). Patterns of reproduction in female diprotodont marsupials. *Symp. Zool. Soc. London* **15**, 205–232.

Shelesnyak, M. C. (1960). Nidation of the fertilized ovum. *Endeavour* **74**, 81–86.

Short, R. V. (1966). Oestrous behaviour, ovulation and the formation of the corpus luteum in the African elephant, *Loxodonta africana. East Afr. Wildl. J.* **4**, 56–68.

Short, R. V., and Hay, M. F. (1965). Delayed implantation in the roe deer. *J. Reprod. Fertil.* **9**, 372–374.

Smith, C. L. (1955). Reproduction in female Amphibia. *Mem. Soc. Endocrinol.* **4**, 39–56.

Smith, P. E. (1946). Non-essentiality of hypophysis for maintenance of pregnancy in rhesus monkeys. *Anat. Rec.* **94**, 497.

Smith, P. E. (1954). Continuation of pregnancy in the rhesus monkey (*Macaca mulatta*) following hypophysectomy. *Endocrinology* **55**, 655–664.

Smith, P. E., and White, W. E. (1931). The effect of hypophysectomy on ovulation and corpus luteum formation in the rabbit. *J. Am. Med. Assoc.* **97**, 1861–1863.

Smithberg, M., and Runner, M. N. (1956). The induction and maintenance of pregnancy in prepubertal mice. *J. Exp. Zool.* **133**, 441–458.

Sobotta, J. (1895). Zur Frage der Wanderung des Säugetiereies durch den Eileiter. *Arch. Mikrosk. Anat.* **45**, 15–93.

Sod-Moriah, U. A., and Schmidt, G. H. (1968). Deoxyribonucleic acid content and proliferative activity of rabbit mammary gland epithelial cells. *Exp. Cell Res.* **49**, 584–597.

Spies, H. G., Zimmerman, D. R., Self, H. L., and Casida, L. E. (1960). Maintenance of early pregnancy in ovariectomized gilts treated with gonadal hormones. *J. Anim. Sci.* **19**, 114–118.

Steinetz, B. G., Beach, V. L., and Kroc, R. L. (1959). The physiology of relaxin in laboratory animals. *In* "Recent Progress in the Endocrinology of Reproduction" (C. W. Lloyd, ed.), pp. 389–423. Academic Press, New York.

Suzuki, Y. (1947). Studies on the physiology of the corpus luteum hormone. 2. The fate of the corpus luteum hormone during pregnancy in mares. *Jpn. J. Vet. Sci.* **9**, 149.

Sykes, J. F., and Wrenn, T. R. (1951). Hormonal development of mammary tissue in dairy heifers. *J. Dairy Sci.* **34**, 1174–1179.

Te Winkel, L. E. (1950). Notes on ovulation, ova and early development in the smooth dogfish, *Mustelus canis. Biol. Bull. (Woods Hole, Mass.)* **99**, 474–486.

Thorburn, G. D., and Schneider, W. (1972). The progesterone concentration in the plasma of the goat during the oestrous cycle and pregnancy. *J. Endocrinol.* **52**, 23–36.

Thorburn, G. D., Nicol, D. H., Bassett, J. M., Shutt, D. A., and Cox, R. I. (1972).

Parturition in the goat and sheep: changes in corticosteroids, progesterone, oestrogens and prostaglandin F. *J. Reprod. Fertil., Suppl.* **16,** 61–84.

Thornton, V. F. (1971). A bioassay for progesterone and gonadotropins based on the meiotic division of *Xenopus* oocytes *in vitro. Gen. Comp. Endocrinol.* **16,** 599–605.

Trentin, J. J., and Turner, C. W. (1948). The experimental development of the mammary gland with special reference to the interaction of the pituitary and ovarian hormones. *Mo., Agr. Exp. Sta., Res. Bull.* **418,** 48.

Tuchmann-Duplessis, H. (1945). Correlations hypophysoendocrines chez le triton. *Actual. Sci. Ind.* **987.**

Tullner, W. W., and Hertz, R. (1966). Normal gestation and chorionic gonadotropin levels in the monkey after ovariectomy in early pregnancy. *Endocrinology* **78,** 1076–1078.

Tulsky, A. S., and Koff, A. K. (1957). Some observations on the role of the corpus luteum in early human pregnancy. *Fertil. Steril.* **8,** 118–130.

Turnbull, A. C., and Anderson, A. B. M. (1969a). The influence of the foetus on myometrial contractility. *Ciba Found. Study Group* **34,** 106–119.

Turnbull, A. C., and Anderson, A. B. M. (1969b). Evidence of a foetal role in determining the length of gestation. *Postgrad. Med. J.* **45,** 65–67.

Turner, C. L. (1938). Adaptations for viviparity in embryos and ovary in *Anableps anableps. J. Morphol.* **62,** 323–342.

Turner, C. W. (1939). The mammary glands. *In* "Sex and Internal Secretions" (E. Allen, ed.), 2nd ed., pp. 740–803. Williams & Wilkins, Baltimore, Maryland.

Turner, C. W., and Gomez, E. T. (1934). The experimental development of the mammary gland. II. The male and female dog. *Mo., Agr. Exp. Sta., Res. Bull.* **207.**

Tyndale-Biscoe, C. H. (1955). Observations on the reproduction and ecology of the brush-tailed possum *Trichosurus vulpecula* Kerr. *Aust. J. Zool.* **3,** 162–184.

Tyndale-Biscoe, C. H. (1963a). Effects of ovariectomy in the marsupial *Setonix brachyurus. J. Reprod. Fertil.* **6,** 25–40.

Tyndale-Biscoe, C. H. (1963b). Blastocyst transfer in the marsupial *Setonix brachyurus. J. Reprod. Fertil.* **6,** 41–48.

Tyndale-Biscoe, C. H. (1963c). The role of the corpus luteum in the delayed implantation of marsupials. *In* "Delayed Implantation" (A. C. Enders, ed.), pp. 15–32. Univ. of Chicago Press, Chicago, Illinois.

Tyndale-Biscoe, C. H. (1966). The marsupial birth canal. *Symp. Zool. Soc. London* **15,** 233–250.

Tyndale-Biscoe, C. H. (1970). Resumption of development by quiescent blastocysts transferred to primed ovariectomized recipients in the marsupial *Macropus eugenii. J. Reprod. Fertil.* **23,** 25–32.

Tyndale-Biscoe, C. H. (1973). "Life of Marsupials." Arnold, London.

Uyldert, I. E., David, K. G., and Freud, J. (1940). Mammary growth in rats. *Acta Brev. Neerl. Physiol. Pharmacol., Microbiol.* **10,** 105–109.

Valle, J. R., and Valle, L. A. R. (1943). Gonadal hormones in snakes. *Science* **97,** 400.

van der Horst, C. J., and Gillman, J. (1940). Ovulation and corpus luteum formation in *Elephantulus. S. Afr. J. Med. Sci.* **5,** 73–91.

van der Horst, C. J., and Gillman, J. (1946). The corpus luteum of *Elephantulus* during pregnancy; its form and function. *S. Afr. J. Med. Sci.* **11,** Biol. Suppl., 87–111.

Velardo, J. T., Olsen, A. G., Hisaw, F. L., and Dawson, A. B. (1953). The influence of decidual tissue upon pseudopregnancy. *Endocrinology* **53**, 216–220.

Venable, J. H., and McDonald, L. E. (1958). Postparturient bovine uterine motility—normal and after experimentally produced retention of the fetal membranes. *Am. J. Vet. Res.* **19**, 308–313.

Venning, E. (1957). The secretion of various hormones and the activity of the adrenal cortex in pregnancy. *In* "Gestation" (C. A. Villee, ed.), pp. 71–89. Josiah Macy, Jr. Found., New York.

Vilter, V., and Lugand, A. (1959a). Trophisme intra-utérin et croissance embryoannaire chez le *Nectophrynoides occidentalis* Ang., Crapaud totalement vivipare du mont Nimba (Haute Guinée). *C. R. Soc. Biol.* **153**, 29–32.

Vilter, V., and Lugand, A. (1959b). Recherches sur le déterminisme interne et externe du corps jaune gestatif chez le crapaud vivipare du mont Nimba, le *Nectophrynoides occidentalis* Ang., de la Haute Guinée. *C. R. Soc. Biol.* **153**, 294–297.

Vilter, V., and Vilter, A. (1960). Sur la gestation de la salamandra noire des Alpes, la *Salamandra atra* Laur. *C. R. Soc. Biol.* **154**, 290–294.

von Mikulicz-Radecki, F. (1930). Untersuchungen über die Tubenkontraktionen mit Hilfe der Pertubation. *Zentralbl. Gynaekol.* **54**, 2183–2191.

von Mikulicz-Radecki, F., and Nahmmacher, W. (1925). Zur Physiologie der Tube. II. Mitteilung. Beobachtung von Fortbewegung korpuskulärer Elemente in der Kaninchentube durch Muskelkontraktionen. *Zentralbl. Gynaekol.* **49**, 2322–2327.

Votquenne, M. (1936). Relations physiologiques hormonales au cours de la gestation chez la chienne. Hypophysectomie. *C. R. Soc. Biol.* **122**, 91–93.

Weekes, H. C. (1934). The corpus luteum in certain oviparous and viviparous reptiles. *Proc. Linn. Soc. N.S.W.* **59**, 380–391.

Weichert, C. K. (1942). Experimental control of prolonged pregnancy in the lactating rat by means of oestrogen. *Anat. Rec.* **83**, 1–17.

Weir, B. J. (1966). Aspects of reproduction in chinchilla. *J. Reprod. Fertil.* **12**, 410–411.

Weir, B. J. (1971a). The reproductive organs of the female plains viscacha, *Lagostomus maximus*. *J. Reprod. Fertil.* **25**, 365–373.

Weir, B. J. (1971b). Some notes on reproduction in the Patagonian mountain viscacha, *Lagidium boxi* (Mammalia: Rodentia). *J. Zool.* **164**, 463–467.

Weir, B. J. (1974). Reproductive characteristics of hystricomorph rodents. *Symp. Zool. Soc. London* **34**, 265–301.

Westman, A., and Jacobsohn, D. (1937). Über Oestrin Wirkungen auf die Corpus luteum Funktion. *Acta Obstet. Gynecol. Scand.* **17**, 1–23.

White, W. E. (1933). The effect of hypophysectomy of the rabbit. *Proc. R. Soc. London, Ser. B* **114**, 64–79.

Whitney, R., and Burdick, H. O. (1939). Effect of massive doses of an estrogen in ova transport in ovariectomized mice. *Endocrinology* **24**, 45–49.

Whitten, W. K. (1958). Endocrine studies on delayed implantation in lactating mice. Role of the pituitary in implantation. *J. Endocrinol.* **16**, 435–440.

Wiest, W. G., and Kidwell, W. R. (1969). The regulation of progesterone secretion by ovarian dehydrogenases. *In* "The Gonads" (K. W. McKerns, ed.), pp. 295–325. Appleton, New York.

Wiest, W. G., Kidwell, W. R., and Balogh, K., Jr. (1968). Progesterone catabolism in the rat ovary: a regulatory mechanism for progestational potency during pregnancy. *Endocrinology* **82**, 844–859.

Wilkinson, J. F., and de Fremery, P. (1940). Gonadotropic hormones in the urine of the giraffe. *Nature* (*London*) **146**, 491.

Williams, F. W. (1952). Pituitary necrosis in diabetics during pregnancy. Houssay phenomenon in man. *Diabetes* **1**, 37–40.

Williamson, H. C. (1911). Report on the reproductive organs of *Sparus centrodontus* (Delaroche); *Sparus cantharus* (L.); *Sebastes dactylopterus* (Delaroche); and the ripe eggs and larvae of *Sparus centrodontus* and *Sebastes marinus*. *Glasgow Rep. Sci. Invest.* **1**, 1–35.

Wislocki, G. B. (1931). Notes on the female reproductive tract (ovaries, uterus and placenta) of the collared peccary (*Pecari angulatus bangsi,* Goldman). *J. Mammal.* **12**, 143–149.

Wislocki, G. B., and Snyder, F. F. (1933). The experimental acceleration of the rate of transport of ova through the Fallopian tube. *Bull. Johns Hopkins Hosp.* **52**, 379–386.

Wolstenholme, G. E. W., and Knight, J., eds. (1969). "Progesterone: its Regulatory Effect on the Myometrium," Ciba Found. Study Group No. 34. Churchill, London.

Wright, P. L. (1966). Observations on the reproductive cycle of the American badger (*Taxidea taxus*). *Symp. Zool. Soc. London* **15**, 27–46.

Xavier, F. (1970). Analyse du rôle des corpora lutea dans le mantien de la gestation chez *Nectophrynoides occidentalis* Ang. *C. R. Acad. Sci. Ser. D* **270**, 2018–2020.

Yeates, N. T. M. (1949). The breeding season of the sheep with particular reference to its modification by artificial means using light. *J. Agric. Sci.* **39**, 1–43.

Yeates, N. T. M. (1954). The effect of high air temperature on reproduction in the ewe. *J. Agric. Sci.* **43**, 199–203.

Yoshimi, T., Strott, C. A., Marshall, J. R., and Lipsett, M. B. (1969). Corpus luteum function in early pregnancy. *J. Clin. Endocrinol. Metab.* **29**, 225–230.

Yoshinaga, K., Hawkins, R. A., and Stocker, J. F. (1969). Estrogen secretion by the rat ovary *in vivo* during the estrous cycle and pregnancy. *Endocrinology* **85**, 103–112.

Zeilmaker, G. H. (1964). Quantitative studies on the effect of the suckling stimulus on blastocyst implantation in the rat. *Acta Endocrinol.* (*Copenhagen*) **46**, 483–492.

Zeiner, F. N. (1943). Studies on the maintenance of pregnancy in the white rat. *Endocrinology* **33**, 239–249.

Zondek, B. (1928). Die Schwangerschaftsdiagnose aus dem Harn durch Nachweis des Hypophysenvorderlappenhormons: Grundlagen und Technik der Methode. *Klin. Wochenschr.* **7**, 1404–1411.

Zuckerman, S., and Parkes, A. S. (1932). The menstrual cycle of the primates. V. The cycle of the baboon. *Proc. Zool. Soc. London* pp. 138–191.

9

The Ovary and Nervous System in Relation to Behavior

J. T. Eayrs, A. Glass, and Heidi H. Swanson

I. INTRODUCTION

The reproductive system is concerned with the survival of the species rather than, as is the case with other bodily systems, with that of the individual. It cannot be considered solely in terms of the gonads, reproductive tracts, and the endocrine organs which regulate fertilization and the transformation of the zygote into the fetus since, in most species, activities involving more than one individual and synchronized with environmental changes are necessary to ensure successful reproduction. The mediation of these activities involves the participation of the nervous system either overtly as manifest by behavior or through the reciprocal control of

399

endocrine activity through the intermediary of the pituitary gland. Neural mechanisms, for example, must be involved in the synchronization of copulatory activity with the state of the gonads not only to ensure fertilization but, in many species, to ensure that the young are born at a time of year favorable for their survival. Such mechanisms may be of considerable complexity as, for instance, are to be found in phenomena of seasonal migration (Rowan, 1931; Farner, 1964) in which the responsiveness of the nervous system to environmental stimuli as well as to feedback control by gonadal hormones has been demonstrated.

Since many reproductive phenomena require the complementary participation of both sexes, it would be artificial to consider the behavior of the female alone. Thus, although emphasis will be placed on interrelations between the ovary and the brain, the interaction between male and female must also be considered. The behavior of the female may be just as much affected by that of a male as by her own hormones, and certain physiological events in the female may be initiated by an action on the part of the male, e.g., the induction of ovulation in the cat and rabbit by the act of coitus (Ross *et al.,* 1963; Sawyer, 1966), while the advent of estrus in the females of several species (e.g., sheep, fox, and chimpanzee) is advanced in the presence of a male (Ford and Beach, 1965).

Most behavior related to survival of the species is influenced by sex hormones. Under the heading of sexual behavior may be included the desire to seek for a mate, the exhibition of various sensory clues, displays designed to attract a mate and to minimize aggression, and, finally, the act of copulation at a time propitious for fertilization. Peripheral behavior patterns associated with the selection and defence of territory and the establishment of hierarchal interrelations between members of a social group ensure adequate dispersal of population to match the food supply and the improvement of genetic stock by favoring reproduction of the best adapted individuals. Thus territorial marking and agonistic behavior usually come under the influence of gonadal secretions (Rothballer, 1967; Guhl, 1961). Finally, there must occur nest building and maternal behavior appropriate to the relative degree of maturity of the infants at birth.

No attempt has been made in this chapter to review completely the very extensive literature dealing with these topics. Rather, the salient features of the relationship between the ovary and brain have been illustrated by choice of examples drawn from mammalian reproductive physiology. Furthermore, although there is no reason to believe that neuroendocrine relationships in man differ fundamentally from those in other mammalian species, the impact of affective social and psychological phenomena, which either play a less significant part or are impossible to study in lower animals, has made it convenient to afford this topic independent consideration in the human being (see Section VI, p. 423).

II. ORGANIZATIONAL ROLE OF SEX HORMONES

The primary basis for sex differentiation is genetic. Both sexes are morphologically undifferentiated in the earliest stages of development and the sex chromosomes determine whether an ovary or a testis will develop. Development is then controlled by secretion from the fetal testis. The presence of androgens during development leads to differentiation of the reproductive tract along the male line. The presence of ovaries is not necessary for differentiation along the female line and there is no evidence that the ovaries secrete significant amounts of hormone (Critchlow and Bar-Sela, 1967; van der Werff ten Bosch, 1966). Chromosomal abnormalities are more often expressed through the lack or abnormal development of the testes than through faulty development of ovaries. In the absence of fetal androgens, whatever the genetic constitution, the organism will present a female phenotype (Gallien, 1965). The details of these processes have been described in other volumes of this work. Other recent reviews include those by Wells (1962), Price and Ortiz (1965), and Jost (1965).

In addition to controlling masculinization of the genital tract, fetal androgens are also active in the organization of brain centers concerned with regulation of the ovarian cycle and sexual behavior (see reviews by: Harris, 1964; Barraclough, 1967; Valenstein, 1968). Again, it appears that the brain is potentially bisexual. In the absence of male hormones (ovarian hormones not being required) certain hypothalamic centers acquire a specific rhythm which becomes manifest at puberty. Stimuli arising from this center lead to cyclic secretion of ovulation-inducing gonadotropic hormones by the pituitary. The action of androgens during fetal development is to suppress this cyclic activity so that the secretion of gonadotropin is tonic in the normal male or in the female masculinized by early administration of testosterone.

The first systematic experiments on the role of the gonads during early life were performed by Pfeiffer (1936) who found that ovaries, implanted into the anterior chamber of the eye of adult male rats castrated at birth, contained a normal complement of follicles and corpora lutea. If castration had been performed later in life, such ovaries contained only follicles, indicating that ovulation did not occur. Transplantation of testes into newborn females also resulted in a failure of ovulation in their own ovaries and persistent vaginal cornification. Removal of ovaries from newborn females or implantation into newborn males had no effect on the function of ovaries implanted into adults. The author suggested that the pituitary of the newborn rat was undifferentiated and that testicular secretions masculinized it so that it secreted gonadotropins in a tonic rather than cyclic fashion. Possibly because this interpretation of the findings could not be substantiated, this elegant work was largely ignored until the 1950's when essentially similar effects were produced by the injection of a single dose of testosterone propionate (TP) into female rats and mice within a few days of birth (Barraclough and Leathem, 1954; Barraclough, 1955, 1961).

The finding that small anovulatory ovaries of females treated with TP in infancy could be restored to normal size and function by transplantation into an untreated female (Bradbury, 1941) showed that the effect on the ovary is reversible and indirect, presumably mediated by the pituitary. Transplantation of hypophyses from untreated females to a site under the median eminence of TP-treated females and vice versa has shown that the TP effect on the pituitary is likewise reversible (Segal and Johnson, 1959). These early experiments have been repeated with many refinements which have led to the specification of the minimal dose of androgens (5–10 μg TP) needed to prevent ovulation in the female rat (Gorski and Barraclough, 1963; Swanson and van der Werff ten Bosch, 1964a,b) and the critical period (0–10 days after birth) (Swanson and van der Werff ten Bosch, 1964a; Gorski, 1968) during which they must be present. Conversely, the testes must be removed within a few days of birth for ovulatory cycles to occur in implanted ovaries (Grady and Phoenix, 1963; Harris, 1964). It is now well established that the cyclic pattern of gonadotropin secretion in the female rat is under hypothalamic control (Harris and Jacobsohn, 1952; Martinez and Bittner, 1956) and that the prevention of cyclic ovarian activity by the administration of androgens to infant female rats is mediated through the hypothalamus rather than by direct action in the pituitary as earlier proposed by Pfeiffer (1936). There is some evidence that the early action of androgens can be blocked by giving reserpine or chlorpromazine, which suggests the participation of higher nervous centers (Kikuyama, 1961, 1962; Ladosky *et al.,* 1968).

The fact that the neuromuscular responses which characterize reproductive behavior in the two sexes differ widely needs no emphasis, but rather raises the question of the extent to which such differences may be attributed to some specific interaction between male and female sex hormones and the nervous system. The possible interrelationship has been subjected to detailed experimental analysis by the procedure of giving, to both male and female, the hormone characteristic of the opposite sex. Such experiments, together with observations (e.g., Hammond, 1927; Beach, 1938, 1942b; Young and

Rundlett, 1939; Beach *et al.,* 1968) to the effect that at least some components of male reproductive behavior are normally exhibited by the female of many species, have made it clear that neural mechanisms responsible for mediating the behavior of either sex are present in both. These mechanisms must be laid down very early in life, possibly before birth (Ball, 1939; Beach, 1942f), for the lordosis reaction typical of the estrous guinea pig to be elicited from newborn young of either sex for a few hours after parturition (Boling *et al.,* 1939), and the emergence of complete sexual behavior can be greatly accelerated by giving either gonadotropic (Cole, 1937) or sex (Beach, 1942a) hormones. Furthermore, electrical stimulation of specific hypothalamic centers elicits the complete pattern of mating behavior of either sex (Roberts, 1970).

The capacity of both sexes to display the behavioral repertoire characteristic of either sex applies not only to overt copulatory behavior but also to more diffuse behavioral differences of temperament, aggression, and exploration, which are sexually dimorphic. Androgens present during development seem to facilitate the expression of masculine and suppress the expression of feminine behavior. Thus female monkeys, whose mothers had received testosterone during pregnancy, were born with masculinized genitalia and showed rough and tumble and chasing play characteristic of young males. They also showed more mounting and less "presenting" than normal females. These patterns of behavior occurred before puberty, at a time when the gonads were nonsecretory (Young *et al.,* 1964; Phoenix *et al.,* 1967; Goy, 1968). Similarly, in the rat and hamster, the level of exploratory activity of each sex is determined by the presence or absence of androgens in the perinatal period of life but is independent of the influence of gonadal hormones at the time of testing (Swanson, 1966, 1967; Valenstein, 1968).

Mating behavior, on the other hand, is activated by the presence of the appropriate sex hormones. Although behavior patterns characteristic of the opposite sex may be present under certain circumstances (the extent and frequency depending on the species), males have a lower threshold for the expression of male behavior and females for female behavior (Beach, 1968; Davidson, 1969). The function of androgens during development seems to be to facilitate the response to male hormones and to suppress the response to female hormones. In the male, the effects may be mediated both on the central nervous system and at the periphery; e.g., the size and sensitivity of the phallus may affect the success of copulation (Beach and Levinson, 1950). In the absence of androgens, the individual will have a female orientation of behavior. The presence or absence of the ovary during development does not influence these events. Ovarian hormones given in pharmacological doses interfere with both male and female behavior (Whalen and Nadler, 1963; Levine and Mullins, 1964; Phoenix *et al.,* 1967).

III. ACTIVATIONAL ROLE OF SEX HORMONES DURING PUBERTY

It has been proposed that the hypothalamus is extremely sensitive before puberty to the negative feedback action of gonadal hormones and that secretion of gonadotropin is inhibited by the presence of very small amounts of hormones secreted by the immature testis or ovary. As the age of puberty approaches, the hypothalamus becomes less sensitive to the inhibiting action of ovarian or testicular hormones and starts inducing the release of gonadotropins. These, in turn, stimulate growth and increased hormone secretion by the gonads until a new steady state is reached (Donovan and van der Werff ten Bosch, 1959). It is not known what developmental factors initiate the change in sensitivity of the hypothalamus. All other morphological, functional, and behavioral changes occurring at puberty are secondary to the secretion of sex hormones by the gonads. The events initiating ovarian development at puberty and the regulation of ovarian secretion are dealt with in detail in other chapters (see reviews by Donovan and van der Werff ten Bosch, 1965; van der Werff ten Bosch, 1966; Critchlow and Bar-Sela, 1967).

IV. ACTIVATIONAL ROLE OF SEX HORMONES IN THE ADULT

The maturation of the germ cells in almost all female mammals occurs at certain more or less regular intervals characteristic of the species. The presence of a mature egg is associated with estrous behavior whether an animal is a seasonal or continuous breeder, and whether ovulation is spontaneous or induced. The function of mating behavior is to achieve the union of male and female gametes at a time when both are ripe for fertilization. There are, therefore, two aspects to sexual behavior. The first is concerned primarily with seeking and attracting a male and the second with consummation of the sexual act (Beach, 1958). The role of the female tends to be more passive so that most of the seeking out and display is done by the male. Lack of attention by the male, however, often induces the female to attract, for example, by "presenting" in the monkey, or ear wiggling and darting in the rat. Further, there is evidence of increased activity and restlessness, as exemplified by increased wheel running activity of rats at estrus (Wang, 1923; Slonaker, 1924; Kennedy and Mitra, 1963) and other changes in behavior (Nalbandov, 1964). Of obviously functional significance is the striking decrease of aggressiveness of the females of certain species, such as the European and golden hamster, which normally prefer to live alone and will not tolerate other individuals near them unless they are in estrus (Eibl-Eibesfeldt, 1953; Payne and Swanson, 1970).

A. Estrous and Mating Behavior

1. General Nature

It is perhaps not surprising that ovarian influence on behavior should be reflected in the events associated with reproduction but, as Young (1941) has pointed out, the changes in the behavior of the female at the time of estrus were at one time thought to be slight and were overlooked during early investigations of the reproductive cycle in favor of morphological criteria. More recent studies, however, have centered upon correlations between the cyclic activity of the ovary and sexual receptivity, and have revealed the vital part played by the ovarian hormones in the elicitation of behavior patterns essential for the process of mating.

Behavioral estrus or "heat" may be taken to signify the stage of the reproductive cycle at which the female will most readily accept the male. The terms "estrus" and "heat" are used synonymously to signify the behavioral changes associated with sexual receptivity rather than physiological changes in the reproductive tract with which correlation is high, but not always complete (Young, 1941). Estrus is marked by changes in behavior which have been studied most extensively in the common laboratory animals, such as the rat, hamster, mouse, rabbit, guinea pig, species used in animal husbandry, and in subhuman primates, notably the baboon, rhesus macaque, and chimpanzee.

The onset of estrus is generally marked by the appearance of stereotyped patterns of behavior which, at least initially, seem little concerned with the copulatory act but which are later succeeded by the more specifically directed forms of activity associated with mating (for literature relating to different species, see Beach, 1948, 1965a, 1968; Young, 1941; Ford and Beach, 1965; Nalbandov, 1964; Cole and Cupps, 1969). Thus there occurs a change in attitude toward the male whose approaches and attentions are usually resisted at other times. These may take the form of an increased interest in the genitalia and attempts to attract attention in a variety of ways such as by approaching and retreating as in the rat and mouse, by licking and nuzzling as in cattle, or, in primates, by gesture or other antics. During this time, postures similar to or identical with those used for copulation may appear either spontaneously, on the approach of the male, or may be elicited by appropriate manual stimulation; in some species, male-like behavior, such as, pursuit of other females, mounting, and attempted copulation may appear. So characteristic are these responses that many have been used as quantitative measures of the duration and intensity of estrus (e.g., Hemmingsen, 1933; Young *et al.,* 1935; Ball, 1937b; Bard, 1940; Blandau *et al.,* 1941; Beach, 1943; Davidson *et al.,* 1968).

The complex patterns of behavior which emerge during estrus depend on the interaction of a number of factors. There is considerable evidence that various components of mating behavior are independently inherited. Two complete sets of alleles exist for components of male and female sexual behavior. Both sets are found in each individual but each set normally only finds expression in one sex. Sexual behavior is a sex-linked trait in which the transmission of the genetic contribution is the same for both sexes but the mode of expression is modified by the sexual milieu (Goy and Jackway, 1962). In the guinea pig, three independent genetic factors have been found to affect the character of estrus: the latency and duration of heat as well as the percentage of responsive animals have shown correlated inheritance associated with sensitivity to estradiol benzoate while the duration of maximum lordosis and the tendency for male-like mounting appear to be independently controlled (Dilger, 1962).

In addition, the vigor of male sexual behavior is inversely related to the vigor of lordosis displayed by females of the same genetic strain (Goy and Jackway, 1962). It is obvious that, like any other character involved in "fitness," sexual behavior appropriate to the socioecological environment is of selective importance in evolution. Thus copulation tends to be quick in animals which are preyed upon in the wild and prolonged in carnivores (Le Boeuf, 1967). Rodents with short estrous cycles of 4 to 5 days (rats, mice, and hamsters) depend upon coital stimulation for the maintenance of the functional corpus luteum necessary for pregnancy. A specific intromission pattern increases the chances of reproductive success between animals of the same strain (Wilson *et al.,* 1965; Diamond, 1970).

Both pre- and postpubertal experience play a role in the development of sexual behavior, the importance of social interaction varying between species and between the sexes (Beach, 1947a,b). Estrous female rats reared in complete isolation will copulate satisfactorily and immediately when placed with a male, this ability being realized at the same age as in animals reared with others of the same or opposite sex (Stone, 1926; Beach, 1942c). In guinea pigs of both sexes, "sex drive" is depressed by social isolation before puberty, remaining unchanged throughout life and unresponsive to hormones above a threshold level (Lehrman, 1962). Female rhesus monkeys raised without physical contact with the mother or other monkeys fail to develop normal sexual or maternal behavior (Harlow and Harlow, 1965).

2. Endocrine Basis of Estrous Behavior

a. Ovarian Cycle. There is abundant evidence correlating the follicular phase of the ovarian cycle with the changes in behavior which characterize estrus. The prepubertal female does not show any readily recognizable signs of estrus and fails to arouse the sexual interest of the male (Beach, 1947b). At puberty, the full pattern of response coincides with the first ovarian cycle, the onset of behavioral estrus being, in general, rapid and reaching a peak at a time when follicular development is maximal and the production of estrogen high (Beach, 1948). In ungulates, however, which have a relatively prolonged period of estrus, the increase in sexual receptivity is more gradual (McKenzie and Andrews, 1937; Altmann, 1939). Sexual drive in the adult rat reaches its peak toward the end of heat (Warner, 1927), just before ovulation occurs (Boling *et al.*, 1941a) and at a time when vaginal cornification is complete (Young *et al.*, 1941). Similar observations have been reported for the guinea pig (Myers *et al.*, 1936), a species in which the number of maturing follicles is a significant factor regulating the volume of mounting activity during heat (Young *et al.*, 1939). In this species, however, correlation between estrus and follicular development is not always complete in so far as heat may not occur even in the presence of large ovarian follicles, a state thought to be attributable to a low individual sensitivity to ovarian hormone (Young *et al.*, 1938).

A similar relationship between the ovarian cycle and estrous behavior is seen in species such as the rabbit, ferret, and cat, which do not ovulate spontaneously. In the rabbit, the ovary regresses during a period of anestrus during October and November. Subsequent follicular growth is accompanied by a recovery of sexual receptivity which lasts throughout the breeding season (Hammond and Marshall, 1925). This receptivity, accompanied by all the signs of estrous behavior, may, however, occur before follicles are ripe for ovulation, and copulations which occur during this time are necessarily sterile. In the ferret, too, estrus is associated with seasonal growth of follicles and once established may continue unabated until terminated by copulation (Hammond and Marshall, 1930). Likewise, estrous behavior in the cat, which comes into heat for a short spell two or three times a year (Bard, 1939), is almost invariably associated with a proestrous or estrous vaginal smear (Michael, 1958).

Correlations between sexual behavior and changes in the reproductive organs of primates have been studied by Ball and Hartman (1935) in the macaque and by Yerkes and Elder (1936) and Young and Orbison (1944) in the chimpanzee. Although social factors tend to obscure endocrine influence, these studies have demonstrated a persistent increase in estrous-like behavior during the follicular phase of the reproductive cycle, reaching a peak shortly before the time of ovulation and coinciding with the corresponding increase in tumescence of the sexual skin.

Implantation of estrogen into various regions of the brain has established the presence of certain centers in the hypothalamus (suprachiasmatic regions) where implants too small to have a systemic effect will induce estrous behavior (Lisk, 1969a,b, 1970). By this procedure sexual receptivity may be divorced from ovulation. Normally the two events are closely associated by the synchronization of stimuli from the hypothalamic centers controlling cyclic release of gonadotropins with the feedback action of ovarian hormones.

Not all sexual behavior, however, is directly related to normally occurring ovarian cycles. Several species, notably the rat (Blandau *et al.,* 1940) and rabbit (Hammond and Marshall, 1925), come into estrus immediately after parturition. The rat (Slonaker, 1934) and the mouse (Rollhäuser, 1949) may come into heat during pregnancy, the former as a result of an ovulation occurring at about the fourteenth or fifteenth postcoital day. The golden hamster will copulate throughout pregnancy up to one day before parturition (Bond, 1945), the mating response occurring at times corresponding to the estrous rhythm had pregnancy not supervened (Krehbiel, 1952).

b. Effect of Ovariectomy. When a male initiates copulatory activity, an estrous female will assume the position of lordosis which is designed to facilitate entry by the male. Under normal circumstances, synchronization of ovarian maturation and sexual receptivity is such that mating is almost always successful and results in pregnancy. Possibly because of the greater necessity for synchronization in the female, sexual behavior is only exhibited under the influence of the correct balance of sex hormones, presumably that associated with ovulation. Although most males require testosterone to initiate copulatory behavior on the first occasion, experienced males will often continue to show various amounts of mating behavior after castration (rat: Bloch and Davidson, 1968; Davidson, 1966; dog: Hart, 1968; cat: Rosenblatt and Aronson, 1958). However, almost all female mammals will fail to show any sexual behavior after castration unless the appropriate hormones are administered. In the cat (Michael, 1958) and the dog (Leathem, 1938) all trace of sexual behavior disappears after ovariectomy, and similar findings have been reported for the guinea pig (Hertz *et al.,* 1937), the mouse (Wiesner and Mirskaia, 1930), and the rat (Nissen, 1929; Beach, 1942c). Ball (1936) reported a decline in the sex interest of the monkey at varying intervals of time after removal of the ovaries and Young and Orbison (1944), who used quantitative measures of sexual behavior in the female chimpanzee, found that behavior was similar to that seen during the luteal phase of the menstrual cycle of the normal animal. The elimination of receptivity as a result of spaying is, however, neither immediate (Hammond and Marshall, 1925; Ball, 1936) nor universal. Both Brambell (1944) and Harris (1952) have observed repeated coitus in spayed rabbits. The spayed female rat continues to attract the male though resisting copulation (Koster, 1943), and female chimpanzees may continue to "present" and to permit coitus (Beach, 1947b). Variations in sexual and emotional behavior during the menstrual cycle in women are discussed in Section VI (p. 424).

c. Effects of Giving Homologous Sex Hormones. The ability of various hormones to accelerate the onset of estrus or to reverse the effects of ovariectomy has been studied in several species. Gonadotropic hormones, since they are ineffective in the ovariectomized animal (e.g., Leathem, 1938; Friedgood, 1939), may be regarded as having no influence on sexual behavior other than that mediated through their control of ovarian secretions (Ball, 1941a). Heat can be induced in the intact dog (Leathem, 1938), mouse (Wiesner and Mirskaia, 1930), cat (Windle, 1939; Friedgood, 1939), and rat (Witschi and Pfeiffer, 1935) by giving chorionic gonadotropin or extracts of menopausal urine which, therefore, may be presumed to possess some follicle-stimulating and luteinizing properties. Rats given follicle-stimulating hormone alone do not exhibit mating behavior, although show a permanently cornified vaginal smear and large ovarian follicles (Witschi and Pfeiffer, 1935), but will mate if subsequently given luteinizing hormone.

Such findings are in keeping with the observed effects of giving ovarian hormones. Administration of estrogen alone will give rise to estrous behavior in the intact cat (Bard, 1936), dog (Hancock, 1952), and sheep (Hammond, 1945) while, in the cow, mounting activity may occur with such vigor as to cause fracture of the pelvic bones (Hammond and Day, 1944; Folley and Malpress, 1944). In this species, similar behavior has been shown to be associated with the occurrence of cystic follicles whose estrogenic content is considerably higher than that present in the normal follicle (Yamanchi and Inni, 1954). In the castrate, however, species differences appear. The cat (Maes, 1939; Michael and Scott, 1957, 1964), dog (Beach *et al.*, 1968), rabbit (Klein, 1952; Sawyer and Everett, 1959; McDonald *et al.*, 1970; Palka and Sawyer, 1966), ferret (Marshall and Hammond, 1945), and monkey (Ball, 1936) all readily respond to small doses of estrogen alone, whereas the reaction of certain rodents such as the guinea pig (Dempsey *et al.*, 1936; Boling *et al.*, 1938; Young and Rundlett, 1939), mouse (Ring, 1944), rat (Nissen, 1929; Hemmingsen and Krarup, 1937; Boling and Blandau, 1939; Beach, 1942c), and hamster (Frank and Fraps, 1945) can be elicited only by relatively large doses and even then has a long latency, tending to be weak, infrequent, and irregular. Estrous behavior indistinguishable from normal can, however, be induced in these species by the subsequent injection of progesterone, although chemically related compounds such as pregnanedione and testosterone have proved ineffective (Hertz *et al.*, 1937).

It thus seems that the hormones necessary to induce estrous behavior vary between species. Except in primates, estrogen is necessary, but its activity may be enhanced or inhibited by the presence of progesterone. The exact balance between estrogens and progesterone is characteristic of the species and depends on whether ovulation is spontaneous or induced by coitus. The refractoriness produced by progesterone in certain species is probably associated with its increased secretion during the luteal phase, when fertilization should already have been achieved or when the egg is no longer available. In spontaneously ovulating rodents, there is a surge of estrogen (Hori *et al.*, 1968) followed 24 hours later by a sharp rise in plasma progesterone, coincident with the appearance of lordosis (Feder *et al.*, 1966, 1967). Experimentally, receptivity may be induced in spayed rats by "priming" with a small dose of estrogen, followed 24 hours later by progesterone; estrous behavior appears 4–6 hours after subcutaneous, but immediately after intravenous, injection of progesterone (Lisk, 1960). On the other hand, in animals which only ovulate after the stimulus of coitus, such as the cat and the rabbit, estrous behavior may be induced by giving estrogen alone. This is not surprising since normally progesterone is not secreted until after the release of LH stimulated by coitus. Furthermore, although progesterone may exert a slightly facilitating effect in the rabbit when given shortly after estrogen priming, delayed administration inhibits lordosis (Sawyer and Everett, 1959). This biphasic action of progesterone, also seen in the guinea pig (Zucker and Goy, 1967; Zucker, 1968), is related to the behavioral refractoriness which occurs during the luteal phase of a normal estrous cycle. In monkeys also, sexual interaction declines markedly during the progestational phase of the sexual cycle (Zuckerman, 1932; Ball, 1936; Michael *et al.*, 1967; Herbert, 1970) and the sexual excitability induced by estrogen in the spayed individual is reduced by giving progesterone (Ball, 1941b). Inhibitory effects of progesterone are less marked in the rat (Zucker, 1967) and gerbil (Kuehn and Zucker, 1968) where there is no luteal phase associated with the normal 4- to 5-day estrous cycle. In pseudopregnancy and pregnancy, however, when progesterone levels are high, these species become sexually unreceptive (Powers and Zucker, 1969).

A striking parallel between natural secretion and exogenous administration of ovarian hormones has been shown in the ewe, a species in which ovarian secretion of progesterone is high until shortly before estrus when it declines sharply to be followed by a surge of estrogen (Moore *et al.*, 1969). Here the induction of estrous behavior is effected by giving ovarian hormones in reversed sequence, i.e., when progesterone is followed rather than preceded by estrogen.

Another aspect of sexual behavior that has been studied most extensively in the primate is the differentiation between sexual attractiveness and receptivity. In rhesus monkeys it has been shown that estrogens, which do not make a castrated female receptive unless small amounts of androgen are present, render her so attractive to the male that he initiates copulation. Conversely, administration of androgens to the female will make her "present" to the male and seek his attention though he will not show interest unless she is also receiving estrogen. Progesterone, on the other hand, makes the female unattractive. Some of these hormonal factors may be important in regulating sexual activity in the human (Herbert, 1967, 1968, 1970; Trimble and Herbert, 1968; Everitt and Herbert, 1969a,b, 1971; Everitt *et al.,* 1972).

d. Effects of Giving Heterologous Sex Hormone. Androgen given to normal adult female rats increases the incidence of male-like behavior without, however, entirely suppressing the female pattern (Ball, 1940). This finding has been confirmed by Koster (1943) who showed that, since both androgen and estrogen given to females from the time of birth will evoke male-like copulatory patterns together with behavior attractive to the male, the neuromuscular organization characteristic of both sexes can function concurrently in the same individual. As Beach (1945, 1948) has pointed out, however, the interpretation of results based on giving sex hormones to the normal individual is complicated by the effects of resultant changes in the secretory activity of the pituitary. More convincing evidence concerning the relationship between individual hormones and the nervous system can be derived from a study of the castrate. Estrogen given to male rats castrated during adult life will partly restore the level of sexual activity (Ball, 1937a) and will elicit a more complete copulatory response than is normally seen in the prepubertal castrate (Beach, 1942e). Estrogen also enables the lordosis reaction to be obtained from the prepubertally castrated male, although considerably larger dosage is required than will produce the same effect in the spayed female (Ball, 1939; Davidson *et al.,* 1968). Prepubertal gonadectomy in the female followed by treatment with androgen is associated both with an increase in male-like behavior and the appearance of some elements of the female copulatory response (Beach, 1942d). Male mating behavior can, however, be induced in the female by giving very large doses of estrogen (Engel, 1949).

These effects, while emphasizing the bisexual nature of neuromuscular organization in both sexes, suggest at the same time that the male pattern is the more dominant. There is additional evidence on this point. Not only does male-like behavior appear in the activity of the normal female but gonadectomy in the female has been shown to eliminate or to prevent the appearance of the female components of sexual behavior while interfering to a considerably lesser extent with the male elements (Beach and Rasquin, 1942; Beach *et al.,* 1968). Furthermore, Beach (1945) has reported a case of a male rat displaying prominent bisexual behavior in which the female reactions disappeared immediately after castration, the elimination of male responses being delayed. Such observations imply that homotypical female behavior owes more to the influence of ovarian hormone than does masculine behavior to androgen. The greater dependence of female behavior on hormonal stimulation is also shown by the rapidity with which it disappears after castration and is restored by ovarian hormones, which is in marked contrast to the delayed reactions of the male.

These considerations suggest that the main organizing action of androgens during perinatal development are not so much to establish the male pattern of behavior as to lower the threshold of response to androgens; the suppression of female elements of behavior seems to be a separate effect (Swanson and Crossley, 1971). In particular, the facilitatory effect of progesterone on lordosis following priming with estrogen is suppressed in rats which have been given androgen neonatally (Clemens *et al.,* 1969); castration of males at birth preserves progesterone facilitation (Davidson and Levine, 1969).

e. Indirect and Nonendocrine Factors. Although it is clear that the secretion of ovarian hormones plays a vital part in eliciting sexual behavior, there is, at the same time, evidence that other factors must be concerned. Such observations as the occurrence of heat in rats despite the failure of ovulation (Boling *et al.,* 1941b), copulation by castrated rabbits (Harris, 1952), and the dissociation of ovarian activity from sexual receptivity (Hammond, 1944; Phillips *et al.,* 1945) all suggest that factors other than cyclic ovarian secretion may be involved. The possibility that one such factor could be the secretion of steroid hormone from some nonovarian source cannot be entirely discounted (Everitt and Herbert, 1969b, 1971; Everitt *et al.,* 1972). A further possibility is that of social interaction between individuals of the same or opposite sex.

The most striking example of social factors affecting the estrous cycle is the action of pheromones in mice. These are external secretions which, presumably through olfactory cues, affect the behavior of other animals in the vicinity. Thus, when female mice are housed together, their estrous cycles are suppressed (van der Lee and Boot, 1956). The introduction of a male precipitates the appearance of synchronous estrous cycles (Whitten, 1956). Finally, exposure to the presence or scent of an alien male prevents implantation in a recently fertilized mouse (Bruce, 1959). Olfactory cues as well as visual and auditory stimuli coming from the boar may elicit lordosis in estrous sows (Signoret, 1970). The onset of estrus in ewes may be accelerated by the presence of rams (Schinkel, 1954; Thibault *et al.*, 1966). Male rats can distinguish between the odors of estrous and anestrous females (Le Magnen, 1952), a capacity which depends on the presence of androgen (Carr and Caul, 1962; Carr *et al.*, 1965). Male dogs prefer the scent of bitches in estrus (Beach and Merari, 1968) and the preference of male rhesus monkeys for spayed females which have been treated with estrogen is related to olfactory cues from vaginal secretions (Michael and Keverne, 1968).

A degree of emancipation from endocrine control is shown by estrous bitches, which show definite preference for some male partners (Le Boeuf, 1967). The preference of male rhesus monkeys for certain females is not only related to her hormonal status, but also to her behaviour and to other more subtle social factors (Herbert, 1968; Everitt and Herbert, 1969a).

Dominance status is known to influence behavior. Thus a female chimpanzee, when housed with a compatible consort, will present for copulation only during the late follicular phase of the menstrual cycle, whereas a timid female, housed with a vigorous male may present, as a submissive gesture, at any time of the cycle (Yerkes and Elder, 1936). Similar observations led Young and Orbison (1944) to conclude that such differences between consorts were more significant for determining behavior than the sexual status of the female. Human sexual behavior would appear to have become even more emancipated from endocrine control (see Section VI, p. 423).

Finally, environmental and sensory stimuli which affect the secretion of sex hormones (via the nervous system and the pituitary) should be mentioned. The effects of light on seasonal breeders is well known and recently the pineal gland has been implicated in the chain of events regulating sexual receptivity (Herbert, 1969). Visual cues, i.e., the sight of another bird, or even a reflection in a mirror, may elicit ovulation in birds (Matthews, 1939). Tactile stimuli derived from the act of mating activate a neuroendocrine reflex in "induced ovulators" and under certain conditions also in spontaneous ovulators (Aron *et al.*, 1966). Another effect which is induced by copulation is that of stimulating, in rodents having a short estrous cycle, the secretion of the prolactin necessary to convert the corpus luteum of the cycle into that of pregnancy. The pattern of intromission typical of a given species is also a factor regulating the formation of a functional corpus luteum (Wilson *et al.*, 1965; Diamond, 1970).

f. Conclusions. The foregoing considerations permit certain generalizations concerning the relationship between ovarian secretion and sexual behavior. In the first place, the central nervous mechanisms which mediate such behavior are innately organized in a manner which differs between species. While laid down early in life, the behavioral patterns involved do not appear prepubertally, but are first elicited at puberty by the activating, triggering, or releasing influence of ovarian hormones whose cyclical fluctuations begin at this time. The manner in which such behavior appears is not "all or none," but one in which the various components of estrous behavior are graded both in probability of occurrence and intensity of reaction, presumably in accordance with the titers of ovarian hormones circulating at different times of the cycle. The patterns of behavior seen at estrus are not normally present at other times and their emergence may be presumed to excite the male to coitus at a time when the probability of fertilization of the ovum is high. The precise part played by the ovarian hormones in the induction of estrus appears to differ between species; but, with certain unexplained exceptions, the presence of estrogen appears to be prerequisite in all. In some, the action of estrogen alone seems adequate to induce estrous behavior; in others the synergistic action of progesterone appears necessary for its full development, while in yet others progesterone seems to inhibit the development of such behavior. The presence of progesterone would seem to be more necessary in species that ovulate spontaneously than in those that ovulate following coitus. There is considerable evidence to suggest that the endocrine control of estrous behavior plays a greater part in reproduction in lower mammals than it does in primates where the influence of social and environmental factors tends to dominate the scene.

B. Maternal Behavior

1. General Nature

A further aspect of reproductive behavior concerns the care of the young. In this, environmental influences seem to play a more significant part than in the mating reactions of the female and stimuli which, in other circumstances, would evoke little response, under the influence of endocrine secretions, elicit a variety of innately organized and stereotyped reactions. Some of these, such as retrieving, nest building, and nursing, may be present in a weak form in nonparturient animals but are greatly accentuated during the period of lactation, waning in intensity as the young mature (e.g., Tietz, 1933; Cooper, 1942), and are found in most mammalian species. A detailed review of parental behavior in mammals has been provided by Lehrman (1961).

2. Endocrine Basis

It is perhaps not surprising that early investigators should have attempted to explain maternal behavior in terms of the possible activating influence of ovarian hormones. Steinach (1912) and Moore (1919) reported strongly marked maternal behavior in castrated male guinea pigs and in rats bearing transplanted ovaries, but Stone (1925) was unable to show, as the result of the investigation of behavior of rats living in parabiosis, that hormones secreted by parturient females were capable of inducing maternal reactions in a nonparturient parabiont. A complete reassessment of the part played by ovarian hormones became necessary in the light of the discovery that the anterior pituitary hormone, prolactin, which is responsible for the initiation of lactation, is a potent activator of maternal behavior (Riddle *et al.*, 1935, 1941–1942, 1942). The role of ovarian hormones and their relationship to the secretion of prolactin is still far from clear, however, and the available evidence is conflicting.

Recent studies suggest that the various components of maternal behavior appear as the result of interaction between the maternal physiological condition and environmental stimuli. The successive phases of the maternal behavior cycle are related to one another through the development of these interrelationships (Rosenblatt and Lehrman, 1963).

One aspect of maternal behavior which seems to be under direct hormonal control is maternal nest building in the rabbit, which has been studied by Zarrow *et al.* (1968). One or 2 days before parturition the rabbit will start to build a "maternal nest" which she lines with fur pulled from her body; characteristically there is a loosening of the hair (Sawin *et al.*, 1960). The same occurs in pseudopregnancy. Interruption of pregnancy by ovariectomy or abortion (including removal of placenta) precipitates hair loosening and nest building. Administration of progesterone near term prolongs pregnancy and inhibits nest building. The induction of maternal nest building in the rabbit seems to require exposure to a combination of estrogen and progesterone, progesterone being initially the dominant hormone (Zarrow *et al.*, 1968) as is the case near the end of normal pregnancy (Schofield, 1957; Ross *et al.*, 1963). Experimentally, nest building can be induced in castrated females maintained on a regimen of estrogen and progesterone either by discontinuing treatment with progesterone or by increasing the dosage of estrogen (Zarrow *et al.*, 1968). It has not proved possible to induce nest building in castrated males by any hormonal treatment, thus suggesting that development of the mechanisms underlying hair loosening and nest building is prevented by the influence of androgens during early development.

The hormonal control of nest building seems to be different in the mouse (Koller, 1952, 1955). Males and nonpregnant females build sleeping nests but much larger brood nests are built by pregnant females. This change in the pattern of nest building coincides with the development of the corpora lutea of pregnancy. Treatment of intact or castrated female mice with estrogen or prolactin has no effect but the administration of progesterone induces the building of a brood nest. Similar results have been obtained in the hamster (Richards, 1965, 1969). In contrast to the rabbit, nest building in these species is induced by an increase rather than a decrease in progesterone and hair loosening does not occur; moreover, estrogen priming is not necessary. Progesterone does not induce nest building in intact or castrated males of either species. The presence of young maintains nest-building activity when the progesterone levels drop after parturition and may even induce it in nonpregnant females. Other behavioral changes occurring during pregnancy, e.g., decreased activity, social isolation, and increased aggressiveness may also be related to high levels of progesterone (Eibl-Eibesfeldt, 1958; Altmann, 1963).

The expulsion of the placenta at parturition drastically alters the hormonal condition of the mother. The reduction in the titer of progesterone leads to the release of prolactin which initiates milk secretion. Ingestion of the placenta by most mammalian species can provide the parturient female with, in addition to estrogen and progesterone, a luteotropic factor, which may produce effects on maternal behavior (Grota and Eik-Nes, 1967; Zarrow et al., 1967). Newborn young have less chance of survival when fostered on a parturient parent before the placenta and membranes have been eaten than after. The injection of estrogen or progesterone into the foster mother has proved adequate substitutes for the ingestion of the placenta in these circumstances (Denenberg et al., 1963; Grota et al., 1967).

It has been suggested that the hormones which control milk production also induce maternal behavior. Lactation has been produced in both castrated male and female rabbits by combined treatment with estrogen and progesterone, followed by prolactin, but nest building has been seen only in a small number of females and never in males (Ross et al., 1963).

The first contact of the mother with the young occurs as an extension of self-licking during parturition. The conditions are highly favorable for the establishment of a behavioral bond between them; licking, nursing, retrieving, and nest building appear very shortly after birth. These activities are minimal in the rat in response to the presence of young before parturition and disappear within a few days if the young are removed at birth and replaced by new young 2-4 days later. However, once maternal behavior has become established removal of young at a later age does not abolish it (Rosenblatt and Lehrman, 1963). There may be a parallel between this behavior and the maintenance of milk secretion, which requires the stimulus of suckling (Meites and Turner, 1942a,b,c). The short period during which the mother is susceptible to bond formation is most marked in herd animals like sheep and goats who will reject their own young after separation of only a few hours (Hersher et al., 1963). Lehrman (1962) and Rosenblatt and

Lehrman (1963) have emphasized that the establishment of appropriate maternal behavior requires a sequential interaction between the mother and her growing litter. Experience during an early stage may change the mother's physiological condition so that, in the first place, she is motivated to perform the behavior characteristic of the next stage and, second, she orientates herself toward the environment in a manner appropriate to the behavior during later stages. In support of this concept it is noteworthy that injection of progesterone and prolactin does not release retrieving activity in virgin rats (Lott, 1962; Lott and Fuchs, 1962) while primiparous females are less likely to accept foster young presented after Caesarian delivery than multiparous mothers (Moltz and Wiener, 1966).

In primates, maternal behavior is influenced by the social experience of the mother during youth. Monkeys brought up in isolation frequently reject their own infants (Harlow *et al.*, 1963). They are more likely to accept them if they are delivered per vaginam than by Caesarian section. In contrast, females raised in the wild and captured as adults accept their young even when delivered surgically (Meier, 1965).

C. Social Behavior

As will have emerged from the foregoing, varying degrees of complexity may be discerned in the influence of hormones on reproductive behavior. First are the effects on the individual; second, effects on the relations between two participants in the sexual act; third, interactions between parents and offspring; and, finally, the progressively more complex relationships between the individual, other individuals, and the environment. As the degree of complexity increases it becomes more and more difficult to assess the role of individual factors. Thus behavior may stimulate the secretion of hormones which then affect other aspects of behavior whose expression may be conditioned by environmental stimuli, the whole process having the characteristics of a chain reaction. In these circumstances the classical approaches appropriate for the study of hormonal action in the individual, such as study of the effects of removing and replacing an endocrine secretion, cannot be expected to yield meaningful results in a study of social behavior. Furthermore the role of specific hormones on the initiation, expression, facilitation, or inhibition of certain aspects of social behavior varies not only between species, but within species living in differing habitats and between individuals who have had different experiences during life. In the past, the more subtle, indirect effects of hormonal balance on mood, motivation, and arousal have been largely overlooked.

With these reservations in mind, a few generalizations can be made. It is clear that, since the function of sexual reproduction is the perpetuation of the species, optimal conditions for survival must be taken into account. These include the provision of sufficient living space in which to make the best use of environmental resources and a consequent appropriate limitation to the growth of the population. It is, therefore, not surprising that the influence of sex hormones can be discerned in the establishment of territories and in the selection of the fittest individuals as the most active breeders (Defries and McClearn, 1970). In some species, males ready to mate establish a territory into which they entice a female. In other species, a hierarchy is set up in which the dominant males have first access to estrous females or collect the largest harems. Examples of various types of social organization throughout the vertebrates have been reviewed by Guhl (1961), Wynne-Edwards (1962), Beach (1965b), Crook (1970), Esser (1971), and many others.

In most mammals, the male is more aggressive than the female. Aggressiveness usually appears at puberty (Seward, 1945; Brain and Nowell, 1969), and qualitative differences in postures may be found between adult fighting and the play fighting of juveniles (Dieterlen, 1959), whose development may be accelerated by giving androgens (Fredericton, 1950; Levy and King, 1953). In many seasonally breeding animals, changes in social structure and heightened aggression accompany increases in testicular weight and secretion (see Altmann, 1952; Lincoln *et al.*, 1970). Castration usually decreases aggressiveness which may result in displacement in the hierarchy of dominance (Edwards, 1968, 1969, 1970; Suchowsky *et al.*, 1969; Sigg, 1969).

The effects of androgens on females are less clear-cut. Although some earlier workers have reported increased pugnacity and irritability in female rats following the administration of androgen (Ball, 1940; Beach, 1942d; Huffman, 1941) more recent workers have found no such increase (Conner *et al.*, 1969; Conner and Levine, 1969). Female mice do not fight when given androgen (Tollman and King, 1956; Edwards, 1968, 1969; Edwards and Herndon, 1970). Evidence seems to be accumulating that androgens may exert an organizational effect on aggressiveness at the same developmental stage in which sexual behavior is affected (Rothballer, 1967; Conner *et al.*, 1969; Money, 1965). Estrogens are usually considered to decrease aggressiveness or conversely to increase submissiveness. In some species, females are aggressive toward the male when sexually unreceptive yet become docile and submissive when in estrus, e.g., guinea pig (Young *et al.*, 1935), oppossum (Hartman, 1945), shrew (Pearson, 1944), and hamster (Kislak and Beach, 1955; Payne and Swanson, 1970). In the ovariectomized hamster, aggressiveness is unaffected by estrogen but increased by progesterone administration (Payne and Swanson, 1971). Suchowsky *et al.* (1969) found that estradiol completely, and progesterone partly, inhibited aggressive behavior in intact male mice; estrogen did not induce fighting in castrated males while progesterone partly increased aggression. Moreover, estrogen suppressed the aggression-enhancing tendencies of testosterone given simultaneously. The increased aggressiveness of females during pregnancy and lactation may be related to a relatively high progesterone:estrogen ratio.

The influence of pheromones on the inhibition of aggression must also be considered. In the male mouse, the presence of a female may lead to sexual arousal which, in turn, stimulates an increase in the secretion of gonadotropin. This results in the release of gonadal steroids, some of which are excreted by way of the urine to which they contribute specific odors (Bruce, 1965). Detection of these odors by the female elicits neuroendocrine reflexes which result in an increased secretion of gonadotropin, followed by the release of estrogen and progesterone in the appropriate sequence to induce ovulation, estrous behavior, and suppression of aggression (Rothballer, 1967). The synchronization of estrous cycles following the introduction of a male into a cage full of female mice has already been described.

The situation is more complicated in primates where dominance–subordination relationships are modified by a variety of factors, one of which is the sexual status of the female. In the chimpanzee, for example, the male is normally the dominant partner of a heterosexual pair and commands priority in obtaining available food except during estrus when priority is conceded to the female (Yerkes, 1939, 1940; Nowlis, 1942). Similar changes occur in the social relationship between pairs of females in so far as a normally subordinate individual may achieve temporary dominance either as a result of the onset of estrus (Nowlis, 1942) or, in the case of the spayed individual, of giving estrogen (Clark and Birch, 1946). There is good evidence, however, that the effect is not so much due to any increase in the capacity of the female actively to command a higher social status, but rather is mediated through the more passive influence of some change in appearance or demeanor upon a previously dominant partner or consort. An important factor in this respect would appear to be the swelling of the sexual skin which occurs during estrus, for the change in status which would normally accompany the administration of estrogen does not arise when the sexual swelling is prevented by giving small doses of progesterone (Birch and Clark, 1950). There is, in fact, reason to believe that any direct effect exerted by estrogen is in the reverse direction. Thus estrogen given to male castrates (Clark and Birch, 1945) or to spayed females in which sexual swelling is inhibited (Birch and Clark, 1950) causes a reduction in dominance, whereas a normally dominant female, while in estrus, will give way to a previously subordinate companion (Yerkes, 1939). Such findings, however, cannot be regarded as universal in their application for Mirsky (1955) was unable to detect any disturbance to a carefully established hierarchy of dominance in the macaque as a result of implanting pellets of estrogen.

Another factor determining the dominance–subordination relationship between heterosexual pairs of monkeys would seem to be associated with olfactory cues emanating from the vaginal secretion (Michael and Keverne, 1968; Herbert, 1970). The female consort of a dominant male usually acquires dominance within the group. If two ovariectomized female rhesus monkeys are given identical doses of estrogen and caged with a single male, the preferred female becomes dominant over the other. If she loses her favored position (as a result of a reduced dosage of estrogen or giving progesterone) the other female will display aggression toward her (Everitt and Herbert, 1969a). There is some evidence that female monkeys are more aggressive toward males during the luteal phase of the normal menstrual cycle (Herbert, 1970) and that this can be duplicated by giving progesterone to spayed females. It is likely that the increased dominance of the female during the follicular phase of the cycle or after giving estrogen is the indirect result of the female's increased sexual attractiveness which acquires for her a favored position in the hierarchy by dominant males. Michael (1969) reports that female monkeys threaten the males during the height of sexual receptivity but that fighting is averted through the greater tolerance of the males at this time (Michael and Zumpe, 1970).

The social implications of the continuous sexual availability of the female primate are of considerable interest, for this imposes a pattern on group behavior and organization. Zuckerman (1932) states: "Both male and female primates are always sexually active, their heterosexual interests providing the bonds that hold them together in permanent bisexual association . . . The primate family consists of male, female or females and young, but the family of the lower mammals consists only of the female and her young." The social and behavioral function, then, of an almost continuous state of female receptivity in the primates is that it fosters the formation of a relatively stable reproductive organization, in which both males and females participate, and to which the young of the species are exposed. Out of this, but outside this primary stable reproductive unit, elaborate nonreproductive organizations have grown. The limits to which the process of female participation in nonreproductive organizations can go are determined by the facts of reproductive physiology, and the extent to which female behavior is controlled by the ovary and its reproductive function.

It is thus clear that the undoubted influence of ovarian secretions upon social interrelations is a complex mechanism. It seems probable that in most, if not all, species they induce, through a direct action upon nervous tissues, changes in aggressive behavior which are compatible with reproductive activity. These changes, which might be expected to reduce the status of dominance in species where a hierarchal structure plays an important part in social organization, are offset by indirect effects such as the evocation under hormonal influence of the overt signs appropriate for releasing a conciliatory type of behavior from other normally dominant members of the community.

In recent years much interest has been shown in mechanisms which regulate the density of population and maintain it at a level below that which would destroy the environment. In closed colonies of mice, crowding leads to an increase in adrenal activity (Brain and Nowell, 1969, 1971a,b). Male mice characteristically hold territories from which they exclude other males. When the number of potential nesting sites is limited, aggressive encounters increase. It has been shown that fighting or even the threat of a fight, such as that arising from the exposure of one mouse, through a barrier, to another mouse known to the former as a good fighter, increases adrenal cortical secretion (Bronson and Eleftheriou, 1965). Subordinate males excluded from a territory are recognizable by their bedraggled appearance, associated with prolonged adrenal oversecretion (Crowcroft and Rowe, 1963; Christian and Davies, 1964; Christian *et al.*, 1965). It has been shown that most fertile matings are with dominant males, the subordinate mice contributing little genetic material to the next generation (Defries and McClearn, 1970). It seems that subordinate males which do not hold a territory provide a surplus population which is capable of replacing a dominant male if a territory becomes vacant; they quickly become aggressive and sleek as their adrenal weight falls (Crowcroft and Rowe, 1963).

The major effect of crowding is the suppression of reproductive activity and the consequent reduction of the population of the crowded region. This results partly from a direct action of corticosteroids which inhibit the production of gonadotropin (Christian *et al.*, 1965). This effect is most marked in females in which puberty is delayed, estrous cycles are disturbed, the frequency of abortion or fetal resorption is increased, and lactation and maternal behavior are impaired. At every stage of the reproductive process there are deleterious effects tending to decrease procreation and increase mortality (Rothballer, 1967). A dramatic account of the deterioration in social behavior in a closed colony of rats allowed to breed freely is given by Broadhurst (1963).

V. MODE OF ACTION OF OVARIAN HORMONES

There remain to be considered the possible mechanisms by which the ovarian secretions exert their modifying influence on behavior. Present knowledge permits little more than speculation concerning the exact nature of the hormonally induced changes occurring somewhere on the pathway linking stimulus and response. Some of the possibilities have been considered by Lashley (1938) and by Beach (1948). They fall into four principal groups, based, respectively, on modification of sensory input, on metabolic factors, on morphological changes within the central nervous system, and on the differential sensitivity to endocrine influence of "centers" controlling the several aspects of innately organized behavior. These are not necessarily mutually exclusive and it is possible that more than one is involved.

The important part played by perception in many aspects of sexual behavior suggests that changes in the nature of the sensory input under the influence of ovarian hormone may be a significant factor controlling the elicitation of the appropriate motor pattern. Only the olfactory modality appears to have been studied in relation to ovarian activity and there is general agreement that fluctuations in olfactory acuity are correlated with the menstrual cycle. Le Magnen (1949, 1950a,b) reports that in the rat sensitivity to a variety of odors, normally detectable with difficulty, is significantly raised both during the follicular phase of the cycle and as a result of giving estrogen, but is lowered by ovariectomy. The finding of Schneider and Wolf (1955) that a reduction in olfactory acuity occurs during menstruation is consistent with these results, but differs from that reported earlier by Elsberg *et al.* (1935). The significance of these fluctuations lies not so much in their direct effect, for olfaction appears to play little part in the sexual behavior of the female rabbit (Brooks, 1937), as in the possibility that similar changes may occur in relation to other receptor mechanisms, such as the visual and tactile, both of which are known to be concerned in the mediation of reproductive behavior in the female. Proprioceptive impulses, originating as a result of morphological changes in the reproductive tract, were long thought to be responsible for estrous behavior (Hammond and Marshall, 1925; Nissen, 1929), and the possible role of such receptors has more recently been discussed by Hammond (1945). However, normal mating behavior in the rat survives removal of both uterus and vagina (Ball, 1934; Kaufman, 1953), and although denervation of the genitalia abolishes the afterreaction of mating in the cat, this procedure has no effect on courtship activities (Bard, 1935). Furthermore, there are many reports of instances in which the onset of estrous behavior is dissociated from any observable change in the reproductive tract (e.g., Alphin and Dey, 1944; Kent and Liberman, 1949; Harris *et al.*, 1958), while the estrous activity cycle in the rat persists after hysterectomy (Wang, 1923). Although, therefore, impulses arising from the reproductive tract may play some part in the female mating reaction, they are by no means essential and an alternative site of influence of ovarian hormone must be sought. It seems certain that such a focus must be within the central nervous system itself. This aspect of the problem is dealt with fully in Volume III, Chapter 3.

VI. BEHAVIOR IN THE HUMAN SPECIES

A. Introduction

Any consideration of ovarian influence on human behavior must necessarily differ in approach from that adopted with lower animals. There are three main reasons for this. In the first place, evidence from experimental work on animals is for the most part controlled, whereas that from human sources is, in many cases, fragmentary and has frequently been gathered for a different purpose. However, much of the more recent work on human behavior, and of human sexual behavior in particular, no longer exhibits this lack of experimental control. As an example, the direct study of human sexual responses both in the normal (Masters and Johnson, 1965, 1966) and less than normal or inadequate individual (Masters and Johnson, 1970) may be cited. The introduction of the oral contraceptive has provided a further source of knowledge of controlled noncyclic behavioral effects in human subjects. In addition, the study of human behavior in general has tended to approximate more closely to the more rigorous criteria adopted for animal experimentation in recent years. Second, there is the phenomenon described by Beach (1948) as "the emancipation of female sexual behavior from ovarian control" which denotes an apparent lack of marked periodicity in human female sexual responsiveness in relation to the cycle of ovulation as compared with the regular estrous cycle of lower mammals and of some nonhuman primates. Here again, however, the development of more precise knowledge of the effect of the ovary on the nervous system (see Beach, 1965b) has thrown light on the mediation of human sexual behavior by the nervous system and has more closely delimited the degree of its emancipation from endocrine control.

Finally, and by contrast, ovarian participation in the regulation of human behavior must include psychological and sociological phenomena that are at once more complex but more accessible to study than in other species. Here, not only is there now more interest in sociological factors than before but paradoxically, through the importance of population control and therefore the dissociation of sexual and reproductive behavior, there is a greater impact of the "ovary" in behavioral terms on society as a whole, typified by the development of and reaction to the "permissive" society.

B. Effects of Cyclic Variations in Ovarian Function

1. Putative Effects

The ovary, as is the case with lower animals, appears to impose a definite psychological rhythm of endocrine origin upon the nonpregnant adult woman with the important difference that, in so far as the study of the impact of this rhythm on behavior is concerned, interest has focused upon the phenomenon of menstruation rather than upon that of ovulation. In this connection, however, the work of Redgrove (1968, 1971) indicates that ovulation is not without effect on performance and suggests that a more accurate indication of the time of human ovulation is required for the objective assessment of behavioral changes which are associated with the menstrual cycle than is provided by the daily oral measurement of basal body temperature. Such conclusions as these clearly complicate any attempt to correlate experimental findings of the type considered elsewhere in this chapter with observations on man since, in nonprimates, ovulation is the most conspicuous behavioral event in the ovarian cycle, whereas in humans, this phase passes relatively unnoticed, by contrast with menstruation itself. Thus, although there is some evidence of a more direct effect of ovarian hormones upon behavioral mechanisms, much of the influence of the ovary on human behavior may be regarded as being mediated secondarily through the profound social and psychological implications of menstruation and especially of the premenstruum.

2. Objectively Assessed Effects

Several investigators have studied the relationship of certain psychological or neurological functions, which often involved the learning and performance of complex tasks, to the rhythm of the menstrual cycle in an attempt to correlate objectively assessed rhythms in central nervous function with different phases of the cycle. Such studies have often formed part of the study of biological rhythms, in general.

a. General Activity. There is some evidence for an activity cycle in human beings comparable with that seen in other animals. Farris (1944), for instance, found variations, not assessed statistically, in the total amount of daily walking which appeared to be related to the ovarian cycle. Three peaks of increased activity were observed, the greatest coinciding with ovulation. Lesser peaks occurred during the latter one-third of the cycle, and immediately after menstruation. The findings agree with the pedometric observations of Billings (1934) whose graphs showed a high postmenstrual burst of activity.

b. Skilled Performance. One of the earliest comprehensive quantitative investigations of the influence of the menstrual period and the events of the menstrual cycle upon accuracy of observation, attention, speech, and precision of movement was reported by Sowton and Myers (1928). No consistent differences could be observed between performance during the menstrual period and that at other stages of the cycle, but such tendencies as did occur suggested that performance was improved during midcycle and lowered during the immediate premenstrual phase. A similar lack of significant correlation between efficiency of performance and stage of the menstrual cycle has been recorded in the output of manual workers (Kirihara, 1926), visual and auditory reaction times (D'Amour and Woods, 1941; Doring, 1954), and vestibular function (Rubin and Winston, 1953). Engel and Hildebrandt (1969) have shown that the reaction time in women varies with menstrual cycle, being lowest about 4 to 6 days after the onset of menstruation. By contrast, Kopell *et al.* (1969) in studying physiological measures of arousal during the menstrual cycle could find no significant correlation between reaction time and phase of the cycle. They also measured galvanic skin potential, two-flash threshold, and time duration estimation. Only ability to estimate durations of 15 and 30 seconds showed significant variation during the menstrual cycle. On the other hand, Johnson (1932), who correlated the learning curves of subjects trained in motor skills, claimed that a decline in learning occurs during the menstrual period followed by a rise which is maximal at the fifteenth day of the cycle.

More recently, Smith (1950a,b) found significant differences in performance related to the menstrual cycle in women employed in a parachute factory among those who were working in jobs demanding a high level of mental activity but not among those who were engaged in what were described as simpler tasks. Sfogliano (1964) also found significant changes during the menstrual cycle in the performance of women assembling transistors, a job requiring very high levels of visual discrimination. Redgrove (1968, 1971), in a careful study of the efficiency of typists, laundry workers, and punch-card operators, observed that the phases of the menstrual cycle of the typists were significantly correlated with the speed but not with the accuracy of typing. The speed of typing increased significantly with the onset of menstruation. Employees in the other two groups, however, showed no consistent changes suggesting that a "menstrual" effect was more likely to be demonstrated in jobs or tests in which subjects were working at the limit of their capacity.

In a study by Dalton (1960a) the average weekly marks of 217 postpubertal schoolgirls showed a pronounced fall premenstrually and during the menstrual period, and a sharp rise in the intermenstruum. Study of Advanced (A) level and Ordinary (O) level examination marks demonstrated significantly lower pass and distinction rates and lower average marks during the premenstruum or during the menstrual period than during the rest of the cycle. This effect was much greater in girls having menstrual cycles which exceeded 31 days in duration and in those whose menstrual periods were longer than 6 days. The percentage failure rate in girls in this group examined during their paramenstruum (the 4 days immediately prior to the menstrual period and the 4 days immediately after it) was over twice that of girls in the same group examined at other stages of their menstrual cycles (Dalton, 1968). This agrees with Wickham's (1958) findings that army personnel had lower intelligence scores during menstruation, although motor aptitude tests were unaffected.

In a recent review of this aspect of this subject, Redgrove (1971) concluded that the menstrual cycle did affect the capacity to carry out certain tasks, but the extent to which this effect was manifest depended upon the degree to which decrements in capacity would be offset by increased effort. It was, therefore, necessary to investigate situations where capacity was as closely as possible reflected by performance. Such conditions were likely to be met when individuals were performing close to their limits.

c. Sleep. Sleep is reported to be disturbed during menstruation, with increased difficulty in getting to sleep, longer duration of sleep, and a statistically significant increased incidence of dreaming during menstruation than on control nights (Kleitman, 1939). These disturbances of sleep may be related to the increased fatigue and irritability reported by McCance *et al.* (1937).

d. Electroencephalographic Effects. Dusser de Barenne and Gibbs (1942) first recorded the EEG during specific phases of the menstrual cycle and showed that it varied. Premenstrually, an increase in the peak frequency of the alpha rhythm occurred with a decline in the "amount" of this rhythm (Lamb *et al.*, 1953). Margerison *et al.* (1964) used low-frequency EEG analysis to demonstrate a decline in mean abundance of alpha rhythm during the premenstrual phase. Differences between male and female subjects have been reported in respect of attenuation of the alpha rhythm during mental arithmetic (Glass, 1968). The greater attenuation of alpha rhythm in females was ascribed to the effect of premenstrual changes. Roubicek *et al.* (1968), in thirty-three healthy females, has shown that the frontal theta/ alpha index was increased during menstruation, the alpha rhythm being retarded and slow waves increased, while Gautray *et al.* (1970) found, by power spectral analysis of the EEG, that the theta (3-7 Hz) component of the EEG attained a maximum in the days preceding ovulation. The alpha rhythm reached its lowest level at the time of ovulation (Gautray, 1969). Ferroni *et al.* (1969) could find no electroencephalographic abnormalities during or after treatment with synthetic estroprogestinic drugs except for two cases treated with Mestranol and Lynestrenol.

e. Abnormal Behavior. There is some statistical evidence for the association of various forms of abnormal behavior with phases of the menstrual cycle. For example, Balazs (quoted by Rosenzweig, 1943) claimed that the highest incidence of attempted suicide was related to the menstrual period. This observation has more recently been confirmed by Dalton (1959) who found that more than half the cases of suicide investigated by her occurred during the premenstrual week or in the menstruum, while the majority of suicides studied by MacKinnon and MacKinnon (1956) were found to have been perpetrated during the middle luteal phase of the cycle. Crimes of violence are also much more common just before menstruation and during the early premenstruum (Cooke, 1945; Morton *et al.,* 1953). Whitehead (1934) found that in air accidents involving women pilots where no other ascertainable cause came to light, the pilots were in the menstrual phase of the cycle. In a study of accident cases, Dalton (1960b) found that, out of a total of eighty-four regularly menstruating women, forty-four accidents occurred during 4 menstrual or 4 premenstrual days. These two 4-day paramenstrual periods contained an accident incidence of twice the expected level (on a uniform expectation hypothesis) whereas all the other 4-day periods of the cycle were below the expected level. Dalton (1960c) likewise found, in data obtained from the school records of 350 schoolgirls, a significant increase in "misbehavior," defined as unpunctuality, forgetfulness, and avoidance of games, during menstruation. The effect was stated to be most marked in those girls exhibiting generally higher levels of misbehavior.

Finally, there is some evidence for a more direct relationship between hormones of ovarian origin and the central nervous system in so far as mental instability is said to be increased, and symptoms of neurological disease such as epilepsy and disseminated sclerosis to be exacerbated, during menstruation (Ask-Upmark, 1955), as also are psychiatric conditions (Dalton, 1959; H. Rey, personal communication).

f. Conclusions. The evidence is sometimes contradictory and hard to summarize and, although earlier work gave little reason to believe that pronounced cyclical changes occur in behavior, some more recent investigations, and perhaps more sensitive tests, have shown some significant associations. There still appears to be a need for experimental investigations specifically designed to elucidate the points which have been uncovered in some of the investigations summarized here. The statistical tests often applied in recent investigations have been neither sensitive enough nor always suitable to detect changes, and have suffered from the disadvantage that there is inevitably a constant long-term improvement in performance due to experience. However, the application of more advanced and recently developed statistical techniques of time series analysis, especially of cross-correlation analysis, to the problem of the cyclical associations, could help.

Most of the experiments suffer from lack of controls, since, ideally, the performance of normal females should be compared with that of ovariectomized individuals. In recent investigations there is an indication that some attention is being given to this point. Although there is little evidence of a fundamental physiological rhythm universally affecting women's working capacity, enough data have been accumulated to support the contention that a fall in functional efficiency and adjustment to the environment occurs during the paramenstruum. Too much of even the more recent evidence has been in anecdotal form; thus the observation "One has only to watch typists in a large pool and notice how on one day some typists are filling their waste paper baskets with spoiled work and on other days it is a quite different group of typists doing the same thing" (Dalton, 1969, p. 129) has been put forward as evidence of an association between the menstrual period and diminished working efficiency.

3. Subjective Effects

The most complete investigation of periodicity in subjectively recorded psychological change is still that of McCance *et al.* (1937). This study leaves little room for doubt that a close association exists between affective states and the menstrual cycle.

a. Fatigue. The incidence of fatigue and headache was high and followed a similar time course, each showing a steady premenstrual rise and postmenstrual fall to a constant intermenstrual level.

b. Emotionality. A "tendency to cry" which was noted on a number of days, showed a definite monthly rhythm, with increased incidence at the menstrual period symmetrically falling and rising about a low point at midcycle. The occurrence of depression and elation were fairly evenly distributed throughout the cycle. The former showed a slight rise in the premenstrual and early menstrual phase; the latter a rise on the first day to a peak on the eighth day when it declined until the twentieth to twenty-third day. It is interesting to note that this parallels the distribution of libido (Section VI,B,6, p. 430), and that the incidence of depression, together with that of many other symptoms, was also low on the twentieth day, suggesting that this period is characterized by emotional stability.

c. Irritability. Reports of irritability revealed a marked fall in incidence from the first to the fifth day of the cycle, followed by a small rise toward the time of ovulation and then a fall during the premenstrual period.

d. Capacity for Intellectual Pursuit. As regards the variation in incidence and intensity of the effort required for intellectual work, the least effort on average was required on the twentieth day followed by a small rise during the premenstruum and a further marked increase on the first day of the menstrual period. This result, therefore, would suggest that there should be an overall fall in cerebral efficiency or alertness on the first day of the menstrual period, associated with generalized feelings of dissatisfaction, maladjustment, and fatigue.

4. Psychosomatic Aspects

The incidence of "psychodynamic" changes during the menstrual cycle has been attributed to fluctuations in ovarian secretion. Benedek and Rubinstein (1939a,b) and Benedek (1952), for instance, studied separately the daily vaginal smear and basal body temperature of women undergoing daily psychoanalysis. Correlations between the two sets of observations were high so that correct predictions could be made from the psychoanalytical data about the state of vaginal smears.

A marked change in the symbolism of the dreams and fantasies of these patients occurred at ovulation. It is not stated if there was any regular difference between the dreams during normal and anovulatory cycles. In the preovulatory phase, corresponding to a rise in estrogen, a marked psychological polarization toward the opposite sex was found, associated with aggressive tendencies or a fearful, defensive attitude if heterosexual gratification was not achieved. Immediately after ovulation the effect was one of "generalized erotization," the heterosexual tension was relaxed but sexual receptivity was often consciously expressed. In the progestational phase following ovulation a tendency to introversion and return to a narcissistic state was noted. This has been interpreted as preparation for motherhood, when consciously or unconsciously the woman is preoccupied with her own body, the "nest of her offspring" (Rosenzweig, 1943; Freed and Kroger, 1950). Finally, if fertilization did not occur, pregnancy fantasies became common, while regressive infantile attitudes were also observed.

5. Premenstrual Tension Syndrome

The premenstrual tension syndrome is a clinical phenomenon affecting the last 10 days of the menstrual cycle. It is characterized by an emotional state which has been described as a tension manifesting itself in a desire to find relief in foolish actions (Frank, 1931). Headache, irritability, instability, and depression are frequently present, together with abnormal hunger and thirst (Morton, 1950), marked emotional instability, lack of initiative, and hypersomnia (Rees, 1958). It thus represents a recurrent impairment of behavioral disturbance for a number of days and occurs in quite a high proportion of the female population.

These symptoms were first ascribed to ovarian influence by Israel (1938a,b) and have since been more specifically attributed to an increase in the ratio of circulating estrogen to progesterone independent of the absolute levels of these hormones (Morton, 1950; Greene and Dalton, 1953, 1955; Rees, 1958). The view is widely held that hydration of the tissues resulting from hormonal imbalance (Long and Zuckerman, 1937; Zuckerman et al., 1950) gives rise to symptoms referable to the central nervous system. However, the correlation between these symptoms and the retention of electrolyte (Swyer, 1958) and water (Thorn et al., 1938; Chesley and Hellman, 1957) is by no means complete, and it would seem that some additional and possibly more direct action of estrogen and progesterone upon the nervous system must be postulated (Greene and Dalton, 1953). Recent evidence derived from, among other sources, single-unit recording from hypothalamic neurons has supported this view (see Volume III, Chapter 3).

6. Libido

Although primates differ from other mammals in so far as they are sexually receptive at all times, there does appear to be a rhythmic flux in the libido of the human female which is related to the ovarian cycle. However, as Forbes (1955) has stated, "whereas in sub-human primates sexual receptivity is greatest during the intermenstruum, in man the periods of greatest sexual receptivity have been less definitely determined." Ellis (1936) quotes a number of authorities to the effect that female libido is greatest in the paramenstruum and that it persists throughout the period itself. This view is supported by the observations of Kinsey et al. (1953), while nymphomania has been reported to recur during the premenstrual week (Greenblatt et al., 1942). McCance et al. (1937), however, found that sexual desire in single and married women tends to fall abruptly on the first day of the menstrual period, rising after the second day to a peak at the eighth day of the cycle and thereafter declining gradually. The curve for the incidence of sexual intercourse among married women followed much the same course. Hart (1960) in a study of 123 healthy women showed that in 59% enhancement of libido occurred paramenstrually, only in 3% did enhancement occur during menstruation, and in only 6% did enhancement of libido occur during midcycle when ovulation could be presumed to occur.

Knowledge of the "safe period" did not appear to be a factor in these results. Udry and Morris (1968), however, have advanced evidence that, in a group of negro women studied, the highest percentage rates of intercourse and orgasm occurred during the putative phase of ovulation. Although they claimed these results were supported by those in a white, middle class group (studied by a different method), the statistical analysis over both groups does not appear convincing. Masters and Johnson (1966), in a novel approach which involved full visualization of the cervix, showed in fifty women what they describe as full "orgasmic phase tension release" during the period of heaviest menstrual flow. Among actively menstruating women, 173 out of 331 expressed personal interest in sexual activity during menstruation and only 33 of their subjects objected to sexual activity during menstruation. Money (1965) has stated that acuity of sense of smell is greater during the intermenstruum than at other times (Schneider and Wolf, 1955), and has attempted to link this to variations in sexual behavior during the cycle. There is experimental and clinical evidence that the "libido" hormone in women as well as men is androgen (Money, 1961; Everitt and Herbert, 1971). Salmon and Geist (1943) have claimed that increased susceptibility to psychosexual stimulation, greater sensitivity of genitalia, and greater orgasmic gratification can be attributed to androgens. The precise role of androgens of ovarian origin in this behavioral reaction is at present obscure. The ovary, however, is known as an important source of these hormones.

While, therefore, many investigations place the peak of libido close to the menstrual period, there is no universal agreement as to the precise stage of the ovarian cycle at which libido is maximal. This lack of a definitive finding, however, fails to accord with the teleological interpretation usually applied to the coincidence of estrus and ovulation in other species. The periodicity of libido in the human female would be expected to influence copulatory activity, but a maximum frequency of intercourse at the eighth day or earlier will not produce maximum fertility since ovulation occurs at about the thirteenth day of the cycle. Thus, although a rhythm of receptivity correlated with the ovarian cycle may be discovered in woman, this would appear to play but a minor role in human fertility. Such observations serve to emphasize the importance of social and psychological factors in human reproductive behavior.

7. Menstruation and Sociological Aspects

Possibly the most severe impact of ovarian physiology on human behavior is mediated through the phenomenon of menstruation. In primitive and semiprimitive societies, a multitude of prohibitions and superstitions attach themselves to a woman during the actual menstrual period. Apart from severe prohibitions against intercourse during the menstruum (Leviticus) common to many codes and societies, a menstruating woman was not supposed to touch anything living, as death emanated from her body at this time (Rosenzweig, 1943). Although such ideas have, with social evolution, largely disappeared, the psychological effect of menstruation upon modern women and the relation of society to the phenomenon remain. The secondary effect of the menstruating woman upon her husband and family (Redgrove, 1971) cannot be ignored.

It is not surprising then that the powers of suggestion and social factors alone, without endocrinological or physiological influences, would make of menstruation something which, at its minimum, is a regularly occurring inconvenience, if not an ordeal. The mingled feelings of shame, impurity, and yet pride described by de Beauvoir (1956) in young girls on attaining the menarche may well be a reaction to, and a reflection of, the traditional beliefs and customs of her society and family, ". . . . so it is the social context that makes menstruation a curse . . . the girl stands abashed before the brutal and prescribed drama that decides her destiny." It symbolizes femininity, ". . . and because femininity signifies alterity and inferiority. . . . its manifestation is met with shame." Rosenzweig (1943) further emphasizes the power of suggestion and other social factors, and observes that whether a young girl goes through the menarche quite normally or is unduly disturbed by it is frequently dependent on the attitudes displayed by older women in her environment. However, it should not be supposed that social factors alone influence feminine attitudes toward the menstrual period, for these may well be the aftermath of accumulated individual somatic and psychological experience. The fact that menstruation is still a considerable inconvenience to the average woman in civilized society is indisputable (Hoskins, 1941).

Theories of the psychological attitudes of a woman in relation to her periods have been many and varied. Thus Morales (quoted by Freed and Kroger, 1950) suggests that increased sexual desire during the menses may be an attempt to provoke the male, while immune to sexual attack, and thus try to seduce him and at the same time deny him. Alternatively, the depression which accompanies menstruation may be regarded as of general biological origin, the process itself being a "frustration of nature," in that fertilization has not occurred and an ovum has been irrevocably wasted, and is felt as such, consciously or unconsciously, by the woman at her periods (Rosenzweig, 1943). On the other hand, if the woman experiences either conscious or unconscious pregnophobia, then the onset of menstruation will be greeted with relief. Menstruation is said to be accompanied by a feeling of symbolic loss of the uterus because it has not received a child. The depression accompanying menstruation leads to fantasies of death, and actual loss of blood is equated symbolically with guilt, and loss of body substance. "Eliminative" psychological trends become pronounced. Menaker (1957) states "To many women menstruation is a sub-conscious monthly reminder of their inferiority and part of the envy of the masculine and rejection of the feminine." Thus, it can be seen that there is ample evidence of ovarian influence, though mediated indirectly, in producing a cyclical change in female psychology, and in the relation of the woman to the society in which she lives.

C. Noncyclical Influences

Noncyclical changes in female behavior can be directly or indirectly related to changes within the ovary. Behavioral changes at puberty, the menopause, and the artificial menopause will be considered briefly and the effect of oral contraceptives on behavioral symptomatology will be evaluated.

1. Menarche

Changes occur in the personality and behavior of girls at the onset of menstruation. Certain of the psychological reactions to the menarche have already been considered in an earlier section. Other aspects of the relationship are exemplified in the work of Stone and Barker (1937) who studied personality and intelligence in a large sample of postmenarcheal and premenarcheal schoolgirls paired for chronological age and matched with respect to upbringing and family origin. They found a definite increase in intelligence, interest attitude, and developmental age in the postmenarcheal group of girls, and, from the study as a whole, concluded that the greatest behavioral differences at this time are found in regard to personal appearance, heterosexual interests and attitudes, avoidance of physical exertion, tendency to daydreaming, and involvements in domestic conflict, all of which are more common among postmenarcheal than premenarcheal individuals.

It may be argued that the onset of menstruation tended to occur earlier in the psychologically more mature schoolgirls, but in the light of the psychoanalytical findings of Benedek and Rubinstein (1939a,b) it seems reasonable to relate heterosexual activities and the more feminine tendencies of the postmenarcheal girls to a rising level of circulating estrogen which has reached a threshold with the onset of menstruation, and to the marked psychological effect of the menarche itself.

2. Menopause

With the decline in secretion of ovarian hormone, women suffer from restlessness, irritability, insomnia, palpitations, and "hot flushes," symptoms which become manifest to a greater or lesser extent in their behavior as a deterioration of character. There is, however, according to Kinsey *et al.* (1953), no abrupt failure in the female libido after the menopause; their records show a gradual decline in all sexual activities. This decrease was thought to be similar to the slight decline found in non-menopausal women of approximately the same age, and primarily dependent on the male's declining interest in sociosexual activity. It must be pointed out that it may also be that the secondary sexual characters of the female no longer provide an object of sociosexual attraction in the male. However, according to Devereux (1950) whose evidence would tend to support the positive aspect of Kinsey's conclusions, the menopause should be regarded as a developmental rather than an involutional phase of a woman's life. In the Indian tribe which he studied there was abundant evidence that sexual life did not terminate at the climacteric.

3. Effect of Ovariectomy

In the light of previously cited evidence linking fluctuations in sexual desire with the ovarian cycle, and by analogy with the corresponding findings in lower mammals, it might be expected that the ovary must, to some extent, be responsible for maintaining sexual behavior and that its removal would result in a marked reduction in libido. In fact, estrogen replacement therapy is at present widely employed after bilateral ovariectomy for the purpose of increasing libido in ovariectomized women, but it has been suggested (Kroger and Freed, 1957) that its effect may be due less to any specific libidinizing action than to general increase in feeling of well being.

The available evidence, whether taken from the earlier literature before estrogen therapy was employed (see Ellis, 1936) or from more recent work, such as that of Filler and Drezner (1944) and Huffman (1950), does not, however, wholly support this view. Although in a few instances sexual behavior was found to decline, in the majority of cases there was little change. Kinsey *et al.* (1953) conclude that the levels of sexual responsiveness and frequency of sexual activity are not reduced in ovariectomized women. A number of their cases in fact showed heightened sexual activity after ovariectomy, and although it is not stated whether or not estrogen substitution therapy was employed in any or all of these cases, it is possible that some of this augmented sexual behavior may have been due to release from pregnophobia.

The fact that ovariectomy does not necessarily modify or impair human female sexual behavior or libido indicates that such functions are not entirely dependent on ovarian hormones for their maintenance. Several hypotheses can be advanced in explanation. First, sexual behavior and the capacity for orgasm is learned, and therefore can become "imprinted" on the central nervous system as other learned responses become habitual. Second, it is possible that pituitary gonadotropins have some effect on behavior, which although unlikely (see Section II) would explain the occasional increase in libido after ovariectomy as being due to a rise in gonadotropic activity. The third, and most likely, possibility is that sex hormones of adrenal origin are secreted in quantities sufficient to maintain sexual activity (see Everitt and Herbert, 1971; Everitt *et al.,* 1972). This hypothesis receives some support from the finding by Sheehan and Summers (1949) that complete and permanent loss of sex function occurs in patients suffering from Simmonds' disease. Furthermore, more exact methods of hormonal analysis have shown that after ovariectomy there is a small constant urinary excretion of pregnanediol by the adrenal (Klopper, 1952). Thus, the adrenal cortex may replace the ovary to some extent as a source of sex hormones at titers sufficient to maintain sexual behavior without causing cyclical menstrual changes (Everitt and Herbert, 1971).

4. Effect of Oral Contraceptives on Behavior

The effects of oral contraceptives may properly be considered noncyclic, although they are used to achieve the normal cycle artificially, and their behavioral effects are mainly known through clinical observation. Herzberg and Coppen (1970) report that 5 to 10% of their series of 153 women starting an estrogen/progestogen oral contraceptive complained of depression and irritability, more than in a control group of forty women. Lewis and Hoghughi (1969) also showed that 13/50 women were mildly depressed and six severely depressed. Of the severely depressed women, two attempted suicide. Patients with an earlier history of mental disturbance were more often affected than those without. There is evidence also that high rates of depression are associated with contraceptives having a high progestogen content (Grant and Mears, 1967). According to Dennis and Jeffery (1968) libido is also affected by the administration of estrogen-progestogen mixtures. In a larger sample (218 women), Herzberg *et al.* (1971) found that 25% of the sample discontinued oral contraceptives because of depression, headaches, and loss of libido.

Thus only in a proportion of women does the administration of ovarian hormones have a clinical level of influence on behavior.

D. Sex Differences in Behavior

Hillgard (1957) draws attention to striking differences between female and male interests, occupations, and achievements in society which may well reflect the influence of reproductive factors. Since, however, the role of woman in social organization is molded by tradition and upbringing rather than by any direct effect of ovarian hormones upon female behavior, these matters cannot be discussed in this chapter.

E. Behavior in the Intersex

It might be expected that some light would be thrown on the influence of the ovary in the control of what has come to be regarded as specifically "feminine" in female behavior from a study of attitudes, and of social and psychological orientation in true and pseudohermaphroditism. A wide variety of such cases has been described and it is only possible to select one or two relevant examples out of a very large number.

Ellis (1945) reviewed eighty-four cases of pseudo- and true hermaphroditism in which information could be obtained on libido and sexual orientation. A large number of genetically male pseudohermaphrodites brought up as females experienced almost exclusively female libido and sexual orientation, while those brought up as males experienced male libido and assumed the male sex role. The same factors of upbringing were found to apply in true hermaphrodites possessing gonads of both sexes. From this study, Ellis infers that while the power of the human sex drive may be dependent on the quantity and type of sex hormone secreted, the direction of this drive depends rather upon the gender of the upbringing. This conclusion has been supported by Hamblen (1957) who writes: "The true sex of an individual is not necessarily that which agrees with his or her chromosomes, gonads, hormones, morphology of internal or external genitalia, but after two and a half years it is definitely the sex of assignment and rearing, fortified by gender role and psycho-sexual orientation." Admittedly the idea underlying this dictum has partially the purpose of discouraging the use of undue psychological disturbance in attempting to "correct" the sex of an individual, but it does point to the importance of learning and conditioning in the development of sexual behavior.

The organizational role of sex hormones in the development of behavior characteristic of male and female has been considered in Section II. The application of this concept to human behavior has been discussed by Money and Ehrhardt (1968).

F. Conclusion

Taking the evidence as a whole it would appear that the role played by the ovary in the direct control of overt human behavior is, by comparison with that found in other mammals, less marked and less readily definable. It is, as is becoming increasingly recognized, a persistent factor in human behavior which must be taken into consideration. This is perhaps to be expected in a species whose function is based on adaptive rather than innate patterns of behavior, and is in harmony with the somewhat equivocal ovarian influence on learning in nonhuman species. It would appear that, in man, hormones associated with the ovary (though they may in certain conditions emanate from alternative sources) are necessary for the maintenance of libido and normal sexual behavior. The effect of cyclical variation in ovarian function upon the behavior and ergonomic capacity of women is a matter of psychosociological importance, in view of the increasing demand for an "equal" role for women in society.

VII. SUMMARY AND CONCLUSIONS

Any attempt to analyze the interaction between biological systems is a difficult task which, in the case of neuroendocrine relationships, is rendered no easier by the wide variety of influences, both from within and without, which can act upon the nervous system and so contribute to changes in behavior.

Among the endocrine organs, the internal secretions of the gonads might, a priori, be expected to exert their greatest influence on those aspects of behavior which are concerned with reproduction and, in lower animals at least, this expectation is borne out by considerable experimental evidence. It is now possible to distinguish an organizational and an activational role for the sex hormones. The secretions of the ovary would seem to play little part in the former, such characteristics as aggressiveness and the suppression of cyclic activity being confined to the influence of androgens during a critical phase of development. However, ovarian secretions are closely concerned with the activation of reproductive behavior, and there is, in most species, a close correlation between the ovarian cycle and estrous behavior. Surgical removal of the ovaries results in abolition or impairment of estrus, and the effects of giving hormones of ovarian origin suggest, in spite of certain unresolved difficulties, that estrus is the outcome of the synergistic activity of estrogen and progesterone. The component patterns involved in reproductive behavior (in which may be included maternal behavior before and during lactation) are innate and peculiar to each species, and it seems probable that such patterns are released as a result of hormonal influence acting directly upon central nervous tissues. The neurological mechanisms responsible for such behavior are frequently regarded as being organized in terms of discrete components whose localization may well coincide with the sites of action of the several hormones involved, and the available evidence, which is far from complete, points to the limbic region and hypothalamus as being particularly involved. Lesions to the neocortex, however, are not without influence on those aspects of sexual behavior which depend on complex perceptual processes, and there is some reason to postulate a relationship between sex hormones and a possible inhibitory function of rhinencephalic structures upon reproductive behavior. In addition to theories based on a specific relationship between hormones and nervous "centers" controlling the various components of reproductive activity, the view has been put forward that a generalized hormonal effect, maintained through the reticular formation, not only increases the alertness of the animal as measured by its cerebral electrical activity but also results in an increase in exploratory behavior, thereby creating a potential increase in social and sexual contact. Evidence from behavioral sources supports this concept in so far as the level of nonspecific activity has been found to depend on the presence of the ovary and to fluctuate in phase with the ovarian cycle.

The triggering of estrous behavior under ovarian control may be taken to serve the purpose of ensuring that coitus is timed to coincide with ovulation when the probability of fertilization of the ovum is at its maximum. This teleological explanation has been successfully applied to studies of human behavior by the psychodynamic school, but in other spheres the evidence is less convincing. Libido in the human female does not appear to reach a maximum at the time of ovulation and, as is well recognized, the female primate is sexually receptive throughout the menstrual cycle. Thus the manifestation of ovarian influence upon primate, and in particular upon human, behavior would seem to reside in the indirect impact of the menstrual cycle upon individual and social aspects of female, as opposed to male, behavior, and in the psychological and sociological implications of menstruation itself for the developing girl and adult woman. Such a finding is in harmony with the relatively minor part played by endocrine factors in the mediation of learned responses, given the tendency for human behavior to be regulated adaptively rather than by the innately organized reactions which dominate the behavior of other species. Ovarian influence would thus seem to impinge more upon the emotional aspects of a behavioral situation than directly upon intellectual function. Indirectly, however, the relative emancipation of behavior from ovarian control, manifest in the continuous receptivity of the human female, may be regarded as providing a stabilizing factor which permits the development of larger nonreproductive organizations than are usually found in nonprimates and so produces a climate favorable for the maturation of intellectual capacity.

In conclusion it must be stressed that much of the evidence bearing on a causative relation between ovarian secretion and behavior is incomplete and equivocal. A more exact delimitation of the role of the ovary and its hormones awaits further investigation of the nervous system with particular reference to its mode of organization from a physicochemical and enzymological point of view. It is from such investigations that the important sociological questions implicit in the differences between female and male behavior, and the changing role of the female in modern society, may derive an answer.

REFERENCES

Alphin, T. H., and Dey, F. L. (1944). Changes in the hypophysis after hypothalamic lesions. *Fed. Proc., Fed. Am. Soc. Exp. Biol.* **3**, 2.

Altmann, M. (1939). Behavior of the sow in relation to the sex cycle. *Am. J. Physiol.* **126**, 421.

Altmann, M. (1952). Social behaviour of elk, *Cervus canadensis* Nelsoni, in the Jackson Hole area of Wyoming. *Behaviour* **4**, 116–143.

Altmann, M. (1963). Naturalistic studies of maternal care in moose and elk. *In* "Maternal Behavior in Mammals" (H. L. Rheingold, ed.), pp. 233–253. Wiley, New York.

Aron, C., Asch, G., and Roos, J. (1966). Triggering of ovulation by coitus in the rat. *Int. Rev. Cytol.* **201**, 139–172.

Ask-Upmark, E. (1955). Monthly periodicity of symptoms from the central nervous system. *Neurology* **5**, 584–586.

Ball, J. (1934). Sex behavior of the rat after removal of the uterus and vagina. *J. Comp. Psychol.* **18**, 419–422.

Ball, J. (1936). Sexual responsiveness in female monkeys after castration and subsequent estrin administration. *Psychol. Bull.* **24**, 811.

Ball, J. (1937a). Sex activity of castrated male rats increased by estrin administration. *J. Comp. Psychol.* **24**, 135–144.

Ball, J. (1937b). A test for measuring sexual excitability in the female rat. *Comp. Psychol. Monogr.* **14**, No. 67, 1–37.

Ball, J. (1939). Male and female mating behavior in prepuberally castrated rats receiving estrogens. *J. Comp. Psychol.* **28**, 273–283.

Ball, J. (1940). The effect of testosterone on the sex behavior of female rats. *J. Comp. Psychol.* **29**, 151–165.

Ball, J. (1941a). Mating behavior induced in hypophysectomized female rats by injected estrogen. *Proc. Soc. Exp. Biol. Med.* **46**, 699.

Ball, J. (1941b). Effect of progesterone upon sexual excitability in the female monkey. *Psychol. Bull.* **38**, 533.

Ball, J., and Hartman, C. G. (1935). Sexual excitability as related to the menstrual cycle in the monkey. *Am. J. Obstet. Gynecol.* **29**, 117–119.

Bard, P. (1935). The effects of denervation of the genitalia on the oestrual behavior of cats. *Am. J. Physiol.* **113**, 5–6.

Bard, P. (1936). Oestrual behavior in surviving decorticate cats. *Am. J. Physiol.* **116,** 4–5.

Bard, P. (1939). Central nervous mechanisms for emotional behavior patterns in animals. *Res. Publ., Assoc. Res. Nerv. Ment. Dis.* **19,** 190–218.

Bard, P. (1940). The hypothalamus and sexual behavior. *Res. Publ., Assoc. Res. Nerv. Ment. Dis.* **20,** 551–579.

Barraclough, C. A. (1955). Influence of age on the response of preweaning female mice to testosterone propionate. *Am. J. Anat.* **97,** 493–521.

Barraclough, C. A. (1961). Production of anovulatory, sterile rats by a single injection of testosterone propionate. *Endocrinology* **68,** 62–67.

Barraclough, C. A. (1967). Modifications in reproductive function after exposure to hormones during the prenatal and early postnatal period. *In* "Neuroendocrinology" (L. Martini and W. F. Ganong, eds.), Vol. 2, pp. 62–100. Academic Press, New York.

Barraclough, C. A., and Leathem, J. W. (1954). Infertility induced in mice by a single injection of testosterone propionate. *Proc. Soc. Exp. Biol. Med.* **85,** 673–674.

Beach, F. A. (1938). Sex reversals in the mating pattern of the rat. *J. Genet. Psychol.* **53,** 329–334.

Beach, F. A. (1942a). Sexual behavior of prepuberal male and female rats treated with gonadal hormones. *J. Comp. Psychol.* **34,** 285–292.

Beach, F. A. (1942b). Execution of the complete masculine copulatory pattern by sexually receptive female rats. *J. Genet. Psychol.* **60,** 137–142.

Beach, F. A. (1942c). Importance of progesterone to induction of sexual receptivity in spayed female rats. *Proc. Soc. Exp. Biol. Med.* **51,** 369–371.

Beach, F. A. (1942b). Male and female mating behavior in prepuberally castrated female rats treated with androgens. *Endocrinology* **31,** 673–678.

Beach, F. A. (1942e). Copulatory behavior in prepuberally castrated male rats and its modification by estrogen administration. *Endocrinology* **31,** 679–683.

Beach, F. A. (1942f). Analysis of factors involved in the arousal, maintenance and manifestation of excitement in male animals. *Psychosom. Med.* **4,** 173–198.

Beach, F. A. (1943). Effects of injury to the cerebral cortex upon the display of masculine and feminine mating behavior by female rats. *J. Comp. Psychol.* **36,** 169–199.

Beach, F. A. (1945). Bisexual mating behavior in the male rat: effects of castration and hormone administration. *Physiol. Zool.* **18,** 390–402.

Beach, F. A. (1947a). A review of physiological and psychological studies of sexual behavior in mammals. *Physiol. Rev.* **27,** 240–307.

Beach, F. A. (1947b). Evolutionary changes in the physiological control of mating behavior in mammals. *Psychol. Rev.* **54,** 297–315.

Beach, F. A. (1948). "Hormones and Behavior." Harper (Hoeber), New York.

Beach, F. A. (1958). Neural and chemical regulation of behavior. *In* "Biological and Biochemical Bases of Behavior" (H. F. Harlow and C. N. Woolsey, eds.), pp. 263–283. Univ. of Wisconsin Press, Madison.

Beach, F. A. (1965a). Experimental studies of mating behavior in animals. *In* "Sex Research—New Developments" (J. Money, ed.), pp. 113–133. Holt, New York.

Beach, F. A. (1965b). Retrospect and prospect. *In* "Sex and Behavior" (F. A. Beach, ed.), pp. 535–568. Wiley, New York.

Beach, F. A. (1968). Factors involved in the control of mounting behavior by female mammals. *In* "Perspectives in Reproduction and Sexual Behavior" (M. Diamond, ed.), pp. 83–131. Indiana Univ. Press, Bloomington.

Beach, F. A., and Levinson, G. (1950). Effects of androgen on the glans penis and mating behavior of castrated male rats. *J. Exp. Zool.* **114**, 159–172.

Beach, F. A., and Merari, A. (1968). Coital behavior in dogs. IV. Effects of progesterone in the bitch. *Proc. Natl. Acad. Sci. U.S.A.* **61**, 442–446.

Beach, F. A., and Rasquin, P. (1942). Masculine copulatory behavior in intact and castrated female rats. *Endocrinology* **31**, 393–409.

Beach, F. A., Rogers, C. M., and Le Boeuf, B. J. (1968). Coital behavior in dogs: effects of estrogen on mounting by females. *J. Comp. Physiol. Psychol.* **66**, 296–307.

Benedek, T. (1952). The functions of the sexual apparatus and their disturbance. *In* "Psychosomatic Medicine" (F. Alexander, ed.), pp. 216–271. Allen & Unwin, London.

Benedek, T., and Rubinstein, B. B. (1939a). The correlations between ovarian activity and psychodynamic process; ovulative phase. *Psychosom. Med.* **1**, 245–270.

Benedek, T., and Rubinstein, B. B. (1939b). The correlations between ovarian activity and psychodynamic process; menstrual phase. *Psychosom. Med.* **1**, 461–485.

Billings, E. G. (1934). Occurrence of cyclic variations in motor activity in relation to the menstrual cycle in the human female. *Bull. Johns Hopkins Hosp.* **54**, 440–454.

Birch, H. G., and Clark, G. (1950). Hormonal modification of social behavior. IV. The mechanism of estrogen-induced dominance in chimpanzees. *J. Comp. Physiol. Psychol.* **43**, 181–193.

Blandau, R. J., Jordan, E. S., and Soderwall, A. L. (1940). Postparturitional heat and time of ovulation in the albino rat. *Anat. Rec.* **78**, Suppl., 1–58.

Blandau, R. J., Boling, J. L., and Young, W. C. (1941). The length of heat in the albino rat as determined by the copulatory response. *Anat. Rec.* **79**, 453–463.

Bloch, G. J., and Davidson, J. M. (1968). Effects of adrenalectomy and prior experience on post-castrational sex behavior in the male rat. *Physiol. Behav.* **3**, 461–465.

Boling, J. L., and Blandau, R. J. (1939). The estrogen-progesterone induction of mating responses in the spayed female rat. *Endocrinology* **25**, 359–371.

Boling, J. L., Young, W. C., and Dempsey, E. W. (1938). Miscellaneous experiments on the estrogen-progesterone induction of heat in the spayed guinea-pig. *Endocrinology* **23**, 182–187.

Boling, J. L., Blandau, R. J., Wilson, J. G., and Young, W. C. (1939). Post-parturitional heat responses of newborn and adult guinea-pigs. Data on parturition. *Proc. Soc. Exp. Biol. Med.* **42**, 128–132.

Boling, J. L., Blandau, R. J., Soderwall, A. L., and Young, W. C. (1941a). Growth of the Graafian follicle and the time of ovulation in the rat. *Anat. Rec.* **79**, 313–331.

Boling, J. L., Blandau, R. J., Rundlett, B., and Young, W. C. (1941b). Factors underlying the failure of mating behavior in the albino rat. *Anat. Rec.* **80**, 155–171.

Bond, C. R. (1945). The golden hamster (*Cricetus auratus*): care, breeding and growth. *Physiol. Zool.* **18**, 52–59.

Bradbury, J. T. (1941). Permanent after-effects following masculinization of the infantile female rat. *Endocrinology* **28**, 101–106.

Brain, P. F., and Nowell, N. W. (1969). Some endocrine and behavioral changes in the development of the albino laboratory mouse. *Commun. Behav. Biol.* **4**, 203–220.

Brain, P. F., and Nowell, N. W. (1971a). Isolation versus grouping effects on adrenal and gonadal function in albino mice. I. The male. *Gen. Comp. Endocrinol.* **16**, 149–154.

Brain, P. F., and Nowell, N. W. (1971b). Isolation versus grouping effects on adrenal and gonadal function in albino mice. II. The female. *Gen. Comp. Endocrinol.* **16**, 155–159.

Brambell, F. W. R. (1944). The reproduction of the wild rabbit. *Proc. Zool. Soc. London* **114**, 1–45.

Broadhurst, P. L. (1963). Anarchy in rat town. *Aspect* **8**, 51–58.

Bronson, F. H., and Eleftheriou, B. E. (1965). Adrenal response to fighting in mice: separation of physical and psychological causes. *Science* **147**, 627–628.

Brooks, C. McC. (1937). The role of the cerebral cortex and of various sense organs in the excitation and execution of mating activity in the rabbit. *Am. J. Physiol.* **120**, 544–553.

Bruce, H. M. (1959). An exteroceptive block to pregnancy in the mouse. *Nature (London)* **184**, 105.

Bruce, H. M. (1965). Olfactory hormones and reproduction in mice. *Proc. Int. Congr. Endocrinol., 2nd, 1964* Excerpta Med. Found. Int. Congr. Ser. No. **83**, pp. 193–197.

Carr, W. J., and Caul, W. F. (1962). The effect of castration in the rat upon the determination of sex odours. *Anim. Behav.* **10**, 20–27.

Carr, W. J., Loeb, L. S., and Dissinger, M. L. (1965). Responses of rats to sex odors. *J. Comp. Physiol. Psychol.* **59**, 370–377.

Chesley, L. C., and Hellman, L. M. (1957). Variations in body weight and salivary sodium in the menstrual cycle. *Am. J. Obstet. Gynecol.* **74**, 582–590.

Christian, J. J., and Davis, D. E. (1964). Endocrines, behavior and population. *Science* **146**, 1550–1560.

Christian, J. J., Lloyd, J. A., and Davis, D. E. (1965). The role of endocrines in the self-regulation of mammalian populations. *Recent Prog. Horm. Res.* **21**, 501–578.

Clark, G., and Birch, H. G. (1945). Hormonal modifications of social behavior. I. The effect of sex-hormone administration in the social status of the male castrate chimpanzee. *Psychosom. Med.* **7**, 321–329.

Clark, G., and Birch, H. G. (1946). The effect of sex hormones on the dominance-subordination relationships of the castrate female chimpanzee. *Fed. Proc., Fed. Am. Soc. Exp. Biol.* **5**, 16–17.

Clemens, L. G., Hiroi, H., and Gorski, R. A. (1969). Induction and facilitation of female mating behavior in rats treated neonatally with low doses of testosterone propionate. *Endocrinology* **34**, 1430–1438.

Cole, H. H. (1937). Superfecundity in rats treated with mare gonadotropic hormone. *Am. J. Physiol.* **119**, 704–712.

Cole, H. H., and Cupps, P. T., eds. (1969). "Reproduction in Domestic Animals," 2nd ed. Academic Press, New York.

Conner, R. L., and Levine, S. (1969). Hormonal influences on aggressive behavior. *Aggressive Behav., Proc. Int. Symp., 1968* pp. 150–163.

Conner, R. L., Levine, S., Wertheim, G. A., and Cummer, J. F. (1969). Hormonal determinants of aggressive behavior. *Ann. N.Y. Acad. Sci.* **159**, 760–776.

Cooke, W. R. (1945). The differential psychology of the American woman. *Am. J. Obstet. Gynecol.* **49**, 457–472.

Cooper, J. B. (1942). An exploratory study on African lions. *Comp. Psychol. Monogr.* **17**, 1–48.

Critchlow, V., and Bar-Sela, M. E. (1967). Control of the onset of puberty. *In* "Neuroendocrinology" (L. Martini and W. F. Ganong, eds.), Vol. 2, pp. 101–162. Academic Press, New York.

Crook, J. H. (1970). "Social Behavior in Birds and Mammals." Academic Press, New York.

Crowcroft, R., and Rowe, F. P. (1963). Social organization and territorial behavior of wild house mice. *Proc. Zool. Soc. London* **140**, 517–531.

Dalton, K. (1959). Menstruation and acute psychiatric illnesses. *Br. Med. J.* **1**, 148–149.

Dalton, K. (1960a). Effect of menstruation on schoolgirls' weekly work. *Br. Med. J.* **1**, 326–328.

Dalton, K. (1960b). Menstruation and accidents. *Br. Med. J.* **2**, 1425–1426.

Dalton, K. (1960c). Schoolgirls' behavior and menstruation. *Br. Med. J.* **2**, 1647–1649.

Dalton, K. (1968). Menstruation and examinations. *Lancet* **2**, 1386–1388.

Dalton, K. (1969). "The Menstrual Cycle." Penguin Books, London.

D'Amour, F. E., and Woods, L. (1941). Sex hormone excretion; sex cycle of normal women. *J. Clin. Endocrinol. Metab.* **1**, 433–435.

Davidson, J. M. (1966). Characteristics of sex behavior in male rats following castration. *Anim. Behav.* **14**, 266–272.

Davidson, J. M. (1969). Effects of estrogen on the sexual behavior of male rats. *Endocrinology* **84**, 1365–1372.

Davidson, J. M., and Levine, S. (1969). Progesterone and heterotypical behaviour in male rats. *J. Endocrinol.* **44**, 129–130.

Davidson, J. M., Smith, E. R., Rodgers, C. H., and Bloch, G. J. (1968). Relative thresholds of behavioral and somatic responses to estrogen. *Physiol. Behav.* **3**, 227–229.

de Beauvoir, S. (1956). "The Second Sex" (transl. by H. M. Parshley). Cape, London.

Defries, J. C., and McClearn, G. E. (1970). Social dominance and Darwinian fitness in the laboratory mouse. *Am. Nat.* **104**, 408–410.

Dempsey, E. W., Hertz, R., and Young, W. C. (1936). The experimental induction oestrus (sexual receptivity) in the normal and ovariectomized guinea-pig. *Am. J. Physiol.* **116**, 201–209.

Denenberg, V. H., Grota, L. J., and Zarrow, M. X. (1963). Maternal behavior in the rat: analysis of cross-fostering. *J. Reprod. Fertil.* **5**, 133–141.

Dennis, K. J., and Jeffery, J. d'A. (1968). Depression and oral contraceptives. *Lancet* **2**, 454–455.

Devereux, G. (1950). Psychology of feminine genital bleeding; analysis of Mohave Indian puberty and menstrual rites. *Int. J. Psycho-Analysis* **31**, 237–257.

Diamond, M. (1970). Intromission pattern and species-specific vaginal code in relation to pseudopregnancy. *Science* **169**, 995–997.

Dieterlen, F. (1959). Das Verhalten des syrischen Goldhamster (*Mesocricetus auratus* Waterhouse). *Z. Tierpsychol.* **16**, 47–103.

Dilger, W. C. (1962). Behavior and genetics. *In* "Roots of Behavior" (E. L. Bliss, ed.), pp. 35–47. Harper, New York.

Donovan, B. T., and van der Werff ten Bosch, J. J. (1959). The hypothalamus and sexual maturation in the rat. *J. Physiol. (London)* **147**, 78–92.

Donovan, B. T., and van der Werff ten Bosch, J. J. (1965). "The Physiology of Puberty." Arnold, London.

Doring, G. K. (1954). Die Reaktion auf Sinnesreize im Verlauf des Menstruations-zyklus. *Dtsch. Med. Wochenschr.* **79**, 885–886.

Dusser de Barenne, D., and Gibbs, F. A. (1942). Variations in the electroencephalo-gram during the menstrual cycle. *Am. J. Obstet. Gynecol.* **44**, 687–690.

Edwards, D. A. (1968). Mice: fighting by neonatally androgenized females. *Science* **161**, 1027–1028.

Edwards, D. A. (1969). Early androgen stimulation and aggressive behavior in male and female mice. *Physiol. Behav.* **4**, 333–338.

Edwards, D. A. (1970). Effects of cyproterone acetate on aggressive behavior and seminal vesicles of male mice. *J. Endocrinol.* **46**, 477–481.

Edwards, D. A., and Herndon, J. H. (1970). Neonatal estrogen stimulation and aggressive behavior in female mice. *Physiol. Behav.* **5**, 993–997.

Eibl-Eibesfeldt, I. (1953). Zur Ethologie des Hamsters (*Cricetus cricetus* L.). *Z. Tierpsychol.* **10**, 204–254.

Eibl-Eibesfeldt, I. (1958). Das Verhalten der Nagetiere. *Handb. Zool.* **10**, 1–88.

Ellis, A. (1945). Sexual psychology of human hermaphrodites. *Psychosom. Med.* **7**, 108–125.

Ellis, H. (1936). "Studies in the Psychology of Sex." Random House, New York.

Elsberg, C. A., Brewer, E. D., and Levy, I. (1935). The sense of smell. IV. Concern-ing conditions which may temporarily affect olfactory acuity. *Bull. Neurol. Inst. N.Y.* **4**, 31–34.

Engel, P. (1949). Male mating behavior shown by female rats treated with enormous doses of estrone. *Endocrinology* **44**, 289–290.

Engel, P., and Hildebrandt, G. (1969). Die rhythmischen Schwankungen der Reak-tionszeit beim Menschen. *Psychol. Forsch.* **32**, 324–336.

Esser, A. H. (1971). "Behavior and Environment." Plenum, New York.

Everitt, B. J., and Herbert, J. (1969a). The role of ovarian hormones in the sexual preference of rhesus monkeys. *Anim. Behav.* **17**, 738–746.

Everitt, B. J., and Herbert, J. (1969b). Adrenal glands and sexual receptivity in female rhesus monkeys. *Nature (London)* **222**, 1065–1066.

Everitt, B. J., and Herbert, J. (1971). The effects of dexamethasone and androgens on sexual receptivity of female rhesus monkeys. *J. Endocrinol.* **51**, 575–588.

Everitt, B. J., Herbert, J., and Hamer, J. D. (1972). Sexual receptivity of bilaterally adrenalectomized female rhesus monkeys. *Physiol. Behav.* **8**, 409–415.

Farner, D. S. (1964). The photoperiodic control of reproductive cycles in birds. *Am. Sci.* **52**, 137–156.

Farris, E. J. (1944). Comparison of patterns of cyclic activity in women and the female albino rat. *Anat. Rec.* **89**, 536.

Feder, H. H., Resko, J. A., and Goy, R. W. (1966). Progestin levels in peripheral plasma of guinea pigs during the estrous cycle. *Am. Zool.* **6**, 597.

Feder, H. H., Goy, R. W., and Resko, J. A. (1967). Progesterone concentrations in the peripheral plasma of cyclic rats. *J. Physiol. (London)* **191**, 136–137.

Ferroni, A., Lauro, V., Marinelli, M., and Bergonzi, P. (1969). Indagini elettroen-cefalografiche in pazienti trattate con estroprogestinici di sintesi. *Arch. Ostet. Ginecol.* **74**, 236–249.

Filler, W., and Drezner, N. (1944). The results of surgical castration in women under 40. *Am. J. Obstet. Gynecol.* **47**, 122–124.

Folley, S. J., and Malpress, F. H. (1944). The artificial induction of lactation in the bovine by the subcutaneous implantation of synthetic oestrogen tablets. *J. Endocrinol.* **4**, 1–22.

Forbes, T. R. (1955). Reproduction in the female. *In* "Textbook of Physiology" (J. F. Fulton, ed.), pp. 1208–1234. Saunders, Philadelphia, Pennsylvania.

Ford, C. L., and Beach, F. A. (1965). "Patterns of Sexual Behaviour." Eyre & Spottiswoode, London.

Frank, A. H., and Fraps, R. M. (1945). Induction of estrus in the ovariectomized golden hamster. *Endocrinology* **37**, 357–361.

Frank, R. T. (1931). The hormonal cause of premenstrual tension. *Arch. Neurol. Psychiatry* **26**, 1053–1057.

Fredericton, E. (1950). The effect of food deprivation upon competitive and spontaneous combat in C57 Black mice. *J. Psychol.* **29**, 89–100.

Freed, S. C., and Kroger, W. S. (1950). Psychologic manifestations of the menstrual cycle. *Psychosom. Med.* **12**, 229–235.

Friedgood, H. B. (1939). Induction of estrous behavior in anestrous cats with the follicle-stimulating and luteinizing hormones of the anterior pituitary gland. *Am. J. Physiol.* **126**, 229–233.

Gallien, L. G. (1965). Genetic control of sexual differentiation in vertebrates. *In* "Organogenesis" (R. L. de Haan and H. Ursprung, eds.), pp. 583–610. Holt, New York.

Gautray, J. P. (1969). Quantitative analysis of EEG variations during spontaneous or restored menstrual cycle. *Neuroendocrinology* **5**, 368–373.

Gautray, J. P., Garrel, S., and Eberard, A. (1970). Variations de l'électroencéphalogramme au cours du cycle menstruel. *Rev. Fr. Gynecol. Obstet.* **65**, 257–262.

Glass, A. (1968). Intensity of attenuation of alpha activity by mental arithmetic in females and males. *Physiol. Behav.* **3**, 217–220.

Gorski, R. A. (1968). Influence of age on the response to paranatal administration of a low dose of androgen. *Endocrinology* **82**, 1001–1004.

Gorski, R. A., and Barraclough, C. A. (1963). Effects of low dosages of androgen on the differentiation of hypothalamic regulatory control of ovulation in the rat. *Endocrinology* **73**, 210–216.

Goy, R. W. (1968). Organizing effect of androgen on the behavior of rhesus monkeys. *In* "Endocrinology and Human Behaviour" (R. P. Michael, ed.), pp. 12–31. Oxford Univ. Press, London and New York.

Goy, R. W., and Jackway, J. S. (1962). Role of inheritance in determination of sexual behavior patterns. *In* "Roots of Behavior" (B. L. Bliss, ed.), pp. 96–122. Harper, New York.

Grady, K. L., and Phoenix, C. H. (1963). Hormonal determinants of mating behavior; the display of feminine behavior by adult male rats castrated neonatally. *Am. Zool.* **3**, 482.

Grant, E. C. G., and Mears, E. (1967). Mental effects of oral contraceptives. *Lancet* **2**, 945–946.

Greenblatt, R. B., Morton, F., and Torpin, R. (1942). Sexual libido in the female. *Am. J. Obstet. Gynecol.* **44**, 658–663.

Greene, R., and Dalton, K. (1953). The premenstrual syndrome. *Br. Med. J.* **1**, 1001–1014.

Greene, R., and Dalton, K. (1955). Discussion on premenstrual syndrome. *Proc. R. Soc. Med.* **48**, 337–341.

Grota, L. J., and Eik-Nes, K. B. (1967). Plasma progesterone concentrations during pregnancy and lactation in the rat. *J. Reprod. Fertil.* **13**, 83–91.

Grota, L. J., Denenberg, V. H., and Zarrow, M. X. (1967). Maternal behavior in the rat: some parameters affecting the acceptance of young delivered by Caesarian section. *J. Reprod. Fertil.* **13**, 405–411.

Guhl, A. M. (1961). Gonadal hormones and social behavior in infrahuman vertebrates. *In* "Sex and Internal Secretions" (W. C. Young, ed.), 3rd ed., Vol. 2, pp. 1240–1267. Williams & Wilkins, Baltimore, Maryland.

Hamblen, E. C. (1957). Assignment of sex to individual. *Am. J. Obstet. Gynecol.* **74**, 1228–1243.

Hammond, J. (1927). "The Physiology of Reproduction in the Cow." Cambridge Univ. Press, London and New York.

Hammond, J., and Marshall, F. H. A. (1925). "Reproduction in the Rabbit." Oliver & Boyd, London.

Hammond, J., and Marshall, F. H. A. (1930). Oestrus and pseudo-pregnancy in the ferret. *Proc. R. Soc. London, Ser. B* **105**, 607–630.

Hammond, J., Jr. (1944). Control of ovulation in farm animals. *Nature (London)* **153**, 702–704.

Hammond, J., Jr. (1945). Induced ovulation and heat in anoestrous sheep. *J. Endocrinol.* **4**, 169–180.

Hammond, J., Jr., and Day, F. T. (1944). Oestrogen treatment of cattle: induced lactation and other effects. *J. Endocrinol.* **4**, 53–82.

Hancock, J. L. (1952). *In Ciba Found. Colloq. Endocrinol. [Proc.]* **3**, 59.

Harlow, H. F., and Harlow, M. K. (1965). The effect of rearing conditions on behavior. *In* "Sex Research—New Developments" (J. Money, ed.), pp. 161–175. Holt, New York.

Harlow, H. F., Harlow, M. K., and Hansen, E. W. (1963). The maternal affectional system of rhesus monkeys. *In* "Maternal Behavior in Mammals" (H. L. Rheingold, ed.), pp. 254–281. Wiley, New York.

Harris, G. W. (1952). *In Ciba Found. Colloq. Endocrinol. [Proc.]* **3**, 58.

Harris, G. W. (1964). Sex hormones, brain development and brain function. *Endocrinology* **75**, 627–647.

Harris, G. W., and Jacobsohn, D. (1952). Functional grafts of the anterior pituitary gland, *Proc. R. London, Ser. B* **139**, 263–276.

Harris, G. W., Michael, R. P., and Scott, P. P. (1958). Neurological site of action of stilboestrol in eliciting sexual behaviour. *Neurol. Basis Behav., Ciba Found. Symp., 1957* pp. 236–251.

Hart, B. (1968). Role of prior experience in the effects of castration on sexual behavior of male dogs. *J. Comp. Physiol. Psychol.* **66**, 719–725.

Hart, R. d'A. (1960). Monthly rhythm of libido in married women. *Br. Med. J.* **1**, 1023–1024.

Hartman, C. G. (1945). The mating of mammals. *Ann. N.Y. Acad. Sci.* **46**, 23–44.

Hemmingsen, A. M. (1933). Studies on the oestrus-producing hormone (oestrin). *Skand. Arch. Physiol.* **65**, 97–250.

Hemmingsen, A. M., and Krarup, N. B. (1937). Rhythmic diurnal variations in the oestrous phenomena of the rat and their susceptibility to light and dark. *D. Kgl. Danske Vidensk. Selsk., Biol. Medd.* **13**, No. 7, 1–61.

Herbert, J. (1967). The social modification of sexual and other behaviour in the rhesus monkey. *Prog. Primatol., Congr. Int. Primatol. Soc., 1st, Frankfurt, 1966* pp. 232–246.

Herbert, J. (1968). Sexual preference in the rhesus monkey, *Macaca mulatta*, in the laboratory. *Anim. Behav.* **16**, 120–128.

Herbert, J. (1969). The pineal gland and light-induced oestrus in the ferret. *J. Endocrinol.* **43**, 625–636.

Herbert, J. (1970). Hormones and reproductive behaviour in rhesus and talapoin monkeys. *J. Reprod. Fertil., Suppl.* **11**, 119–140.

Hersher, L., Richmond, J. B., and Ulric Moore, A. (1963). Maternal behavior in sheep and goats. *In* "Maternal Behavior in Mammals" (H. L. Rheingold, ed.), pp. 203–232. Wiley, New York.

Hertz, R., Meyer, R. K., and Spielman, M. A. (1937). The specificity of progesterone in inducing receptivity in the ovariectomized guinea pig. *Endocrinology* **21**, 533–535.

Herzberg, B. N., and Coppen, A. (1970). Changes in psychological symptoms in women taking oral contraceptives. *Br. J. Psychiatry* **116**, 161–164.

Herzberg, B. N., Draper, C., Johnson, A. L., and Nicol, C. (1971). Oral contraceptives, depression and libido. *Br. Med. J.* **3**, 495–500.

Hillgard, E. R. (1957). "Introduction to Psychology." Methuen. London.

Hori, T., Ide, M., and Miyake, T. (1968). Ovarian estrogen secretion during the estrous cycle and under the influence of exogenous gonadotropins in rats. *Endocrinol. Jpn.* **15**, 215–222.

Hoskins, R. G. (1941). "Endocrinology." Norton, New York.

Huffman, J. W. (1941). Effect of testosterone propionate upon reproduction in the female. *Endocrinology* **29**, 77–79.

Huffman, J. W. (1950). Effect of gynecologic surgery on sexual reactions. *Am. J. Obstet. Gynecol.* **59**, 915–917.

Israel, S. L. (1938a). Treatment of dysfunctional menstrual disorders. *Endocrinology* **22**, 253–261.

Israel, S. L. (1938b). Premenstrual tension. *J. Am. Med. Assoc.* **110**, 1721–1723.

Johnson, G. B. (1932). The effects of periodicity on learning to walk a tight wire. *J. Comp. Psychol.* **13**, 133–141.

Jost, A. (1965). Gonadal hormones in the sex differentiation of the mammalian fetus. *In* "Organogenesis" (R. L. de Haan and H. Ursprung, eds.), pp. 611–628. Holt, New York.

Kaufman, K. S. (1953). Effects of preventing intromission upon sexual behavior in rats. *J. Comp. Physiol. Psychol.* **46**, 209–211.

Kennedy, G. C., and Mitra, J. (1963). Hypothalamic control of energy balance and the reproductive cycle in the rat. *J. Physiol. (London)* **166**, 395–407.

Kent, G. C., and Liberman, M. J. (1949). Induction of psychic estrus in the hamster with progesterone administered via the lateral brain ventricle. *Endocrinology* **45**, 29–32.

Kikuyama, S. (1961). Inhibitory effect of reserpine on the induction of persistent estrus by sex steroids in the rat. *Annot. Zool. Jpn.* **34**, 111–116.

Kikuyama, S. (1962). Inhibition of induction of persistent estrus by chlorpromazine in the rat. *Annot. Zool. Jpn.* **35**, 6–11.

Kinsey, A. C., Pomeroy, W. B., Marten, C. E., and Gebhard, P. H. (1953). "Sexual Behavior of the Human Female." Saunders, Philadelphia, Pennsylvania.

Kirihara, H. (1926). Quoted by Sowton and Myers (1928). From *Stud. Sci. Lab.* **3**, 265–328.

Kislak, J. W., and Beach, F. A. (1955). Inhibition of aggressiveness by ovarian hormones. *Endocrinology* **56**, 684–692.

Klein, M. (1952). Uterine distension, ovarian hormones and maternal behavior in rodents. *Ciba Found. Colloq. Endocrinol.* [*Proc.*] **3**, 84–88.

Kleitman, N. (1939). "Sleep and Wakefulness as Alternating Phases in the Cycle of Existence." Univ. of Chicago Press, Chicago, Illinois.

Klopper, A. I. (1952). The excretion of pregnanediol during the normal menstrual cycle. *J. Obstet. Gynaecol. Br. Emp.* **64**, 504–511.

Koller, G. (1952). Das Nestbau des weissen Maus und seine hormonale Auslösung. *Zool. Anz.,* Suppl. **17**, 160–168.

Koller, G. (1955). Hormonal und psychische Steurung beim Nestbau weisser Mäuse. *Zool. Anz.,* Suppl. **19**, 123–132.

Kopell, B. S., Lunde, D. T., Clayton, R. B., and Moos, R. H. (1969). Variations in some measures of arousal during the menstrual cycle. *J. Nerv. Ment. Dis.* **148**, 180–187.

Koster, R. (1943). Hormone factors in male behavior of the female rat. *Endocrinology* **33**, 337–348.

Krehbiel, R. H. (1952). Mating of the golden hamster during pregnancy. *Anat. Rec.* **113**, 117–121.

Kroger, W. S., and Freed, S. C. (1957). "Psychosomatic Gynecology." Saunders, Philadelphia, Pennsylvania.

Kuehn, R. E., and Zucker, I. (1968). Reproductive behavior of the Mongolian gerbil (*Meriones unguiculatus*). *J. Comp. Physiol. Psychol.* **66**, 747–752.

Ladosky, W., Kesikowski, W. M., Momoli, D. M. M., and Felipe, I. (1968). Corpora lutea formation in male rats injected neonatally with chlorpromazine. *Proc. Int. Congr. Endocrinol., 3rd, 1968* Excerpta Med. Found. Int. Congr. Ser. No. 157, p. 28.

Lamb, W. M., Ulett, G. A., Masters, W. H., and Robinson, D. W. (1953). Premenstrual tension: EEG, hormonal and psychiatric evaluation. *Am. J. Psychiatry* **109**, 840–848.

Lashley, K. S. (1938). Experimental analysis of instinctive behavior. *Psychol. Rev.* **45**, 445–471.

Leathem, J. H. (1938). Experimental induction of estrus in the dog. *Endocrinology* **22**, 559–567.

Le Boeuf, B. J. (1967). Inter-individual association in dogs. *Behaviour* **24**, 268–295.

Lehrman, D. S. (1961). Hormonal regulation of parental behaviour in birds and infra-human mammals. *In* "Sex and Internal Secretions" (W. C. Young, ed.), 3rd ed., Vol. 2, pp. 1268–1382. Williams & Wilkins, Baltimore, Maryland.

Lehrman, D. S. (1962). Interaction of hormonal and experiential influence on development of behavior. *In* "Roots of Behavior" (E. L. Bliss, ed.), pp. 142–156. Harper, New York.

Le Magnen, J. (1949). Travaux récents de psycho-physiologie et physiologie de l'olfaction et de la gustation. *Année Psychol.* **51**, 189–199.

Le Magnen, J. (1950a). Nouvelles données sur le phénomène de l'exaltolide. *C. R. Acad. Sci.* (*Paris*) **230**, 1103–1105.

Le Magnen, J. (1950b). L'odeur des hormones sexuelles. *C. R. Acad. Sci.* (*Paris*) **230**, 1367–1369.

Le Magnen, J. (1952). Les phénomènes olfacto-sexuels chez le rat blanc. *Arch. Sci. Physiol.* **6**, 295–332.

Levine, S., and Mullins, R., Jr. (1964). Estrogen administered neonatally affects adult sexual behavior in male and female rats. *Science* **144**, 185–187.

Levy, J. V., and King, J. A. (1953). The effects of testosterone propionate on fighting behavior in young male C57 BL/10 mice. *Anat. Rec.* **117**, 562–563.

Lewis, A., and Hoghughi, M. (1969). An evaluation of depression as a side effect of oral contraceptives. *Br. J. Psychiat.* **115,** 697–701.

Lincoln, G. A., Youngson, R. W., and Short, R. V. (1970). The social and sexual behaviour of the red deer stag. *J. Reprod. Fertil., Suppl.* **11,** 71–103.

Lisk, R. D. (1960). A comparison of the effectiveness of intravenous, as opposed to subcutaneous injection of progesterone for the induction of estrous behavior in the rat. *Can. J. Biochem. Physiol.* **38,** 1381–1383.

Lisk, R. D. (1969a). Progesterone: biphasic effects on the lordosis response in adult or neonatally gonadectomized rats. *Neuroendocrinology* **5,** 149–160.

Lisk, R. D. (1969b). Progesterone: role in limitation of ovulation and sex behavior in mammals. *Trans. N.Y. Acad. Sci. [2]* **31,** 593–601.

Lisk, R. D. (1970). Mechanisms regulating sexual activity in mammals. *J. Sex Res.* **6,** 220–228.

Long, C. N. H., and Zuckerman, S. (1937). Relation of the adrenal cortex to cyclical changes in the female accessory reproductive organs. *Nature (London)* **139,** 1106–1107.

Lott, D. F. (1962). The role of progesterone in the maternal behavior of rodents. *J. Comp. Physiol. Psychol.* **45,** 610–613.

Lott, D. F., and Fuchs, S. S. (1962). Failure to induce retrieving by sensitization or the injection of prolactin. *J. Comp. Physiol. Psychol.* **55,** 1111–1113.

McCance, R. A., Luff, M. D., and Widdowson, E. E. (1937). Physical and emotional periodicity in women. *J. Hyg.* **37,** 571–605.

McDonald, P. G., Vidal, N., and Beyer, C. (1970). Sexual behavior in the ovariectomized rabbit and treatment with different amounts of gonadal hormones. *Horm. Behav.* **1,** 161–172.

McKenzie, F. F., and Andrews, F. N. (1937). Estrus and ovulation in the mare. Quoted by Young (1941).

MacKinnon, P. C. B., and MacKinnon, I. L. (1956). Hazards of the menstrual cycle. *Br. Med. J.* **1,** 555.

Maes, J. P. (1939). Neural mechanisms of sexual behavior in the female cat. *Nature (London)* **144,** 598–599.

Margerison, J. H., Anderson, W. McC., and Dawson, J. (1964). Plasma sodium and the EEG during the menstrual cycle of normal human females. *Electroencephalogr. Clin. Neurophysiol.* **17,** 540–544.

Marshall, F. H. A., and Hammond, J. (1945). Experimental control by hormone action of the oestrous cycle in the ferret. *J. Endocrinol.* **4,** 159–168.

Martinez, C., and Bittner, J. J. (1956). A non-hypophyseal sex difference in estrous behavior of mice bearing pituitary grafts. *Proc. Soc. Exp. Biol. Med.* **91,** 506–509.

Masters, W. H., and Johnson, V. E. (1965). The sexual response cycles of the human male and female: comparative anatomy and physiology. *In* "Sex and Behavior" (F. A. Beach, ed.), pp. 512–534. Wiley, New York.

Masters, W. H., and Johnson, V. E. (1966). "Human Sexual Response." Churchill, London.

Masters, W. H., and Johnson, V. E. (1970). "Human Sexual Inadequacy." Churchill, London.

Matthews, L. H. (1939). Visual stimulation and ovulation in pigeons. *Proc. R. Soc. London, Ser. B* **126,** 557–560.

Meier, G. W. (1965). Maternal behaviour of feral and laboratory-reared monkeys following surgical delivery of their infants. *Nature (London)* **206,** 492–493.

Meites, J., and Turner, C. W. (1942a). Studies concerning the mechanism controlling the initiation of lactation at parturition. I. Can estrogen suppress the lactogenic hormone of the pituitary? *Endocrinology* **30**, 711–718.

Meites, J., and Turner, C. W. (1942b). Studies concerning the mechanism controlling the initiation of lactation at parturition. II. Why lactation is not initiated during pregnancy. *Endocrinology* **30**, 719–725.

Meites, J., and Turner, C. W. (1942c). Studies concerning the mechanism controlling the initiation of lactation at parturition. III. Can estrogen account for the precipitous increase in the lactogen content of the pituitary following parturition? *Endocrinology* **30**, 726–733.

Menaker, J. S. (1957). Psychosomatic problems in gynecologic practice. *Obstet. Gynecol.* **10**, 176–180.

Michael, R. P. (1958). Sexual behaviour and the vaginal cycle in the cat. *Nature (London)* **181**, 567–568.

Michael, R. P. (1969). Effects of gonadal hormones on displaced and and direct aggression in pairs of rhesus monkey of opposite sex. *Aggressive Behav., Proc. Int. Symp., 1968* pp. 172–178.

Michael, R. P., and Keverne, E. B. (1968). Pheromones in the communication of sexual status in primates. *Nature (London)* **218**, 746–749.

Michael, R. P., and Scott, P. P. (1957). Quantitative studies on mating behaviour of spayed female cats stimulated by treatment with oestrogens. *J. Physiol. (London)* **138**, 46–47.

Michael, R. P., and Scott, P. P. (1964). The activation of sexual behaviour in cats by the subcutaneous administration of oestrogen. *J. Physiol. (London)* **171**, 254–274.

Michael, R. P., and Zumpe, D. (1970). Aggression and gonadal hormones in captive rhesus monkeys. *Anim. Behav.* **18**, 1–10.

Michael, R. P., Saayman, G., and Zumpe, D. (1967). Inhibition of sexual receptivity by progesterone in rhesus monkeys (*Macaca mulatta*). *J. Endocrinol.* **39**, 309–310.

Mirsky, A. F. (1955). The influence of sex hormones on social behavior in monkeys. *J. Comp. Physiol. Psychol.* **48**, 327–335.

Moltz, H., and Wiener, E. (1966). Effects of ovariectomy on maternal behavior of primiparous and multiparous rats. *J. Comp. Physiol. Psychol.* **62**, 382–387.

Money, J. (1961). Sex hormones in human eroticism. *In* "Sex and Internal Secretions" (W. C. Young, ed.), 3rd ed., Vol. 2, pp. 1383–1400. Williams & Wilkins, Baltimore, Maryland.

Money, J. (1965). Psychosexual differentiation. *In* "Sex Research—New Developments" (J. Money, ed.), pp. 3–23. Holt, New York.

Money, J., and Ehrhardt, A. A. (1968). Prenatal hormonal exposure: possible effects on behaviour in man. *In* "Endocrinology and Human Behaviour" (R. P. Michael, ed.), pp. 32–48. Oxford Univ. Press, London and New York.

Moore, C. R. (1919). On the physiological properties of the gonads as controllers of somatic and psychical characteristics. I. The rat. *J. Exp. Zool.* **28**, 137–160.

Moore, N. W., Barrett, S., Brown, J. B., Schindler, I., Smith, M. A., and Smyth, B. (1969). Oestrogen and progesterone content of ovarian vein blood of the ewe during the oestrous cycle. *J. Endocrinol.* **44**, 55–62.

Morton, J. H. (1950). Premenstrual tension. *Am. J. Obstet. Gynecol.* **60**, 343–352.

Morton, J. H., Additon, H., Addison, R. G., Hunt, L., and Sullivan, J. J. (1953). A clinical study of premenstrual tension. *Am. J. Obstet. Gynecol.* **65**, 1182–1191.

Myers, H. I., Young, W. C., and Dempsey, E. W. (1936). Graafian follicle development throughout the reproductive cycle in the guinea-pig with special reference to changes during oestrus (sexual receptivity). *Anat. Rec.* **65**, 381–401.

Nalbandov, A. V. (1964). "Reproductive Physiology," 2nd ed. Freeman, San Francisco, California.

Nissen, H. W. (1929). The effects of gonadectomy, vasectomy, and injections of placental and orchid extracts on the sex behavior of the white rat. *Genet. Psychol. Monogr.* **5**, 451–547.

Nowlis, V. (1942). Sexual status and degree of hunger in chimpanzee competitive interaction. *J. Comp. Psychol.* **34**, 185–194.

Palka, Y. S., and Sawyer, C. H. (1966). The effects of hypothalamic implants of ovarian steroids on oestrous behaviour in rabbits. *J. Physiol. (London)* **135**, 251–269.

Payne, A. P., and Swanson, H. H. (1970). Agonistic behaviour between pairs of hamsters of the same and opposite sex in a neutral observation area. *Behaviour* **36**, 259–269.

Payne, A. P., and Swanson, H. H. (1971). Hormonal control of aggressive dominance in the female hamster. *Physiol. Behav.* **6**, 355–357.

Pearson, O. P. (1944). Reproduction in the shrew (*Blarina brevicauda* Say). *Am. J. Anat.* **75**, 39–93.

Pfeiffer, C. A. (1936). Sexual differences of the hypophyses and their determination by the gonads. *Am. J. Anat.* **58**, 195–227.

Phillips, R. W., Fraps, R. M., and Frank, A. H. (1945). Hormonal stimulation of estrus and ovulation in sheep and goats: a review. *Am. J. Vet. Res.* **6**, 165–179.

Phoenix, C. H., Goy, R. W., and Young, W. C. (1967). Sexual behavior: general aspects. *In* "Neuroendocrinology" (L. Martini and W. F. Ganong, eds.), Vol. 2, pp. 163–196. Academic Press, New York.

Powers, J. B., and Zucker, I. (1969). Sexual receptivity in pregnant and pseudopregnant rats. *Endocrinology* **84**, 820–827.

Price, D., and Ortiz, E. (1965). The role of fetal androgen in sex differentiation in mammals. *In* "Organogenesis" (R. L. de Haan and H. Ursprung, eds.), pp. 629–652. Holt, New York.

Redgrove, J. A. (1968). "Work and the menstrual cycle." Ph.D. Thesis, University of Birmingham, Birmingham, England.

Redgrove, J. A. (1971). Menstrual cycles. *In* "Biological Rhythms and Human Performance" (W. P. Colquhuon, ed.), pp. 211–240. Academic Press, New York.

Rees, L. (1958). The premenstrual tension syndrome in relation to personality, neurosis, certain psychosomatic disorders and psychotic states. *In* "Psychoendocrinology" (M. Reiss, ed.), pp. 82–95. Grune & Stratton, New York.

Richards, M. P. M. (1965). The behaviour of the pregnant golden hamster. *J. Reprod. Fertil.* **10**, 285–286.

Richards, M. P. M. (1969). Effects of oestrogen and progesterone on nest building in the golden hamster. *Anim. Behav.* **17**, 356–361.

Riddle, O., Bates, R. W., and Lahr, E. L. (1935). Maternal behavior induced in virgin rats by prolactin. *Proc. Soc. Exp. Biol. Med.* **32**, 730–734.

Riddle, O., Hollander, W. F., Miller, R. A., Lahr, E. L., Smith, G. C., and Marivin, H. M. (1941–1942). Endocrine studies. *Carnegie Inst. Washington, Yearb.* **41**, 203–211.

Riddle, O., Lahr, E. L., and Bates, R. W. (1942). The role of hormones in the initiation of maternal behavior in rats. *Am. J. Physiol.* **137**, 200–317.

Ring, J. R. (1944). The estrogen-progesterone induction of sexual receptivity in the spayed female mouse. *Endocrinology* **34**, 269–275.

Roberts, W. W. (1970). Hypothalamic mechanism for motivational and species-typical behaviour. *In* "The Neural Control of Behaviour" (R. E. Whalen *et al.*, eds.), pp. 175–206. Academic Press, New York.

Rollhäuser, A. R. (1949). Superfetation in a mouse. *Anat. Rec.* **105**, 657–663.

Rosenblatt, J. S., and Aronson, L. R. (1958). The decline in sexual behaviour in male cats after castration with special reference to the role of prior sexual experience. *Behaviour* **12**, 285–338.

Rosenblatt, J. S., and Lehrman, D. S. (1963). Maternal behavior in the laboratory rat. *In* "Maternal Behaviour in Mammals" (H. L. Rheingold, ed.), pp. 8–57. Wiley, New York.

Rosenzweig, S. (1943). Psychology of the menstrual cycle. *J. Clin. Endocrinol. Metab.* **3**, 296–300.

Ross, S., Sawin, P. B., Zarrow, M. X., and Denenberg, V. H. (1963). Maternal behaviour in the rabbit. *In* "Maternal Behavior in Mammals" (H. L. Rheingold, ed.), pp. 94–121. Wiley, New York.

Rothballer, A. B. (1967). Aggression and defence, neural mechanisms and social patterns. *In* "Brain Function" (C. D. Clemente and D. B. Lindsley, eds.), Vol. V, pp. 135–170. Univ. of California Press, Berkeley.

Roubicék, J., Tachezy, R., and Matousék, M. (1968). Elektrická cinnost mosková v prubĕhu menstruačm'ho cyklu. *Cesk. Psychiatr.* **64**, 90–95.

Rowan, D. (1931). "The Riddle of Migration." Williams & Wilkins, Baltimore, Maryland.

Rubin, A. D., and Winston, J. (1953). Effects of menstruation upon vestibular function in normal women. *J. Aviat. Med.* **24**, 234–236.

Salmon, U. J., and Geist, S. H. (1943). The effect of androgens upon libido in women. *J. Clin. Endocrinol. Metab.* **3**, 235–238.

Sawin, P. B., Denenberg, V. H., Ross, S., Hafter, E., and Zarrow, M. X. (1960). Maternal behavior in the rabbit; hair loosening during gestation. *Am. J. Physiol.* **198**, 1099–1102.

Sawyer, C. H. (1966). Neural mechanisms in the steroid feed-back regulation of sexual behavior and pituitary gonad function. *Brain Gonadal Funct., Proc. Conf. Brain Behav., 3rd, 1963* Vol. 3, pp. 221–256.

Sawyer, C. H., and Everett, J. W. (1959). Stimulatory and inhibitory effects of progesterone on the release of pituitary ovulating hormone in the rabbit. *Endocrinology* **65**, 644–651.

Schinkel, P. G. (1954). The effect of the ram on the incidence and occurrence of oestrus in ewes. *Aust. Vet. J.* **30**, 189–195.

Schneider, R. A., and Wolf, S. (1955). Olfactory perception thresholds for citral utilizing a new type olfactorium. *J. Appl. Physiol.* **8**, 337–342.

Schofield, B. (1957). The hormonal control of myometrial function during pregnancy. *J. Physiol. (London)* **138**, 1–10.

Segal, S. J., and Johnson, D. C. (1959). Inductive influence of steroid hormones on the neural system: ovulation controlling mechanisms. *Arch. Anat. Microsc. Morphol. Exp.* **48**, 261–274.

Seward, J. P. (1945). Aggressive behavior in the rat. I. General characteristics; age and sex differences. *J. Comp. Psychol.* **38**, 175–197.

Sfogliano, C. (1964). Menstrual cycle and work production of young workers employed in the manufacture of transistors in an electronics factory. *Rass. Med. Ind. Ig. Lav.* **32**, 218–221.

Sheehan, H. L., and Summers, V. K. (1949). Syndrome of hypopituitarism. *Q. J. Med.* **18**, 314–378.

Sigg, E. B. (1969). Relationship of aggressive behaviour to adrenal and gonadal function in male mice. *Aggressive Behav., Proc. Int. Symp., 1968* pp. 143–149.

Signoret, J. P. (1970). Reproductive behaviour of pigs. *J. Reprod. Fertil., Suppl.* **11**, 105–118.

Slonaker, J. R. (1924). The effect of pubescence, oestruation and menopause on the voluntary activity in the albino rat. *Am. J. Physiol.* **68**, 294–315.

Slonaker, J. R. (1934). Superfetation in the albino rat. *Am. J. Physiol.* **108**, 322–323.

Smith, A. J. (1950a). Menstruation and industrial efficiency. I. Absenteeism and activity level. *J. Appl. Psychol.* **34**, 1–5.

Smith, A. J. (1950b). Menstruation and industrial efficiency. II. Quality and quantity of production. *J. Appl. Psychol.* **34**, 148–152.

Sowton, S. C. M., and Myers, C. S. (1928). Two contributions to the experimental study of the menstrual cycle. I. Its influence on mental and muscular efficiency. *Med. Res. Counc. (G.B.), Ind. Fatigue Res. Board Rep.* No. 45.

Steinach, E. (1912). Willkürliche Umwandlung von Säugetier-Männchen in Tiere mit ausgeprägt weiblichen Geschlechtscharakteren und weiblicher Psyche. *Pflugers Arch. Gesamte Physiol. Menschen Tiere* **144**, 71–108.

Stone, C. P. (1925). Preliminary note on maternal behavior of rats living in parabiosis. *Endocrinology* **9**, 505–512.

Stone, C. P. (1926). The initial copulatory response of female rats reared in isolation from the age of 20 days to puberty. *J. Comp. Psychol.* **6**, 73–83.

Stone, C. P., and Barker, R. G. (1937). Aspects of personality and intelligence in post-menarcheal and pre-menarcheal girls of the same chronological age. *J. Comp. Psychol.* **23**, 439–455.

Suchowsky, G. K., Pegrassi, L., and Bonsignori, A. (1969). The effect of steroid on aggressive behaviour in isolated male mice. *Aggressive Behav., Proc. Int. Symp., 1968* pp. 164–171.

Swanson, H. H. (1966). Sex differences in behaviour of hamsters in open field and emergence tests: effects of pre- and post-pubertal gonadectomy. *Anim. Behav.* **14**, 522–529.

Swanson, H. H. (1967). Alteration of sex-typical behaviour of hamsters in open field and emergence tests by neonatal administration of androgen or oestrogen. *Anim. Behav.* **15**, 209–216.

Swanson, H. H., and Crossley, D. A. (1971). Sexual behaviour in the golden hamster and its modification by neonatal administration of testosterone propionate. *In* "Hormones in Development" (M. Hamburgh and E. J. W. Barrington, eds.), pp. 677–687. Appleton, New York.

Swanson, H. H., and van der Werff ten Bosch, J. J. (1964a). The "early-androgen" syndrome; its development and the response to hemi-spaying. *Acta Endocrinol. (Copenhagen)* **45**, 1–12.

Swanson, H. H., and van der Werff ten Bosch, J. J. (1964b). The "early-androgen" syndrome; differences in response to pre-natal and post-natal administration of various doses of testosterone propionate in female and male rats. *Acta Endocrinol. (Copenhagen)* **47**, 37–50.

Swyer, G. I. M. (1958). Hormonal aspects of water and electrolyte metabolism in relation to age and sex. *Ciba Found Colloq. Ageing* **4**, 78–98.

Thibault, C., Courit, M., Martinet, L., Mauleon, P., du Mesnil du Buisson, F., Ortavant, P., Pelletier, J., and Signoret, J. P. (1966). Regulation of breeding season and estrous cycles by light and external stimuli in some mammals. *J. Anim. Sci.* **25**, Suppl., 119–142.

Thorn, G. W., Nelson, K. R., and Thorn, D. W. (1938). A study of the mechanism of edema associated with menstruation. *Endocrinology* **22**, 155–163.

Tietz, E. G. (1933). The humoral excitation of the nesting instincts in rabbits. *Science* **78**, 316.

Tollman, J., and King, J. A. (1956). The effects of testosterone propionate on aggression in male and female C57 BL/10 mice. *Br. J. Anim. Behav.* **4**, 147–149.

Trimble, M., and Herbert, J. (1968). Effects of testosterone or oestradiol upon the sexual and associated behaviour of the adult female rhesus monkey. *J. Endocrinol.* **42**, 171–185.

Udry, J. R., and Morris, N. M. (1968). Distribution of coitus in the menstrual cycle. *Nature (London)* **220**, 593–396.

Valenstein, E. S. (1968). Steroid hormones and the neuropsychology of development. *In* "Neuropsychology of Development" (R. L. Isaacson, ed.), pp. 1–40. Wiley, New York.

van der Lee, S., and Boot, L. M. (1956). Spontaneous pseudopregnancy in mice. *Acta Physiol. Pharmacol. Neerl.* **5**, 213–214.

van der Werff ten Bosch, J. J. (1966). Anterior pituitary function in infancy and puberty. *In* "The Pituitary Gland" (G. W. Harris and B. T. Donovan, eds.), Vol. 2, pp. 324–345. Butterworth, London.

Wang, G. H. (1923). Relationship between "spontaneous" activity and estrous cycle in the white rat. *Comp. Psychol. Monogr.* **2**, 1–27.

Warner, L. H. (1927). A study of sex behavior in the white rat by means of the obstruction method. *Comp. Psychol Monogr.* **4**, 1–68.

Wells, L. J. (1962). Experimental studies of the role of the developing gonads in mammalian sex differentiation. *In* "The Ovary" (S. Zuckerman, ed.), 1st ed., Vol. 2, pp. 131–153. Academic Press, New York.

Whalen, R. E., and Nadler, R. D. (1963). Suppression of the development of female mating behavior by estrogen administered in infancy. *Science* **141**, 273–275.

Whitehead, R. E. (1934). Forces which produced changes in physical examinations for aircraft pilots for Aeronautics Branch of Department of Commerce. *J. Aviat. Med.* **5**, 24–25.

Whitten, W. K. (1956). Modification of the oestrous cycle of the mouse by external stimuli associated with the male. *J. Endocrinol.* **13**, 399–404.

Wickham, M. (1958). The effects of the menstrual cycle on test performance. *Br. J. Psychol.* **49**, 34–41.

Wiesner, B. P., and Mirskaia, L. (1930). On the endocrine basis of mating in the mouse. *Q. J. Exp. Physiol.* **20**, 274–279.

Wilson, J. R., Adler, N., and Le Boeuf, B. J. (1965). The effects of intromission frequency on successful pregnancy in the female rat. *Proc. Natl. Acad. Sci. U.S.A.* **53**, 1392–1395.

Windle, W. F. (1939). Induction of mating and ovulation in the cat with pregnancy urine and serum extracts. *Endocrinology* **25**, 365–371.

Witschi, E., and Pfeiffer, C. A. (1935). The hormonal control of oestrus, ovulation and mating in the female cat. *Anat. Rec.* **64**, 85–99.

Wynne-Edwards, V. C. (1962). "Animal Dispersion in Relation to Social Behaviour." Oliver & Boyd, Edinburgh and London.

Yamanchi, M., and Inni, S. (1954). Studies on the ovarian cyst in the cow. II.

Endocrinological and histological studies on the correlation between the ovarian cyst and the symptoms. *Jpn. J. Vet. Sci.* **16**, 27–36.

Yerkes, R. M. (1939). Social dominance and sexual status in the chimpanzee. *Q. Rev. Biol.* **14**, 115–136.

Yerkes, R. M. (1940). Social behavior of chimpanzees. Dominance between mates in relation to sexual status. *J. Comp. Psychol.* **30**, 147–186.

Yerkes, R. M., and Elder, J. H. (1936). Oestrus, receptivity and mating in chimpanzee. *Comp. Psychol. Monogr.* **13**, 1–39.

Young, W. C. (1941). Observations and experiments on mating behavior in female mammals. *Q. Rev. Biol.* **16**, 135–156 and 311–335.

Young, W. C., and Orbison, W. W. (1944). Changes in selected features of behavior in pairs of oppositely sexed chimpanzees during the sexual cycle and after ovariectomy. *J. Comp. Psychol.* **37**, 107–143.

Young, W. C., and Rundlett, B. (1939). The hormonal induction of homosexual behavior in the spayed female guinea-pig. *Psychosom. Med.* **1**, 449–460.

Young, W. C., Dempsey, E. W., and Myers, H. I. (1935). Cyclic reproductive behavior in the guinea-pig. *J. Comp. Psychol.* **19**, 313–335.

Young, W. C., Dempsey, E. W., Myers, H. I., and Hagquist, C. W. (1938). The ovarian condition and sexual behavior in the female guinea-pig. *Am. J. Anat.* **63**, 457–483.

Young, W. C., Dempsey, E. W., Hagquist, C. W., and Boling, J. L. (1939). Sexual behavior and sexual receptivity in the female guinea-pig. *J. Comp. Psychol.* **27**, 49–68.

Young, W. C., Boling, J. L., and Blandau, R. J. (1941). The vaginal smear picture, sexual receptivity and time of ovulation in the albino rat. *Anat. Rec.* **80**, 37–45.

Young, W. C., Goy, R. W., and Phoenix, C. H. (1964). Hormones and sexual behavior. *Science* **43**, 212–217.

Zarrow, M. X., Grota, L. J., and Denenberg, V. H. (1967). Maternal behavior in the rat: survival of newborn fostered young after hormonal treatment of the foster mother. *Anat. Rec.* **156**, 13–18.

Zarrow, M. X., Brody, P. N., and Denenberg, V. H. (1968). The role of progesterone in behavior. *In* "Reproduction and Sexual Behavior" (M. Diamond, ed.), pp. 363–389. Indiana Univ. Press, Bloomington.

Zucker, I. (1967). Progesterone in the experimental control of the behavioural sex cycle in the female rat. *J. Endocrinol.* **38**, 269–277.

Zucker, I. (1968). Biphasic effects of progesterone on sexual receptivity in the female guinea-pig. *J. Comp. Physiol. Psychol.* **65**, 472–478.

Zucker, I., and Goy, R. W. (1967). Inhibition and facilitation of sexual receptivity in female guinea pigs by progesterone and related compounds. *J. Comp. Physiol. Psychol.* **64**, 378–383.

Zuckerman, S. (1932). "The Social Life of Monkeys and Apes." Kegan Paul, London.

Zuckerman, S., Palmer, A., and Hanson, D. A. (1950). The effect of steroid hormones on the water content of tissues. *J. Endocrinol.* **6**, 261–276.

10

External Factors and Ovarian Activity
in Mammals

J. Herbert

457

I. INTRODUCTION

A comprehensive account of the evidence for the proposition that changes in the environment can profoundly influence ovarian activity would be impossible within the bounds of a single chapter. References to various reviews of the subject are given in the text which follows. Since a notable account principally, but by no means exclusively, concerned with the effect of natural changes in the environment on reproduction has recently appeared (Sadleir, 1969), this chapter is particularly concerned with experimental approaches that attempt to isolate, define, and specify the nature of environmental variables that alter ovarian function (see also Perry and Rowlands, 1973). Attention is particularly directed toward the mammals; certain phenomena (e.g., delayed implantation) which lie on the border line between the function of the ovaries and that of other organs, and others which might qualify as "external events" (e.g., behavioral and social factors), require a separate chapter, and have been omitted or merely mentioned in passing.

Although, in this chapter, the different external factors have been treated separately, it has to be recognized that under natural conditions, and even under some experimental ones, an animal is subject to combinations of changes in the different parameters of its environment. It is therefore necessary to synthesize from an account such as this the way in which these various components interact with each other and with the animal to influence the activity of the ovary.

II. LIGHT

A. Light and the Breeding Season

1. The Breeding Season

The majority of mammals living in temperate climates breed only in part of the year (for reviews, see Amoroso and Marshall, 1960; Clegg and Ganong, 1969). Many insectivores, carnivores, nonruminating ungulates, and some rodents start to breed in the spring; rather fewer species, such as deer, sheep and other ruminants, badgers, seals, and some primates breed in the autumn. Some species that breed all the year round show annual fluctuations in fertility (e.g., the rabbit), and "domestication" seems to have been associated, in some cases (e.g., cows, pigs, cats, and dogs), with loss of the pronounced breeding season which characterizes the wild population.

It is customary to distinguish between breeding seasons and breeding peaks, the former implying that reproductive activity is entirely limited to part of the year, and the latter that most breeding occurs during a particular season, though possible throughout the year. It is doubtful whether such a distinction is still tenable. All gradations between the two are observed, and the ovaries of even a "true" seasonal breeder are not completely quiescent even during anestrus. Species in either category show, in common, a seasonal decrement in reproductive function which, in the case of one, prevents breeding altogether.

The breeding season is best regarded as a rhythm. It has, therefore, various components: periodicity, the duration of various parts (e.g., estrus, anestrus), and timing relative to an external signal ("Zeitgeber:" Aschoff, 1960). Moreover, there are seasonally polyestrous animals (e.g., some rodents) which exhibit estrous cycles for only part of the year. In such cases, light can have a twofold effect upon reproduction: not only can it affect the timing and duration of the breeding season, but it may also influence events, such as ovulation, which concern the estrous cycle itself (see Section II,B). Although it is clear that many or all of the components of the mammalian breeding season may be affected by light, experimental analyses have not yet attained the sophistication of those applied to the breeding season of birds (see Marshall, 1960; Lofts, 1970; Meier and MacGregor, 1972) or of those attempted of the 24-hour (or circadian) rhythm which occurs in a large number of physiological processes (Aschoff, 1960; Pittendrigh, 1960; Bunning, 1967).

2. Habitat and Breeding Seasons

Animals transferred from one hemisphere to the other adjust their breeding season to correlate with their new environment, though this may take several years (Bedford and Marshall, 1942). Sheep kept at increasing distances from the equator (and, therefore, exposed to greater annual changes in environmental day length) show correspondingly more restricted breeding seasons. Both sheep and deer originating from near the equator (where there is little change in daylength throughout the year) have more prolonged breeding seasons in temperate climates than nonequatorial species under the same conditions (Yeates, 1954; R. E. Chaplin, personal communication), indicating an interaction between an animal's genetic constitution and its environment in determining breeding seasons.

3. Light and the Onset of Breeding Seasons

Although other factors play a role (see below), changes in daylength are predominant in initiating breeding in mammals. In general, species that breed in the spring ("long-day" species) can be made to breed earlier than usual by being put into photoperiods resembling those of late spring or summer (e.g., ferrets: Bissonnette, 1932; field mice: Baker and Ranson, 1932), whereas autumn breeders ("short-day" species) can be induced to breed early by a reduction in light during the summer (e.g., sheep: Sykes and Cole, 1944; Yeates, 1949; goat: Yoshioka *et al.,* 1952). However, species normally living near the equator may not be sensitive to variations in photoperiod; for example, sheep native to Rhodesia are not affected by artificial light (Symington and Oliver, 1966). It is not necessary to mimic the gradual change in daylength which occurs naturally in order to induce premature estrus. Putting ferrets or voles abruptly into long-day regimens (e.g., 14L:10D) or sheep into short days (e.g., 8L:16D) results in estrus occurring earlier than normal (Hammond, 1952; Lecyk, 1962; Breed and Clarke, 1970; Ducker and Bowman, 1970).

The induction of the breeding condition in response to light is not abrupt. Both long-day and short-day breeders take several weeks to come into estrus after being put into the appropriate photoperiod and this, of course, is far slower than the visual response to light. It is also much slower than, for example, skin color changes induced in amphibians and reptiles by alterations in light; these occur in a matter of hours or, at the most, days. Little is known about the physiological processes occurring during the time between the onset of artificial light and the induction of estrus. A prolonged period of experimental treatment is evidently necessary, since ferrets require at least 30 days in artificial long days if the onset of estrus is to be advanced. Less exposure results in the animals coming into estrus at the same time as controls (Vincent, 1970). This suggests that a threshold amount of light, measured as the number of days for which the animal is exposed to it, is necessary for estrus to be retimed in this species.

4. Light and the Duration of the Breeding Season

Once estrus is initiated by light, it may come to an end either because the animal eventually becomes refractory to continued stimulation (see Section II,A,6), or because the duration of estrus is, in some way, predetermined so that once initiated it is independent of light. Alternatively, seasonal changes in illumination opposite to those which initiate estrus may bring it to an end. Lengthening the photoperiod during estrus hastens anestrus in the ewe (Ducker and Bowman, 1970), and the ferret (a long-day species) can be driven out of estrus prematurely by artificially shortened days (Thorpe and Herbert, 1977).

Estrus may come to an end at a time when the daylength seems similar to, or even more stimulating than, that prevailing at the start of the breeding season. This suggests that the characteristics of the light required for initiating the onset of estrus may be substantially different from those determining its duration. There is also the problem of how an animal's neuroendocrine system defines a "long" or "short" photoperiod. For example, an animal in a photoperiod of, say, 20L:4D may treat a subsequent one of 12L:12D as a "short" day, whereas animals kept in 8L:16D may react to a subsequent photoperiod of 12L:12D as if it were a "long" day. Thus, the midsummer solstice may render a given photoperiod immediately following it nonstimulatory (or even inhibitory); yet the same photoperiod might be stimulatory when it follows the winter solstice. Factors such as these require much more investigation (see Yeates, 1949). Furthermore, the earliest changes in the reproductive system may occur before the solstice preceding estrus (see Watson and Radford, 1955). This may only indicate, however, that a particular rhythm in the estrous cycle has been established by the animals being exposed to changes in daylength in previous years (see Section II,A,5). Ferrets kept in artificial long days for prolonged periods show an initial period of estrus which is greater than the normal summer breeding season (Herbert, 1972; see also Donovan, 1967). This suggests that certain conditions may prolong estrus.

5. Light and Annual Periodicity

Light can also determine the periodicity of the breeding season as well as accelerating its onset. The best evidence for this proposition comes from experiments in which animals have been put into artificial light cycles of periodicities less than 1 year. Sheep exposed to changes in light mimicking those occurring naturally but concentrated in a period of 6 months (instead of 12) come into estrus twice a year (Thibault *et al.*, 1966). Ferrets behave in the same way and can even be made to breed three times a year by being exposed to 2-monthly alternating regimens of long (i.e., 14L:10D) and short (i.e., 8L:16D) days (Vincent, 1970).

Another way of testing the periodicity of the annual rhythm is to keep animals in unvarying conditions of illumination. In these circumstances, provided that other environmental factors (for example, temperature) are not acting as secondary "Zeitgebers," annual breeders lose their annual periodicity. Mice kept in continuous light or darkness (Whitaker, 1940) show recurrent periods of estrus and anestrus unrelated to the seasons, and ferrets become mutually asynchronous (Thorpe, 1967). Exceptional findings have been reported in sheep (Radford, 1961; Thibault *et al.*, 1966) but seasonal fluctuations in temperature, which were not controlled, may have been responsible for the timing of estrus in these experiments (see Sadleir, 1969).

It is important to recognize that there are fundamental differences to be expected from putting animals into constant conditions that are themselves stimulatory from those that do not, per se, influence the onset of estrus. If ferrets ("long-day" animals) are blinded or put into constant "short" days, they come into estrus at fairly regular intervals which do not correspond to an annual rhythm (Thomson, 1954; Vincent, 1970). However, should they be exposed to continuous "long" days, a prolonged period of estrus results, which is followed, in most cases, by anestrus. The latter persists indefinitely unless the animals are transferred to short days for a time, after which a further period of long-day stimulation induces estrus once more (Donovan, 1967; Vincent, 1970). These findings strongly suggest that a condition of photorefractoriness analogous to that known to occur in certain birds (see Lofts, 1970) exists in some mammals. The period of short days which is required to dispel the refractory state may act in some positive way to resensitize the animal to light, or it may simply represent a period during which the stimulus (in this case long days) is absent, and the animal is able to recover its photosensitivity spontaneously.

Long days, which are ordinarily thought of as stimulating estrus in ferrets, can also inhibit breeding in ways other than that described above. Ferrets born and reared in long days show greatly delayed first estrus (i.e., puberty). Animals born into short days and put into a long-day regimen before they are 100 days old show a similar effect. Ferrets older than this, exposed to long photoperiods, exhibit the usual stimulating effect (D. H. Thorpe, unpublished observations). Thus there seem to be maturational changes within the light-sensitive neuroendocrine system that alter its response to experimental lighting regimens. It is still not known precisely how such changes can themselves be affected by particular light conditions experienced by an animal during only postnatal life.

6. Is There an Endogenous Rhythm?

Animals kept in conditions in which they no longer receive changing light stimuli (e.g., blinded or in constant illumination) may still come into estrus. This indicates that the increase or decrease in the hours of daylight which would precede the onset of the normal mating season are not a condition necessary for its occurrence. Although the evidence is still incomplete, it is becoming more likely that there is, in some animals at least, an endogenous rhythm which has a cycle greater or less than a year and which is normally synchronized with the seasons by changes in daylength (Gwinner, 1973; Pengelley, 1974). Endogenous rhythms of about 24 hours, which are synchronized by diurnal variations in daylength, are well known (see Pittendrigh, 1960), and an analogous process may exist in some seasonal

breeders. From the above discussion (Section II,A,5), it is clear that keeping animals in constant lighting conditions that are stimulatory (e.g., sheep
in "short" days, ferrets in "long" days or continuous light) have not
contributed to our understanding of the existence of such an underlying
rhythm.

7. Variations in the Parameters of the Light Stimulus

a. Single Photoperiods. The effects of a light stimulus upon the timing of
estrus seem to depend upon its duration, provided that this occurs as a
single period, each of 24 hours. A period greater than about 11 hours is
required to advance the onset of estrus in the ferret (D. S. Vincent,
unpublished observations); 14 hours seems to be about the optimal stimulus,
since longer and shorter periods are less effective (Hart, 1951; Hammond,
1952). Voles are said to be equally stimulated by different photoperiods
exceeding 12 hours (Breed and Clarke, 1970), but inspection of these data
suggests that longer photoperiods may, in fact, induce estrus more rapidly.
Thibault *et al.* (1966) report that 15 hours of light daily stimulates more
follicular growth in voles than does 5, 10, or 20 hours. In general, single
periods shorter than 12 hours are effective in initiating estrus in sheep,
though a thorough study of the consequence of exposing them to different
categories of short days has not been reported (but see Means *et al.,* 1959).
There has also been considerable discussion of whether it is the duration of
the dark phase, rather than the light, which is the more important factor in
long- and short-day species (see below).

b. Multiple Photoperiods. An amount of light which is not stimulatory
when presented in a single period, may initiate estrus if allotted in divided
doses. Anestrous ferrets kept in single photoperiods of 7 hours a day do not
show premature estrus; however, they do so when exposed to two $3\frac{1}{2}$-hour
photoperiods separated by a 5-hour period of darkness (Hammond, 1951).
Conversely, a regimen consisting of a single 8-hour period stimulates estrus
in sheep, whereas the same amount of light is not as effective when given as
two periods of 4 hours separated by a 2-hour period of darkness. If the two
light periods are separated by 4 hours of darkness, estrus is advanced even
less (Hart, 1950; Hafez, 1952). Thus, the total amount of light given over a
24-hour period is not a critical parameter in these conditions.

Experiments on birds have suggested that there is, during each 24 hours,
a light-sensitive ("scòtophil") period which is determined by the first light
stimulus (e.g., dawn). If the bird experiences a light stimulus during the scotophil phase then reproduction is stimulated (Follett and Sharp, 1969;
Follett, 1973). Thus in the experiments of Hart (1950), Hammond (1951),
and Hafez (1952), the second light stimulus may have occurred during a
light (or dark) sensitive period.

In certain circumstances, the light-sensitive system may react to two periods of light separated by a period of darkness as if there had been a single light stimulus, starting at the onset of the first photoperiod and ending at the conclusion of the second (for further discussion, see Farner, 1965; Menaker, 1965; Follett and Sharp, 1969; Follett, 1973). Much discussion has taken place on the relative roles played by light or dark periods, scotophil or photophil (i.e., light-insensitive) phases, light/dark ratios, and "on–off" stimuli. Experiments such as those described above (Section II,A,7) show that many parameters can affect the breeding season in different conditions, though some (e.g., the duration of the photoperiod) are likely to be more significant in natural conditions than others (e.g., "on–off" stimuli). Not enough attention, however, has been paid to determining how far an animal's neuroendocrine system defines the duration of a photoperiod, either in absolute terms or relative to the amount of light the animal has experienced immediately before (see Yeates, 1949; and Section II,A).

c. **Wavelength.** Information is extremely scanty on the relative effectiveness of light of different wavelengths. Marshall and Bowden (1934) claim to have shown that ultraviolet light is particularly stimulating in the ferret, but their experimental groups were very small and each animal received different light treatments in successive years. Lecyk (1962) exposed voles for 16 hours/day to either blue–violet (390 to 500 nm) or red–orange (570–760 nm) light of equal intensity and found the two regimens equally stimulating.

d. **Intensity.** It is well established that, in some birds, stimulation of reproductive activity is proportional to the intensity of the light to which they are exposed (Bissonnette, 1931). Although there is less information on mammals, the same general conclusions seem likely to apply. Earlier workers (Bissonnette, 1935; Marshall, 1940) suggested that higher intensities were more effective, but their experiments have been criticized because of inadequate controls or the small number of animals used (see Wurtman, 1967; Sadleir, 1969). Marshall and Thomson (1951) come to the opposite conclusion, since estrus occurred at about the same time in ferrets exposed to between 10 and 125 ft-c. More recently, it has become apparent that these, and other, negative results are a consequence of the extreme sensitivity of the light-responsive neuroendocrine system in mammals, and variations in response to different intensities can thus only be demonstrated if the values used are very low. Lecyk (1962) showed that estrus is not evoked in voles exposed to only 0.02 ft-c, whereas those receiving 1 ft-c were only slightly less stimulated than others kept in 20 ft-c. Whitaker (1940) reported that mice could be entrained to artificial light cycles if the intensity lay between 15 and 32 ft-c, but not if it was between 1 and 2 ft-c. In the latter condition, the estrous cycle in mice was the same as that in hamsters kept in continuous darkness. A logarithmic positive relationship between the light intensity and onset of estrus has been demonstrated in anestrous ferrets exposed to long days (14L:10D) with intensities ranging from 0.01 to 5 ft-c, although differences between groups became practically imperceptible at intensities higher than 0.42 ft-c (Vincent, 1970).

Periods of low illumination at dawn or dusk could thus be of importance in regulating ovarian function and the breeding season. Furthermore, light sufficient to influence breeding may still be received even when the animal closes its eyes, or if it lives in a very dimly lit habitat.

B. Light and the Estrous Cycle

1. Sequence of Events during the Estrous Cycle

Schwartz and her colleagues have shown that, in the rat at least, there is an orderly sequence of events occurring during the estrous cycle. Each of these is related in some way to the photoperiod (Schwartz, 1969). For example, release of LH and uterine ballooning occur, in that order, during the light period on the day of proestrus, whereas estrus and ovulation occur during the subsequent dark period. It is becoming clear that this sequence can be disrupted by experimental lighting schedules, although there is still considerable doubt about the way this is brought about (Lawton and Schwartz, 1967).

2. Light and the Timing of Ovulation

Nocturnal rodents, such as mice, rats, and hamsters, normally ovulate early in the morning (during the dark) on the day of estrus in response to the pulse of LH secreted during the preceding afternoon during the light phase. Originally, this LH pulse was demonstrated by finding that certain drugs (e.g., pentobarbitone, atropine, and Dibenamine) prevented ovulation if given during a certain critical period on the afternoon of proestrus (see Everett, 1964). Assays of blood and pituitary levels of LH have shown that the hormone is discharged at about this time (Schwartz and Caldarelli, 1965). The precise timing of the ovulatory discharge of LH depends upon the species and upon the lighting conditions in which the animals are kept. It has long been known that light determines the timing of ovulation in rats. Hemmingsen and Krarup (1937) found that the time of ovulation shifts by 12 hours in rats kept in reversed light/dark regimens (i.e., in darkness during the day and vice versa). Altering the phase of the light/dark cycle by 3 hours causes a corresponding change in ovulation (Hoffmann, 1969). Similar responses have been found in mice and hamsters (Alleva *et al.*,

1968). The time of day that rodents ovulate may alter during the year, probably because of the change in daylength, and this effect has been, in part, replicated by exposing animals to photoperiods of different lengths (Austin, 1956; Braden, 1957). However, Hoffmann (1969) found that rats in photoperiods of either 12, 14, or 16 hours showed a critical period for ovulation which was not related to the onset or duration of the respective light periods, but to the midpoints of the light or dark phases.

Because ovulation represents an easily definable process, the effect of light on ovulation rather than on other parts of the estrous cycle has been studied most extensively. Whether ovulation is the central point in the cycle, upon which other components depend, or whether the several components can be affected independently by different light treatments, is still undecided (for further discussion, see Schwartz, 1969; Neguin *et al.*, 1975).

3. Effect of Gonadotropin, Estrogen, and Progesterone on the Timing of Ovulation Relative to Light

If an imminent ovulation in the rat is prevented by administering a drug (such as phenobarbitone) during the "critical phase," an ovulation will occur almost exactly 24 hours later (Everett, 1964). It is now thought that every 24 hours there is a phase during which the secretion of estrogen, if it reaches a high enough value, provokes the discharge of LH from the pituitary. The timing of this estrogen-sensitive phase is determined by the duration of the light or dark period. If a device releasing constant amounts of estradiol is implanted subcutaneously in an ovariectomized rat then LH surges occur every 24 hours in the second half of the photoperiod (Hery *et al.*, 1975). Administration of PMSG (pregnant mare's serum) can evoke LH release (and hence ovulation) in the immature rat, presumably because of an induced release of estrogen, but the ovulation will occur at the normal time 2 days later, regardless of the timing of the PMSG, since LH is only released during the estrogen-sensitive critical phase (Strauss and Meyer, 1962a,b; McCormack and Bennin, 1970). A similar effect can be obtained by administering estrogen to adult rats (Everett, 1964, 1969). Treatment with estrogen or PMSG mimics the surge of estrogen secretion and thereby determines the day on which ovulation will occur, although the time of day is determined by light.

The time of ovulation can be advanced by progesterone even in animals kept in light regimens which otherwise prevent ovulation from occurring (Section II,B,5). It is possible that progesterone can also alter the estrogen-sensitive period of the LH release mechanism, since the administration of this hormone at different times may alter the timing of LH release relative to the photoperiod (see Everett, 1969).

There has been comparatively little study of the effects of light on the secretion of FSH, though diurnal changes of secretion are known to exist and rats exposed to constant light reach puberty earlier than normal (Fiske, 1941).

4. Light and the Duration of the Estrous Cycle

The estrous cycle of the rat normally lasts for 4 or 5 days. If rats are kept in a 12L:12D photoperiod, most of them (70%) have cycles lasting 4 days, whereas many (46%) of those kept in a 16-hour photoperiod (16L:8D) have 5-day cycles. Those in 14L:10D show intermediate results (Hoffmann, 1969). This effect of light on the duration of the rat's cycle has been ascribed to a delay, induced by the longer photoperiod, in the secretion of estrogen. This, in turn, is presumably due to a corresponding decline in FSH secretion. The result may be that on the day of proestrus, the estrogen peak is delayed and "misses" the estrogen-sensitive phase of the neural structures responsible for LH secretion. Individual differences in secretion rates are suggested as an explanation for the fact that some animals, even in 12L:12D, experience 5-day cycles (Hoffmann, 1969). A long photoperiod would need to delay FSH secretion, a process analogous to that which may occur in short-day seasonal breeders such as the deer (Lincoln *et al.,* 1970), but differs from that in the young prepubertal rat, in which prolonged illumination stimulates the onset of puberty (Fiske, 1941). However, measurements of plasma estrogen in 4- or 5-day cycles have failed to show significant differences. Instead, it has been proposed that differences in progesterone may be responsible (Neguin *et al.,* 1975).

5. Effects of Continuous Light

Keeping rats and other rodents in continuous light has been frequently used to study the interaction between light and the estrous cycle. It is, however, a highly abnormal procedure, inducing severe atrophy of the retina in rats (O'Steen, 1970; Reiter and Klein, 1971), with little obvious relevance to the study of diurnal changes in lighting and has yielded little information about the nature of the control effected by light.

Browman (1937) found that rats kept in constant illumination eventually experienced persistent vaginal cornification (constant estrus). The ovaries of such animals lack corpora lutea, indicating that ovulation is no longer occurring. Constantly estrous rats, when returned to alternating periods of light and dark, resume normal estrous cycles (Takahashi and Suzuki, 1969).

Lawton and Schwartz (1967), in an elegant and detailed study (see also Schwartz, 1969), have shown that the initial effect of continuous light is to desynchronize the various components of the estrous cycle; ovulation is delayed (but not prevented) during the first two or three cycles under these conditions, and vaginal cornification, uterine ballooning, and ovulation no longer appear in the normal sequence (see Section II,B,2). Eventually ovulation ceases, though it may still be evoked by giving estrogen or progesterone or by electrical or mechanical stimulation of the cervix (Everett, 1964). This implies that the pituitary still contains enough LH to initiate ovulation, but that the stimulus necessary for LH release is lacking. Why this is so is still not clear. It may reflect a suppression of the estrogen surge which ordinarily provokes LH release, or the rhythms in estrogen secretion may move out of phase with the period of steroid sensitivity that exists in the brain. Moreover, it is not known whether continuous light simply allows the components of the cycles to disappear (if they are exogenous) or to become free-running and hence mutually desynchronous (if they are endogenous), or whether there is actual suppression or stimulation of different components, for example, FSH and LH secretion. Strauss and Meyer (1962a) report that plasma LH levels are very high during the first 60 days of exposure to continuous light, but this is disputed (Daane and Parlow, 1971). It would be interesting to determine the way in which diurnal rhythms in hormone secretion change under such conditions.

Continuous illumination also affects the time of PMSG-induced ovulation in immature rats. Young females given PMSG after being kept in continuous light for 3, 6, or 9 days ovulated later than controls maintained in 14-hour photoperiods (McCormack and Bennin, 1970), presumably as the periodicity in activity of the structures responsible for releasing LH lengthened.

After prolonged periods in continuous light, the ovaries of the rat involute, hormone secretion declines, and the appearance of the vaginal smear resembles that of anestrus (Lawton and Schwartz, 1967). This sequence of events recalls that observed in long-day annual breeders put into artificial long days for extended periods and may be similarly related to the process of photorefractoriness recognized in birds (see Section II,A,5). Rats born and reared in continuous illumination continue to experience estrous cycles, though they may be less regular than normal. If they are transferred to alternating photoperiods and then back to continuous light, the rats become continuously estrous (Takahashi and Suzuki, 1969). Thus diurnal lighting has some conditioning effect on the light-sensitive system.

Although studied less extensively, continuous light has been shown to have similar effects on the estrous cycle of other rodents. Mice may experience persistent vaginal cornification and their ovaries lack corpora lutea (Murthy and Russfield, 1970) though the effects of light are less pronounced than they are in the rat (Campbell *et al.,* 1976). Hamsters normally exhibit 4-day cycles of great regularity; Greenwald (1963) found that these continued in constant light, but became slightly less regular with occasional intervals of 5 days between successive estrous periods. Moreover, animals with cycles that were synchronous before the experimental treatment began became asynchronous. In contrast, Kent *et al.* (1968) reported that a single day's exposure of hamsters to continuous light delayed ovulation by 6 hours. Thereafter, animals became mutually asynchronous and finally stopped ovulating, though this took longer to come about than in rats (Alleva *et al.,* 1968). These findings indicate that the sequence of events within the estrous cycle becomes erratic in constant light and that the timing of ovulation is regulated by light in mice and hamsters as well as in the rat.

Old rats may experience persistent vaginal cornification even when kept in photoperiods of 12L:12D. Those that continue to cycle regularly enter constant estrus more rapidly than young animals when both ages are kept in continuous light. This suggests that the mechanism regulating the estrous cycle of the rat becomes more sensitive to light as the animals age, so that shorter photoperiods, initially ineffective in disturbing the cycle, become capable of so doing. Wild rats probably spend far more time in the dark than in the light, and it would therefore be interesting to see whether the incidence of "spontaneous" constant estrus could be reduced by keeping laboratory rats in very short photoperiods.

6. Effect of Continuous Darkness or Blinding

Rats which are blinded or kept continuously in the dark maintain the daily critical period for LH discharge for at least 1 week, though its periodicity may no longer be exactly 24 hours (Strauss and Meyer, quoted by McCormack and Bennin, 1970). Thereafter, the estrous cycle becomes longer with protracted periods of anestrus. The ovaries of such animals contain small follicles and few corpora lutea (Hoffmann, 1967). The surge of LH which is induced by treating immature rats with PMSG is also delayed by darkness (McCormack, 1971). The estrous cycle of rats kept in darkness is no longer synchronized with the time of day. For example, the onset of vaginal cornification, which normally occurs toward the end of the light phase, becomes randomly distributed throughout the 24 hours (Hoffmann, 1967). Thus, ovarian cyclicity continues, to some degree, in continuous darkness in contrast to the effects of constant light described above, a finding which emphasizes the importance of the nature of the photic stimulus, as well as its constancy.

The ovaries of the hamster involute if the animal is maintained in very short days (e.g., 1L:23D) or if blinded (Hoffman and Reiter, 1966). Presumably the decrease in the size of the testes of the male hamster which occurs in the winter is correspondingly related to the reduction in the hours of daylight (Czyba *et al.*, 1964).

Compensatory hypertrophy of one ovary, which normally follows removal of the other, seems to depend partly upon light since rats kept in darkness showed only 25% increase in the residual gonad, compared with 63% in controls in 14L:10D (Dickson *et al.*, 1971). Thus light is required to promote FSH secretion in these conditions.

III. TEMPERATURE

A. Homeothermy and Reproduction

Mammals have evolved efficient ways of keeping their body temperature constant and it is sometimes maintained that changes in temperature do not have any effect on reproduction. In fact, mammals possess highly efficient detectors of changes in external temperature which could also function to induce specific responses in the reproductive system. The property of homeothermy has another corollary. Because mammals have to keep their body temperature relatively constant, elaborate mechanisms exist to compensate or adjust for changes in external temperature and the operation of these may alter ovarian activity as a secondary consequence. Moreover, if homeostatic mechanisms prove insufficient, the widespread metabolic disturbances that result may affect reproduction in a relatively nonspecific manner. It is therefore important to distinguish between experiments in which unacclimatized animals are suddenly exposed to very high or low temperatures well outside their normal range, from those in which there is consideration of the temperature characteristics of the normal habitat of the animals. The results of the first are likely to be of as much interest to the pathologist as to the physiologist.

B. Ecological Evidence

Sadleir (1969) reviews evidence from studies of natural populations which indicates that there may be correlations between changes in temperature and reproduction. Fluctuations in temperature, though to some extent predictable, are much more variable from one year to the next than those in light. Many field workers have therefore attempted to relate variations in temperature from year to year, or from habitat to habitat, with breeding performance. Earlier work seemed to show little correlation; wild mice bred as freely in severe as in mild winters (Baker, 1930). However, more recent studies suggest that in ewes, rabbits, and hares there is a predictable relationship between environmental temperature and breeding (Brambell, 1944; Wight and Conaway, 1961; Lees, 1966; Flux, 1967). A cool summer hastens the onset of estrus in ewes, whereas a severe winter reduces the breeding performance of the other species. The problem, as Sadleir (1969) points out, is to distinguish the effects of temperature on reproduction from those of other factors like rainfall and food which can themselves depend upon climatic variables such as the prevailing temperature.

C. High Temperatures

There is little doubt that exposure of animals to high temperatures can disturb ovarian function, though interpretation of such experiments is difficult (see Section III,A). There is a wide variety of metabolic and endocrine consequences to exposure to high temperature (see MacFarlane *et al.*, 1959; MacFarlane, 1963; Collins and Weiner, 1968). Rats and mice kept at high temperature are less fertile than animals maintained in normal conditions and ovarian function is depressed (Ogle, 1934; Chang and Fernandez-Cano, 1959). Rats maintained at high temperatures (25°–30°C) gradually show a decreased incidence of estrus and of ovulations. If the animals remain in these conditions, some degree of acclimatization occurs and reproductive function then improves (Chang and Fernandez-Cano, 1959). The restoration of heat-treated mice to a temperate environment results in a rapid return of normal cycles (Ogle, 1934). Sheep do not breed well after a hot summer (Moule, 1950) and man is also less fertile in hot environments (MacFarlane, 1963; Fernandez-Cano, 1959). Many authors (e.g., Fernandez-Cano, 1959; MacFarlane, 1963) consider these ovarian effects to be aspects of a generalized response to a hot environment. The fact that thyroxine could restore the fecundity of sheep during unusually hot summers (Moule, 1950) seems to support this conclusion. However, small increases 3°–13°C) in temperature reduced ovulation in cows (Miller and Alliston, 1974).

There seems to be no convincing experimental evidence that increase of temperature can promote ovarian activity in mammals in a manner analogous to that of increased light. Cowan and Arsenault (1954) reported that either increased temperature or increased light could induce winter breeding in captive creeping voles (a fossorial species), but their controls were taken from the wild population (which does not breed in winter) and significant differences in food supply or other factors could not be excluded.

D. Low Temperatures

In many studies on the effects of low temperature, animals have been exposed to conditions which approach those found in the natural environment, and therefore have more interest for reproductive physiologists than the work so far reported on effects of high temperature. Nevertheless, the same provisos concerning more generalized metabolic effects apply to low as well as to high temperatures. Moore *et al.* (1934) showed that female ground squirrels, whose reproduction was unaffected by experimental lighting, could be made to breed at a temperature of 4°C. The collared lemming (an arctic species) is similarly unaffected by light, but temperatures between $-9°$ and $-3°C$ resulted in reductions in size of the follicles and anestrus (Quay, 1960).

The stimulatory effects of low temperature on reproduction have been shown most clearly in the ewe. Sheep kept in natural daylengths during the summer but at 7° to 9°C (about 20°C less than the natural temperature then prevailing) came into estrus 7–8 weeks earlier than those in entirely normal conditions (Dutt and Bush, 1955). Godley *et al.* (1966) also found that reduced daylight, or reduced temperature, or a combination of both could increase the number of ewes in estrus as well as their fertility, but whether the onset of estrus was advanced in these experiments is not clear.

Rats kept in low temperatures respond in various ways. Most workers have found that reproductive activity decreases, ovarian weight declines, and the estrous cycle becomes prolonged because of an extended anestrus. Rats exposed to 7.5°C became anestrus, but after about 80 days ovarian function began to recover, presumably as the animals became acclimatized. Similar observations have been made on hamsters (Grindeland and Folk, 1962). Mice kept at 3°C showed delayed puberty and, when adult, anestrus was prolonged (Barnett and Coleman, 1959). After about 8 weeks, however, these abnormalities tended to diminish, and were ascribed to metabolic changes due to cold and reduced food intake (Barnett and Coleman, 1959; Barnett and Mount, 1967). However, Dennison and Zarrow (1955) reported that cold temperatures (1°–2°C) increased the incidence of estrus and proestrus in rats (an effect reminiscent of that of continuous illumination) and that this effect could be counteracted by treating the animals with thyroxine. Brolin (1945) maintained rats in various temperatures from about 1°C upward and found no sign of ovarian malfunction.

E. Hibernation and Reproduction

The physiological changes occurring during hibernation and estivation in rodents, insectivores, and bats have been reviewed by Wimsatt (1969, 1975).

He shows that there is a general decrease in most metabolic functions, including the responsiveness of target organs to gonadotropic and steroid hormones. In some species (e.g., the thirteen-lined ground squirrel), the ovaries appear stimulated during hibernation, and breeding commences almost immediately after emergence. In temperate zones, vespertilionid bats copulate just before hibernation, but ovulation, fertilization, and implantation are delayed until the animal emerges from hibernation, except for *Miniopterus* which enters hibernation pregnant. In the hedgehog, the activity of the ovaries diminishes, and anestrus becomes accentuated during hibernation.

Estivation, a period of torpor which some ground squirrels and pocket mice undergo during summer, has no clear effect on reproduction.

IV. OLFACTION

Changes in light and temperature act as signals from the animal's environment, but smell transmits information of a quite different category, that from conspecifics. The importance of such information depends, therefore, upon the social structure of the animal's habitat rather than its physical characteristics.

A. Pheromones

Chemical substances which are released by one animal and, after being received and detected by another, result in one or more specific physiological changes, are called pheromones, a term introduced by Karlsson and Butenandt (1959). This definition, therefore, depends upon the substance released being specific and having a direct effect upon a dependent physiological or behavioral process. As we shall see, accepting this definition may oversimplify the role of pheromones. Pheromones may be detected through an animal's olfactory system, or by ingestion, or by absorption through the skin (Bruce, 1970). All mammalian pheromones so far studied belong to the first category (but see Weir, 1973; Vandenbergh *et al.*, 1975), and there are considerable differences, between species, in their role. Information is particularly plentiful for mice, which seem peculiarly susceptible to pheromonal action (Bronson, 1971).

Pheromones may have an immediate and rapid effect upon the recipient, in which case they are classified as "releasers." These have particular relevance to studies of social, aggressive, or sexual behavior and are not considered further here. "Primer" pheromones require a more prolonged action to be effective, and response to them is slower; it is these that are important in the present context. Pheromones which, by acting on the young animal during some critical phase of development, produce permanent or long-lasting effects, are called "imprinting" pheromones; other environmental factors (e.g., visual input) can have similar effects (see Bateson, 1966). It can be seen that this classification defines pheromones according to the nature of the responses they elicit. Though these definitions are widely accepted, the distinction between them is becoming blurred. For example, pheromones may induce sexual or aggressive behavior, which can result in rapid changes in hormone levels (including gonadal steroids) (E. B. Keverne and R. E. Meller, unpublished).

B. Males and the Estrous Cycle (Whitten Effect)

Whitten (1958, 1959) described two experiments which suggested that the presence of a male mouse stimulates ovarian function in the female. First, the addition of a male to a group of females results in an unexpectedly high proportion coming into estrus 3 days later, and does not depend upon physical contact between male and female. Second, groups of females housed together may show prolonged estrous cycles because diestrus is lengthened (see Section IV,C). Adding a male to such groups shortened the estrous cycles and made them more regular (see also Dominic, 1969). The duration of exposure and its time relative to the female cycle was critical; males had to be introduced for at least 48 hours immediately after estrus. Castrated males had little or no effect but androgen-treated females were able to synchronize the estrous cycles of normal females (Dominic, 1969).

Since removal of the olfactory bulbs itself induces anestrus (see Section IV,F), evidence that an olfactory mechanism was operating was determined by observing the same response to male urine placed in the cage (Bronson and Marsden, 1964), and after putting females into a cage recently vacated by a male (Whitten, 1966).

Some other rodents, such as the prairie vole (Richmond and Conaway, 1969) and the wild mouse (Chipman and Fox, 1966), show similar responses. In deer mice, there is an interaction between the size of the group and the effects of a male; males introduced to single females mated with equal frequency on the following nights; the effect described by Whitten was seen only if the females were kept in groups of ten. Groups of four gave results intermediate between single animals and those in the larger groups (Bronson and Marsden, 1964).

Similar observations have been made in sheep and goats, in which estrus and ovulation are stimulated if a male is introduced into a flock (Schinkel, 1954; see Bruce, 1970). For this to occur, males must be introduced shortly before the breeding season is due to start. There is no effect if the ewes are in breeding condition and showing regular estrous cycles. The implication is that there is an interaction between the presence of a male and the environmental factors that determine the onset of the breeding season (see Section II).

The estrous cycle of the rat, however, is much less sensitive to pheromones, although some shortening may occur (Hughes, 1964). Rats fed half their normal food supply tend to become anestrous (see Section V) but the introduction of a male partly counteracts the result of underfeeding (Cooper and Haynes, 1967). However, puberty is delayed in rats after removal of the olfactory bulbs (Sato *et al.,* 1974).

A more extreme example of the effect of the male on the estrous cycle has been described in the cuis (*Galea musteloides*), a species found in South America and related to the guinea pig. The majority (over 85%) of females in captivity caged singly or with other females fail to experience estrous cycles and are permanently anestrous. In the presence of a vasectomized male, however, cycles become regular. This finding differs from those for mice in that physical contact is necessary between male and female and behavioral as well as pheromonal mechanisms may be involved (Weir, 1971, 1973), though even in mice tactile cues may act synergistically with pheromones in accelerating puberty (Bronson and Maruniak, 1975).

C. Females and the Estrous Cycle (Lee–Boot Effect)

Female mice caged in fours show an unusually high incidence of pseudo-pregnancy, whereas those kept in larger groups of thirty or so tend to have long cycles with prolonged periods of anestrus and reduced ovulation rates (van der Lee and Boot, 1956). These changes are reversible since estrous cycles become regular again if the females are caged singly, although occasional "spontaneous" pseudopregnancies may still occur (Parkes and Bruce, 1961). Physical contact between mice is not needed to bring about the suppression of estrus since females in cages divided by perforated partitions show similar responses (Whitten, 1959). As we have seen, the introduction of a male prevents the effect described by van der Lee and Boot (see Section IV,B). Removal of the olfactory bulbs, but not blinding, reduces the incidence of pseudopregnancy or anestrus caused by caging mice in groups of different sizes (Lamond, 1959; Whitten, 1966; Bruce, 1967). The Lee–Boot effect is not seen in rats (Hughes, 1964).

D. Males and the Onset of Puberty

Immature female mice housed in the presence of, but not necessarily in physical contact with, a male, reach puberty earlier than those housed without a male (Vandenbergh, 1967). This observation is consistent with Whitten's findings in adult mice. Preliminary results suggest that the active substance may be a polypeptide of some sort since dialysis of urine removes its activity (Vandenbergh *et al.,* 1972). Conversely, urine from other females may delay puberty in weanling female mice (Colby and Vandenberg, 1974).

The opposite effects of males or females on the ovarian cycle has led to suggestions that pheromones from the former stimulate, and from the latter inhibit, the secretion of FSH from the anterior pituitary (Whitten, 1966). If FSH secretion were increased, then the resulting estrogen surge could, by acting on the hypothalamus during its estrogen-sensitive phase, potentiate the release of LH, and hence ovulation (see Section II,B).

E. The Male and Luteal Function (Bruce Effect)

Neither pregnancy nor pseudopregnancy will occur in mice which are mated and then, 3 to 4 days later, put into a cage containing a strange male (i.e., one other than the stud) or into a cage marked by the odor of his urine (Bruce, 1959; Parkes and Bruce, 1961). They do occur, despite the presence of a strange male, if the stud male remains with the female, or the stranger is castrated, or the females are made anosmic. Experiments with different strains of deer mice demonstrate that the less closely the stranger and the female are related, the more consistently is pseudopregnancy prevented (Bronson *et al.,* 1964). Females can also block pregnancy if they belong to a different strain (Bronson *et al.,* 1964) or are given androgen (Bruce, 1970; Hoppe, 1975). Although this phenomenon has been observed in various strains of mouse, it has not been reproduced in either rats or cattle (Hughes, 1964; Bruce, 1970). Bruce and Parrott (1960) have proposed that, in mice, the pregnancy-blocking effect aids exogamy, though whether this is so or why it should be limited to mice and some other myomorphs is still not clear.

Injections of progesterone or prolactin (LTH) into recently mated female mice exposed to the odor of strange males causes pregnancy or pseudopregnancy to develop normally (Bruce, 1970). It thus appears that the suppression of pregnancy reflects some effect on the corpus luteum caused by the presence of a strange male.

Whitten (1966) points out that the putative pheromones from a male mouse can either block pregnancy or stimulate the estrous cycle, depending upon conditions. He suggests that there may be several pheromones with different functions, or that an "identifying" pheromone, indicating to the female whether the male is strange or not, may modify the action of a second one which can either promote FSH secretion or inhibit the secretion of prolactin (but see Section IV,H).

F. Anosmia and the Estrous Cycle

In some species the estrous cycle is disturbed when the olfactory bulbs are removed. Female guinea pigs, mice, and pigs all exhibit prolonged anestrus, whereas in the rat, rabbit, and ewe there is little or no detectable effect (Whitten, 1966; Aron *et al.*, 1970). Clearly, this finding complicates investigations of olfactory stimuli since procedures specifically designed to eliminate olfactory input can themselves alter reproductive function. An interesting interaction between olfactory and visual stimuli has been described in the female rat. If an animal is made blind and anosmic, there is marked atrophy of the ovaries and reproductive tracts; yet either operation alone results in very little change in ovarian function (Reiter and Sorrentino, 1971).

G. Pheromones in Primates

There is, as yet, no satisfactory evidence that olfactory stimuli can directly modify ovarian function in primates. The menstrual cycle may, however, alter olfactory sensitivity in women. Le Magnen (1952), following his work on rats, found that adult women were able to detect a musklike compound called "exaltolide." Vierling and Rock (1967) demonstrated that sensitivity to this substance was maximal at midcycle and again before menstruation, findings which they suggested might be correlated with those phases of the cycle characterized by a high secretion rate of estrogen. There are some reports that human, as well as nonhuman, primates that live together or spend much time together may undergo synchronous reproductive cycles (*Lemur catta*: Evans and Goy, 1968; baboons: Kummer and Kurt, 1963; talapoin monkeys: Scruton and Herbert, 1970; man: McLintock, 1971). Male company is said to shorten and regularize human menstrual cycles, comparable to the effect described by Whitten in mice (McLintock, 1971). Whether these effects are transmitted by olfactory stimuli is not known. There is a congenital condition in man (Kallman's syndrome) in which anosmia is combined with gonadal hypofunction. It is not known whether this is a result of deficiencies in hypothalamic gonadatropin-releasing factors (Naftolin *et al.,* 1971).

H. Specificity of Pheromonal Effects on the Estrous Cycle

The possibility that some of the effects now ascribed to direct actions of pheromones are, in fact, less specific has to be recognized. Whitten (1966) briefly considered behavioral mechanisms and, surprisingly, found no evidence of social dominance in cages containing thirty mice. However, prevention of physical contact between the mice reduced suppression of estrus (Whitten, 1959). Female mice reared together do not become anestrous, but if they are separated and then reassembled anestrus occurs (Lamond, 1959). Aggressive interaction in groups of mice familiar with one another can be induced by similar maneuvers (Edwards, 1970; Brain *et al.,* 1971). Quantitative studies on the possible relationships between aggressive interactions and disturbances of the estrous cycle in mice are still lacking. Pseudopregnancy in small groups could equally be the result of homosexual mounting activity which occurs in female rodents, as well as in most other mammals (Young, 1961), and anosmia can abolish mounting behavior in some rodents (Murphy and Schneider, 1970). It is not yet known what effect the odors of males have on this or other categories of behavioral interaction between females.

The Bruce effect (pregnancy blocking by strange males) is not observed in adrenalectomized female mice (Snyder and Taggart, 1967) possibly because they can no longer respond fully to the "stress" of encountering a strange male. The administration of reserpine to female mice also counteracts the effect of strange males, a finding which has been related to the ability of drugs of this kind to induce prolactin secretion (Dominic, 1969). However, it must be remembered that reserpine is also a tranquilizing drug which may, for example, mitigate the effects of aggressive interaction. Moreover, simply putting a female into a different cage can block pregnancy (Eleftheriou *et al.,* 1962) and this indicates that female mice are highly susceptible to a number of environmental variables other than the odor of strange males. The effect of the male can be limited by placing the two sexes in a large cage (Eleftheriou *et al.,* 1962) though whether this is simply the result of less behavioral interaction between them is not yet known. Recent findings indicate that male mice produce pheromones that elicit aggression from other males, and that females produce substances that diminish the male's aggression; both are influenced by gonadal hormones (Archer, 1968; Mugford and Nowell, 1971; Dixon and Mackintosh, 1971). Castrating a male mouse thus not only diminishes his effectiveness in blocking pregnancy but also his aggression; treatment of female mice during the neonatal period with androgen has the converse effect (Edwards, 1970). It is clear that much more information is required about the interaction of behavioral and endocrinological phenomena in these rodents.

V. COITUS

It is known that mating in some species modifies ovarian function by inducing ovulation and activating the corpora lutea. If the mating is fertile, pregnancy supervenes; if not, pseudopregnancy. Whether sexual interaction can modify ovarian function in other, less obvious, ways in a manner comparable, for example, to that demonstrated in the bull in which the secretion of androgen is markedly stimulated (Katongole *et al.*, 1971), it not yet known.

A. Reflex Ovulation

Members of some species usually ovulate only after mating. It has been suggested that reflex ovulation is particularly prevalent in animals that are widely dispersed, whereas spontaneous ovulation is found in gregarious species (Zarrow and Clark, 1968). The rabbit, the cat and, more recently, the vole (Breed, 1967), have been extensively studied. Coitus-induced ovulation can be replicated by electrical stimulation of the cervix, or (in the cat) with a glass rod introduced into the vagina. The same stimulus is not effective in the untreated rabbit (review by Everett, 1964). The act of coitus results in a surge of LH being released from the anterior pituitary, causing ovulation (Everett, 1964; Hilliard *et al.*, 1967), which can be blocked by those drugs (for example, atropine or pentobarbitone) that are effective in preventing ovulation in "spontaneous" ovulators such as the rat. In the rabbit the stimulus of coitus does not lead to the immediate release of much LH but what is released is sufficient to induce secretion of 20α-dihydroprogesterone from the ovary. This, in turn, seems to prolong and amplify the release of LH: progesterone can have the same effect (Hilliard *et al.*, 1967; Hilliard and Eaton, 1971).

The distinction, however, between species that do, or do not ovulate spontaneously is not clear-cut. Mice may ovulate spontaneously or after coitus (Allen, 1922), and rats, treated with estrogen to induce behavioral estrus, ovulate earlier if mated than unmated (Aron *et al.*, 1966). Rats which are either in "spontaneous" or light-induced constant estrus, or in which ovulation is inhibited by chlorpromazine, will ovulate if they are mated (Everett,

1939; Dempsey and Searles, 1943; Zarrow and Clark, 1968). Reflex ovulators may, on occasion, ovulate apparently "spontaneously" (Everett, 1964), though the possibility that the animal is receiving stimuli of some kind from the environment has to be carefully excluded. Ovulation is not dependent on vaginal stimuli since rabbits mounted by other females will sometimes ovulate (Brooks, 1937). Just as certain treatments (e.g., continuous light) elevate the threshold stimulus required for ovulation (see above), so others can lower it. Vaginal stimulation with glass rods will lead to ovulation in rabbits treated with estradiol but not in untreated animals (Sawyer and Markee, 1959).

Multiple copulations are necessary to induce ovulation in the female mink and short-tailed weasel, though a single brief intromission in the cat and rabbit (Aron *et al.,* 1966) and a more protracted one in the ferret is sufficient. The presence of epidermal penile spines, found in many rodents and carnivores, has been correlated with the male's ability to induce reflex ovulation (Clark and Zarrow, 1971), but guinea pigs have such spines and ovulate spontaneously. Penile spines are also found in many male primates, though not in man. The evidence that has been presented for the view that induced ovulation occurs in primates is not convincing (Clark and Zarrow, 1971).

B. Coitus and Luteal Function

The corpus luteum formed during the estrous cycle in certain rodents, such as mice, rats, and hamsters, is activated to secrete more progesterone if the animal copulates. The multiple intromissions that characterize the mating behavior of rodents such as the rat seem to be required for the maximal release of prolactin and, therefore, for the optimal function of the corpus luteum (Ball, 1934). If the mating is sterile, as after copulation with a vasectomized male, a period of pseudopregnancy results. Electrical or mechanical stimulation of the cervix replicates the effect of mating (see Clegg and Doyle, 1967). In these species, the hormone responsible for the induction of pseudopregnancy is prolactin (LTH) which is released from the pituitary after mating or stimulation of the cervix (Everett, 1968). Stimulation during estrus is the most effective way of eliciting prolactin secretion (Staples, 1965), but the same stimuli during diestrus may result in a period of estrus which is then followed by pseudopregnancy (Everett, 1939, 1968). Thus it seems that a comparatively long-lasting sensitivity to circulatory hormones is imparted to the neural structures concerned in prolactin release by cervical stimulation. Secretion of prolactin occurs only after ovulation when the corpus luteum (and, therefore, a minimal degree of progesterone secretion) is present in the ovary (Everett, 1968). The induction of a hormone-sensitive condition in the nervous system by coitus resulting in prolactin secretion recalls the mechanism activated by light for the secretion of LH and, hence, the timing of ovulation (see Section II,B). In rats which are "spontaneously" in persistent estrus, the stimulus from a single coital episode can induce ovulation and the subsequent pseudopregnancy (Everett, 1939).

The induction of pseudopregnancy has been used as a model to investigate the neural connections between the genitalia and the anterior pituitary and is discussed below (Section VIII).

VI. ALTITUDE

A. Physiological Consequences

The conspicuous physiological effects of transporting animals to high altitudes are related to the fall in barometric pressure, and, hence, to the decline in the partial pressure of oxygen in the lungs and in the level of oxygenation of the blood. Alteration in such a basic metabolic parameter has a widespread effect on many physiological processes (Monge, 1943; Stickney and van Liere, 1953), of which reproduction is but one. Animals exposed to high altitudes for long periods, or to intermittent or prolonged low pressures in the laboratory, gradually become acclimatized as their metabolism (particularly that of the cardiovascular and respiratory systems) adapts (Monge, 1943; Stickney and van Liere, 1953).

B. Ovarian Function

There is no doubt that high altitude adversely affects fertility. Fernandez-Cano (1959) tells the story of Potosí, a Bolivian city at 14,000 feet. While the indigenous population of Indians were extremely fertile, the immigrant Spaniards remained infertile for 53 years. In 1535 the capital of Peru was moved from Cuzco at 11,500 feet to its present sea level site at Lima because men, cattle, and pigs could not breed (Monge, 1943). While it is well established that gonadal function of the male is sensitive to altitude and that libido may decline (Steele, 1971), ovarian activity appears to be less sensitive (Moore and Price, 1948). The estrous cycle remained normal in rats, mice, hamsters, or guinea pigs exposed to simulated altitudes of 18,000 feet for several months (Nelson and Barrill, 1944; Moore and Price, 1948). Greater altitudes (up to 27,000 feet) lead to a reduction in uterine and ovarian weight in rats as well as a loss in body weight. The estrous cycle and ovulation continued normally. The testes of male rats kept at these altitudes involuted relatively more. Puberty may be delayed (Altland, 1949). "Castration" cells appeared in the anterior pituitary (Gordon *et al.*, 1943; Altland, 1949).

Although lower altitudes seem to be relatively innocuous, mice kept for several years at 10,000 to 14,000 feet initially showed little change but gradually became infertile. Fertility was rapidly restored when the animals were returned to sea level (Cook and Krum, 1955). It should be noted that, in contrast to the estrous cycle, pregnancy is very sensitive to changes in altitude, and abortion or resorption of fetuses is frequent (Baird and Cook, 1962). Thus low fertility at high altitudes is due to interruptions of pregnancy, failure of implantation, possibly some decrease in ovulation rate, infertility of the male, and reduced sexual behavior (Chang and Fernandez-Cano, 1959).

VII. NUTRITION

A. Study of Nutrition and Reproduction

A great deal of our knowledge about the relationship of an animal's nutrition and its reproductive capacity derives from studies on economically important species, particularly cattle and sheep. Here, the problem is to define a diet that will yield maximum reproductive efficiency at the lowest cost, or to determine why certain diets impair fertility or growth. There is evidence that either under- or overfeeding may alter reproduction.

Biological necessity seems to require that shortage of food should lower fertility, since it seems advantageous to spare a malnourished female the additional burdens of pregnancy and lactation, and to prevent her young being born into conditions hazardous to their survival. In fact, while some species are fairly sensitive to food shortage, others are not. Undernutrition may affect reproduction indirectly as well as directly. For example, deer entering their first breeding season are preferentially affected in a general food shortage (Verme, 1965); food intake during a particular season may determine reproductive performance in the next (Smith, 1964). Shortage of water may also inhibit reproduction (Yahr and Kessler, 1975).

It is often difficult to separate the effects of variations in food supply per se from other associated environmental variables, such as rainfall or temperature, and this applies, of course, particularly to studies of populations in the wild (see Sadleir, 1969). There are, however, circumstances in which food supply can have major effects on breeding, for example, in wild rodents, particularly those living in deserts, in lagomorphs, and in marsupials. The quokka kept in captivity in conditions of temperature and light comparable to those of natural environment, but fed a high protein diet, breeds continuously. The wild population ceases to reproduce when food is scanty (Shield, 1964). Bodenheimer (1949) found that voles living near irrigated fields bred during the dry season; others in more arid conditions did not. These observations show that, in some circumstances, food supply can be an important factor in reproduction.

The effects of altering food supply on reproductive activity may be due to direct effects upon the ovary, to changes in gonadotropin secretion, to changes in fertilization rate, implantation, or pregnancy, to interference with the response of target tissues to hormones, or secondary to changes in other endocrine organs such as the adrenal or thyroid (Leathem, 1961). In many cases, while results have been documented, the underlying mechanisms remain obscure.

B. Changes in Energy Intake

Reduced food intake retards puberty in female rats (Asdell and Crowell, 1935), though pigs fed two-thirds of a "normal" diet actually reached puberty earlier (Self et al., 1955) which suggests that the "controls" were, in fact, being overfed. Starvation may induce permanent damage to the pituitary (Lamond, 1970).

The characteristic effect of underfeeding the adult is a gradual lengthening of the cycle, prolonging anestrus and leading to ovarian atrophy. Eventually, estrous cycles may cease altogether (Leathem, 1961; Bellows et al., 1966; Cooper and Haynes, 1969; Cooper et al., 1970). The principal effect seems to be upon ovulation. Feeding between 24 and 35% of their normal diet reduces ovulation rates in rats (Cooper et al., 1970), sows (Zimmerman et al., 1960; Lodge and Hardy, 1968), mice (McClure, 1967), and hamsters (Morin, 1975). The converse effect of stimulating ovulation by unusually rich feeding is known as "flushing" and is used in animal husbandry, though the effectiveness of flushing in well-fed animals is disputed (Thomson and Aitken, 1959; Moustgaard, 1969).

There is increasing evidence that the incidence of ovulation can be altered by changes in food supply of quite short duration. Feeding a full diet for only 12 hours to rats which were previously maintained on 75% of their normal intake increased their ovulation rate (Cooper et al., 1970). Such procedures are particularly effective if instituted just before the expected surge in LH (Cooper and Haynes, 1969). Similar effects have been observed in underfed sows (Zimmermann et al., 1960; Lodge and Hardy, 1968). Superovulation is not induced by these methods; the ovulation rate simply returns to normal.

Food intake can also affect the duration of the breeding season. Sheep fed extra rations during the winter experience a shorter anestrous period (Smith, 1965), while green food is said to stimulate or prolong breeding in the rabbit, although this is disputed (Stodart and Myers, 1966; Sadleir, 1969; Section VIII,A).

There have been only a few systematic attempts to correlate the rate of synthesis or secretion of gonadotropins with food intake. Assays of gonadotropin in underfed animals have given very variable results (see Leathem, 1961, 1966; Bellows *et al.*, 1966); pituitary levels of FSH or LH tend to rise (Nakanishi *et al.*, 1976), presumably reflecting reduced rates of secretion. There is some evidence that the response of the ovary to gonadotropin is impaired in starved animals (see Lutwak-Mann, 1962).

Acute starvation in women generally causes greater disturbance to the menstrual cycle than does chronic malnutrition (Lutwak-Mann, 1962), although the latter has recently been thought to reduce fertility (Gopalan and Naidu, 1972). Severe starvation may cause amenorrhea. In the condition known as anorexia nervosa, patients (nearly always female) refuse to eat and show a variety of metabolic abnormalities including amenorrhea; this may be associated with low urinary levels of estrogen and gonadotropin (Russell, 1965; Starkey and Lee, 1969). Patients who begin to eat and regain weight, also menstruate again and become fertile (Starkey and Lee, 1969). Although the malnutrition may contribute to ovarian dysfunction, it is also possible that the anorexia and the amenorrhea stem from a common lesion in some part of the brain, possibly the hypothalamus (Russell, 1965).

Overfeeding can also reduce ovarian activity. Obese mice, rats, and overfed cattle are subfertile and rats show ovarian atrophy (Leathem, 1961; Hellman, 1965; Smith, 1965; Moustgaard, 1969). There is a negative correlation between body fat and ovulation rate in cattle (Self *et al.*, 1955). The effects of obesity are not understood; infiltration of the ovarian bursa with fat, thus preventing access of the ova to the Fallopian tubes, has been suggested (Verme, 1965; Maynard and Loosli, 1969) but hardly seems to explain changes in ovulation rate.

The definition of overfeeding and underfeeding is necessarily arbitrary, and depends upon the criteria used. Whether the effects described above act directly upon the pituitary or ovary, or are part of a more general metabolic disturbance, can be determined only by further study.

C. Proteins

The effects of protein deficiency on reproduction have been studied extensively by Leathem (1961, 1966). In general, low protein consumption results in ovarian atrophy in adults, and to a failure of maturation of the reproductive organs in young animals. Species differ, however, and rabbits, for example, seem less sensitive than rats to protein deficiency. The ovaries of heifers fail to develop normally but recover as adequate protein is supplied (Moustgaard, 1969; Maynard and Loosli, 1969). Rats may cease ovulating or undergo continuous vaginal cornification (Leathem, 1966).

Changes in pituitary FSH during protein deficiency are variable and inconspicuous, but the increase in FSH, which ordinarily occurs after unilateral ovariectomy, is prevented (Leathem, 1966). Pituitary LH is reduced. Adequate studies of the turnover of FSH and LH in protein-deprived animals are still needed. The response of the ovaries to gonadotropins is unimpaired in protein-deficient animals, but that of the uterus to estrogen and progesterone is reduced.

D. Carbohydrates

Reducing carbohydrate intakes to very low levels has little effect on ovarian function, probably because glucose can be produced from fat or protein precursors. Overfeeding with carbohydrate results in precocious puberty and later obesity, which, in turn, reduces fertility (see Section VII,B). The condition of diabetes mellitus, in which there is an absolute or relative deficiency of insulin secretion, gives rise to widespread metabolic disturbances, particularly in glucose metabolism. Rodents and women with untreated diabetes are usually infertile and may not ovulate (Leathem, 1961; Hellman, 1965; Weir, 1974).

E. Fat

Rats and pigs fed either fat-free or low-fat diets exhibit abnormalities of their reproductive systems, principally failure to ovulate, and prolonged estrous cycles (Burr and Burr, 1929; Wirz and Beeson, 1951). This appears to be due to lack of one or more of the "essential" fatty acids such as linoleic or arachidonic acids (Maynard and Loosli, 1969). Deficiency in these substances results in numerous abnormalities, of which reproductive failure is only one (Wirz and Beeson, 1951). Overfeeding with fat results in obesity, the effects of which have already been discussed (see Section VII,B). In addition, high-fat diets may introduce relative deficiencies in vitamin E (Maynard and Loosli, 1969). This, in itself, may result in changes in reproductive function.

F. Vitamins

The vitamins are, by definition, substances that are essential in the diet, and it is not surprising, therefore, that vitamin deficiency leads to reproductive failure. Certain vitamins seem to be more closely concerned with ovarian function than others, but it is important to interpret the results of experiments with vitamin-deficient diets according to the species being studied, since some can synthesize various vitamins to differing degrees. For example, rabbits can synthesize members of the vitamin B group (Aitken and Wilson, 1962) which are not, therefore, required in their diet; this is not so for man. It is also important to be able to distinguish specific effects of deficiency in vitamins from those following a generally deficient food intake. Certain vitamins (e.g., A and D, A and E) may interact so that a deficiency or excess in one may precipitate symptoms actually due to a relative deficiency of the other (see Maynard and Loosli, 1969).

Vitamin A is necessary for normal functioning of epithelia, and deficiency results in abnormal keratinization of the vagina, which then becomes an unreliable indicator of ovarian function. Vitamin A-deficient rats, however, continue to ovulate (see Leathem, 1961). Follicular atresia is induced in cattle after long periods of vitamin A deficiency (Leathem, 1961; Lutwak-Mann, 1962). Abnormally high levels of vitamin A are said to prolong the estrous cycle (Masin, 1950).

Of the vitamin B group, deficiencies in pantothenic acid, folic acid, thiamine, riboflavin, and pyridoxine can all cause ovarian atrophy and reproductive failure (Leathem, 1961; Lutwak-Mann, 1962; Maynard and Loosli, 1969). Pyridoxine deficiency has the interesting result of increasing the level of FSH within the pituitary, while markedly impairing the ovarian response to gonadotropins (Leathem, 1966).

Despite the large amount of vitamin C in the ovaries, no specific disturbance has resulted from a diet deficient in this vitamin (Aitken and Wilson, 1962). The reputed efficacy of green food in stimulating reproduction (see Section VII,A) is not, therefore, a consequence of the vitamin C contained in it. Vitamin D is particularly important during pregnancy, when considerable demands are made on a mother's calcium metabolism. However, the ovary seems to be little affected by vitamin D-deficient diets (Maynard and Loosli, 1969).

For many years it was thought that vitamin E played some special role in reproduction. Implantation may be prevented (Leathem, 1961) but there is little evidence of specific ovarian abnormalities in vitamin E-deficient rats. Some reports suggest that the estrous cycle becomes prolonged and that some degree of ovarian atrophy occurs in rats (Lutwak-Mann, 1962).

However, vitamin E has no clearly defined role in the reproduction of cattle or sheep, nor does the supply of extra vitamin E improve fertility (Maynard and Loosli, 1969).

The remaining vitamins have not been implicated in ovarian function. It is clear that information about the role of vitamins is still incomplete; in particular, little is known about the specific role played by them in the various components of ovarian or pituitary activity.

G. Minerals

Information about the role of minerals in the function of the ovaries is scanty. Diets deficient in calcium or phosphorus, zinc, and, possibly, magnesium result in low fertility in cows, sheep, and rats (Hignett and Hignett, 1952; Lutwak-Mann, 1962; Moustgaard, 1969; Gombe et al., 1973) which can be partly compensated for by a high vitamin D intake (Hignett, 1959). Why these effects occur is not known. Iodine deficiency can induce hypothyroidism, with consequent disturbance in reproduction. Manganese deficiency is said to be particularly liable to cause infertility in cattle (Hignett, 1959; Moustgaard, 1969). This has been correlated with the unusually high content and turnover of this element in the ovary, reproductive tract, and thyroid (Englund, 1960).

VIII. NEURAL PATHWAYS

It is possible to discuss the neural pathways linking the environment with ovarian activity only in the case of specific and discrete stimuli such as those provided by light, smell, or coitus. Since none of these has been shown to affect the ovary directly, changes in its activity are secondary to antecedent alterations in the anterior pituitary and the hypothalamus (for reviews see McCann et al., 1968; Yates et al., 1971). The problem, therefore, is to identify the receptors sensitive to each stimulus in its neuroendocrine context and to trace the pathways through the brain to their common termination. Although the pathways mediating the effects of temperature on ovarian function are not considered further (there is practically no evidence for their existence), it should be noted that the hypothalamus contains neurons which react to changes in body temperature, and are presumed to be concerned with homeostasis, though whether they can also transmit impulses to the pituitary is not known.

A. Receptors

1. Light

Since removal of the eyes or section of the optic nerve prevents the effects of artificial light on the reproductive cycle of the ferret (Thomson, 1954) and rat (review by Wurtman, 1967), the eyes evidently contain the receptors transmitting the effects of light to the reproductive system. Although light can enter the brain directly even through skulls as thick as those of the sheep (Ganong *et al.,* 1963) and, it is claimed, can modify directly the reproductive functions of the hypothalamus (Lisk and Kannwischer, 1964), nothing yet points to these mechanisms playing a significant role in the effect of light on reproduction.

It has been suggested that young rats may possess extraretinal receptors which allow light to modify the enzymes in the pineal (Zweig *et al.,* 1966; and Section VIII,C). The Harderian glands, which are compound tubular–alveolar structures containing, in some species, much porphyrin (a light-absorbing substance) and lying behind the eye, have been postulated as receptors (Wetterberg *et al.,* 1970). However, removal of these glands in adult rats (Reiter and Klein, 1971) does not prevent the effects of constant light on the ovaries or the degeneration of the retina with the disappearance of all rods and cones (see Section II,B), which occurs in response to continuous light. It is not clear whether this degeneration itself contributes to the effects of light on the reproductive cycle, or whether the neuroendocrine receptors in the retina persist and are, in some way, different from those mediating vision. It is interesting, in this context, that rats returned to alternating photoperiods from continuous light resume their former estrous rhythms (Takahashi and Suzuki, 1969), and that low intensities of continuous light may induce prolonged estrous cycles without retinal degeneration (Lambert, 1975).

2. Olfaction

The experiments described above (see Section IV,F) show that, in cases where this is feasible, the effects of pheromones are prevented by removal of the olfactory bulbs and that these structures are stimulated by receptors in the olfactory mucosa. There seems little evidence, so far, that pheromones can act in any way (e.g., by ingestion or absorption) other than through these receptors.

3. Coitus

The receptors lying within the vagina or cervix are not the only ones that are involved in the pathways evoking ovulation. It is important to distinguish experiments in which localized stimulation of the cervix or vagina (e.g., with a glass rod) is used from those in which ovulation is induced by mating, during which tactile, visual, and olfactory stimuli are also operating and which, separately or together, can result in ovulation. Denervation of the vagina or uterus in the cat reduces the incidence of ovulation induced by mechanical stimulation (Diakow, 1971), but neither extensive denervation of the genital tract nor application of local anesthetics has any effect on coitus-induced ovulation in either rabbits or cats (Friedman, 1929; Fee and Parkes, 1930; Brooks, 1937; see also Clegg and Doyle, 1967). However, rats treated with chlorpromazine no longer ovulate after mating if the pelvic nerve, which supplies sensory fibers to the genitalia, is cut (Zarrow and Clark, 1968).

In some animals (e.g., rabbits) stimulation only of receptors in the vagina (e.g., by a glass rod) is not effective; others (e.g., cats) can ovulate after inputs through receptors either in the vagina or elsewhere. The rat seems to be more dependent on vaginal stimulation, at least in some contexts, but removal of the cervix does not prevent activation of the corpus luteum and the development of pseudopregnancy by coitus (Ball, 1934; Krehbiel, 1948).

B. Pathways to the Hypothalamus

1. Light

Pathways mediating vision or visual reflexes are not concerned in the neuroendocrine response to light since removal of the optic tracts, lateral geniculate bodies, superior coliculi, or occipital cortex does not prevent the effects of artificial illumination (Le Gros Clark *et al.,* 1939; Critchlow, 1963). The accessory optic tracts, which leave the main pathway at various points, have been implicated, since section of these wholly crossed tracts prevents the effects of light on the enzymes in the pineal (Moore *et al.,* 1967; and Section VIII,C), as do lesions in the median forebrain bundle (Wurtman *et al.,* 1967), a tract running through the hypothalamus and containing fibers from a number of sources. Accessory optic fibers terminate in two nuclei in the midbrain tegmentum and then may relay, in some as yet unexplained way, to the sympathetic neurons in the spinal cord and to the pineal gland (see Section VIII,C). However, lesions of the terminal nuclei do not prevent light-induced estrus in ferrets (Thorpe and Herbert, 1977).

Whether fibers originating from the retina synapse directly with hypothalamic neurons is disputed. Earlier workers claimed evidence for their existence (e.g., Kreig, 1932; Knocke, 1957) but this was not substantiated by later studies using more sensitive and specific neurohistological methods (see Hayhow, 1959). Recent electron microscope and autoradiographic studies have suggested that there are synapses within the hypothalamus derived from fibers originating in the eye (Sousa-Pinto, 1970; Moore and Lenn, 1972), which are found in the posterior part of the suprachiasmatic nucleus. Lesions of this structure prevent both estrous rhythms (i.e., ovulation) and the gonadal response of hamsters to short days (Stetson and Watson-Whitmyre, 1976).

2. Olfaction

The olfactory bulbs do not project directly to the hypothalamus, but mainly to the piriform lobe or to the underlying corticomedial nuclei of the amygdala. Both these structures, however, project to the hypothalamus either by fibers running directly to it, or by way of the stria terminalis, which connects the amygdala with the preoptic area and the hypothalamus (Powell *et al.,* 1965; Heimer, 1968). Electrical recording from the medial forebrain bundle and preoptic area confirms that olfactory stimuli reach this part of the brain (Scott and Pfaffmann, 1967). However, whether the destruction or stimulation of these structures can modify the effects of pheromones is still largely unexplored, though the same procedures can, in themselves, alter the reproductive functions of the anterior pituitary (reviews by Everett, 1964; McCann *et al.,* 1968; Yates *et al.,* 1971).

3. Coitus

In those instances in which the effective stimulus for ovulation derives presumably from stimulation of the cervix or vagina, it seems that the afferent impulses enter the spinal cord via the pelvic nerves. These nerves, which also supply parasympathetic fibers to the genitalia (Clegg and Doyle, 1967), are branches of the pudendal nerve. Their division prevents pseudopregnancy (Kollar, 1953) or copulation-induced ovulation in rats (Zarrow and Clark, 1968). Division of the sympathetic nerves, however, is ineffective (Brooks, 1937; Labate, 1940). Destruction of the lumbar or thoracic spinal cord, but not the sacral, prevents copulation-induced ovulation in rabbits (Brooks, 1937), although it is likely that behavioral mechanisms are also considerably disturbed by these procedures. The pathway from the spinal cord to the hypothalamus is still largely unknown (the anterolateral system seems likely), though that involved in the release of prolactin during lactation may run with those implicated in oxytocin secretion (Tindal, 1967).

C. The Pineal Gland

It now seems certain that the pineal gland is a component of the link between the eyes and the anterior pituitary. This conclusion follows from the finding that, in its absence or if its autonomic innervation is removed, the presence or absence of light may not have its usual effect upon reproduction (see Wurtman *et al.,* 1968; Reiter and Fraschini, 1969). Blinding hamsters, or putting them into very short days, no longer induces gonadal atrophy if the pineal is removed (Reiter, 1968). The same procedure prevents gonadal regression occurring during winter, presumably in response to the shorter daylengths (Czyba *et al.,* 1964). In this nocturnal rodent, the "positive" effect of the pineal occurs in response to "short" photoperiods and induces anestrus.

In the ferret, pinealectomy prevents the effect of artificial long days in stimulating precocious estrus (Herbert, 1969). In these diurnal animals, therefore, a positive action of the pineal is to induce estrus in response to long photoperiods. Removal of the superior cervical ganglia, which deprives the pineal of its extensive autonomic nerve supply (Kappers, 1960; Trueman and Herbert, 1970), has the same effect in the two species as pinealectomy. Furthermore, in the ferret the premature termination of estrus by short days is also prevented by pinealectomy (Thorpe and Herbert, 1976).

However, neither pinealectomy nor ganglionectomy prevents continuous illumination from inducing persistent estrus in the rat; nor is this induced simply by removing the pineal without exposing the animal to constant illumination (see Wurtman, 1967). There is still no evidence that the pineal is concerned in the timing of ovulation by light in this species, but pinealectomized rats no longer adjust the length of the estrous cycles in accordance with the prevailing photoperiod (Hoffmann and Pomerantz, 1971; Wallen and Yochim, 1974; see Section II,B). Reiter and co-workers have demonstrated that, in certain circumstances (e.g., anosmia, neonatal testosterone treatment, or malnutrition), the reproductive system of the rat becomes more sensitive to blinding, and ovarian atrophy occurs. This, too, is prevented by pinealectomy (Reiter and Sorrentino, 1971) and it seems that in this context the function of the pineal in the rat resembles that in hamsters (though in hamsters, the suppressive effect of starvation on ovulation is not prevented by pinealectomy; Morin, 1975). The prevention by blinding of normal degrees of compensatory ovarian hypertrophy after unilateral ovariectomy (see Section II,B,6) also seems to require the pineal, since pinealectomy counteracted the effect of blinding (Dickson *et al.,* 1971).

In all these species there is evidence that at least some of the effects of light are transmitted to the pituitary through the pineal gland. In no species, however, is the pineal essential for estrus to occur. Pinealectomized ferrets and rats still come into estrus (Herbert, 1971) and hamsters kept in very short photoperiods (in which the pineal is activated), or blinded, eventually recover from gonadal atrophy (Reiter, 1968). Further information comes from the findings that pinealectomized ferrets kept entirely in natural daylight eventually come into estrus at abnormal times of the year, such as the autumn (Herbert, 1972), though these effects only become apparent during the second year after operation. This, together with the evidence summarized above, suggests that light acts through the pineal gland to synchronize reproduction with the seasons. Without the pineal, the animal is no longer sensitive to those parameters of light that normally regulate the onset of breeding; the animal's pituitary, in fact, is "blind." There may be alternative pathways (e.g., via the suprachiasmatic nucleus) which mediate some other effects of light (e.g., the results of continuous illumination in rats). It is an oversimplification to say that the pineal inhibits or stimulates ovarian function; rather, its effect is to allow light to modify secretory activity of the pituitary. The way that this occurs depends upon the species, the particular light treatment, and the context in which light treatment is given (see Section II). How far the pineal itself determines these differences, or whether they represent variations in the way that neural structures respond to the pineal's secretions, remains unknown.

Melatonin (5-methoxy-N-acetyl serotonin) is formed in the pineal (see Wurtman *et al.*, 1968). The enzymes which convert serotonin to melatonin (N-acetyltransferase and hydroxyindole-O-methyltransferase) are markedly altered by light; in rodents they are decreased by continuous illumination. Injection of melatonin caused reduced ovarian function in rats (see reviews by Wurtman *et al.*, 1968; Reiter and Fraschini, 1969). Though melatonin does not replicate the action of the pineal activated by long photoperiods in ferrets (Herbert, 1971), it may do so in the case of short photoperiods (Turek *et al.*, 1975; Thorpe and Herbert, 1976). Conversely, melatonin counteracts the effects of artificial photoperiods under some conditions. Thus, it delays the induction of estrus by long photoperiods in the ferret (Herbert, 1971) and diminishes the gonadal atrophy caused in hamsters by restricted light (Hoffmann, 1974; Reiter *et al.*, 1974). Experiments in which melatonin has been given to rats have not yet allowed a role to be postulated for melatonin in the mechanism whereby light alters the estrous cycle in that species, although melatonin can reproduce the effect of blinding on compensatory ovarian hypertrophy. In fact, though the function of the pineal in the control by light of the breeding season is well established, there is, as yet, no known role for the gland in the regulation of the estrous cycle (see Tigghelaar and Nalbandov, 1975). While melatonin can change the concentration of gonadotropin in the pituitary (Fraschini *et al.*, 1971) it is still undecided whether this substance forms the link between pineal and pituitary in the effects described above or acts upon the pineal gland itself. Other nonindolic substances including many peptides (e.g., arginine vasotocin, releasing factors) have been isolated from the pineal (Moskowska *et al.*, 1971; Benson *et al.*, 1971; Pavel, 1973), but whether they can fill this role has not yet been demonstrated.

REFERENCES

Aitken, F. C., and Wilson, W. K. (1962). Rabbit feeding for meat and fur. *Commonw. Bur. Anim. Nutr., Tech. Commun.* **12**, 1–63.

Allen, E. (1922). The estrous cycle in the mouse. *Am. J. Anat.* **30**, 297–371.

Alleva, J. J., Waleski, M. V., Alleva, F. R., and Umburger, E. J. (1968). Synchronizing effect of photo-periodicity on ovulation in hamsters. *Endocrinology* **82**, 1227–1235.

Altland, P. D. (1949). Effect of discontinuous exposure to 25,000 feet simulated altitude on growth and reproduction of the albino rat. *J. Exp. Zool.* **110**, 1–18.

Amoroso, E. C., and Marshall, F. H. A. (1960). External factors in sexual periodicity. *In* "Marshall's Physiology of Reproduction" (A. S. Parkes, ed.), 3rd ed., Vol. 1, Part 2, pp. 707–831. Longmans Green, London and New York.

Archer, J. (1968). The effect of strange male odor on aggressive behaviour in male mice. *J. Mammal.* **149**, 572–575.

Aron, C., Asch, G., and Roos, J. (1966). Triggering of ovulation by coitus in the rat. *Int. Rev. Cytol.* **20**, 139–172.

Aron, C., Roos, J., and Asch, G. (1970). Effect of removal of the olfactory bulbs on mating behaviour and ovulation in the rat. *Neuroendocrinology* **6**, 109–117.

Aschoff, J. (1960). Exogenous and endogenous components in circadian rhythms. *Cold Spring Harbor Symp. Quant. Biol.* **25**, 11–27.

Asdell, S. A., and Crowell, M. F. (1935). The effect of retarded growth upon the sexual development of rats. *J. Nutr.* **10**, 13–23.

Austin, C. R. (1956). Ovulation, fertilization, and early cleavage in the hamster (*Mesocricetus auratus*). *J. R. Microsc. Soc.* [3] **75**, 141–154.

Baird, B., and Cook, S. F. (1962). Hypoxia and reproduction in Swiss mice. *Am. J. Physiol.* **202**, 611–615.

Baker, J. R. (1930). The breeding season in British wild mice. *Proc. Zool. Soc. London* **100**, 113–126.

Baker, J. R., and Ranson, R. M. (1932). Factors affecting the breeding of the field mouse (*Microtus agrestis*). Part 1. Light. *Proc. R. Soc. London, Ser. B* **110**, 313–322.

Ball, J. (1934). Determination of a quantitative relation between stimulus and response in pseudopregnancy in the rat. *Am. J. Physiol.* **107**, 698–703.

Barnett, S. A., and Coleman, E. L. (1959). The effect of low environmental temperature on the reproductive cycle of female mice. *J Endocrinol.* **19**, 232–240.

Barnett, S. A., and Mount, L. E. (1967). Resistance to cold in mammals. *In* "Thermobiology" (A. Rose, ed.), pp. 411–477. Academic Press, New York.

Bateson, P. P. G. (1966). The characteristics and contexts of imprinting. *Biol. Rev. Cambridge Philos. Soc.* **41**, 177–220.

Bedford, Duke of, and Marshall, F. H. A. (1942). On the incidence of the breeding season in mammals after transference to a new latitude. *Proc. R. Soc. London, Ser. B* **130**, 396–399.

Bellows, R. A., Meyer, R. K., Hoekstra, W. C., and Casida, L. E. (1966). Pituitary potency and ovarian activity in rats on two levels of dietary energy. *J. Anim. Sci.* **25**, 381–385.

Benson, B., Matthews, M., and Rodin, A. E. (1971). A melatonin-free extract of bovine pineal with autogonadotropic activity. *Life Sci.* **10**, 609–612.

Bissonnette, T. H. (1931). Studies on the sexual cycle in birds. V. Effect of light of different intensities upon the testis activity of the European starling (*Sturnus vulgaris*). *Physiol. Zool.* **4**, 542–574.

Bissonnette, T. H. (1932). Modification of mammalian sexual cycles: reactions of ferrets (*Putorius vulgaris*) of both sexes to electric light added after dark in November and December. *Proc. R. Soc. London, Ser. B* **110**, 322–336.

Bissonnette, T. H. (1935). Modification of mammalian sexual cycles. IV. Delay of oestrus and induction of anoestrus in female ferrets by reduction of intensity and duration of daily light periods in the normal oestrous season. *J. Exp. Biol.* **12**, 315–320.

Bodenheimer, F. S. (1949). Quoted by Sadleir (1969).

Braden, A. W. H. (1957). The relationship between the diurnal light cycle and the time of ovulation in mice. *J. Exp. Biol.* **34**, 177–188.

Brain, P. F., Nowell, N. W., and Wouters, A. (1971). Some relationships between adrenal function and the effectiveness of a period of isolation in inducing inter-male aggression in albino mice. *Physiol. Behav.* **6**, 27–29.

Brambell, F. W. R. (1944). The reproduction of the wild rabbit, *Oryctolagus cuniculus* (L.). *Proc. Zool. Soc. London* **114**, 1–45.

Breed, W. G. (1967). Ovulation in the genus *Microtus*. *Nature (London)* **214**, 826.

Breed, W. G., and Clarke, J. R. (1970). Effect of photoperiod on ovarian function in the vole, *Microtus agrestis*. *J. Reprod. Fertil.* **23**, 189–192.

Brolin, S. E. (1945). A study of the structural and hormonal reactions of the pituitary body of rats exposed to cold. *Acta Anat., Suppl.* **3**, 1–165.

Bronson, F. H. (1971). Rodent pheromones. *Biol. Reprod.* **4**, 344–357.

Bronson, F. H., and Marsden, H. M. (1964). Male-induced synchrony of oestrus in deer mice. *Gen. Comp. Endocrinol.* **4**, 434–437.

Bronson, F. H., and Maruniak, J. A. (1975). Male-induced puberty in female mice: evidence for a synergistic action of social cues. *Biol. Reprod.* **13**, 94–98.

Bronson, F. H., Eleftheriou, B. E., and Garick, E. I. (1964). Effects of intra- or inter-specific social stimulation on implantation in deer mice. *J. Reprod. Fertil.* **8**, 23–27.

Brooks, C. M. (1937). Role of the cerebral cortex and of various sense organs in the excitation and execution of mating activity in the rabbit. *Am. J. Physiol.* **120**, 544–553.

Browman, L. G. (1937). Light in its relation to activity and oestrous rhythms in the albino rat. *J. Exp. Zool.* **75**, 375–388.

Bruce, H. M. (1959). An exteroceptive block to pregnancy in the mouse. *Nature (London)* **184**, 105.

Bruce, H. M. (1967). Pheromonal mechanisms regulating mammalian reproduction. *Gen. Comp. Endocrinol., Suppl.* **2**, 260–267.

Bruce, H. M. (1970). Pheromones. *Br. Med. Bull.* **26**, 10–13.

Bruce, H. M., and Parrott, D. M. V. (1960). Role of olfactory sense in pregnancy block by strange males. *Science* **131**, 1526.

Bunning, E. (1967). "The Physiological Clock." Longmans Green, London and New York.

Burr, G. O., and Burr, M. M. (1929). A low deficiency disease produced by the rigid exclusion of fat from the diet. *J. Biol. Chem.* **82**, 345–367.

Campbell, C. S., Ryan, K. D., and Schwartz, N. B. (1976). Estrous cycles in the mouse: relative influences of continuous light and the presence of a male. *Biol. Reprod.* **14**, 292–299.

Chang, M., and Fernandez-Cano, L. (1959). Effects of short changes of environmental temperature and low atmospheric pressure on the ovulation of rats. *Am. J. Physiol.* **196**, 653–655.

Chipman, R. K., and Fox, K. A. (1966). Oestrous synchronisation and pregnancy blocking in wild house mice (*Mus musculus*). *J. Reprod. Fertil.* **12**, 233–236.

Clark, J. H., and Zarrow, M. X. (1971). Influence of copulation on time of ovulation in women. *Am. J. Obstet. Gynecol.* **109**, 1083–1085.

Clegg, M. T., and Doyle, L. L. (1967). Role in reproductive physiology of afferent impulses from the genitalia and other regions. *In* "Neuroendocrinology" (L. Martini and W. G. Ganong, eds.), Vol. 2, pp. 1–17. Academic Press, New York.

Clegg, M. T., and Ganong, W. F. (1969). Environmental factors affecting reproduction. *In* "Reproduction in Domestic Animals" (H. H. Cole and P. T. Cupps, eds.), 2nd ed., pp. 473–488. Academic Press, New York.

Colby, D. R., and Vandenbergh, J. G. (1974). Regulatory effects of urinary pheromones in the mouse. *Biol. Reprod.* **11**, 268–279.

Collins, K. J., and Weiner, J. S. (1968). Endocrinological aspects of exposure to high environmental temperatures. *Physiol. Rev.* **48**, 785–839.

Cook, S. F., and Krum, A. H. (1955). Deterioration of mouse strains exposed for long periods to low atmospheric pressure. *J. Exp. Zool.* **128**, 561–572.

Cooper, K. J., and Haynes, N. B. (1967). Modification of the oestrous cycle of the underfed rat associated with the presence of the male. *J. Reprod. Fertil.* **14**, 317–320.

Cooper, K. J., and Haynes, N. B. (1969). Modification of the oestrous cycle of the underfed rat associated with the time of feeding. *J. Reprod. Fertil.* **19**, 577–579.

Cooper, K. J., Haynes, N. B., and Lamming, G. E. (1970). Effects of unrestricted feeding during oestrus on reproduction in the underfed female rat. *J. Reprod. Fertil.* **22**, 293–301.

Cowan, I. McF., and Arsenault, M. G. (1954). Reproduction and growth in the creeping vole (*Microtus oregoni serpens Meriam*). *Can. J. Zool.* **32**, 198–208.

Critchlow, V. (1963). The role of light in the neuroendocrine system. *Adv. Neuroendocrinol., Proc. Symp., 1961* pp. 377–402.

Czyba, J. C., Girod, C., and Durand, N. (1964). Sur l'antagonisme épiphysiohypophysaire et les variations saissonière de la spermatogénèse chez le hamster doré (*Mesocricetus auratus*). *C.R. Soc. Biol.* **158**, 742–745.

Daane, T. A., and Parlow, A. F. (1971). Serum FSH and LH in continuous light-induced persistent estrus: short-term and long-term studies. *Endocrinology* **88**, 964–968.

Dempsey, E. W., and Searles, J. F. (1943). Experimental modification of certain endocrine phenomena. *Endocrinology* **32**, 119–128.

Dennison, M. E., and Zarrow, M. Y. (1955). Changes in estrous cycles of rats during prolonged exposure to cold. *Proc. Soc. Exp. Biol. Med.* **89**, 632–634.

Diakow, C. (1971). Effects of genital desensitisation on mating behaviour and ovulation in the female cat. *Physiol. Behav.* **7**, 47–54.

Dickson, K., Benson, B., and Tate, C. (1971). The effect of blinding and pinealectomy in unilaterally ovariectomised rats. *Acta Endocrinol. (Copenhagen)* **66**, 177–182.

Dixon, A. K., and Mackintosh, J. H. (1971). Effects of female urine upon the social behaviour of adult male mice. *Anim. Behav.* **19**, 138–140.

Dominic, C. J. (1969). Pheromonal mechanisms regulating mammalian reproduction. *Gen. Comp. Endocrinol., Suppl.* **2**, 206–267.

Donovan, B. T. (1967). Light and the control of the oestrous cycle in the ferret. *J. Endocrinol.* **39**, 105–115.

Ducker, M. J., and Bowman, J. C. (1970). Photo-periodism in the ewe. 4. A note on the effect on onset of oestrus in Clun Forest ewes of applying the same decrease in daylight at two different times of the year. *Anim. Prod.* **12**, 513–514.

Dutt, R. H., and Bush, L. F. (1955). The effect of low environmental temperature in initiation of the breeding season and fertility in sheep. *J. Anim. Sci.* **14**, 885–896.

Edwards, D. A. (1970). Post-neonatal androgenisation and adult aggressive behaviour in female mice. *Physiol. Behav.* **5**, 465–467.

Eleftheriou, B. E., Bronson, F. H., and Zarrow, M. X. (1962). Interaction of olfactory and other environmental stimuli on implantation, in the deermouse. *Science* **132**, 764.

Englund, S. E. (1960). Studies on the metabolism of Mn^{54} in some endocrine and reproductive organs in the female mouse. *Acta Obstet. Gynecol. Scand.* **39**, 575–585.

Evans, C. S., and Goy, R. W. (1968). Social behaviour and reproductive cycles in captive ring-tailed lemurs (*Lemur catta*). *J. Zool.* **156**, 181–197.

Everett, J. W. (1939). Spontaneous persistent estrus in a strain of albino rats. *Endocrinology* **25**, 123–127.

Everett, J. W. (1964). Central neural control of reproductive functions of the adenohypophysis. *Physiol. Rev.* **44**, 373–431.

Everett, J. W. (1968). Delayed pseudopregnancy in the rat, a tool for the study of central nervous mechanisms in reproduction. *In* "Reproduction and Sexual Behavior" (M. Diamond, ed.), pp. 25–31. Univ. of Indiana Press, Urbana, Illinois.

Everett, J. W. (1969). Neuroendocrine aspects of mammalian reproduction. *Annu. Rev. Physiol.* **31**, 383–416.

Farner, D. S. (1965). Circadian systems in the photoperiodic response of vertebrates. *In* "Circadian Clocks" (J. Aschoff, ed.), pp. 357–369. North-Holland Publ., Amsterdam.

Farner, D. S., and Lewis, R. A. (1973). Field and experimental studies of the annual cycles of white-crowned sparrows. *J. Reprod. Fert., Suppl.* **19**, 35–50.

Fee, A. R., and Parkes, A. S. (1930). Studies on ovulation. III. Effect of vaginal anaesthesia on ovulation in the rabbit. *J. Physiol. (London)* **70**, 385–388.

Fernandez-Cano, L. (1959). The effect of increase or decrease of body temperature or of hypoxia on ovulation and pregnancy in the rat. *In* "Recent Progress in the Endocrinology of Reproduction" (C. W. Lloyd, ed.), pp. 97–106. Academic Press, New York.

Fiske, V. M. (1941). Effect of light on sexual maturation, estrous cycles and anterior pituitary of the rat. *Endocrinology* **29**, 187–196.

Flux, J. E. C. (1967). Reproduction and body weights of the hare (*Lepus europeus*) in New Zealand. *N.Z. J. Sci.* **10**, 357–401.

Follett, B. K. (1973). Circadian rhythms and photoperiodic time measurement in birds. *J. Reprod. Fert., Suppl.* **19**, 5–18.

Follett, B. K., and Sharp, P. J. (1969). Circadian rhythmicity in photoperiodically-induced gonadotrophin release and gonadal growth in the quail. *Nature (London)* **223**, 968–971.

Fraschini, F., Collu, R., and Martini, L. (1971). Mechanisms of inhibitory action of pineal principles on gonadotrophin secretion. *Pineal Gland, Ciba Found. Symp., 1970* pp. 254–273.

Friedman, M. H. (1929). The mechanisms of ovulation in the rabbit. I. The demonstration of a humoral mechanism. *Am. J. Physiol.* **89**, 438–442.

Ganong, W. F., Shepherd, M. D., Wall, J. R., van Brunt, E. E., and Clegg, M. T. (1963). Penetration of light into the brain of mammals. *Endocrinology* **72**, 962–963.

Godley, W. C., Wilson, R. L., and Hurst, V. (1966). Effect of controlled environment on the reproductive performance of ewes. *J. Anim. Sci.* **25**, 212–216.

Gombe, S., Apgar, J., and Hansel, W. (1973). Effect of zinc deficiency and restricted food intake on plasma and pituitary LH and hypothalamic LRF in female rats. *Biol. Reprod.* **9**, 415–419.

Gopalan, C., and Naidu, N. (1972). Nutrition and fertility. *Lancet* **2**, 1077–1079.

Gordon, A. S., Tornetta, F. J., D'Angelo, S. A., and Charippa, H. A. (1943). Effect of low atmospheric pressures on the activity of the thyroid, reproductive system and anterior lobe of the pituitary in the rat. *Endocrinology* **38**, 366–383.

Greenwald, G. S. (1963). Failure of continuous light to induce constant oestrus in the hamster. *J. Endocrinol.* **28**, 123–124.

Grindeland, R. E., and Folk, G. E. (1962). Effects of cold exposure on the oestrous cycle of the golden hamster (*Mesocricetus auratus*). *J. Reprod. Fertil.* **4**, 1–6.

Gwinner, E. (1973). Circannual rhythms in birds: their interaction with circadian rhythms and environmental photoperiod. *J. Reprod. Fertil., Suppl.* **19**, 51–65.

Hafez, E. S. E. (1952). Studies on the breeding season and reproduction of the ewe. Part III. The breeding season in artificial light. *J. Agric. Sci.* **42**, 232–265.

Hammond, J. (1951). Control by light of reproduction in ferrets and minks. *Nature (London)* **167**, 150–151.

Hammond, J. (1952). Control of reproductive and pelt changes in ferrets: some experiments with animals kept entirely upon artificial light. *J. Agric. Sci.* **42**, 293–303.

Hart, D. S. (1950). Photoperiodicity in Suffolk sheep. *J. Agric. Sci.* **40**, 143–149.

Hart, D. S. (1951). Photoperiodicity in the female ferret. *J. Exp. Zool.* **28**, 1–12.

Hayhow, W. R. (1959). An experimental study of the accessory optic fiber system in the cat. *J. Comp. Neurol.* **113**, 281–313.

Heimer, L. (1968). Synaptic distribution of centripetal and centrifugal nerve fibres in the olfactory system of the rat. An experimental anatomical study. *J. Anat.* **103**, 413–432.

Hellman, B. (1965). Studies in obese hyperglycemic mice. *Ann. N.Y. Acad. Sci.* **131**, 541–558.

Hemmingsen, A. M., and Krarup, N. B. (1937). Rhythmic diurnal variations in the oestrous phenomena of the rat and their susceptibility to light and dark. *K. Dan. Vidensk. Selsk., Biol. Medd.* **13**, No. 7.

Herbert, J. (1969). The pineal gland and light-induced oestrus in ferrets. *J. Endocrinol.* **43,** 625–636.

Herbert, J. (1971). The role of the pineal gland in the control by light of the reproductive cycle of the ferret. *Pineal Gland, Ciba Found. Symp., 1970* pp. 303–320.

Herbert, J. (1972). Initial observations on pinealectomized ferrets kept for long periods in either daylight or artificial illumination. *J. Endocrinol.* **55,** 591–597.

Hery, M., Laplante, E., and Kordon, C. (1975). Role of pituitary sensitivity and adrenal secretion on the effect of serotonin depletion on luteinizing hormone regulation. *J. Endocrinol.* **67,** 463–464.

Hignett, S. L. (1959). Some nutritional and other interesting factors which may influence the fertility of cattle. *Vet. Rec.* **71,** 217–256.

Hignett, S. L., and Hignett, P. G. (1952). The influence of nutrition on reproductive efficiency in cattle. II. The effect of the phosphorus intake on ovarian activity and fertility of heifers. *Vet. Rec.* **64,** 203–206.

Hilliard, J., and Eaton, L. W. (1971). Estradiol-17β, progesterone and 20α-hydroxy-pregn-4-en-3-one in rabbit ovarian venous plasma. II. From mating through implantation. *Endocrinology* **89,** 522–527.

Hilliard, J., Penardi, R., and Sawyer, C. H. (1967). A functional role for 20α-hydroxy-pregn-4-en-3-one in the rabbit. *Endocrinology* **80,** 901–909.

Hoffmann, K. (1974). Testicular involution in short photoperiods inhibited by melatonin. *Naturwissenschaften* **61,** 364–365.

Hoffman, R. A., and Reiter, R. J. (1966). Response of some endocrine organs of female hamsters to pinealectomy and light. *Life Sci.* **5,** 1147–1151.

Hoffmann, J. C. (1967). Effects of light deprivation on the rat estrous cycle. In "Neuroendocrinology" (L. Martini and W. F. Ganong, eds.), Vol. 2, pp. 1–10. Academic Press, New York.

Hoffmann, J. C. (1969). Light and reproduction in the rat: effect of lighting schedule on ovulation blockade. *Biol. Reprod.* **1,** 185–188.

Hoffmann, J. C., and Pomerantz, D. K. (1971). Effect of pinealectomy on photoperiodic control of estrous cycle length in the rat. *Fed. Proc., Fed. Am. Soc. Exp. Biol.* **30,** 363.

Hoppe, P. C. (1975). Genetic and endocrine studies of the pregnancy-blocking pheromone of mice. *J. Reprod. Fertil.* **45,** 109–115.

Hughes, R. L. (1964). Effect of changing cages, introduction of the male and other procedures on the oestrous cycle of the rat. *Wildl. Res.* **9,** 145–122.

Kappers, J. A. (1960). The development, topographical relations and innervation of the epiphysis cerebri in the rat. *Z. Zellforsch. Mikrosk. Anat.* **2,** 162–215.

Karlsson, P., and Butenandt, A. (1959). Pheromones (ectohormones) in insects. *Annu. Rev. Entomol.* **4,** 39–59.

Katongole, C. B., Naftolin, F., and Short, R. V. (1971). Relationship between blood levels of lutenizing hormone and testosterone in bulls and the effects of sexual stimulation. *J. Endocrinol.* **50,** 457–466.

Kent, G. C., Ridgway, P. M., and Stroebel, E. F. (1968). Continual light and constant estrus in hamsters. *Endocrinology* **82,** 699–703.

Knocke, H. (1957). Die retino-hypothalamische Bahn von Mensch, Hunt, und Kaninchen. *Z. Mikrosk.-Anat. Forsch.* **63,** 461–486.

Kollar, E. J. (1953). Reproduction in the female rat after pelvic nerve neurectomy. *Anat. Rec.* **115,** 641–658.

Krehbiel, R. H. (1948). Reproduction in the rat after cervical bypassing and after cervectomy. *Anat. Rec.* **10,** 299–318.

Kreig, W. J. S. (1932). The hypothalamus of the albino rat. *J. Comp. Neurol.* **55,** 19–89.

Kummer, H., and Kurt, K. (1963). Social units of free-living Hamadryas Baboons. *Folia Primatol.* **1,** 4–19.

Labate, J. S. (1940). Influence of uterine and ovarian nerves on lactation. *Endocrinology* **27,** 342–344.

Lambert, H. (1975). Intensity of continuous light: threshold lower for persistent estrus than for retinal degeneration. *Biol. Reprod.* **13,** 576–580.

Lamond, D. R. (1959). Effect of stimulation derived from other animals of the same species on oestrous cycles in mice. *J. Endocrinol.* **18,** 343–349.

Lamond, D. R. (1970). Effect of PMSG on ovarian function of beef heifers, as influenced by progestins, plane of nutrition and fasting. *Aust. J. Agric. Res.* **21,** 153–161.

Lawton, I., and Schwartz, N. B. (1967). Pituitary-ovarian function in rats exposed to constant light: a chronological study. *Endocrinology* **81,** 497–508.

Leathem, J. H. (1961). Nutrition effects on endocrine secretions. *In* "Sex and Internal Secretions" (W. C. Young, ed.), 3rd ed., pp. 666–704. Williams & Wilkins, Baltimore, Maryland.

Leathem, J. H. (1966). Nutritional effects on hormone production. *J. Anim. Sci.* **25,** 68–82.

Lecyk, M. (1962). The dependence of breeding in the field vole (*Microtus arvalis*) on light intensity and wave length. *Zool. Pol.* **12,** 255–268.

Lees, J. L. (1966). Variations in the time of onset of the breeding season in Clun ewes. *J. Agric. Sci.* **67,** 173–179.

Le Gros Clark, W. E., McKeown, T., and Zuckerman, S. (1939). Visual pathways concerned in gonadal stimulation in ferrets. *Proc. R. Soc. London, Ser. B* **126,** 449–468.

Le Magnen, J. (1952). Nouvelles données sur le phénomène de l'exaltolide. *C.R. Acad. Sci. (Paris)* **230,** 1103–1105.

Lincoln, G. A., Youngson, R. W., and Short, R. V. (1970). The social and sexual behaviour of the red deer stag. *J. Reprod. Fertil., Suppl.* **11,** 71–103.

Lisk, R. D., and Kannwischer, L. R. (1964). Light: evidence for its direct effect on hypothalamic neurons. *Science* **146,** 272–273.

Lodge, G. A., and Hardy, B. (1968). The influence of nutrition during oestrus on ovulation rates in the sow. *J. Reprod. Fertil.* **15,** 329–332.

Lofts, B. (1970). "Animal Photoperiodism." Arnold, London.

Lutwak-Mann, C. (1962). The influences of nutrition on the ovary. *In* "The Ovary" (S. Zuckerman, ed.), 1st ed., Vol. 2, Chapter 18, pp. 291–315. Academic Press, New York.

McCann, S. M., Dhariwal, A. P. S., and Porter, J. C. (1968). Regulation of the adenohypophysis. *Physiol. Rev.* **30,** 589–640.

McClure, T. J. (1967). Infertility in mice caused by fasting at about the time of mating. *J. Reprod. Fertil.* **13,** 387–391.

McCormack, L. E. (1971). Timing of ovulation in rats as influenced by the onset of continuous light (LL) or darkness (DD). *Fed. Proc., Fed. Am. Socs Exp. Biol.* **30,** 310.

McCormack, L. E., and Bennin, B. (1970). Delay of ovulation caused by exposure to continuous light in immature rats treated with pregnant mare's serum gonadotrophin. *Endocrinology* **86,** 611–619.

MacFarlane, W. V. (1963). Endocrine function in hot environments. *In* "Environmental Physiology and Psychology in Arid Conditions: Reviews of Research," Vol. 22, pp. 153–232. UNESCO, Paris.

MacFarlane, W. V., Pennycuik, P. R., Yeath, N. T. M., and Thrift, E. (1959). Reproduction in hot environments. *In* "Recent Progress in the Endocrinology of Reproduction" (C. W. Lloyd, ed.), pp. 81–95. Academic Press, New York.

McLintock, M. C. (1971). Menstrual synchrony and suppression. *Nature (London)* **229**, 244–245.

Marshall, A. J. (1960). Annual periodicity in the migration and reproduction of birds. *Cold Spring Harbor Symp. Quant. Biol.* **25**, 499–506.

Marshall, F. H. A. (1940). The experimental modification of the oestrous cycle in the ferret by different intensities of light irradiation and other methods. *J. Exp. Biol.* **17**, 139–146.

Marshall, F. H. A., and Bowden, F. P. (1934). The effect of irradiation with different wave lengths on the oestrous cycle of the ferret with remarks on the factors controlling sexual periodicity. *J. Exp. Zool.* **11**, 409–422.

Marshall, W. A., and Thomson, A. P. D. (1951). The effect of increasing intensities of light on the rate of appearance of light-induced oestrus in normal ferrets and in ferrets after removal of both superior ganglia of the cervical sympathetic chain. *J. Anat.* **91**, 600.

Masin, F. (1950). Quoted by Leathem (1961).

Maynard, L. A., and Loosli, J. E., eds. (1969). "Animal Nutrition," 6th ed. McGraw-Hill, New York.

Means, T. M., Andrews, F. M., and Fontaine, W. E. (1959). Environmental factors in the induction of estrus in sheep. *J. Anim. Sci.* **18**, 1388–1396.

Meier, A. H., and MacGregor, R. (1972). Temporal organisation in avian reproduction. *Am. Zool.* **12**, 257–271.

Menaker, M. (1965). Circadian rhythms and photoperiodism in *Passer domesticus*. *In* "Circadian Clocks" (J. Aschoff, ed.), pp. 385–395. North-Holland Publ., Amsterdam.

Miller, H. L., and Alliston, C. W. (1974). Influence of programmed circadian temperature changes upon levels of luteinising hormone in the bovine. *Biol. Reprod.* **11**, 187–190.

Monge, C. (1943). Chronic mountain sickness. *Physiol. Rev.* **23**, 146–184.

Moore, C. R., and Price, D. (1948). A study at high altitude of reproduction, growth, sexual maturity and organ weights. *J. Exp. Zool.* **108**, 171–216.

Moore, C. R., Simmons, G. F., Wells, L. J., Zalesky, M., and Nelson, W. O. (1934). On the control of reproductive activity in an annual breeding mammal (*Citellus tridecemlineatus*). *Anat. Rec.* **60**, 279–289.

Moore, R. Y., and Lenn, N. J. (1972). A retino-hypothalamic projection in the rat. *J. Comp. Neurol.* **146**, 1–14.

Moore, R. Y., Heller, A., Wurtman, R. J., and Axelrod, J. (1967). Visual pathway mediating pineal response to environmental light. *Science* **158**, 220–223.

Morin, L. P. (1975). Effects of various feeding regimens and photoperiods or pinealectomy on ovulation on the hamster. *Biol. Reprod.* **13**, 99–103.

Moskowska, A., Kordon, C., and Ebels, I. (1971). Biochemical functions and mechanisms involved in the pineal modulation of pituitary gonadotrophin release. *Pineal Gland, Ciba Found. Symp., 1970* pp. 241–255.

Moule, G. R. (1950). Some problems of sheep breeding in semi and tropical Queensland. *Aust. Vet. J.* **26**, 29–37.

Moustgaard, J. (1969). Nutritional influences upon reproduction. *In* "Reproduction in Domestic Animals" (H. H. Cole and P. T. Cupps, eds.), 2nd ed., pp. 489–516. Academic Press, New York.

Mugford, R. A., and Nowell, N. W. (1971). Endocrine control over production and activity of the anti-aggression pheromone from female mice. *J. Endocrinol.* **49**, 225–232.

Murphy, M. R., and Schneider, G. E. (1970). Olfactory bulb removal eliminates mating behavior in the male golden hamster. *Science* **167**, 302–304.

Murthy, A. S. K., and Russfield, A. B. (1970). Endocrine changes in two strains of mice exposed to constant illumination. *Endocrinology* **86**, 914–917.

Naftolin, F., Harris, G. W., and Bobrow, M. (1971). Effect of purified luteinising hormone releasing factor in normal and hypogonadotropic anosmic men. *Nature (London)* **32**, 496–497.

Nakanishi, Y., Mori, J., and Nagasawa, H. (1976). Recovery of pituitary secretion of gonadotropins and prolactin during refeeding after chronic restricted feeding in female rats. *J. Endocrinol.* **69**, 329–339.

Neguin, L. G., Alvarez, J., and Schwartz, N. B. (1975). Steroid control of gonadotropin release. *J. Steroid Biochem.* **6**, 1007–1012.

Nelson, D., and Barrill, M. W. (1944). Repeated exposure to simulated high altitude: estrous cycles and fertility of the white rat. *Fed. Proc., Fed. Am. Soc. Exp. Biol.* **3**, 34.

Ogle, L. (1934). Adaptation of sexual activity to environmental stimulation. *Am. J. Physiol.* **107**, 628–634.

O'Steen, W. K. (1970). Retinal and optic nerve serotonin and retinal degeneration as influenced by photoperiod. *Exp. Neurol.* **27**, 194–205.

Parkes, A. S., and Bruce, H. M. (1961). Olfactory stimuli in mammalian reproduction. *Science* **134**, 1049–1054.

Pavel, S. (1973). Arginine vasotocin release into cerebrospinal fluid of CNS induced by melatonin. *Nature (London) New Biol.* **246**, 183–184.

Pengelley, E. T. (Ed.) (1974). "Circannual Clocks." Academic Press, New York.

Perry, J. S., and Rowlands, I. W. (eds). (1973). The environment and reproduction in mammals and birds. *J. Reprod. Fertil., Suppl.* **19**, 768.

Pittendrigh, C. S. (1960). Circadian rhythms and the circadian organization of living systems. *Cold Spring Harbor Symp. Quant. Biol.* **25**, 154–182.

Powell, T. P. S., Cowan, W. C., and Raisman, G. (1965). The central olfactory connexions. *J. Anat.* **99**, 791–813.

Quay, W. B. (1960). The reproductive organs of the collared lemming under diverse temperature and light conditions. *J. Mammal.* **41**, 74–89.

Radford, H. M. (1961). Photoperiodism and sexual activity in Merino ewes. I. The effect of continual light on the development of sexual activity. II. The effect of equinotal light on sexual activity. *Aust. J. Agric. Res.* **12**, 139–153.

Reiter, R. J. (1968). Pineal function on long-term blinded male and female hamsters. *Gen. Comp. Endocrinol.* **12**, 860–468.

Reiter, R. J. (1971). In discussion of Reiter and Sorrentino (1971, p. 343).

Reiter, R. J., and Fraschini, F. (1969). Endocrine aspects of the pineal gland: a review. *Neuroendocrinology* **5**, 219–255.

Reiter, R. J., and Klein, D. C. (1971). Observations on the pineal gland, the Harderian glands, the retina, and the reproductive organs of adult female rats exposed to continuous light. *J. Endocrinol.* **51**, 119–125.

Reiter, R. J., and Sorrentino, S. (1971). Factors influential in determining the gonad-inhibiting activity of the pineal gland. *Pineal Gland, Ciba Found. Symp., 1970* pp. 329–344.

Reiter, R. J., Vaughan, M. K., Black, D. E., and Johnson, L. Y. (1974). Melatonin: its inhibition of pineal antigonadotrophic activity in male hamster. *Science* **185,** 1169–1171.

Richmond, M., and Conaway, C. H. (1969). Induced ovulation and oestrus in *Microtus ochrogaster. J. Reprod. Fertil., Suppl.* **6,** 357–376.

Russell, G. F. M. (1965). Metabolic effects of anorexia nervosa. *Proc. R. Soc. Med.* **58,** 811–814.

Sadleir, R. F. M. S. (1969). "The Ecology of Reproduction in Wild and Domestic Mammals." Methuen, London.

Sato, N., Haller, E. W., Powell, R. D., and Henkin, R. I. (1974). Sexual maturation in bulbectomised female rats. *J. Reprod. Fertil.* **36,** 301–309.

Sawyer, C. H., and Markee, J. E. (1959). Estrogen facilitation of release of pituitary ovulating hormone in the rabbit in response to vaginal stimulation. *Endocrinology* **65,** 614–621.

Schinkel, P. G. (1954). The effect of the ram on the incidence and occurrence of oestrus in ewes. *Aust. Vet. J.* **30,** 189.

Schwartz, N. B. (1969). A model for the regulation of ovulation in the rat. *Recent Prog. Horm. Res.* **25,** 1–55.

Schwartz, N. B., and Caldarelli, D. (1965). Plasma LH in cyclic female rats. *Proc. Soc. Exp. Biol. Med.* **119,** 16–20.

Scott, J., and Pfaffmann, C. (1967). Olfactory input to the hypothalamus: electrophysiological evidence. *Science* **158,** 1592–1594.

Scruton, D. M., and Herbert J. (1970). The menstrual cycle and its effect on behaviour in the Talapoin monkey (*Miopithecus talapoin*). *J. Zool.* **162,** 419–436.

Self, H. L., Grummer, R. H., and Casida, L. E. (1955). The effect of various sequences of full and limited feeding on the reproductive phenomena in Chester White and Poland China gilts. *J. Anim. Sci.* **14,** 573–592.

Shield, J. W. (1964). A breeding season difference in two populations of the Australian macropod marsupial (*Setonix brachyurus*). *J. Mammal.* **45,** 616–625.

Smith, J. D. (1964). Reproduction in Merino sheep in tropical Australia. *Aust. Vet. J.* **40,** 156–160.

Smith, J. D. (1965). The influence of nutrition during winter and spring upon oestrous activity in the ewe. *World Rev. Anim. Prod.* **4,** 95–102.

Snyder, R. L., and Taggart, N. E. (1967). Effects of adrenalectomy on male-induced pregnancy block in mice. *J. Reprod. Fertil.* **14,** 451–455.

Sousa-Pinto, A. (1970). Electron-microscopic observations on the possible retinohypothalamic projection in the rat. *Exp. Brain Res.* **11,** 528–538.

Staples, R. E. (1965). Induction of pseudo-pregnancy in the rat by vaginal stimulation at various stages of the oestrus cycle. *Anat. Rec.* **152,** 499–502.

Starkey, J. A., and Lee, R. A. (1969). Menstruation and fertility in anorexia nervosa. *Am. J. Obstet. Gynecol.* **105,** 370–379.

Steele, P. (1971). Medicine on Mount Everest, 1971. *Lancet* **2,** 32–39.

Stetson, M. H., and Watson-Whitmyre, M. (1976). Nucleus suprachiasmaticus: the biological clock in the hamster? *Science* **19,** 197–199.

Stickney, J. C., and van Liere, E. J. (1953). Acclimatization to low oxygen tension. *Physiol. Rev.* **33,** 13–34.

Stodart, E., and Myers, K. (1966). The effects of different foods on confined populations of hill rabbits, *Oryctolagus cuniculus* (L.). *CSIRO Wildl. Res.* **11,** 111–124.

Strauss, W. F., and Meyer, R. K. (1962a). Neural timing of ovulation in immature rats treated with gonadotropin. *Science* **137**, 860–861.

Strauss, W. F., and Meyer, R. K. (1962b). Neural timing of ovulation in immature rats treated with gonadotropin: effect of light. *Am. Zool.* **2**, 219.

Sykes, J. F., and Cole, C. L. (1944). Modification of mating season in sheep by light treatment. *Mich., Agric. Exp. Stn., Q. Bull.* **26**, 250–256.

Symington, R. B., and Oliver, J. (1966). Observations on the reproductive activity of tropical sheep in relation to photoperiod. *J. Agric. Sci.* **67**, 7–12.

Takahashi, M., and Suzuki, Y. (1969). The dependence of the rat estrous cycle on the daily alternation of light and dark. *Endocrinol. Jpn.* **16**, 87–102.

Thibault, C., Courot, M., Martinet, L., Mauléon, P., du Mesnil du Buisson, F., Ortavant, R., Pelletier, J., and Signoret, J. P. (1966). Regulation of breeding season and oestrous cycles by light and external stimuli in some mammals. *J. Anim. Sci.* **25**, 119–139.

Thomson, A. P. D. (1954). The onset of oestrus in normal and blinded ferrets. *Proc. R. Soc. London, Ser. B* **142**, 126–135.

Thomson, W., and Aitken, F. C. (1959). Diet in relation to reproduction and the viability of the young. III. Sheep: world survey of reproduction and review of feeding experiments. *Commonw. Bur. Anim. Nutr., Tech. Commun.* **20**, 1–93.

Thorpe, D. H. (1967). Basic parameters in the reactions of ferrets to light. *Ciba Found. Study Group* **26**, 33–66.

Thorpe, P. A., and Herbert, J. (1976). Studies on the duration of the breeding season and photorefractoriness in female ferrets pinealectomised or treated with melatonin. *J. Endocrinol.* **70**, 255–262.

Thorpe, P. A., and Herbert, J. (1977). The effect of lesions of the accessory optic tract terminal nuclei on the gonadal response to light in ferrets. *Neuroendocrinology* (in press).

Tigghelaar, P. V., and Nalbandov, A. V. (1975). The effect of the pineal gland on ovulation and pregnancy in the rat. *Biol. Reprod.* **13**, 461–469.

Tindal, J. S. (1967). Studies on the neuro-endocrine control of lactation. *In* "Reproduction in the Female Mammal" (G. E. Lamming and E. C. Amoroso, eds.), pp. 79–109. Butterworth, London.

Trueman, T., and Herbert, J. (1970). Monoamines and acetyl-cholinesterase in the pineal gland and habenula of the ferret. *Z. Zellforsch. Mikrosk. Anat.* **109**, 83–100.

Turek, F., Desjardins, C., and Menaker, M. (1975). Melatonin: antigonadal and progonadal effects in male golden hamsters. *Science* **190**, 280–281.

Vandenbergh, J. G. (1967). Effect of the presence of a male on the sexual maturity of female mice. *Endocrinology* **81**, 345–349.

Vandenbergh, J. G., Drickamer, L. C., and Colby, D. R. (1972). Social and dietary factors in the seasonal maturation of female mice. *J. Reprod. Fertil.* **28**, 397–405.

Vandenbergh, J. G., Whitsett, J. M., and Lombardi, J. R. (1975). Partial isolation of a pheromone accelerating puberty in female mice. *J. Reprod. Fertil.* **43**, 515–523.

van der Lee, S., and Boot, L. M. (1956). Spontaneous pseudo-pregnancy in mice. *Acta Physiol. Pharmacol. Neerl.* **5**, 213–215.

Verme, L. J. (1965). Reproduction studies on penned white-tailed deer. *J. Wildl. Manage.* **29**, 74–79.

Vierling, J. S., and Rock, J. (1967). Variation of olfactory sensitivity to Exaltolide during the menstrual cycle. *J. Appl. Physiol.* **22**, 311–315.

Vincent, D. S. (1970). Modification of the annual oestrous cycle of the ferret by various regimes of artificial light. *J. Endocrinol.* **48**, iii.

Wallen, E. P., and Yochim, J. M. (1974). Photoperiodic regulation of the estrous cycle of the rat: role of the pineal gland. *Biol. Reprod.* **11**, 117–124.

Watson, R. H., and Radford, H. M. (1955). A note on the hours of daylight associated with the seasonal increase on sexual activity in Merino ewes. *Aust. Vet. J.* **31**, 31–32.

Weir, B. J. (1971). The evocation of oestrus in the cuis, *Galea musteloides*. *J. Reprod. Fertil.* **26**, 405–408.

Weir, B. J. (1973). The rôle of the male in the evocation of oestrus in the cuis, *Galea musteloides*. *J. Reprod. Fertil., Suppl.* **19**, 421–432.

Weir, B. J. (1974). The development of diabetes in the tuco-tuco (*Ctenomys talarum*). *Proc. R. Soc. Med.* **67**, 843–846.

Wetterberg, L., Geller, E., and Yuwiler, A. (1970). Harderian gland: an extraretinal photoreceptor influencing the pineal gland in neonatal rats? *Science* **167**, 884–885.

Whitaker, W. L. (1940). Some effects of artificial illumination in the white-footed mouse, *Peromyscus leucopus noroboracensis*. *J. Exp. Zool.* **83**, 33–60.

Whitten, W. K. (1958). Modification of the oestrous cycle of the mouse by external stimuli associated with the male. *J. Endocrinol.* **17**, 307–313.

Whitten, W. K. (1959). Occurrence of anoestrus in mice caged in groups. *J. Endocrinol.* **18**, 343–349.

Whitten, W. K. (1966). Pheromones and mammalian reproduction. *Adv. Reprod. Physiol.* **1**, 155–177.

Wight, H. M., and Conaway, C. H. (1961). Weather influences on the onset of breeding in Missouri cottontails. *J. Wildl. Manage.* **25**, 87–89.

Wimsatt, W. A. (1969). Some inter-relationships of reproduction and hibernation in mammals. *Symp. Soc. Exp. Biol.* **23**, 511–549.

Wimsatt, W. A. (1975). Some comparative aspects of implantation. *Biol. Reprod.* **121**, 40.

Wirz, N. M., and Beeson, W. M. (1951). The physiological effects of a fat deficient diet in the pig. *J. Anim. Sci.* **10**, 112–128.

Wurtman, R. J. (1967). Effect of light and visual stimuli on endocrine function. *In* "Neuroendocrinology" (L. Martini and W. G. Ganong, eds.), Vol. 2, pp. 19–59. Academic Press, New York.

Wurtman, R. J., Axelrod, J., Chu, E. W., Heller, A., and Moore, R. Y. (1967). Medial forebrain bundle lesions: blockage of effects of light on rat gonads and pineal. *Endocrinology* **89**, 509–514.

Wurtman, R. J., Axelrod, J., and Kelly, D. E. (1968). "The Pineal Gland." Academic Press, New York.

Yahr, P., and Kessler, S. (1975). Suppression of reproduction in water-deprived mongolian gerbils (*Meriones unguiculatus*). *Biol. Reprod.* **12**, 249–254.

Yates, F. E., Russell, S. M., and Maran, J. W. (1971). Brain-adenohypophysial communication in mammals. *Physiol. Rev.* **33**, 393–444.

Yeates, N. T. M. (1949). The breeding season of sheep with particular reference to its modification by artificial means using light. *J. Agric. Sci.* **39**, 1–42.

Yeates, N. T. M. (1954). Daylight changes. *In* "Progress on the Physiology of Farm Animals" (J. Hammond, ed.), Vol. 1, Chapter 8, pp. 363–392. Butterworth, London.

Yoshioka, Z., Awasawa, T., and Suzuki, S. (1952). Effect of the short day treatment on the modification of the breeding season in goats. *Bull. Anat. Inst. Agr. Soc., Ser. 9* **111**, 105–111.

Young, W. C. (1961). The hormones and mating behavior. *In* "Sex and Internal Secretions" (W. C. Young, ed.), 3rd ed., Vol. 2, pp. 1173–1239. Williams & Williams, Baltimore, Maryland.

Zarrow, M. Y., and Clark, J. H. (1968). Ovulation following vaginal stimulation in a spontaneous ovulator and its implications. *J. Endocrinol.* **40**, 343–352.

Zimmerman, D. R., Spies, H. G., Self, H. C., and Casida, L. E. (1960). Ovulation rate in swine as affected by increased energy intake fat prior to ovulation. *J. Anim. Sci.* **19**, 295–301.

Zweig, M., Snyder, S. H., and Axelrod, J. (1966). Evidence for a non-retinal pathway of light to pineal gland in newborn rats. *Proc. Natl. Acad. Sci. U.S.A.* **56**, 515–520.

Author Index

Numbers in italics refer to the pages on which the complete references are listed.

Subject Index

A

Actinomycin D, effect on initiation of ovarian tumors, 154
Activity cycle, in humans, 425
Adenoma
    chromophobe, 203
    tubular, induced, 172, 173, 174
Adrenal cortex
    fetal, influence on myometrial contractility, in women, 358
    influence on reproduction, 302
    overactivity
        causing primary amenorrhea, 199
        causing secondary amenorrhea, 204
        in subordinate mice, 421
    postovariectomy role, 435
    source of prepubertal hormone secretion, 297
Adrenal hyperplasia, congenital: differential diagnosis from true hermaphroditism, 194
Adrenalectomized pregnant animals, ovarian activity, 355
Adrenogenital syndrome, 200, 204
Adreno-ovarian relationships, 301
Aflatoxin B1, effect on initiation of ovarian tumors, 154
Age, effect on ovarian steroidogenesis, 289
Aging reproductive powers, effects of ovarian transplants, 116
Aggressive sexual behavior
    hormonal influence, 418
    pheromonal factors, 478
Agonadism, 186
Alertness, hormonal influence, 438
Allatectomy, in Insecta, effect on vitellogenesis, 25
Allograft, see Grafts
Altitude, effect on ovarian activity, 481
Amenorrhea
    emotional, 202
    environmental, 202
    primary, 196
        idiopathic, 200
    secondary, 196, 200
        idiopathic, 206
        ovarian causes, 204
        physiological causes, 200
        pituitary causes, 203
        psychogenic causes, 202
        systemic disease, 201
Amino acid, dietary, and hypothyroidism, 300
Amphibians
    corpus luteum and maintenance of gestation, 367
    ovarian cycle, 224
Amphineura, endocrine control of ovarian development, 13
Androgenic glands, in vitellogenesis, of Crustacea, 22
Androgens
    administration, effect on pituitary function, in transplantation, 120
    cellular origin, 275
    effect on aggressive sexual behavior, 418
    effect on sexual behavior, 403
    fetal, organizational role, 401
    libido hormone, 431
    neonatal influence on ovarian function, 297
    neuromuscular organization, 411
    peripheral conversion to estrogens, 289
    sterilizing action, mechanism, 298
    stromal production, 286
Androstenedione, 277
    ovarian venous blood concentration, 287, 289
    plasma, conversion to estrone, 290
    role as precursor, 290
    Δ^4-, of polycystic ovary, 205
Anencephaly, and adrenocortical hypoplasia, 358